# Circularly Polarized Luminescence of Isolated Small Organic Molecules

D1824395

Tadashi Mori
Editor

# Circularly Polarized Luminescence of Isolated Small Organic Molecules

 Springer

*Editor*
Tadashi Mori
Osaka University
Suita, Japan

ISBN 978-981-15-2311-3     ISBN 978-981-15-2309-0   (eBook)
https://doi.org/10.1007/978-981-15-2309-0

© Springer Nature Singapore Pte Ltd. 2020
This work is subject to copyright. All rights are reserved by the Publisher, whether the whole or part of the material is concerned, specifically the rights of translation, reprinting, reuse of illustrations, recitation, broadcasting, reproduction on microfilms or in any other physical way, and transmission or information storage and retrieval, electronic adaptation, computer software, or by similar or dissimilar methodology now known or hereafter developed.
The use of general descriptive names, registered names, trademarks, service marks, etc. in this publication does not imply, even in the absence of a specific statement, that such names are exempt from the relevant protective laws and regulations and therefore free for general use.
The publisher, the authors, and the editors are safe to assume that the advice and information in this book are believed to be true and accurate at the date of publication. Neither the publisher nor the authors or the editors give a warranty, expressed or implied, with respect to the material contained herein or for any errors or omissions that may have been made. The publisher remains neutral with regard to jurisdictional claims in published maps and institutional affiliations.

This Springer imprint is published by the registered company Springer Nature Singapore Pte Ltd.
The registered company address is: 152 Beach Road, #21-01/04 Gateway East, Singapore 189721, Singapore

# Preface

Circularly polarized luminescence (CPL) is one of the optical properties of chiral materials that measures the intensity difference between the left and right circularly polarized spontaneous emission from an intrinsically chiral fluorophore or fluorophore in a chiral environment. Therefore, CPL can be considered as an emission analog of circular dichroisms (CD). While tremendous efforts have been devoted to characterize the observed CD based on the structure of the molecule, successfully generating the so-called structure–chiroptical property relationships, such study on the CPL has been hitherto rather limited, due primarily to the instrumental limitations. As the CPL measurement becomes more accessible, the related study has been more concerned, particularly in this decade. Because the CPL responses are fundamentally based on the (emissive) excited state of the fluorophore, this property can be also used to effectively probe the configurational and/or conformational features of a chiral molecule in its electronic excited states. The degree of circular polarization is usually quantified by the dissymmetry factor, i.e., the relative intensity difference of left and right circularly polarized absorption (in CD) or emission (in CPL), respectively. Theoretically, the degrees of these are characterized by the intensity and orientation of electric and magnetic transition dipole moments of the relevant transition. These features are not easily deduced from the molecular structure alone, but the quantum chemical calculations, especially with the ansatz equivalent or superior to the cost-efficient time-dependent density functional theory, are ordinarily respectable for accurately reproducing these values and better understanding of the relationship, although a reliable yet accessible general calculation is still in the developing stage. The differences between CPL and CD (of the emissive state) are often small, as the structural difference between the thermally equilibrated ground state and the excited state, particularly of fluorescent $\pi$-systems, is relatively minor, and can also be significant in some cases, where nature of emission state, their structural relaxation in the excited state, effect of forbidden state, vibrational coupling, as well as excitonic coupling may play a substantial role. A complete understanding of all these and other factors are definitely required in order to fully understand the CPL characteristics of the chiral substances. While a number of investigations have addressed and disclosed some of these issues, more studies on the appropriate model systems are certainly anticipated to better comprehend the structure–property relationships for the CPL response.

This book comprehensively collects leading research on recently advanced CPL of small organic molecules, mostly in their isolated states, providing a status quo of the CPL chemistry that will be useful for scientists, researchers, and engineers in general, and is particularly useful for photophysical chemists, organic chemists, supramolecular chemists, spectroscopists, chemical engineers, and others in chemistry-related fields. As compared to lanthanide-based fluorophores, the studies on the organic molecule-based CPL behavior and preparations of CPL-responsive organic molecule are in emerging phase with the relatively smaller dissymmetric factors, but has been extensively attracted in recent years, since chiral organic molecules are potentially more useful in practical applications, as their fluorescence intensity, wavelength, and degree of dissymmetry can be rationally modulated through structural modifications.

This book begins with a short introductory chapter on molecular CPL, and the following chapters of this book consist of detailed descriptions of state-of-the-art advancement of all convincing CPL-responsive organic molecules, classified by the type of inherent chirality as axial chirality in biaryls (by Imai), planar chirality in cyclophanes (by Morisaki), helical chirality in helicenes (by Crassous and Hasobe), distorted chirality in dipyrromethene-related dyes (by Hall and de la Moya Cerero), and other relatively new and unique chiral molecules (by Mori). The book also covers future applications in areas such as advanced imaging and information technologies (by Nakashima). Most of the studies described here are rather focused on the structure–property relationship of relatively simple molecules investigated in non-aggregated form in solution, and also includes some recent progresses on supramolecular behaviors in host–guest interaction and aggregation formation (by Haino, Abbate, and Liu). The extended photophysical processes such as excimer formation as well as delayed fluorescence are also thought-provoking in terms of the CPL responses, which are also addressed in this book (by Lacour and Pieters). Last but not least, this book also highlights the recent development of commercially available CPL instrument (by Suzuki) as well as time-resolved CD spectroscopy (by Araki), to facilitate the further development and future design of CPL molecules. In order to make each contribution complete in itself, there is some inevitable overlap among the chapters.

Reports and studies on CPL materials, especially of organic systems, are rapidly increasing. Accordingly, it is difficult to cover all aspects of this ongoing active research in one book. However, this book on the CPL of isolated organic molecules serves the purpose of providing valuable information and some insights into the industry, academics, and researchers who are searching for current state of this emerging CPL research area, hoping that the book will stimulate further academic and applied research and promote the industrial applications of organic molecule-based CPL-response materials.

The editor of this book really enjoyed working with the authors of all chapters and would like to thank all authors for their excellent contributions to this exciting book. The editor is grateful to the colleagues who reviewed the chapters, as well as the staff of the Springer-Nature Japan for making this valued book possible.

Osaka, Japan                                                                                     Tadashi Mori
2020

# Contents

1   Frontiers of Circularly Polarized Luminescence Chemistry
    of Isolated Small Organic Molecules . . . . . . . . . . . . . . . . . . . . . . .   1
    Tadashi Mori

2   Circularly Polarized Luminescence of Axially Chiral Binaphthyl
    Fluorophores . . . . . . . . . . . . . . . . . . . . . . . . . . . . . . . . . . . .   11
    Yoshitane Imai

3   Circularly Polarized Luminescence from Planar Chiral
    Compounds Based on [2.2]Paracyclophane . . . . . . . . . . . . . . . . . .   31
    Yasuhiro Morisaki

4   Circularly Polarized Luminescence in Helicene and Helicenoid
    Derivatives . . . . . . . . . . . . . . . . . . . . . . . . . . . . . . . . . . . . .   53
    Jeanne Crassous

5   Structural Control of Fluorescent Helicates for Improved
    Circularly Polarized Luminescence Properties . . . . . . . . . . . . . . . .   99
    Taku Hasobe

6   BODIPY Based Emitters of Circularly Polarized Luminescence . . .   117
    Michael John Hall and Santiago de la Moya

7   Propeller Chirality: Circular Dichroism and Circularly Polarized
    Luminescence . . . . . . . . . . . . . . . . . . . . . . . . . . . . . . . . . . . .   151
    Tadashi Mori

8   Photo-Switching of Circularly Polarized Luminescence . . . . . . . . . .   177
    Takuya Nakashima and Tsuyoshi Kawai

9   Circularly Polarized Luminescence of Chirally Arranged
    Achiral Organic Luminophores by Covalent and Supramolecular
    Methods . . . . . . . . . . . . . . . . . . . . . . . . . . . . . . . . . . . . . . . .   197
    Toshiaki Ikeda and Takeharu Haino

10   **Structural and Electronic Information Drawn from the Circularly
     Polarized Luminescence Spectra: Many Questions and Some
     Answers for Simple Organic Molecules, Polymers, and Molecular
     Aggregates** . . . . . . . . . . . . . . . . . . . . . . . . . . . . . . . . . . . . . . . . . . . . . . . .   219
     Giovanna Longhi and Sergio Abbate

11   **Circularly Polarized Luminescence from Gelator Molecules:
     From Isolated Molecules to Assemblies** . . . . . . . . . . . . . . . . . . . . . .   249
     Tonghan Zhao, Pengfei Duan, and Minghua Liu

12   **Circularly Polarized Luminescence from Intramolecular
     Excimers** . . . . . . . . . . . . . . . . . . . . . . . . . . . . . . . . . . . . . . . . . . . . . . . .   273
     Francesco Zinna, Elodie Brun, Alexandre Homberg,
     and Jérôme Lacour

13   **Design of Circularly Polarized Thermally Activated Delayed
     Fluorescence Emitters** . . . . . . . . . . . . . . . . . . . . . . . . . . . . . . . . . . . . .   293
     Gregory Pieters and Lucas Frederic

14   **Principles and Applications of Circularly Polarized Luminescence
     Spectrophotometer** . . . . . . . . . . . . . . . . . . . . . . . . . . . . . . . . . . . . . . . .   309
     Satoko Suzuki

15   **Transient Circular Dichroism Approach to Chirality Detection
     in Dark Photo-Excited States** . . . . . . . . . . . . . . . . . . . . . . . . . . . . . . .   327
     Yasuyuki Araki

# Chapter 1
# Frontiers of Circularly Polarized Luminescence Chemistry of Isolated Small Organic Molecules

**Tadashi Mori**

**Abstract** Lately, circularly polarized luminescence (CPL), differential left- and right-polarized emission from chiral materials, has been attracted great attention, especially that of small organic molecules (SOMs). Despite the fact that the luminescence dissymmetry factor ($g_{lum}$) of SOMs, a measure of degree of chirality in CPL, is typically in a range of $10^{-3}$ to $10^{-5}$, considerably smaller than those based on materials with forbidden transition, the value of SOM-based CPL materials are progressively revised upwards. This trend is primarily due to the fact that a structural modification of SOMs is permanently rational and relatively easy, allowing a straightforward control of absorption and/or emission wavelengths as well as luminescence intensity. Biological compatibility and ease of device fabrication are additional advantages of SOM-based CPL materials. In this chapter, we outline the basics of CPL such as a definition and a quantification. Then, we briefly discuss about an information attained by the CPL measurement of SOMs. Finally, a perspective on the rapid progress of SOM-based CPL materials is provided.

## 1.1 Introduction

In whole of the following chapters of this book, you will soon find the recent advances of circularly polarized luminescence (CPL) behaviors of isolated small organic molecules (SOMs) and some related phenomena. Before going into individual topic in details, we briefly outline in this first chapter the basic concept and technical details of CPL. These include a definition of CPL, a quantification of CPL intensity, and an information attained by the CPL measurement. This chapter may be skipped for the specialists; however, we provide this fundamental information focusing on for those who are relatively new to this field, to facilitate a further reading. At the end of this chapter, a blight perspective on latest development of

T. Mori (✉)
Department of Applied Chemistry, Graduate School of Engineering, Osaka University, Suita, Japan
e-mail: tmori@chem.eng.osaka-u.ac.jp

© Springer Nature Singapore Pte Ltd. 2020                                                    1
T. Mori (ed.), *Circularly Polarized Luminescence of Isolated Small Organic Molecules*, https://doi.org/10.1007/978-981-15-2309-0_1

future CPL materials is also provided, demonstrating the rapid evolution of SOMs-based CPL chemistry in the last 5 years.

## 1.2 Background

Circularly polarized luminescence (CPL) represents a difference of incidental emission of left- and right-handed circularly polarized light from chiral materials (Fig. 1.1). The phenomenon can be observed, in principle, at the atomic, molecular, as well as supramolecular levels, necessarily involving chiral substances and/or surroundings. Such a polarized luminescence may be induced even for achiral systems by the external applied field (such as magnetic field), but we only focus on in this book the "*natural*" CPL that occurs through the electronic transition in the absence of external field. As a whole, emissions from both singlet and triplet (as well as other) electronically excited states are similarly treated without distinction between fluorescence and phosphorescence, unless the cases when a particular emphasis is placed. First of all, the CPL measurement is attractive and powerful means to elucidate the chiral information in their emissive excited states, complemental to its ground-state analog, circular dichroism (CD). Recently, the materials with strong CPL intensity (usually defined by a dissymmetry factor, vide infra) has been attracting great attention due to their promising applications in advanced photonic materials for chiral devices, enantioselective sensing systems, and biomedical applications, more specifically 3D displays, OLED materials, optoelectronic devices, spintronics devices, security painting, information storage, and so forth. More recently, functional CPL materials responsive to various stimuli such as pH, chemicals, temperature changes, mechanical forces, or light have been also developed as smarter chiroptical switching devices.

At the early stage of CPL chemistry, the CPL spectrum was only measured on a custom-made instrument, limiting such measurement only to the spectroscopist. This instrumental limitation unquestionably hampered the progress of CPL chemistry for a certain period. Now, the instrument becomes commercially available, facilitating to obtain reliable and reproducible CPL spectra expanding into other relevant fields

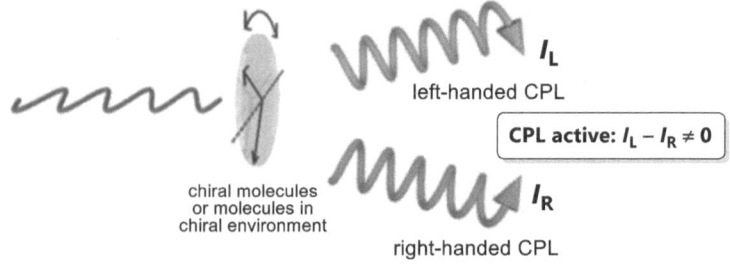

**Fig. 1.1** Circularly polarized luminescence (CPL) obtained from or with chiral substances

(including organic chemistry) and to systematically attain the structure-property relationship. As a consequence, the development of new CPL emitters becomes an active field of research, particularly for the last 5 years, by rationally designing materials with improved dissymmetry factors. Amongst the CPL active compounds, lanthanide complexes, especially those based on europium or terbium, are normally reported for greater CPL responses. Recently, self-assembled materials, liquid crystals, and polymers provide more efficient CPL responses, thanks to a supramolecular chirality, typically with at least one order of magnitude improved luminescence dissymmetry factors than simple SOMs. Nevertheless, the study on the CPL responses of SOMs becomes more and more attractive, probably because SOMs have some intrinsic advantage over the other materials. Hypothetically, their fluorescence efficiency, absorption and emission wavelengths and the bandwidths, and degree of dissymmetry can be rationally controlled by logical and relatively facile structural modifications. Additional advantages may be biological compatibility for in vivo sensing and ease of fabrication and manufacturing for electronic and photonic devices. You will soon find a variety of such efforts on SOMs and related systems in the following chapters, by focusing on the nature of respective structural motifs. In recent CPL research on SOMs, special emphases are placed on the $\pi$-$\pi^*$ transition of extended aromatic systems.

In addition to the materials chemistry standpoint, the CPL spectroscopy is valuable to elucidate the configurational and/or conformational information of optical active molecules in their emissive excited states, although the molecules under investigation must be reasonably luminescent. On the contrary, the information obtained through the CD spectroscopy is based on the thermally equilibrated electronic ground state. In principle, CD and CPL are mutually complementary probes for the structural features of chiral molecules in the different electronic states. When the structural differences and the vibrational contributions can be negligible, the structural information provided by the two chiroptical methods becomes similar. In practice, combination of CD and CPL spectroscopies provides a wealth of information concerning the structural differences in their ground and excited states. The CPL spectroscopy also provides information concerning the excited-state dynamics and energetics along the photophysical consequences from the initial absorption to the emission event. A necessity of fluorophore seems disadvantage at a glance, but can be beneficial for its selectivity and specificity. The CPL active emissive state may be accessible either by direct excitation, by indirect energy transfer process, or by relaxation from the higher excited state, which is in sharp contrast to the CD spectral technique, where the band overlap or small absorption may disturb the accurate analysis.

As stated by the Frank-Condon principle, electronic transition (both absorption and emission) always occurs instantly without accompanying geometrical change, represented as the vertical transition. According to the Kasha's rule, luminescence occurs from the lowest electronically excited state, typically in the $S_1$ state (or $T_1$ state in the triplet manifold). Therefore, information obtained by the CPL is responsible to $S_1$ to $S_0$ electronic transition, parallels to that of the lowest-energy CD, where

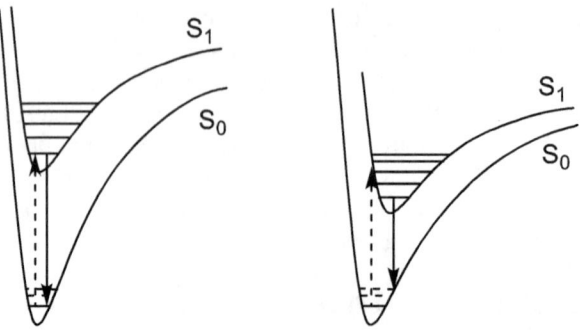

**Fig. 1.2** Potential energy surfaces in the ground ($S_0$) and excited ($S_1$) states with negligible (left) and typical (right) geometry change in a rigid aromatic system, exemplified by the transverse $S_1$ surface displacement. Absorption (CD) and emission (CPL) processes are indicated by dashed and solid arrows, respectively

$S_0$ to $S_1$ transition is involved (Fig. 1.2). Emission in condensed phase, however, generally occurs from the vibrationally relaxed state, which is different from the Frank-Condon state. In other words, depending on the nature of the excited-state potential surface of molecules under study, the degree of relaxation (or structural change) is different to a certain degree. Nevertheless, the absorption and emission spectra are usually mirror-imaged each other with a small energy gap called the Stokes shift, particularly in the rigid π-systems (for which most of the chapters focusing on in this book). In such systems, the potential energy surfaces and the vibrational frequency and spacing are comparable between the $S_1$ and $S_0$ states. As a result, the same sign and comparable magnitudes of dissymmetry factors are anticipated in the CPL and CD spectra. Still, the measurement of CPL spectrum, in comparison with CD spectrum, characterizes the valuable structural differences between the ground and electronically excited states of the chiral molecules. Readers are directed to further discussion on this issue in "Perspective" section.

## 1.3   Definition

When chiral molecules or molecules in chiral environment are excited with an unpolarized light in a typical CPL spectroscopy, the observed intensity of light emission is oscillated between ($I + \Delta I$) and ($I - \Delta I$) by a polarizing modulator. Thus, total intensity

$$I = I_L + I_R$$

and the difference

$$\Delta I = I_L - I_R$$

are simultaneously determined, where $I_L$ and $I_R$ are the emission intensities of left- and right-handed circularly polarized light. Because the evaluation of absolute $I$ and $\Delta I$ values is generally difficult, degree of chirality is generally discussed in terms of dissymmetry factor of luminescence, $g_{lum}$, which is defined by the difference emission divided by an averaged emission intensity, as follows:

$$g_{lum} = \frac{I_L - I_R}{\frac{1}{2}(I_L + I_R)} = 2\frac{\Delta I}{I}$$

The index $g_{lum}$ is occasionally called as luminescence anisotropy factor or more simply $g$ factor. Note that the emission anisotropy is also used to quantify a linearly polarized luminescence; therefore, dissymmetry is a probably more favorable term. By definition, minimum and maximum $g_{lum}$ factors are $-2$ and $+2$.

In most of recent studies to explore the CPL responses in molecular and supramolecular systems primarily concern the $g_{lum}$ value, as this parameter has been always the most limiting factor, especially in small organic molecules where $g_{lum}$ values are typically in an order of $10^{-3}$ to $10^{-5}$ range. However, in order to fully compare the overall CPL efficiency, it should be more appropriate to consider other photophysical parameters beside the polarization efficiency (i.e., dissymmetry factor), such as efficiencies of light absorption (oscillator strength, $f$) and emission intensity (quantum yield, $\Phi$). The magnitude of $f$ for an electronic transition is related to the maximum molar absorption coefficient ($\varepsilon$) and full width at half maximum of the absorption band ($\Delta\nu$):

$$f \propto \varepsilon \times \Delta\nu$$

Accordingly, a circular polarization luminosity ($\Lambda_{CPL}$) from the single chiral molecule in the excited state is described as follows, defined as an intrinsic CPL efficiency based on a single chiral molecule or matter:

$$\Lambda_{CPL} = f \times \Phi \times \frac{|g_{lum}|}{2} \propto \varepsilon \times \Phi \times |g_{lum}|$$

By definition, the $\Lambda_{CPL}$ values should be in the range between 0 and 1. Assuming comparable band-shape ($\Delta\nu$), $\Lambda_{CPL}$ becomes proportional also to $\varepsilon$. The latter term, $\varepsilon \times \Phi$, is commonly referred to as molecular brightness $B$. In principle, the materials with larger $\Lambda_{CPL}$ (or $g_{lum}$ within the similar systems assuming the brightness is comparable) values at the desired excitation and emission wavelengths are considered as favorable chiroptical materials.

From a theoretical point of view, the emission intensities (as well as absorption intensities) $f$ are proportional to the dipole strength $D$ produced by the action of electromagnetic radiation on an electric dipole of substances. The value $D$ is defined as the square of the electric transition dipole moment (etdm, $\mu$) for the electronic transition between the emissive state $j$ and ground state $i$, where $\mu$ is a real vector:

$$D = \langle \Psi_j|\mu|\Psi_i \rangle^2$$

As stated by Rosenfeld equation, the CPL (as well as CD) spectra are characterized by an analogous parameter, which is referred to as a rotational (or rotatory) strength $R$. To describe a rotation of the coordinate system, a magnetic transition dipole moment (mtdm, $m$) is given as a purely imaginary vector. From a product of wavefunction overlap integrals, $R$ can be expressed as follows:

$$R = \text{Im}\left[\langle \Psi_j|\mu|\Psi_i \rangle \cdot \langle \Psi_i|m|\Psi_j \rangle\right]$$

where Im refers to an imaginary component of the scalar product between $\mu$ and $m$. In most situations, it can be also stated that:

$$R = |\mu| \cdot |m| \cos \theta$$

where $\theta$ is the angle between the two dipole moments. Thus, in the CPL active materials, the magnetic and electric transition dipole moments must not be orthogonal to each other. It is to note that the order of magnitude of $\mu$ is typically Debye (1 D $= 3.3 \times 10^{-30}$ C m), while that of $m$ is approximately Bohr magneton (1 $\mu_B = 9.3 \times 10^{-24}$ J T$^{-1}$). In cgs unit, CPL and fluorescence bands as function of transition energy ($E$) in spectra can be obtained through the following equations:

$$\Delta I(E) = \frac{16 \, E^3 \, \rho(E)}{3 \, c^3 \, \hbar^4} R$$

$$I(E) = \frac{4 \, E^3 \, \rho(E)}{3 \, c^3 \, \hbar^4} D$$

where $\hbar$ is the reduced Planck's constant, $c$ is the speed of light, and $\rho(E)$ is a Gaussian band shape. The $E^3$ dependence is explained as the total luminescence and CPL is measured by counting the number of photons in space.

The degree of chirality in CPL, quantified by the dissymmetry factor $g_{\text{lum}}$, is a function of both strengths $D$ and $R$. Strictly speaking, $g_{\text{lum}}$ values are also dependent on the ratio of shape factors of the CPL and emission spectra, refractive index of medium (solvent) $n$, and inverse of internal field correction factor $\beta$. In isotropic solutions, it is approximated by assuming that these factors are mutually cancelled out. The dissymmetric factor $g_{\text{lum}}$ is consequently given by:

$$g_{\mathrm{lum}} = 4 \times \frac{R}{D} = 4 \times \frac{|\boldsymbol{m}|}{|\boldsymbol{\mu}|} \cos\theta$$

Thus, the luminescence dissymmetry factor $g_{\mathrm{lum}}$ is inversely proportional to the amplitude of $\boldsymbol{\mu}$ and directly proportional to that of $\boldsymbol{m}$, as well as cosine of their relative angle ($\theta$).

The rotational and dipole strengths, and thus the dissymmetric factor, can be directly estimated by the quantum chemical calculations. The time-dependent density functional theory (TD-DFT) has been successfully employed for the evaluation of the CPL spectra of distorted chiral ketones and some π-systems. Superior theoretical methods such as coupled-cluster type theory have been also applied recently. The accuracy of predicted values, however, depends on the ansatz used in the calculations as well as the geometrical features (size, flexibility, etc.) of chiral systems under study. Compared with the calculation of the CD spectra, theoretical investigation of CPL counterparts are still in the preliminary stage, which is however most likely dissolved very soon.

## 1.4  Measurement

To the best of our knowledge, very first CPL was reported by Samoilov in 1948 for chiral crystals of sodium uranyl acetate at the liquid helium temperature. In 1967, the first example of CPL from small organic molecule was reported by Emeis and Oosterhof, employing cyclic chiral ketone, *trans-β*-hydrindanone (*trans*-bicyclo [4.3.0]nonan-8-one). Then for a while, the target organic molecules for SOMs-based CPL study have been restricted to optically active cyclic ketones, almost nearly until this century. This is because these intrinsically chiral ketones exhibit relatively large $g_{\mathrm{lum}}$ values due to the magnetically allowed electronically forbidden n-π* transition of distorted carbonyl group, which appears as an isolated band at ~300 nm. In all these measurements, custom-built instrument has been employed, which required in-depth technical attentions in order to avoid any possible artifacts. Recently, measurements became considerably easier since a recent appearance of commercial CPL spectrofluorimeter such as JASCO CPL-300. In this decade, a rapid growth was realized in the CPL chemistry as the CPL spectra of most of chiral π-systems, typically bearing $g_{\mathrm{lum}}$ in an order of $10^{-3}$ or less, can be reliably obtained. Nevertheless, additional difficulties exist for obtaining reliable and reproducible CPL due to several experimental limitations, which are briefly described below.

In addition to the common precautions taken care of in the measurement of UV-vis, CD, fluorescence, and phosphorescence spectra such as medium/solvent and sample cell, amount of sample, temperature dependence, and light intensity (slit width), special attention should be required for the reliable CPL measurement. At first, light beam, polarizer/depolarizer, photoelastic modulator, and monochromator should be properly aligned and carefully calibrated (even for commercial instrument). This task is not trivial as the instrumental standard is not established thus far, although some suggestions are available. In practice, precautions are needed when the fluorescence and CPL is close to the absorption wavelength. In such cases, the observed CPL signals should be properly corrected, as a part of the emitted radiation may be re-absorbed in different degrees for left and right circularly polarized light. Concentration of samples may also matter in several ways. Aggregation leads to the supramolecular effect for CPL measurement, which gives signals typically at least one order of magnitude stronger in $g_{lum}$ than the isolated molecule. Depending on the conditions, an artifact may be also obtained through the photoselection and birefringent. When scattering phenomena become important, chiral scattering may manifest itself as apparent CPL. Last but not the least, the degradation of the sample may occur during the measurement, due to an inevitable exposure to a relatively strong excitation light during the CPL measurement. It is thus important to verify the stability (and recovery) of the sample at the end of the measurement. Unfortunately, above precautions do not always appear to have been completed in the literature examples. More serious artifact can appear for the CPL spectra of solid-state samples. One should consult the literature to understand the possible sources of artifacts before attempting to measure the CPL in the solid state or in any anisotropic conditions.

## 1.5 Perspective

Progressively large numbers of successful examples of SOMs-based CPL-active materials have been appeared in the literature. Nevertheless, total amount of study is still limited. Therefore, it is natural that a solid structure-property (i.e., CPL) relationship has not yet been established. Recent investigations are mostly focusing on the $\pi$-$\pi^*$ transition of rigid aromatic systems, as their fluorescence quantum yields are normally moderate to high. The ideal CPL-active materials are expected to simultaneously have high luminescence quantum yield and high (absolute) $g_{lum}$ value, and if possible that also function with some desired stimuli. It is certainly a great advantage that rational modification on the structure of SOMs, particularly of aromatic systems, can effectively modulate the various physical parameters in the CPL-responsible materials, while the luminescence dissymmetry factors for SOMs still remain unsatisfactory; further progresses are certainly desired.

We have recently examined quantitative relationship between the dissymmetric factors of CPL ($g_{lum}$) and CD ($g_{abs}$) for such $\pi$-$\pi^*$ transitions of extended aromatic systems. It was the first systematic investigation to experimentally elucidate the relationship between the dissymmetry factors in order to strategically design the CPL response through the rational molecular modification on the aromatic systems. Although the number of the available data was still limited to around 100 examples, they unexpectedly afforded a good empirical linear correlation as a global fit: $|g_{lum}| = 0.81 \times |g_{abs}|$ ($r^2 = 0.60$) (Fig. 1.3). It is also to highlight that all the CPL data used in the study have been published in 2011 or later, except for two cases,

**Fig. 1.3** The $g_{lum}$-$g_{abs}$ correlation for of the $\pi$-$\pi^*$ transition in chiral aromatic molecules. Square, cyclophanes; triangle, biaryls; open circle, helicene-like molecules; open inverted triangle, BODIPY derivatives. Reproduced with permission [1]. Copyright 2018, Wiley

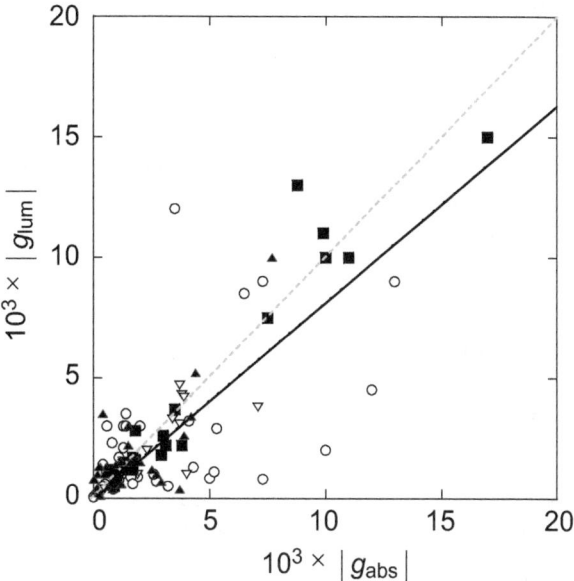

implying the innovative aspect of this study area. Such correlation can be easily understood as follows. In the rigid aromatic systems, minimal structural changes are expected in the $S_1$ state for the relaxation to the emissive state, as they always provide mirror-imaged absorption and emission spectra. Accordingly, the influence of the vibrational and structural relaxation in the excited state can be negligible, and thus the chiroptical properties obtained from the CD and CPL spectra become consistent with each other, as they originate from the upward and downward transitions between the same two states. When more carefully analyzed, distinct (but still linear) correlations are more likely, depending on the molecular structure, i.e., type of chirality, varying from 0.99 for planar chiral cyclophanes, to 0.93 for axially chiral biaryls, to 0.77 for chiral BODIPY dyes, and then to 0.61 for helicenes and related molecules. Nevertheless, the slope of $\approx$0.8 obtained of all the $g_{lum}$-$g_{abs}$ data reported by 2017 clearly suggests that the dissymmetry of chiral extended $\pi$ molecules does not significantly deteriorate on average after the structural relaxation upon the photoexcitation. Such structural considerations are of paramount importance and would make us available to more sensibly design the further advanced CPL materials in the future.

Finally, we draw attention to the recent stimulating discoveries such as the enhanced CPL response through fluorescence resonance energy transfer and for up-converted emission, as well as the CPL associated with delayed and dual fluorescence. All these possibilities show that chiral SOMs, especially those of the extended $\pi$-systems, are the most promising candidates for more innovative CPL materials that are readily accessible, theoretically predictable, rationally designable and modifiable.

Considering the nature of this chapter, we did not cite various original papers as a reference. We suggest the following review articles for a further reading for those more interested in the CPL phenomena, where you will find better and more detailed information and explanation on mathematical expression, historical view, critical issues, and additional examples [1–13]. Also note that photophysical aspect on CPL, CPL under external fields, CPL from metal complexes or polymers, and additional recent findings besides the SOMs-based CPL phenomena are all outside the scope of this book.

# References

1. Tanaka H, Inoue Y, Mori T (2018) Circularly polarized luminescence and circular dichroisms in small organic molecules: correlation between excitation and emission dissymmetry factors. ChemPhotoChem 2:386–402. https://doi.org/10.1002/cptc.201800015
2. Richardson FS, Riehl JP (1977) Circularly polarized luminescence spectroscopy. Chem Rev 77:773–792. https://doi.org/10.1021/cr60310a001
3. Riehl JP, Richardson FS (1986) Circularly polarized luminescence spectroscopy. Chem Rev 86:1–16. https://doi.org/10.1021/cr00071a001
4. Brittain HG (1987) Excited state optical activity, 1983–1986. Photochem Photobiol 46:1027–1034. https://doi.org/10.1111/j.1751-1097.1987.tb04889.x
5. Brittain HG (1996) Excited-state optical activity, 1987–1995. Chirality 8:357–363. https://doi.org/10.1002/(SICI)1520-636X(1996)8:5<357::AID-CHIR1>3.0.CO;2-B
6. Riehl JP, Muller G (2011) Circularly polarized luminescence spectroscopy and emission-detected circular dichroism. Compr Chiropt Spectrosc 1:65–90. https://doi.org/10.1002/9781118120187.ch3
7. Maeda H, Bando Y (2013) Recent progress in research on stimuli-responsive circularly polarized luminescence based on π-conjugated molecules. Pure Appl Chem 85:1967–1978. https://doi.org/10.1351/PAC-CON-12-11-09
8. Sanchez-Carnerero EM, Agarrabeitia AR, Moreno F, Maroto BL, Muller G, Ortiz MJ, de la Moya S (2015) Circularly polarized luminescence from simple organic molecules. Chem Eur J 21:13488–13500. https://doi.org/10.1002/chem.201501178
9. Kumar J, Nakashima T, Kawai T (2015) Circularly polarized luminescence in chiral molecules and supramolecular assemblies. J Phys Chem Lett 6:3445–3452. https://doi.org/10.1021/acs.jpclett.5b01452
10. Longhi G, Castiglioni E, Koshoubu J, Mazzeo G, Abbate S (2016) Circularly polarized luminescence: a review of experimental and theoretical aspects. Chirality 28:696–707. https://doi.org/10.1002/chir.22647
11. Han J, Guo S, Lu H, Liu S, Zhao Q, Huang W (2018) Recent progress on circularly polarized luminescent materials for organic optoelectronic devices. Adv Opt Mater 6:1800538. https://doi.org/10.1002/adom.201800538
12. Chen N, Yan B (2018) Recent theoretical and experimental progress in circularly polarized luminescence of small organic molecules. Molecules 23:3376. https://doi.org/10.3390/molecules23123376
13. Ma J-L, Peng Q, Zhao C-H (2019) Circularly polarized luminescence switching in small organic molecules. Chem Eur J 25:15441–15454. https://doi.org/10.1002/chem.201903252

# Chapter 2
# Circularly Polarized Luminescence of Axially Chiral Binaphthyl Fluorophores

**Yoshitane Imai**

**Abstract** Axially chiral fluorophores bearing a variety of functionalities have been successfully developed using chiral binaphthyl units. These axially chiral binaphthyl fluorophores emit circularly polarized luminescence (CPL) in their solution-dissolved states, organic polymer-film-dispersed states, and inorganic pellet-dispersed states. The CPL emitted from an axially chiral binaphthyl fluorophore is easily tuned by (1) adjusting the dihedral angle in the binaphthyl unit, (2) employing the neighboring group effect between binaphthyl units, and (3) controlling their external environments, without the need for the enantiomeric compound.

## 2.1 Introduction

Organic compounds with chiroptical properties are very important for the development of new functional organic materials.

Axially chiral binaphthyl is a significant fundamental chiral unit that is used to introduce chirality into molecules and materials. An axially chiral binaphthyl organic fluorophore that can emit circularly polarized luminescence (CPL) is therefore very useful, since the binaphthyl unit exhibits both chirality and fluorescence [1–20].

In this chapter, the solution- and solid-state CPL properties of various axially chiral binaphthyl fluorophores are introduced.

Y. Imai (✉)
Kindai University, Osaka, Japan
e-mail: y-imai@apch.kindai.ac.jp

© Springer Nature Singapore Pte Ltd. 2020
T. Mori (ed.), *Circularly Polarized Luminescence of Isolated Small Organic Molecules*, https://doi.org/10.1007/978-981-15-2309-0_2

## 2.2 Controlling Circularly Polarized Luminescence (CPL) Through the Dihedral Angle of the Axially Chiral Binaphthyl [21]

In this section, we report the chiroptical properties of axially chiral binaphthyl compounds in relation to the dihedral angle in the binaphthyl unit. For this purpose, two almost identical binaphthyl derivatives with the same (S)-axial chirality, namely, (S)-2,2'-diethoxy-1,1'-binaphthyl [(S)-1], as an open-type binaphthyl, and (S)-2,2'-(1,4-butylenedioxy)-1,1'-binaphthyl [(S)-2], as a closed-type, were chosen (Fig. 2.1).

The solution-state circularly polarized luminescence (CPL) and unpolarized photoluminescence (PL) spectra of the open-type (S)-1 and the closed-type (S)-2 in chloroform (CHCl$_3$) are shown in Fig. 2.2. (S)-1 and (S)-2 exhibit similar PL in CHCl$_3$; PL maxima ($\lambda_{PL}$) for (S)-1 and (S)-2 were observed at 365 and 368 nm, respectively, and the absolute values of their PL quantum yields ($\Phi_F$s) were 19% and 25%, respectively. The $\Phi_F$ value of (S)-2 is greater than that of (S)-1 because the ro-vibrational modes of (S)-2 are limited compared to those of (S)-1, and the binaphthyl unit of (S)-2 is more planar than that of (S)-1.

**Fig. 2.1** Chiral binaphthyl fluorophores (S)-1 and (S)-2

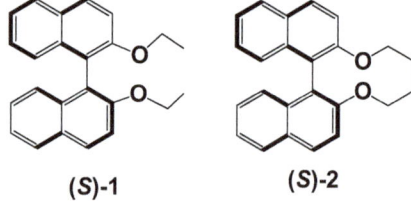

**Fig. 2.2** CPL (upper) and PL (lower) spectra of (S)-1 (black) and (S)-2 (red) in CHCl$_3$ (1.0 × 10$^{-3}$M)

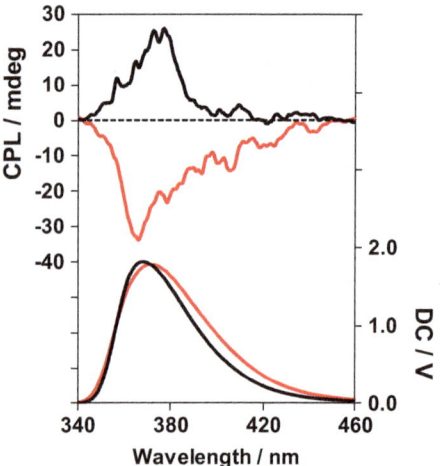

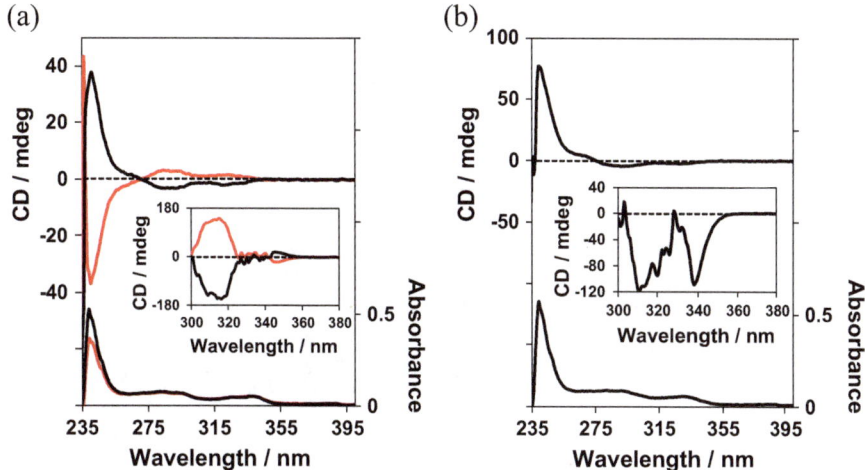

**Fig. 2.3** CD (upper) and UV-Vis absorption (lower) spectra for (**a**) (*S*)-**1** (black) and (*R*)-**1** (red) and (**b**) (*S*)-**2** (black) in CHCl$_3$ (1.0 × 10$^{-5}$ and 1.0 × 10$^{-3}$M (insets))

(*S*)-**1** and (*S*)-**2** emit CPL in CHCl$_3$. Surprisingly, although (*S*)-**1** and (*S*)-**2** have the same axial chirality, their CPL spectra show opposite signs; a positive (+) sign is observed for (*S*)-**1**, while (*S*)-**2** exhibits a negative (−) sign. To quantitatively compared the degrees of CPL and PL, we used the dimensionless Kuhn's anisotropy factor in the photoexcited state, which is defined as $g_{CPL} = 2(I_L − I_R)/(I_L + I_R)$, where $I_L$ and $I_R$ are the intensities of the left- and right-handed CPLs upon excitation with unpolarized light, respectively. The values of $g_{CPL}$ were found to be about $+1.0 \times 10^{-3}$ for (*S*)-**1** and about $−1.4 \times 10^{-3}$ for (*S*)-**2**. These results suggest that the sign of the CPL depends on the nature of the binaphthyl ring (i.e., open or closed) in addition to axial chirality. To discuss the origin of the reversal in the sign of the CPL, solution-state circular dichroism (CD) and unpolarized UV-Vis absorption spectra for (*S*)-**1** and (*R*)-**1** were acquired, as shown in Fig. 2.3.

Several characteristic vibronic UV-Vis bands between 270 and 360 nm that arise from the $^1B_b$ transition moments of (*S*)-**1** and (*S*)-**2** were commonly observed. To quantitatively discuss the magnitude of the CD amplitude, we used the dimensionless Kuhn's anisotropy factor in the ground state, which is defined as $g_{CD} = \Delta\varepsilon/\varepsilon$. Values of $g_{CD}$ for (*S*)-**1** and (*S*)-**2** at their first Cotton CD bands are about $+2.0 \times 10^{-4}$ (345 nm) and about $−9.6 \times 10^{-4}$ (338 nm), respectively. Evidently, the first Cotton CD band of (*S*)-**1** has opposite sign to that of (*S*)-**2**. This reversal in the CD sign, despite the same axial chirality, is possibly due to the difference in their chemical structures, i.e., whether the substituents are in the open or closed form. The opposite signs of the first Cotton CD bands are responsible for the opposite sign of the CPL bands observed for (*S*)-**1** and (*S*)-**2**.

To investigate the origins of the opposing CPL and CD spectral signs of (*S*)-**1** and (*S*)-**2**, CD spectra for (*S*)-**1** were determined computationally. Simulated

**Fig. 2.4** CD and UV-Vis
absorption spectra of (*S*)-**1**
calculated as functions of
the dihedral angle *θ* [+50°
(—), +60° (—), +70° (—),
+80° (—), +90° (—), +100°
(—), +110° (—), +120° (—)]
in the binaphthyl unit

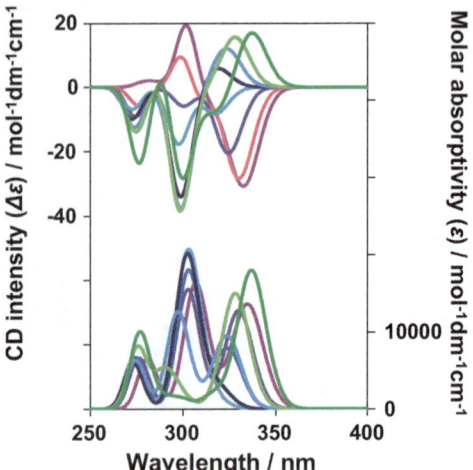

dihedral-angle-dependent ($\theta = (O)C$-$C$-$C$-$C(O)$) CD and UV-Vis absorption spectra of (*S*)-**1** are presented in Fig. 2.4. These simulations reveal that the sign of the first Cotton CD band of (*S*)-**1** is negative ($-$) for dihedral angles ($\theta$) between about +50° and +85°, whereas the sign is positive (+) for $\theta$ between +85° and +120°.

These results suggest that the opposite first CD and CPL bands observed for (*S*)-**1** and (*S*)-**2** are attributable to differences in the $\theta$ values of the binaphthyl units in the ground and photoexcited states. In fact, the calculated equilibrium structures of (*S*)-**1** and (*S*)-**2** were observed to have $\theta$ values of +89.6° and +73.8° respectively, which suggests that the former has a positive first Cotton CD band (~340 nm) and the latter has a negative first Cotton CD band (~340 nm); these simulated results are consistent with experimental observations.

In conclusion, the CPL properties of an axially chiral binaphthyl fluorophore can be controlled by adjusting the dihedral angle of the binaphthyl unit in addition to its chirality. In general, the enantiomeric organic fluorophore is usually required to invert the sign of the CPL of a chiral fluorophore; however, the enantiomeric organic molecule is sometimes difficult to obtain. Therefore, controlling the sign of the CPL of axially chiral fluorophores through dihedral angle, without the need for the enantiomer, is a very useful technique.

## 2.3  Controlling the Sign of the Circularly Polarized Luminescence (CPL) from an Axially Chiral Binaphthyl Fluorophore by Solvent [22]

In Sect. 2.2, we reported that the sign of the CPL from a chiral binaphthyl fluorophore can be controlled by tuning the dihedral angle of the binaphthyl unit. In this section, solvent polarity control of the CPL sign of an axially chiral

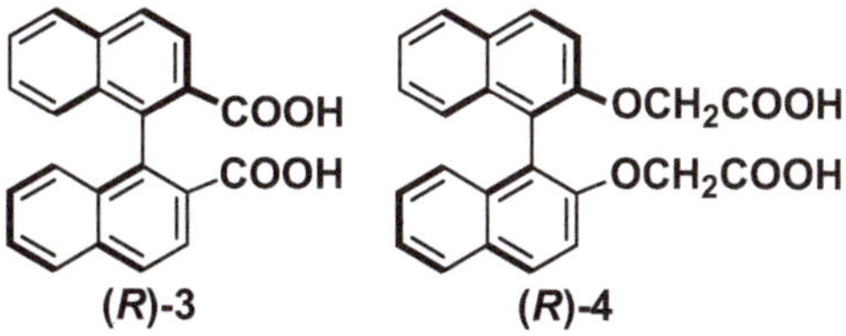

**Fig. 2.5** Chiral binaphthyl fluorophores (*R*)-**3** and (*R*)-**4**

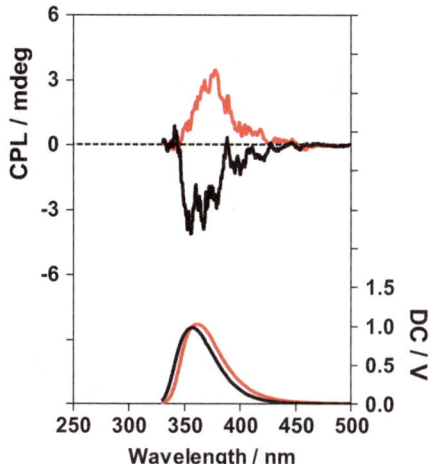

**Fig. 2.6** CPL (upper) and PL (lower) spectra of (*R*)-**4** in CHCl$_3$ (black) and DMF (red) ($1.0 \times 10^{-4}$ M)

binaphthyl fluorophore in solution is demonstrated. As axially chiral binaphthyl fluorophore models, (*R*)-1,1'-binaphthyl-2,2'-dicarboxylic acid [(*R*)-**3**] and (*R*)-2,2'-(1,1'-binaphthyl-2,2'-diylbis(oxy))diacetic acid [(*R*)-**4**] were examined (Fig. 2.5).

The PL spectra of (*R*)-**3** and (*R*)-**4** were acquired in CHCl$_3$ and dimethylformamide (DMF) solutions, owing to their differing polarities. Dicarboxylic acid (*R*)-**3** did not exhibit PL in CHCl$_3$ and DMF solutions. On the other hand, (*R*)-**4** exhibited PL in both CHCl$_3$ and DMF, with PL maxima ($\lambda_{PL}$) observed at 356 and 361 nm, respectively, as shown in Fig. 2.6 (lower panel, black lines for CHCl$_3$ and red lines for DMF). The PL quantum yield ($\Phi_F$) of (*R*)-**4** in CHCl$_3$ was found to be 25%. On the other hand, its $\Phi_F$ value increased to 39% in DMF, possibly due to the suppression of the rotational freedom of the binaphthyl unit as a result of hydrogen-bonding interactions with DMF. (*R*)-**4** exhibited a CPL signal in CHCl$_3$ and DMF, as shown in Fig. 2.6 (upper). Interestingly, although significant changes in $\lambda_{em}$ were not observed in the two solvents, (*R*)-**4** exhibited opposite CPL signs in the two solvents: negative (−) in CHCl$_3$ and positive (+) in DMF. Their $g_{CPL}$ values were found to be similar at about $-0.18 \times 10^{-3}$ for CHCl$_3$ and about $+0.19 \times 10^{-3}$ for DMF.

In order to elucidate the ground state chirality of **4**, CD and UV-Vis absorption spectra in CHCl$_3$ and DMF were acquired, as shown in Fig. 2.7a, b, respectively. The CD spectra of **4** exhibited significant differences in the first Cotton CD bands in

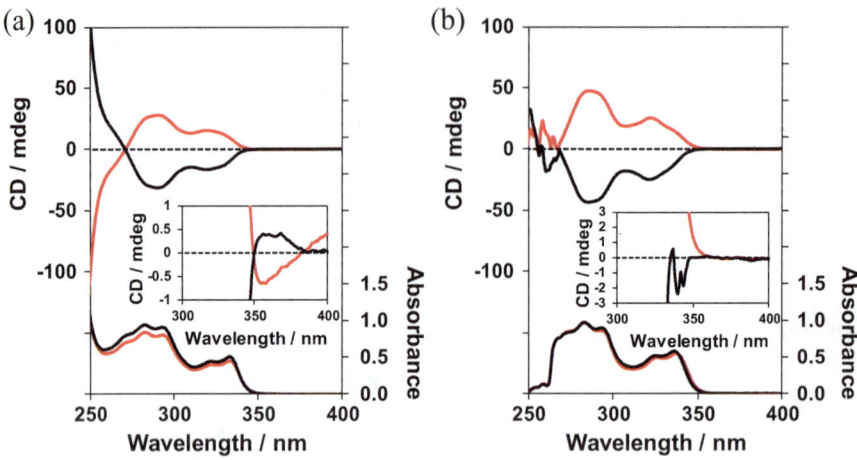

**Fig. 2.7** CD (upper) and UV-Vis absorption (lower) spectra of ($R$)-**4** (red) and ($S$)-**4** (black) in (**a**) CHCl$_3$ and (**b**) DMF ($1.0 \times 10^{-4}$ and $1.0 \times 10^{-3}$M (insets))

CHCl$_3$ and DMF (Fig. 2.7; upper). The negative ($-$) sign ($\lambda_{CD} = 356$ nm) for ($R$)-**4** in CHCl$_3$ became positive (+) ($\lambda_{CD} = 337$ nm) in DMF. The $g_{CD}$ value at the first Cotton CD band of ($R$)-**4** in CHCl$_3$ was about $-0.33 \times 10^{-3}$, while a value of about $+0.38 \times 10^{-4}$ was observed in DMF.

The observed reversals of CPL and CD signs are possibly ascribable changes in the dihedral angles of the binaphthyl units. In DMF, the dihedral angle in the binaphthyl unit is smaller since the carboxylic acid groups in **4** interact efficiently with DMF molecules. Dipole-dipole and/or hydrogen-bonding interactions between DMF and the carboxylic acid groups on the binaphthyl units may also be responsible for the observed CD and CPL inversions in the ground and photoexcited states.

In conclusion, an axially chiral binaphthyl fluorophore can selectively emit positive or negative CPL through control of solvent type.

## 2.4 Controlling Circularly Polarized Luminescence (CPL) Through Neighboring Group Effects Involving Binaphthyl Units [23]

In Sect. 2.2, the signs of the CPL from chiral binaphthyl fluorophores were shown to be controllable by tuning the dihedral angle of the binaphthyl unit. As a more facile method for controlling the sign of the CPL from a chiral binaphthyl moiety while maintaining the same axial chirality, CPL sign control through neighboring-group effects between binaphthyl units has been reported. To enable this form of control, two binaphthyl derivatives with different numbers of chiral binaphthyl units that

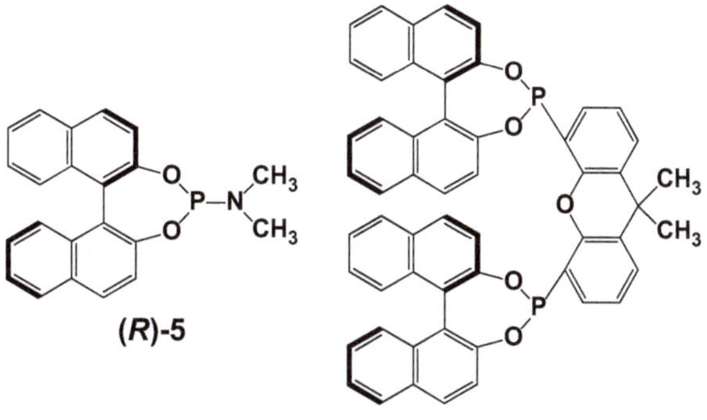

**Fig. 2.8** Chiral binaphthyl
fluorophores (*R*)-**5** and (*R*,
*R*)-**6**

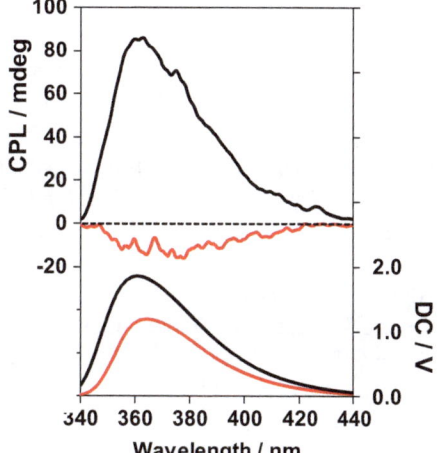

(*R*)-**5**

(*R*,*R*)-**6**

**Fig. 2.9** CPL (upper) and
PL (lower) spectra of (*R*)-**5**
(black) and (*R*,*R*)-**6** (red) in
CHCl$_3$ (1.0 × 10$^{-3}$M)

have the same (*R*)-axial chirality were investigated, namely, (*R*)-(−)-(3,5-dioxa-4-phosphacyclohepta[2,1-a:3,4-a′]dinaphthalen-4-yl)dimethylamine [(*R*)-**5**], with a single binaphthyl unit, and (11b*R*,11′b*R*)-4,4′-(9,9-dimethyl-9*H*-xanthene-4,5-diyl)-bis-di-naphtha[2,1-*d*:1′,2′-f][1,3,2]dioxaphosphepin [(*R*,*R*)-**6**], with two binaphthyl units (Fig. 2.8).

The CPL and PL properties of (*R*)-**5** and (*R*,*R*)-**6** were compared in CHCl$_3$. (*R*)-**5** and (*R*,*R*)-**6** exhibited PL maxima (λ$_{PL}$) at 361 and 362 nm in CHCl$_3$, respectively, as shown in Fig. 2.9. (*R*)-**5** and (*R*,*R*)-**6** emitted CPL in CHCl$_3$. Surprisingly, although (*R*)-**5** and (*R*,*R*)-**6** contain the same axially chiral binaphthyl unit, they exhibited oppositely signed CPL spectra, with a positive (+) sign observed for (*R*)-**5** and a negative (−) sign observed for (*R*,*R*)-**6**. The values of g$_{CPL}$ were determined to be about +3.4 × 10$^{-3}$ for (*R*)-**5** and −1.3 × 10$^{-3}$ for (*R*,*R*)-**6**, where the |g$_{CPL}$| value of **6** is smaller than that of **5** by a factor of three.

The solution-state CD and UV-Vis absorption spectra of (*R*)-**5** and (*R*,*R*)-**6** were acquired (Fig. 2.10). Several characteristic UV-Vis bands arising from π-π*

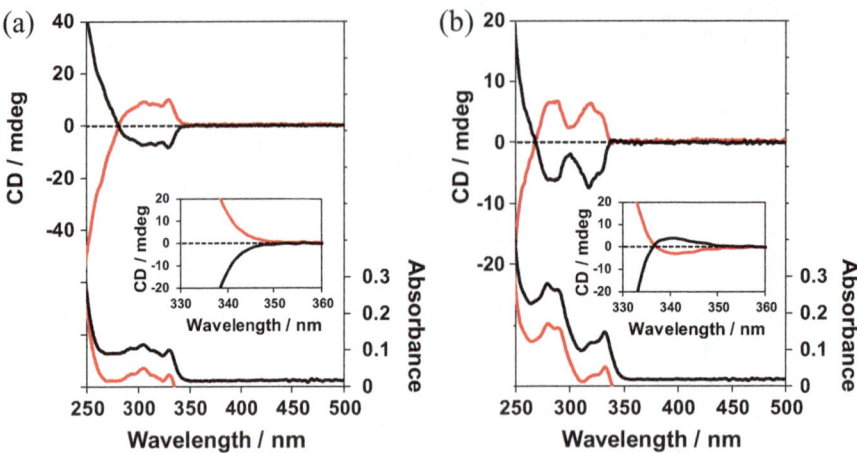

**Fig. 2.10** CD (upper) and UV-Vis absorption (lower) spectra for (**a**) (*R*)-**5** (red) and (*S*)-**5** (black) and for (**b**) (*R,R*)-**6** (red) (*S,S*)-**6** (black) in CHCl$_3$ (1.0 × 10$^{-5}$ and (insets) 1.0 × 10$^{-4}$M)

transitions involving the naphthyl groups in **5** and **6** were observed at approximately 270 and 360 nm, respectively. The signs of the CPL and the first Cotton CD band of (*R*)-**5** are both positive (+). Although the CD spectrum of (*R,R*)-**6** is similar to that of (*R*)-**5**, a weak negative (−) Cotton band was observed at the longest-wavelength edge in the spectrum of (*R,R*)-**6**. Because the CPL sign of (*R,R*)-**6** is negative (−), this weak negative (−) Cotton CD band may be responsible for the negative (−) sign of the CPL band. The |$g_{CD}$| value at the first Cotton band ($\lambda_{CD}$ = 330 nm) of **5** was determined to be about 4.2 × 10$^{-3}$; on the other hand, the |$g_{CD}$| value of **6** at the first Cotton CD band ($\lambda_{CD}$ = 341 nm) is about 2.5 × 10$^{-4}$, which is smaller than that of **5** by a factor of 17. Since the dihedral angles of the binaphthyl units in **5** and **6** are fixed, the opposite CD and CPL signs observed for (*R*)-**5** and (*R,R*)-**6**, despite having the same axial chirality, are due to the effects of the neighboring chiral binaphthyl units in (*R,R*)-**6**.

In conclusion, the CPLs of chiral binaphthyl fluorophores can be controlled by the number of neighboring chiral binaphthyl units, that is, by the effects of the neighboring binaphthyl units.

## 2.5 Dependence of the Circularly Polarized Luminescence (CPL) from Binaphthyl Units on External Environmental Factors [24]

In Sects. 2.2 and 2.4, the CPL signs of chiral binaphthyl luminophores with the same axial chirality were reported to be controllable by tuning the dihedral angle of the binaphthyl unit and by employing the neighboring group effect between fluorescent

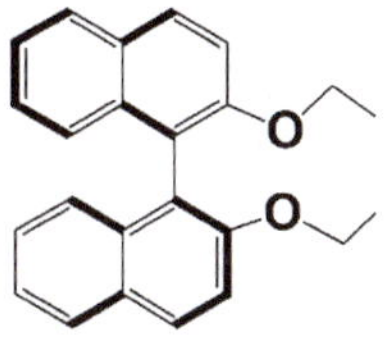

**Fig. 2.11**  Chiral binaphthyl
fluorophore (S)-**1**

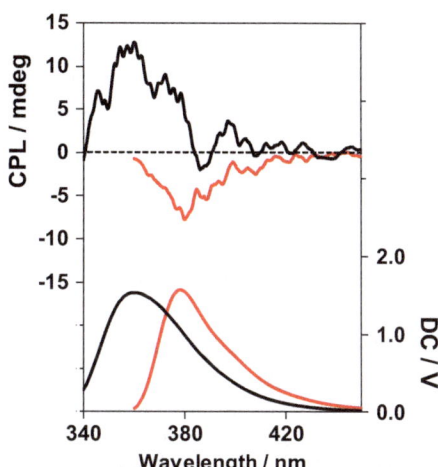

**Fig. 2.12**  CPL (upper) and
PL (lower) spectra of (S)-**1**
in PMMA film-dispersed
(black) and KBr pellet-
dispersed (red) states

binaphthyl units. In this section, an approach for the facile control of the CPL signs
of chiral binaphthyl fluorophores involving the external solid-state environment is
reported; (S)-2,2'-diethoxy-1,1'-binaphthyl [(S)-**1**] is used in this section as a model
optically active binaphthyl fluorophore (Fig. 2.11). As the solid-state environment,
organic poly(methyl methacrylate) (PMMA) and inorganic KBr are used.

The PL and CPL spectra of (S)-**1** in the PMMA film-dispersed state and the KBr
pellet-dispersed state were acquired, as shown in Fig. 2.12. (S)-**1** dispersed in
PMMA and KBr exhibited PL maxima ($\lambda_{PL}$) at 373 and 377 nm, respectively,
with corresponding PL quantum yields ($\Phi_F$) of 45% and 54%. Characteristically,
the values of $\Phi_F$ for (S)-**1** were higher in the solid state than in CHCl$_3$ solution,
which is possibly due to the suppression of vibrational deactivation of the molecule
in the solid state.

(S)-**1** emits CPL both in the PMMA film-dispersed state and the KBr pellet-
dispersed state. Surprisingly, (S)-**1** exhibited opposite CPL signs in these two
states: positive (+) for the PMMA state and negative (−) for the KBr state. (S)-**1**
has $g_{CPL}$ values in the PMMA film and KBr pellet states of about $+7.9 \times 10^{-4}$ and
$-4.4 \times 10^{-4}$, respectively. The $|g_{CPL}|$ values of (S)-**1** in the PMMA and KBr states
were slightly smaller than that of the CHCl$_3$ solution ($g_{CPL} = +1.0 \times 10^{-3}$).

The CD and UV-Vis absorption spectra of (S)-**1** in the PMMA film- and KBr
pellet-dispersed states were also acquired (Fig. 2.13). Several UV-Vis bands were

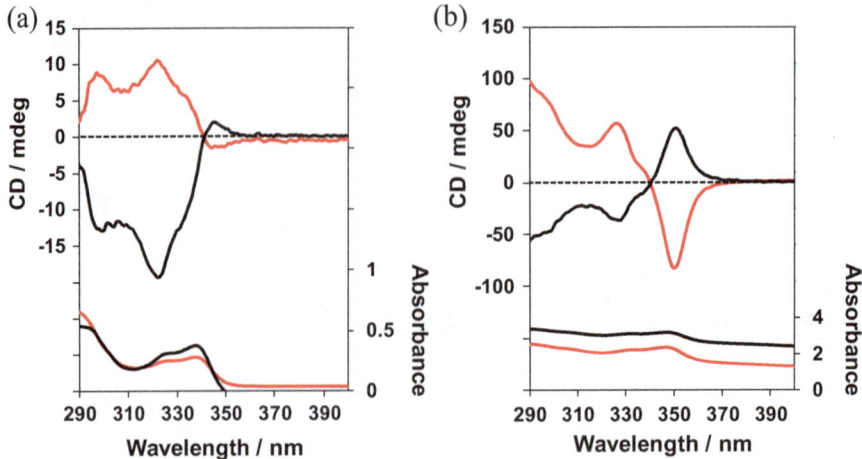

**Fig. 2.13** CD (upper) and UV-Vis absorption (lower) spectra for ($S$)-**1** (black) and ($R$)-**1** (red) in (**a**) the PMMA film-dispersed state and (**b**) the KBr pellet-dispersed state

also observed at ~290 and ~380 nm in the solid state, which correspond to characteristic $\pi$-$\pi*$ transitions in the naphthyl groups of **1**. The CD spectra of ($S$)-**1** in two states are very similar in their long-wavelength tails. In contrast to the CPL spectra, the long-wavelength CD Cotton bands of ($S$)-**1** were both positive (+); the $g_{CD}$ values of first Cotton CD band of ($S$)-**1** were $+3.5 \times 10^{-4}$ at $\lambda_{CD} = 345$ nm in the PMMA film-dispersed state and $+7.6 \times 10^{-4}$ at $\lambda_{CD} = 351$ nm in the KBr pellet-dispersed state. These $g_{CD}$ values are slightly larger than that of the CHCl$_3$ solution of ($S$)-**1** ($g_{CD} = $ ~$+2.0 \times 10^{-4}$). The CD amplitudes of ($S$)-**1** in PMMA and KBr were evaluated as follows: $g_{CD} = ((Abs(L) - Abs(R))/|(Abs(L) + Abs(R))/2|$.

To determine the origins of the CPL and CD spectra of ($S$)-**1** in its PMMA-dispersed and KBr-dispersed states, the equilibrium structures of single molecules of **1** in solution (or PMMA) and KBr were compared. The equilibrium dihedral angle ($\theta = $ (O)C-C-C-C(O)) of ($S$)-**1** in CHCl$_3$ solution was determined to be 89.6°, while the structure of a single ($S$)-**1** molecule in the PMMA film-dispersed state was identical to that in solution. X-ray crystallography revealed that the dihedral angle $\theta$ of ($S$)-**1** in the solid state is 117.1°. Although these angles in the two states differ, the signs of the first Cotton CD bands of **1** in each state should be the same based on the simulated CD spectra reported in Sect. 2.2. These results suggest that the inversion of the CPL bands observed for **1** is determined by whether or not there are intermolecular neighboring group effects operating between chiral binaphthyl units in the photoexcited states.

In conclusion, the solid-state CPL of an axially chiral binaphthyl fluorophore can be controlled by changing its external solid-state matrix environment, from a PMMA film-dispersed state to a KBr pellet-dispersed state.

## 2.6   Dependence of Circularly Polarized Luminescence (CPL) on the Structure of the Tether Connecting Binaphthyl and Fluorescent Units [25]

The binaphthyl family contains one of the best chirality-inducing building blocks because binaphthyl units act as hinges that enable connections to a variety of molecular components, in addition to being a fluorescence source. In this section, control of the sign of the CPL of a chiral non-luminophoric binaphthyl-driven pyrene excimer system is demonstrated with the help of freely rotating alkyl ether tethers. Binaphthyl-pyrene **7** containing long oligoether tethers composed of ten single bonds, and binaphthyl-pyrene **8** with shorter (six single bonds) alkyl ether tethers were used (Fig. 2.14). In these systems, the binaphthyl moiety acts as a molecular hinge.

The PL spectra of (*R*)-**7** and (*R*)-**8** in CHCl$_3$ solutions are provided in the lower parts of Fig. 2.15. Both compounds exhibit almost identical PL maxima ($\lambda_{PL}$: 465 and 480 nm for (*R*)-**7** and 462 and 480 nm for (*R*)-**8**). Although these PL bands are intense, the monomeric PL bands at ~380 nm are very weak. This feature is ascribable to intramolecularly sterically constrained dimeric pyrenes in their $S_0$ states.

In contrast to their PLs, (*R*)-**7** and (*R*)-**8** exhibit opposite CPL signs in their $S_1$-state chiralities; they have the same binaphthyl (*R*)-chiralities as the $S_1$ state chirality: a negative (−) sign for (*R*)-**7** and a positive (+) sign for (*R*)-**8**. The excimeric origin of the CPLs of (*R*)-**7** and (*R*)-**8** in CHCl$_3$ are attributable to long-distance intramolecular chiral interactions between the two remote pyrenes attached to the chiral binaphthyl backbones through reorganization of the two photoexcited pyrenes in (*R*)-**7** and (*R*)-**8**. The $g_{CPL}$ values of (*R*)-**8** (+1.2 × 10$^{-3}$ at 462 nm and +1.0 × 10$^{-3}$ at 480 nm) are larger than those of (*R*)-**7** (−0.41 × 10$^{-3}$ at 465 nm and −0.46 × 10$^{-3}$ at 480 nm). In this case, not only is the axial chirality of the binaphthyl in the $S_0$ state a critical factor that determines the CPL sign of the $S_1$ state, but so is the structure of the tether.

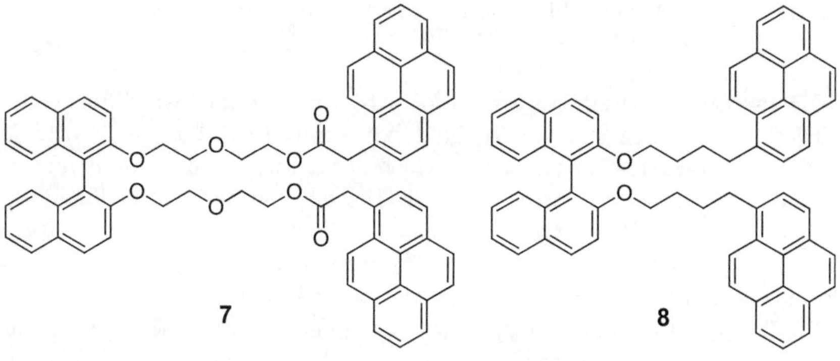

**Fig. 2.14**  Chiral binaphthyl-pyrene fluorophores **7** and **8**

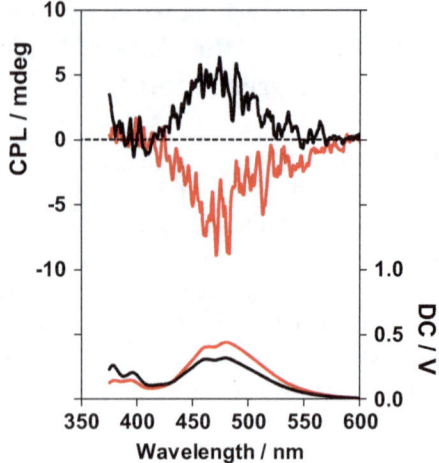

**Fig. 2.15** CPL (upper) and PL (lower) spectra of (R)-**7** (red) and (R)-**8** (black) in CHCl₃ (1.0 × 10⁻⁵M)

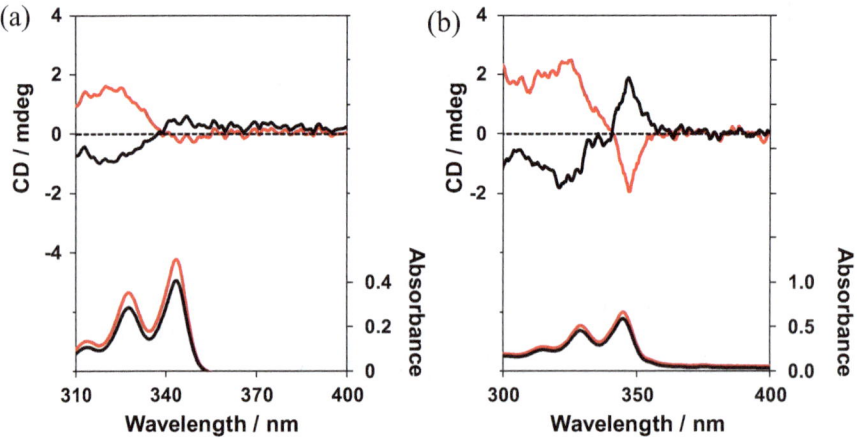

**Fig. 2.16** CD (upper) and UV-Vis absorption (lower) spectra of (**a**) (R)-**7** (red) and (S)-**7** (black lies) and (**b**) (R)-**8** (red) and (S)-**8** (black) in CHCl₃ (1.0 × 10⁻⁵M)

CD and UV-Vis absorption spectra of (R)-**7** [or (S)-**7**] and (R)-**8** [or (S)-**8**] in CHCl₃ were acquired (Fig. 2.16a, b, respectively). Several $\pi$-$\pi*$ vibronic UV transitions in the 315–355 nm range, which are characteristic of the pyrene and binaphthyl moieties, are clearly observed. The three well-resolved vibronic UV bands of (R)-**7** and (R)-**8**, which correspond to 0-0′ (~345 nm), 0-1′ (~329 nm), 0-2′ (~315 nm) bands, are observed; these three vibronic bands originate from the allowed $^1L_a$ transitions of the pyrene moieties in (R)-**7** and (R)-**8**.

The UV-Vis absorption spectra of (R)-**7** and (R)-**8** are almost identical and their CD spectra are also similar. The $g_{CD}$ values of the 0-0′ band of (R)-**7** (at 343 nm) and (R)-**8** (at 345 nm) are similar, at only $-0.22 \times 10^{-4}$ and $-0.67 \times 10^{-4}$, respectively. Surprisingly, the CD signals at the 0-0′ band of (R)-**7** and (R)-**8** both exhibit negative

$(-)$ Cotton signs (Fig. 2.16, upper), even though $(R)$-**7** and $(R)$-**8** display opposite CPL signs (Fig. 2.15, upper). Characteristically, these $g_{CD}$ values are smaller than the corresponding $g_{CPL}$ values.

In this system, intermolecular associations of $(R)$-**7** and $(R)$-**8** in the $S_0$ state are neglected. In addition, an odd-even effect involving the number of atoms in the tethers may be a crucial factor that defines the signs of the CD and CPL signals; however, both tethers in **7** and **8** have odd numbers of atoms. These results lead to the conclusion that, although the chirality-oriented conformations of the two pyrenes in **7** and **8** in the $S_0$ states are similar, their relative orientations in the $S_1$ states may be very different, possibly with opposite handedness, which causes opposite CPL signs to be observed for **7** and **8**.

In conclusion, in axially chiral binaphthyls that connect fluorescent units, the CPL signs, which are generated by the two photoexcited fluorophores, can be controlled through tethered structures involving the axially chiral binaphthyl.

## 2.7 Cryptochiral Circularly Polarized Luminophores Based on Axially Chiral Binaphthyls [26]

In this section, the novel functional binaphthyl fluorophores designed using the CPL-controlling concepts discussed in Sects. 2.2 and 2.6, involving the 5,5′,6,6′,7,7′,8,8′-octahydro-1,1′-bi-2-naphthyl fluorophore **9** connected to two pyrenes through flexible ester tethers, is reported (Fig. 2.17).

The PL and CPL spectra of $(R)$-**9** and $(S)$-**9** in CHCl$_3$ are shown in Fig. 2.18. $(R)$-**9** exhibits an intense excimer PL band at 468 nm ($\lambda_{PL}$) of pyrene origin. In addition, a weak monomeric PL band at 396 nm is observed, which is ascribable to the sterically constrained intramolecular dimeric pyrene moieties in their $S_0$ states. $(R)$-**9** and $(S)$-**9** emit CPL and exhibit positive $(+)$ and negative $(-)$ CPL signs in their $S_1$-state structures, respectively. The absolute value of $g_{CPL}$ for $(R)$-**9** in CHCl$_3$ is $+2.5 \times 10^{-3}$ at 454 nm.

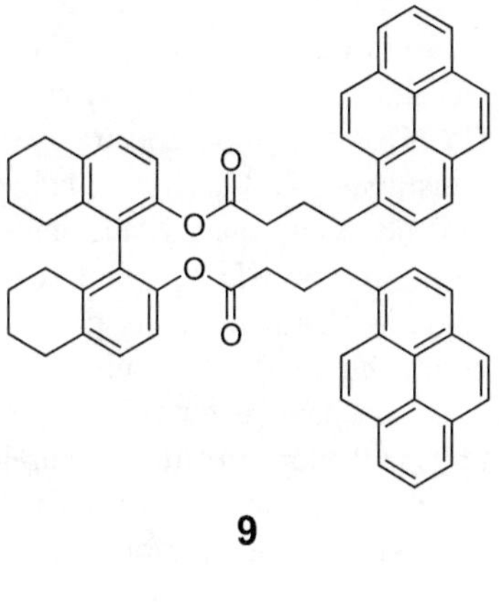

**Fig. 2.17** Chiral binaphthyl-pyrene fluorophore **9**

**9**

**Fig. 2.18** CPL (upper) and
PL (lower) spectra of (*R*)-**9**
(red) and (*S*)-**9** (black) in
CHCl$_3$ (1.0 × 10$^{-5}$M)

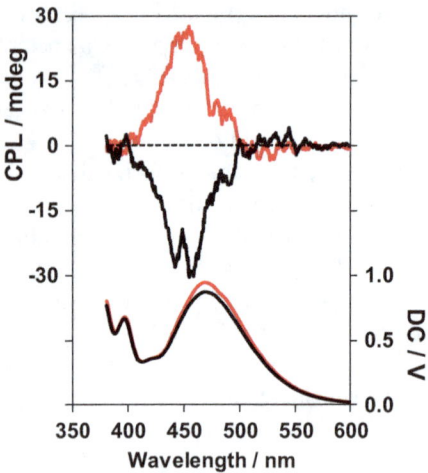

**Fig. 2.19** CD (upper) and
UV-Vis absorption (lower)
spectra of (*R*)-**9** (red) and
(*S*)-**9** (black) in CHCl$_3$ (1.0
× 10$^{-5}$M)

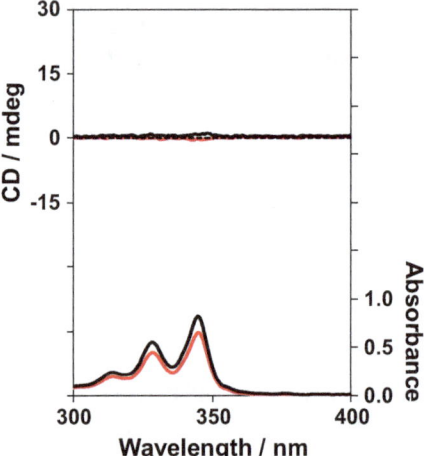

To evaluate the chiroptical properties of compounds **9** in their $S_0$-state structures, CD and UV-Vis absorption spectra of (*R*)-**9** and (*S*)-**9** were acquired (Fig. 2.19). The UV-Vis absorption spectra of (*R*)- and (*S*)-**9** exhibit three main $\pi$-$\pi*$ vibronic transitions ($^1L_a$ transitions) between 315 and 360 nm that correspond to the pyrene moieties. Surprisingly, the intensities of the CD bands of (*R*)-**9** were noticeably weak, and no CD bands of (*R*)-**9** were detected in the 250–300 nm range, which suggests that, due to its pivotal framework, chiral molecular system **9** acts as a cryptochiral CPL fluorophore that has no detectable CD signals.

A mechanism for this cryptochirality has been hypothesized. The two pyrene units in **9** adopt an almost achiral T-shaped structure in the ground $S_0$ state, resulting in a $\theta$ value of about 80–90° in the axially chiral octahydrobinaphthyl unit, resulting in an almost undetectable CD spectrum. On the other hand, in the $S_1$ state, the two pyrene units transform themselves into a chirally π-stacked arrangement. These changes in configuration are responsible for the cryptochiral behavior observed, namely, the ultra-weak CD and intense CPL bands of excimer origin that are due to the two floppy pyrene units that sandwich the octahydrobinaphthyl moiety.

In conclusion, a CD-silent/CPL-active cryptochiral binaphthyl fluorophore was successfully created by connecting an axially chiral binaphthyl and two fluorescent pyrene units through flexible tethers. This system can concurrently control two chiral points (binaphthyl and pyrene) in the photoexcited state.

## 2.8   Controlling Circularly Polarized Luminescence (CPL) of Binaphthyl-Pyrene Fluorophores Using Fluidic and Glassy Media [27]

In Sects. 2.3 and 2.5, the CPL signs of chiral binaphthyl fluorophores with the same axial chirality were reported to be controllable, both in solution and in the solid state, by manipulating their external environments. In this section, a non-classical approach for the dual control of the $\lambda_{CPL}$ and the CPL sign of a chiral organic fluorophore, by varying the external environment, e.g., by choosing $CHCl_3$ solution as the fluidic medium and PMMA film as the glassy solid medium, is reported. To that end, chiral binaphthyl-pyrene fluorophore 7 with oligoether tethers was used again as the model system (Fig. 2.20).

Interestingly, although the $CHCl_3$ solution-state PL of (R)-7 shows the typical excimeric emission of its pyrene units, the PL of (R)-7 dispersed in PMMA film shows a $\lambda_{PL}$ shifted to a significantly shorter wavelength ($\lambda_{PL} = 393$ nm), while its $\Phi_F$ value was higher, at 46%, due to the suppression of thermal vibrational deactivation (Fig. 2.21, lower).

As expected, (R)-7 exhibited CPL signals in $CHCl_3$ and PMMA film (Fig. 2.21, upper panel). The CPL wavelength also changed, as was observed for PL. In addition, (R)-7 exhibited CPL signs that depended on the fluidic and glassy states; it was negative (−) in $CHCl_3$ solution (Fig. 2.15) and positive (+) in the PMMA film.

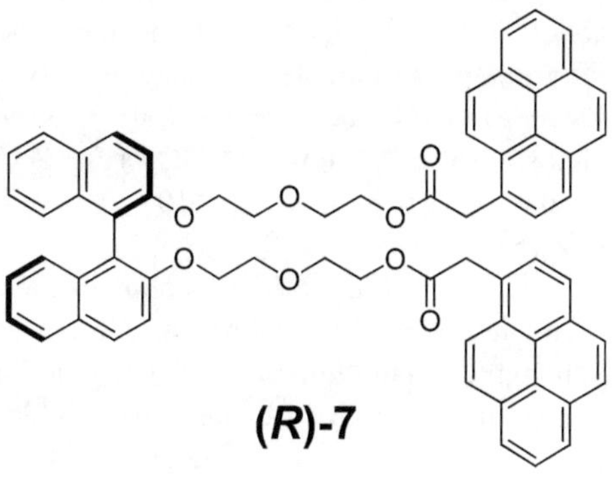

**Fig. 2.20** Chiral binaphthyl-pyrene fluorophore (R)-7

**(R)-7**

**Fig. 2.21** CPL (upper) and
PL (lower) spectra of (R)-**7**
in CHCl₃ (1.0 × 10⁻³ M)
(red) and in the PMMA film-
dispersed state (black)

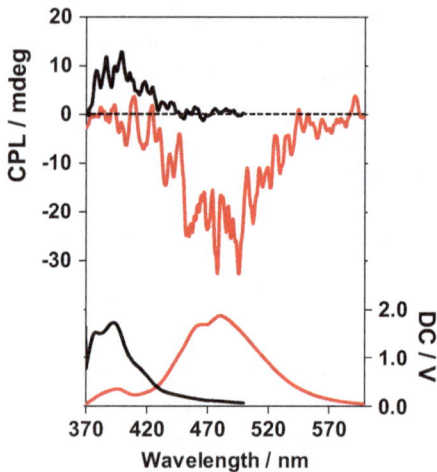

**Fig. 2.22** CD (upper) and
UV-Vis (lower) absorption
spectra of (R)-**7** (red) and
(S)-**7** (black) in the PMMA
film-dispersed state

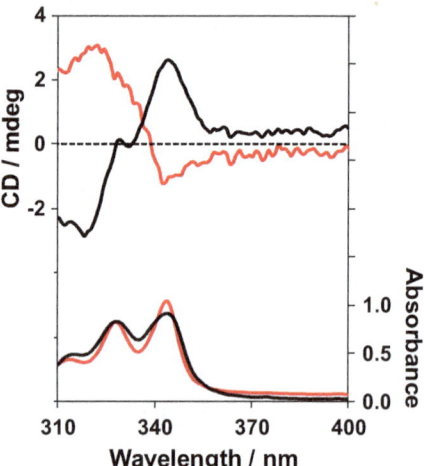

The $g_{CPL}$ values were determined to be about $-7.8 \times 10^{-4}$ in CHCl₃ solution and about $+3.6 \times 10^{-4}$ in PMMA film. Hence, both $\lambda_{CPL}$ and CPL sign of a chiral organic fluorophore are tunable by changing between fluidic CHCl₃ solution and a glassy PMMA film. On the other hand, the CD spectra of **7** in PMMA film (Fig. 2.22) are almost identical to those in CHCl₃ [Fig. 2.16a]; the $g_{CD}$ value of the first Cotton CD band of (R)-**7** is about $-3.8 \times 10^{-5}$ at 344 nm in PMMA film ($-5.9 \times 10^{-5}$ in CHCl₃ solution).

In CHCl₃ solution, the two photoexcited pyrene moieties in (R)-**7** can approach each other more closely through reorganization. On the other hand, in PMMA film, although a remote intramolecular pyrene interaction produce chiral environment of pyrene units, the PMMA glassy solid inhibits the formation of a closer intramolec- ular pyrene excimer. The intramolecular reoriented mode involving the two pyrene units is difficult to achieve upon photoexcitation; as a result, (R)-**7** exhibits monomer-like excimer CPL in PMMA film.

In conclusion, the CPL/PL characteristics of an axially chiral binaphthyl fluorophore connected to fluorescent units can be doubly controlled by selecting a fluidic $CHCl_3$ solution and a glassy PMMA film.

## 2.9  The Appearance of Circularly Polarized Luminescence (CPL) from a Chiral Binaphthyl-Terthiophene Fluorophore in the Solid State [28]

If chirality transfer from the chiral binaphthyl to a remote achiral fluorescent unit in a fluorescent binaphthyl system is not efficient in the solution state, CPL signals due to two remote fluorescent moieties may be not observed. In this section, the appearance of CPL from a dual-tandem fluorophoric molecular system, namely, (R)-**10** and (S)-**10** (Fig. 2.23), which consist of an axially chiral binaphthyl and two achiral terthiophenes, is reported. Terthiophene is a basic $\pi$-conjugated thiophene unit with superior fluorescence properties due to its long wavelength emission.

(R)-**10** exhibited the marked PL associated with terthiophene units in $CHCl_3$ solution, with a PL maximum ($\lambda_{PL}$) observed at 437 nm. Unfortunately, neither (R)- nor (S)-**10** exhibited any meaningful CPL spectrum in $CHCl_3$, presumably due to the highly flexible linkers between the terthiophenes and binaphthyl.

On the other hand, (R)-**10** exhibited marked PL with a $\lambda_{PL}$ at 413 nm in the KBr solid state (Fig. 2.24, lower). Expectedly, (R)-**10** in KBr exhibited CPL with a $g_{lum}$ value of about $+5.0 \times 10^{-4}$.

The solid-state CD and UV-Vis absorption spectra of (R)-**10** in the KBr pellet-dispersed state are shown in Fig. 2.25. Several $\pi$-$\pi*$ transitions associated with the terthiophene groups of **10**, in the 260–360 nm region are observed. The $g_{CD}$ value at the Cotton CD band ($\lambda_{CD} = 342$ nm) of (R)-**10** is about $-3.2 \times 10^{-5}$. The difference in the values of $g_{CD}$ and $g_{CPL}$ is possible ascribable to conformational changes in the ground and photoexcited states.

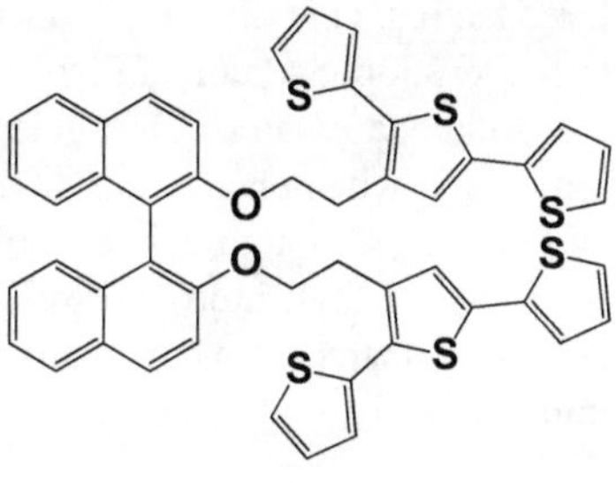

**Fig. 2.23**  Chiral binaphthyl-terthiophene fluorophore **10**

**10**

**Fig. 2.24** CPL (upper panel) and PL (lower panel) spectra of (R)-**10** (red) and (S)-**10** (black) in the KBr pellet-dispersed state

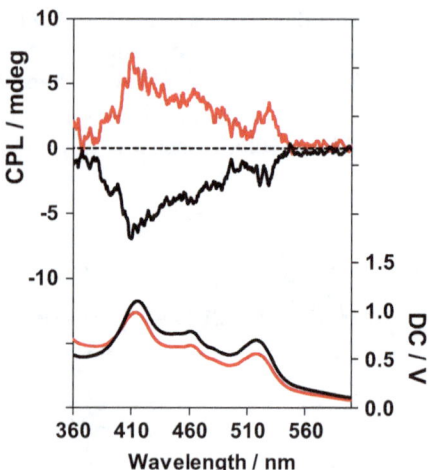

**Fig. 2.25** CD (upper) and UV-Vis absorption (lower) spectra of (R)-**10** (red) and (S)-**10** (black) in the KBr pellet-dispersed state

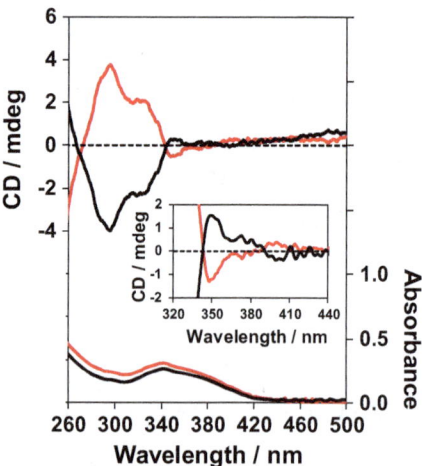

Molecules densely packed in the solid state are surrounded by neighboring molecules. The appearance of solid-state CPL signals may be due to the following three factors: (1) efficient chirality transfer from the chirally rigid binaphthyl moiety to the terthiophene units in the solid state; (2) the very limited molecular motion in the solid state; and (3) efficient photoexcited-state chirality transfer from the binaphthyl when excited at 297 nm to certain chirally arranged terthiophene units.

In conclusion, unlike in the solution state, solid-state CPL is emitted from axially chiral binaphthyl fluorophores bearing remote fluorescent units aided by effective chirality transfer from the axially chiral binaphthyl moiety to the tethered fluorescent units.

## 2.10 Conclusion

Axially chiral binaphthyl is one of the most significant fundamental sources of chirality and fluorescence. An axially chiral binaphthyl fluorophore can emit circularly polarized luminescence (CPL) in the solution-dissolved state, PMMA film-dispersed state, and KBr pellet-dispersed state. The CPL emitted from an axially chiral binaphthyl fluorophore can easily be tuned by (1) tuning the dihedral angle in the binaphthyl unit, (2) employing the neighboring group effect that operates between binaphthyl units, and (3) controlling its external environment, without the need for the enantiomer. This chapter provides novel ideas for the design of novel CPL materials in the solution and solid state.

## References

1. Bian Z, He Y-B, Gao L-X (2003) Circular dichroism and circularly polarized photoluminescence spectra of a new chiral conjugated polymers based on 1,1'-binaphthyl structure. Gaodeng Xuexiao Huaxue Xuebao 24:559–561
2. Sundar MS, Talele HR, Mande HM, Bedekar AV, Tovar RC, Muller G (2014) Synthesis of enantiomerically pure helicene like bis-oxazines from atropisomeric 7,7'-dihydroxy BINOL: preliminary measurements of the circularly polarized luminescence. Tetrahedron Lett 55:1760–1764
3. Park J, Yu T, Inagaki T, Akagi K (2015) Helical network polymers exhibiting circularly polarized luminescence with thermal stability. Synthesis via photo-cross-link polymerizations of methacrylate derivatives in a chiral nematic liquid crystal. Macromolecules 48:1930–1940
4. Ye Q, Zhu D, Xu L, Lu X, Lu Q (2016) The fabrication of helical fibers with circularly polarized luminescence via ionic linkage of binaphthol and tetraphenylethylene derivatives. J Mater Chem C 4:1497–1503
5. Uchida Y, Hirose T, Nakashima T, Kawai T, Matsuda K (2016) Synthesis and photophysical properties of a 13,13'-bibenzo[b]perylenyl derivative as a π-extended 1,1'-binaphthyl analog. Org Lett 18:2118–2121
6. Hirata S, Vacha M (2016) Circularly polarized persistent room-temperature phosphorescence from metal-free chiral aromatics in air. J Phys Chem Lett 7:1539–1545
7. Shi L, Zhu L, Guo J, Zhang L, Shi Y, Zhang Y, Hou K, Zheng Y, Zhu Y, Lv J et al (2017) Self-assembly of chiral gold clusters into crystalline nanocubes of exceptional optical activity. Angew Chem Int Ed 56:15397–15401
8. Shintani R, Misawa N, Takano R, Nozaki K (2017) Rhodium-catalyzed synthesis and optical properties of silicon-bridged arylpyridines. Chem Eur J 23:2660–2665
9. Nishikawa H, Mochizuki D, Higuchi H, Okumura Y, Kikuchi H (2017) Reversible broad-spectrum control of selective reflections of chiral nematic phases by closed-/open-type axially chiral azo dopants. Chem Open 6:710–720
10. Koiso N, Kitagawa Y, Nakanishi T, Fushimi K, Hasegawa Y (2017) Eu(III) chiral coordination polymer with a structural transformation system. Inorg Chem 56:5741–5747
11. Aoki R, Toyoda R, Kogel JF, Sakamoto R, Kumar J, Kitagawa Y, Harano K, Kawai T, Nishihara H (2017) Bis(dipyrrinato)zinc(II) complex chiroptical wires: exfoliation into single strands and intensification of circularly polarized luminescence. J Am Chem Soc 139:16024–16027

12. Song J, Wang M, Zhou X, Xiang H (2018) Unusual circularly polarized and aggregation-induced near-infrared phosphorescence of helical platinum(II) complexes with tetradentate salen ligands. Chem Eur J 24:7128–7132
13. Nishigaki S, Murayama K, Shibata Y, Tanaka K (2018) Rhodium-mediated enantioselective synthesis of a benzopicene-based phospha[9]helicene: the structure-property relationship of triphenylene- and benzopicene-based carbo- and ' phosphahelicenes. Mater Chem Front 2:585–590
14. Fujiki M, Koe JR, Mori T, Kimura Y (2018) Questions of mirror symmetry at the photoexcited and ground states of non-rigid luminophores raised by circularly polarized luminescence and circular dichroism spectroscopy: part 1. Oligofluorenes, oligophenylenes, binaphthyls and fused aromatics. Molecules 23:2606/1–2606/36
15. Wang Y, Harada T, Phuong LQ, Kanemitsu Y, Nakano T (2018) Helix induction to polyfluorenes using circularly polarized light: chirality amplification, phase-selective induction, and anisotropic emission. Macromolecules 51:6865–6877
16. Liu D, Zhou Y, Zhang Y, Li H, Chen P, Sun W, Gao T, Yan P (2018) Chiral BINAPO-controlled diastereoselective self-assembly and circularly polarized luminescence in triple-stranded europium(III) podates. Inorg Chem 57:8332–8337
17. Zhang X, Zhang Y, Zhang H, Quan Y, Li Y, Cheng Y, Ye S (2019) High brightness circularly polarized organic light-emitting diodes based on nondoped aggregation-induced emission (AIE)-active chiral binaphthyl emitters. Org Lett 21:439–443
18. Deng W-T, Qu H, Huang Z-Y, Shi L, Tang Z, Cao X, Tao J (2019) Facile synthesis of homochiral compounds integrating circularly polarized luminescence and two-photon excited fluorescence. Chem Commun 55:2210–2213
19. Maeda C, Ogawa K, Sadanaga K, Takaishi K, Ema T (2019) Chiroptical and catalytic properties of doubly binaphthyl-strapped chiral porphyrins. Chem Commun 55:1064–1067
20. Sun Z-B, Liu J-K, Yuan D-F, Zhao Z-H, Zhu X-Z, Liu D-H, Peng Q, Zhao C-H (2019) 2,2'-Diamino-6,6'-diboryl-1,1'-binaphthyl: a versatile building block for temperature-dependent dual fluorescence and switchable circularly polarized luminescence. Angew Chem 58:4840–4846
21. Kimoto T, Tajima N, Fujiki M, Imai Y (2012) Control of circularly polarized luminescence by using open- and closed-type binaphthyl derivatives with the same axial chirality. Chem Asian J 7:2836–2841
22. Okazaki M, Mizusawa T, Nakabayashi K, Yamashita M, Tajima N, Harada T, Fujiki M, Imai Y (2016) Solvent-controlled sign inversion of circularly polarized luminescent binaphthylacetic acid derivative. J Photochem Photobiol A 331:115–119
23. Amako T, Kimoto T, Tajima N, Fujiki M, Imai Y (2013) Dependence of circularly polarized luminescence due to the neighboring effects of binaphthyl units with the same axial chirality. RSC Adv 3:6939–6944
24. Kimoto T, Amako T, Tajima N, Kuroda R, Fujiki M, Imai Y (2013) Control of solid-state circularly polarized luminescence of binaphthyl organic fluorophores through environmental changes. Asian J Org Chem 2:404–410
25. Nakabayashi K, Kitamura S, Suzuki N, Guo S, Fujiki M, Imai Y (2016) Non-classically controlled signs in a circularly polarised luminescent molecular puppet: The importance of the wire structure connecting binaphthyl and two pyrenes. Eur J Org Chem 2016:64–69
26. Hara N, Yanai M, Kaji D, Shizuma M, Tajima N, Fujiki M, Imai Y (2018) A pivotal biaryl rotamer bearing two floppy pyrenes that exhibits cryptochiral characteristics in the ground state. Chem Select 3:9970–9973
27. Nakabayashi K, Amako T, Tajima N, Fujiki M, Imai Y (2014) Nonclassical dual control of circularly polarized luminescence modes of binaphthyl–pyrene organic fluorophores in fluidic and glassy media. Chem Commun 50:13228–13230
28. Taniguchi N, Nakabayashi K, Harada T, Tajima N, Shizuma M, Fujiki M, Imai Y (2015) Circularly polarized luminescence of chiral binaphthyl with achiral terthiophene fluorophores. Chem Lett 44:598–600

# Chapter 3
# Circularly Polarized Luminescence from Planar Chiral Compounds Based on [2.2] Paracyclophane

Yasuhiro Morisaki

**Abstract** In this chapter, recent development on molecules emitting intense circularly polarized luminescence (CPL) based on planar chiral [2.2]paracyclophane is described. Optical resolution routes of the planar chiral [2.2]paracyclophane compounds, optically active π-stacked molecules, and the CPL profiles are discussed. It is suggested that the optically active higher-ordered structures, such as V-, X-, triangle-, propeller-shaped structures, and so on, in the excited state are important for intense CPL with large dissymmetry factors ($g_{lum}$ values).

## 3.1 Introduction

Cyclophanes are cyclic compounds containing at least one aromatic ring (arylene) in the cyclic skeleton. They have been well studied, particularly in the field of organic chemistry [1, 2]. A typical example of a cyclophane is [2.2]paracyclophane, which was synthesized for the first time in 1949 by the pyrolysis of *para*-xylene [3]. Later, in 1951, it was synthesized directly by the Wurtz-type intramolecular cyclization of 1,4-bis-bromomethylbenzene [4]. [2.2]Paracyclophane has a unique structure consisting of two π-stacked benzene rings that are fixed in the para-position with two ethylene chains. There are many studies on its synthetic routes, reactivities, and physical properties, in the field of organic chemistry [1]. However, there are not many examples where [2.2]paracyclophane is effectively utilized in the fields of polymer chemistry and materials chemistry [5–16].

The two face-to-face benzene rings in close proximity (~3.0 Å apart) in [2.2] paracyclophane completely suppress their rotational movement. The resulting [2.2] paracyclophane becomes a planar chiral compound (Fig. 3.1) upon the introduction of substituent(s) at appropriate position(s) of the benzene ring(s) [17–22]. Planar chirality attained by [2.2]paracyclophane makes it structurally stable, because of which this compound has been utilized as a chiral auxiliary or chiral ligand in the

Y. Morisaki (✉)
School of Science and Technology, Kwansei Gakuin University, Sanda, Hyogo, Japan
e-mail: ymo@kwansei.ac.jp

© Springer Nature Singapore Pte Ltd. 2020
T. Mori (ed.), *Circularly Polarized Luminescence of Isolated Small Organic Molecules*, https://doi.org/10.1007/978-981-15-2309-0_3

31

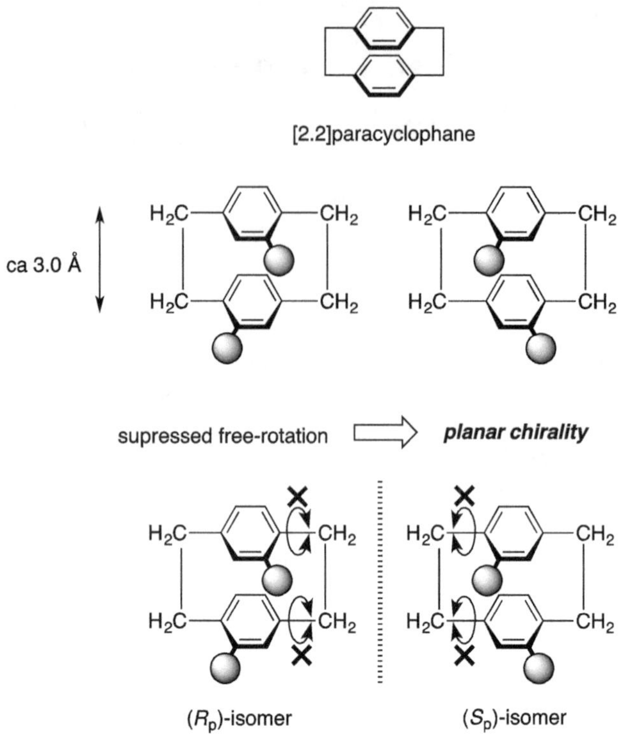

[2.2]paracyclophane

ca 3.0 Å

supressed free-rotation ⟹ *planar chirality*

($R_p$)-isomer                    ($S_p$)-isomer

**Fig. 3.1** Structure of [2.2]paracyclophane and its planar chirality

fields of synthetic organic chemistry and organometallic chemistry. Despite this, it has not been used in the fields of polymer chemistry and materials chemistry, as well.

In this chapter, the author has focused on the planar chirality of [2.2] paracyclophane. Optical resolution of disubstituted and tetrasubstituted [2.2] paracyclophane and their use as chiral building blocks for the syntheses of optically active π-stacked compounds are described. Application of molecules containing planar chiral [2.2]paracyclophane in materials emitting circularly polarized luminescence (CPL) is also shown.

## 3.2 Optical Resolution of [2.2]Paracyclophane Compounds

Optical resolutions of mono-substituted [2.2]paracyclophanes are well established, and various enantiopure *ortho*-, pseudo-*geminal*-, and *syn-latero*-disubstituted [2.2] paracyclophanes have been synthesized [17–22].

Several methods for the optical resolution of pseudo-*ortho*-disubstituted [2.2] paracyclophanes have been reported [23–32]. Optical resolution of *rac*-pseudo-*ortho*-bis(diarylphosphino)[2.2]paracyclophane    (*rac*-[2.2]PHANEPHOS)    by

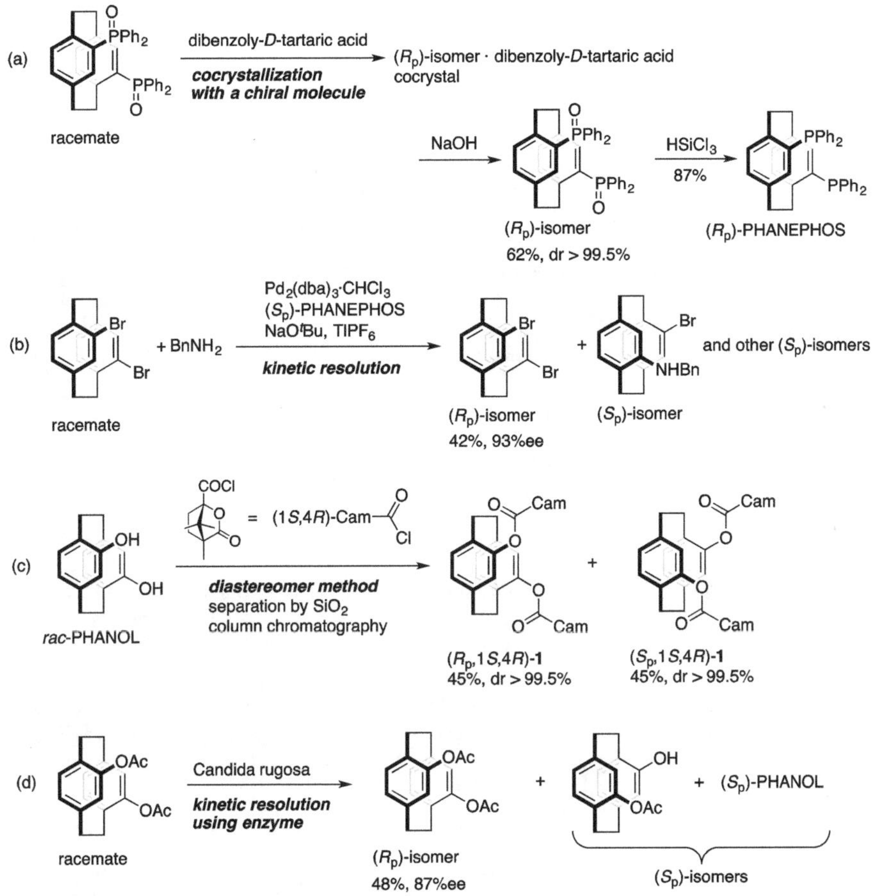

**Fig. 3.2** Representative optical resolutions of pseudo-*ortho*-disubstituted [2.2]paracyclophanes

co-crystallization with a tartaric acid derivative is a successful examples (Fig. 3.2a) [23], and planar chiral ($S_p$)- and ($R_p$)-[2.2]PHANEPHOS are the commercially available chiral ligands for the transition metal-catalyzed asymmetric reactions. $Pd_2(dba)_3$/[2.2]PHANEPHOS-catalyzed amination of *rac*-pseudo-*ortho*-dibromo [2.2]paracyclophane enabled the kinetic resolution [24] for obtaining the enantioenriched pseudo-*ortho*-dibromo[2.2]paracyclophane (Fig. 3.2b). Optical resolutions of *rac*-4-bromo-12-hydroxy[2.2]paracyclophane [25], *rac*-pseudo-*ortho*-dihydroxy[2.2]paracyclophane (*rac*-PHANOL) [26], and *rac*-pseudo-*ortho*-dihydroxymethyl[2.2]paracyclophane [27] were achieved by a diastereomer method using chiral camphanic acid chloride as the chiral auxiliary; the optical resolution of PHANOL is shown in Fig. 3.2c as a representative example. The enzyme-catalyzed kinetic resolutions of *rac*-pseudo-*ortho*-disubstituted [2.2]paracyclophanes were also reported [28–30], and the representative example is shown in Fig. 3.2d. Optical resolution of *rac*-pseudo-*ortho*-dibromo[2.2]paracyclophane was carried out in

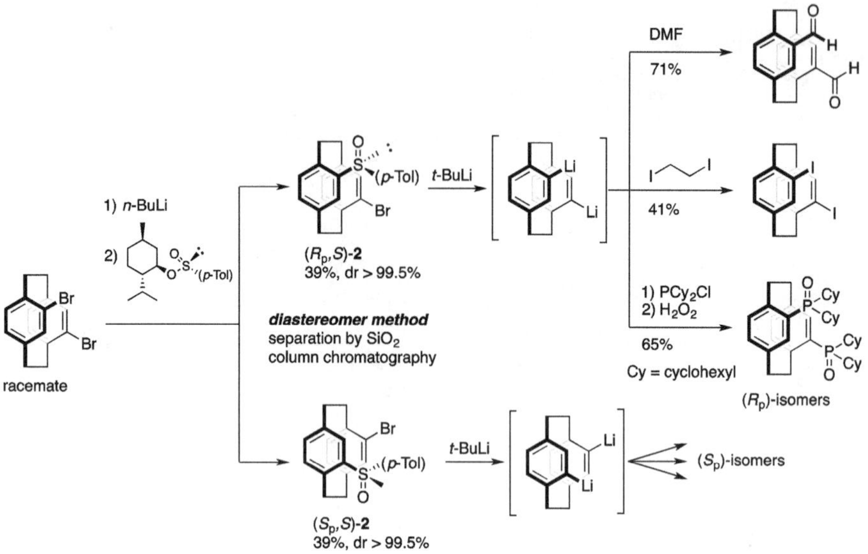

**Fig. 3.3** Optical resolution of pseudo-*ortho*-disubstituted [2.2]paracyclophane using (1*R*,2*S*,5*R*)-(−)-menthyl (*S*)-*p*-toluenesulfinate as a chiral auxiliary

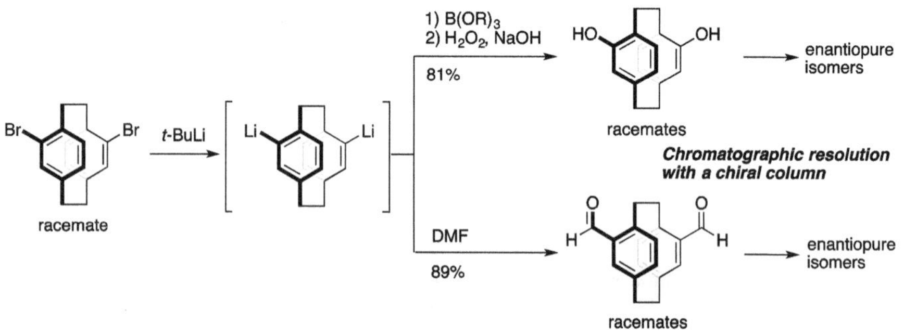

**Fig. 3.4** Optical resolution of *rac*-pseudo-*meta*-disubstituted [2.2]paracyclophane

2012 using (1*R*,2*S*,5*R*)-(−)-menthyl (*S*)-*p*-toluenesulfinate as a chiral auxiliary [31], as shown in Fig. 3.3, and the obtained diastereomers could be used as the parent compounds to produce a wide variety of [2.2]paracyclophane-based chiral molecules. Optical resolution of *rac*-pseudo-*ortho*-disubstituted [2.2]paracyclophanes by chiral columns was reported [32]. The chromatographic optical resolution of *rac*-pseudo-*meta*-disubstituted [2.2]paracyclophanes was also reported by Lutzen, and several enantiopure pseudo-*meta*-disubstituted [2.2]paracyclophanes were produced [33]. Figure 3.4 shows the representative examples of the successful optical resolutions obtained using a chiral column.

In 2008, Hopf and coworkers reported the optical resolutions of bis-(*ortho*)-pseudo-*meta*-4,5,15,16-tetrasubstituted [2.2]paracyclophane and 4,5,15-

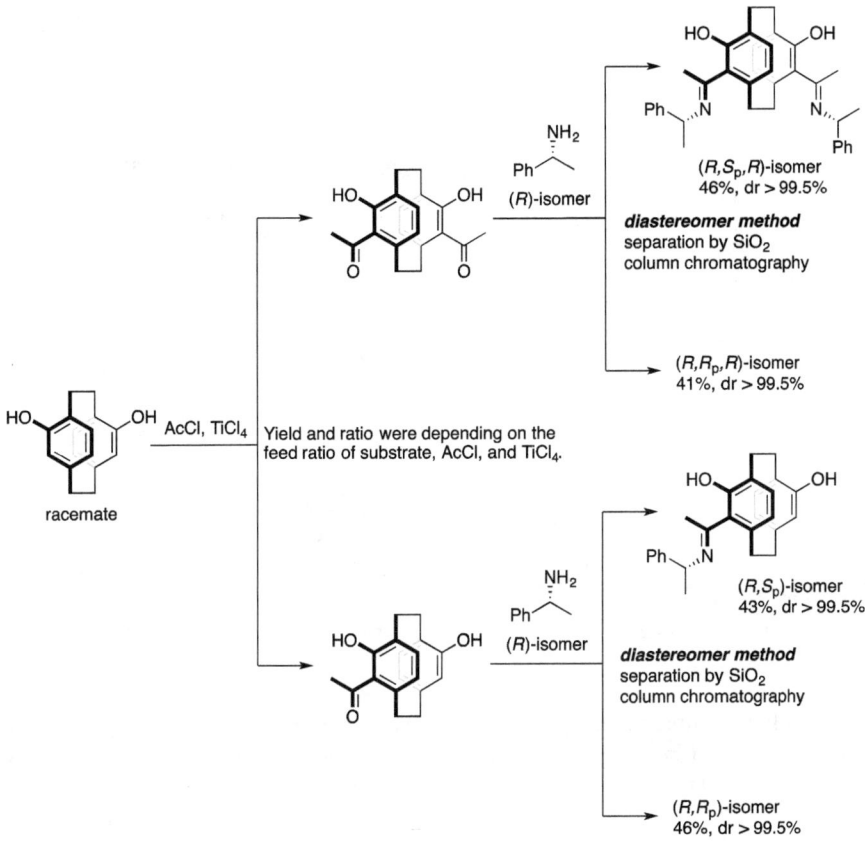

**Fig. 3.5** Optical resolutions of bis-(*ortho*)-pseudo-*meta*-4,5,15,16-tetrasubstituted [2.2] paracyclophane and 4,5,15-trisubstituted [2.2]paracyclophane

trisubstituted [2.2]paracyclophane (Fig. 3.5) by diastereomer methods [34]. The optical resolution of 4,7,12,15-tetrasubstituted [2.2]paracyclophane was achieved in 2014 [35], as shown in Fig. 3.6. The racemate 4,7,12-tribromo-15-hydroxy[2.2] paracyclophane was synthesized from the corresponding tetrabromide, and the reaction with (1*S*,4*R*)-(-)-camphanic chloride afforded the diastereomers (*R*$_p$,1*S*,4*R*)- and (*S*$_p$,1*S*,4*R*)-isomers. The diastereomers could be separated in gram scale using the simple silica gel column chromatography, and the diastereomer ratios were over 99.5%. The hydroxy group was converted to the trifluoromethanesulfonyl group, which is an active site for the Pd-catalyzed cross-coupling, in addition to the bromo groups.

Optical resolution of *rac*-PHANOL [26] was applied to produce enantiopure bis-(*para*)-pseudo-*ortho*-4,7,12,15-tetrasubstituted [2.2]paracyclophanes [36], as shown in Fig. 3.7. The obtained diastereomers were reacted with bromine, in the presence of iron, to afford bis-(*para*)-pseudo-*ortho*-4,7,12,15-tetrasubstituted [2.2] paracyclophanes (*R*$_p$,1*S*,4*R*)- and (*S*$_p$,1*S*,4*R*)-isomers, wherein bromine was

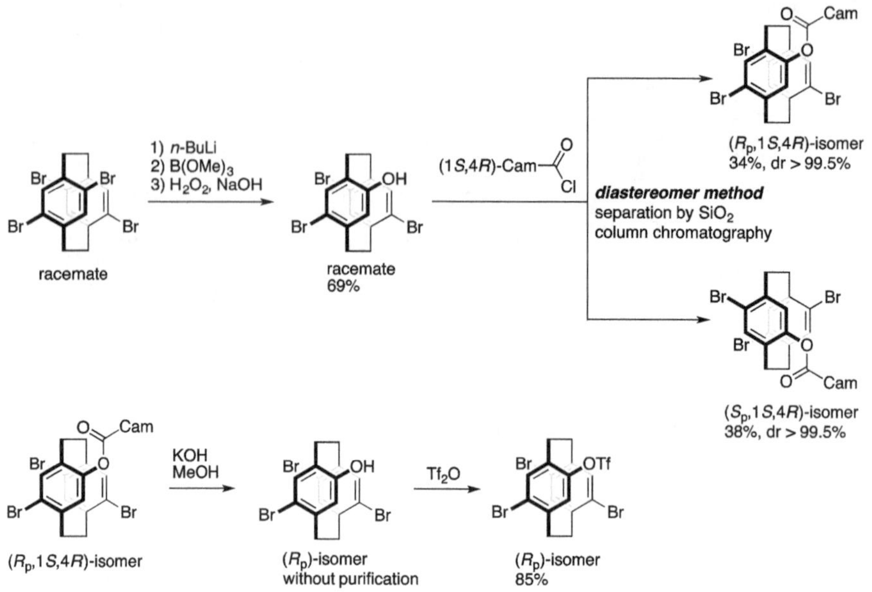

**Fig. 3.6** Optical resolution of 4,7,12,15-tetrasubstituted [2.2]paracyclophane and the transformation

selectively substituted to the *para*-position with respect to their oxygen substituent. The chiral auxiliary groups were removed using KOH to obtain the enantiopure dibrominated PHANOL. The reaction with trifluoromethanesulfonic anhydride as well as MeI affords bis-(*para*)-pseudo-*ortho*-type tetrasubstituted [2.2] paracyclophanes. Optical resolution could also be carried out after the bromination of the mixture of diastereomers, as shown in Fig. 3.8. This modified route involved a single bromination step to produce both the diastereomers [37].

Enantiopure bis-(*para*)-pseudo-*meta*-tetrasubstituted [2.2]paracyclophane derivatives were successfully synthesized (Fig. 3.9) [38]. The starting material, racemic 4,7,12-tribromo-15-hydroxy[2.2]paracyclophane, was reacted with *n*-BuLi. First, the phenol in *rac*-**2** was reacted with *n*-BuLi to form lithium phenoxide. Then, *n*-BuLi attacked selectively at the corresponding pseudo-*meta*-position because of the electronic effect. The optical resolution was achieved by the diastereomer method using camphanoyl chloride as the chiral auxiliary. The camphanoyl groups could be removed by saponification, and the obtained 4,15-dibromo-7,12-dihydroxy[2.2] paracyclophane was converted to ($R_p$)- and ($S_p$)-4,15-dibromo-7,12-trifluoromethanesulfonyloxy[2.2]paracyclophanes, for using them as chiral building blocks.

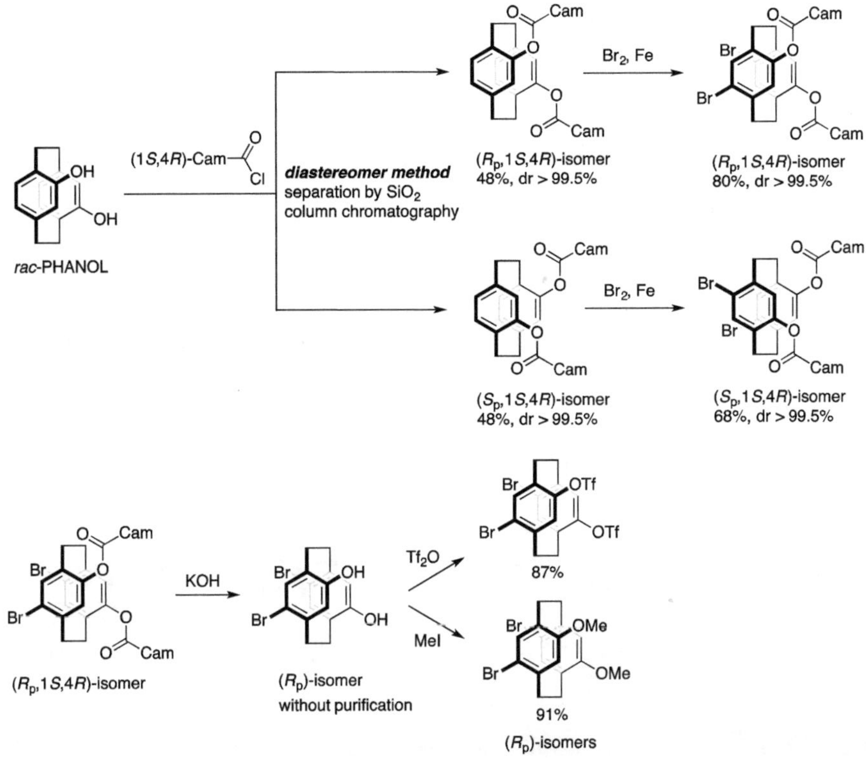

**Fig. 3.7** Synthesis of enantiopure bis-(*para*)-pseudo-*ortho*-4,7,12,15-tetrasubstituted [2.2] paracyclophanes

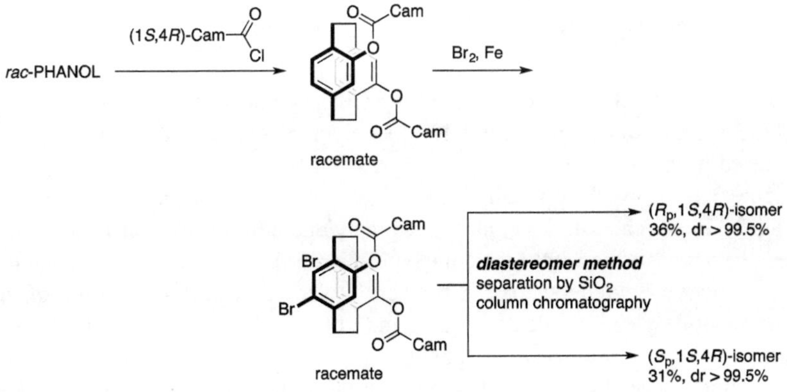

**Fig. 3.8** Modified optical resolution of *rac*-PHANOL

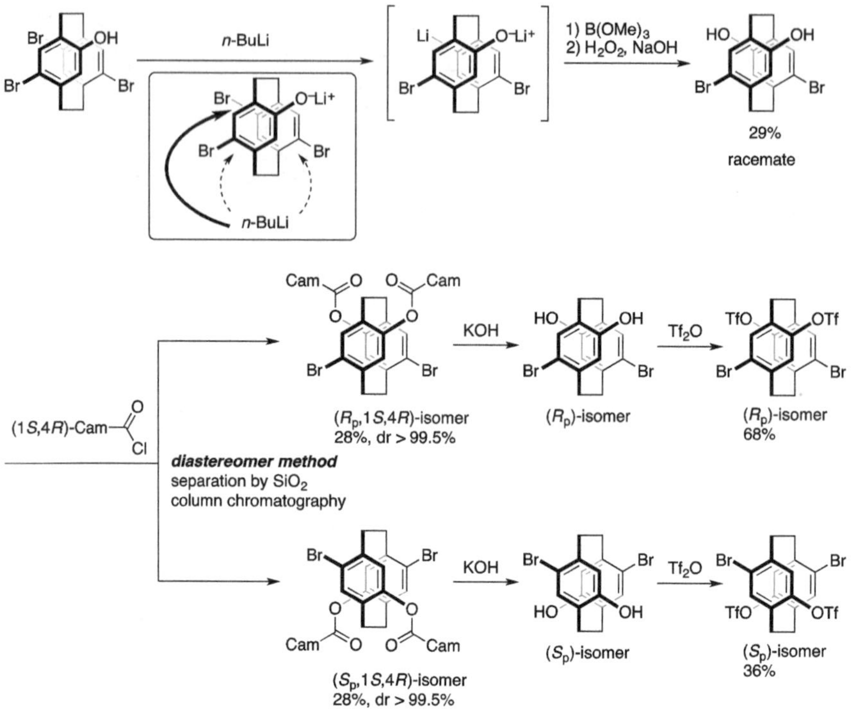

**Fig. 3.9** Synthesis of enantiopure bis-(*para*)-pseudo-*meta*-4,7,12,15-tetrasubstituted [2.2] paracyclophanes

## 3.3 Optically Active Molecules Based on Planar Chiral [2.2] Paracyclophane and Their CPL Profiles

Various optically active molecules emitting CPL, synthesized from the chiral building blocks based on planar chiral [2.2]paracyclophane described above, will be introduced in this section.

Figure 3.10 shows the optically active V-, N-, and W-shaped π-stacked molecules ((*R_p*)-**1-3**), in which two, three, and four phenylene-ethynylene π-electron systems, respectively, are stacked at the terminal benzene ring(s) [39]. Figure 3.10 contains an optically active triangle-shaped π-stacked molecule (*R_p*)-**4** consisting of three phenylene-ethynylene π-electron systems [39]. They were prepared from enantiopure pseudo-*ortho*-disubstituted [2.2]paracyclophane. Their photoluminescence (PL) and CPL spectra in dilute CHCl$_3$ and the optical data are shown in Fig. 3.11a.

The PL spectra of (*R_p*)-**1-3** were gradually red-shifted, depending on the number of stacked π-electron systems. Vibronic structures were observed in the PL spectra, with good PL quantum efficiencies ($\Phi_{PL}$) of approximately 0.8. The decay curves could be fitted with a single exponential function, indicating that the emission from

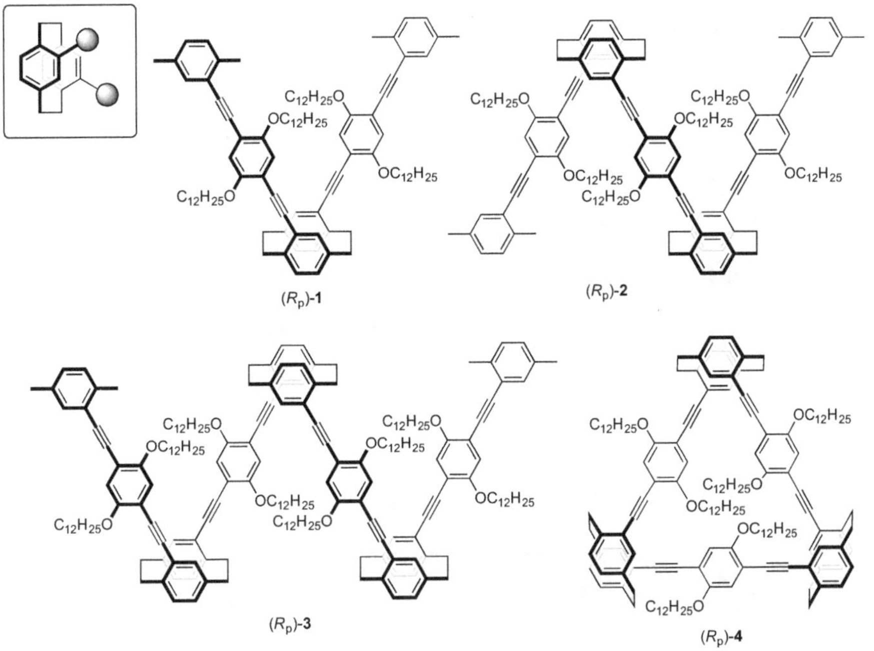

**Fig. 3.10** Optically active V-, N-, and W-shaped π-stacked molecules

$(R_p)$-**1-3** occurred from the chromophore state rather than from the phane state [40–44]. CPL spectra for $(R_p)$-**1-3** and $(S_p)$-**1-3** in dilute CHCl$_3$ (1.0 × 10$^{-5}$M) were obtained. Intense CPL signals of $I_L - I_R$, where $I_L$ and $I_R$ are the intensities of left- and right-handed PL, respectively, were obtained. The CPL signals of the two compounds were the mirror images of each other. Interestingly, the CPL dissymmetry factors $g_{lum} = 2(I_L - I_R)/(I_L + I_R)$ increased gradually with increasing number of the stacked π-electron systems. This result suggests the emergence of additional chirality, besides the planar chirality of the [2.2]paracyclophane skeleton.

Figure 3.11b shows the PL and CPL spectra of the linear trimer **2** (N-shaped molecule) and cyclic trimer **4** (chiral triangle-shaped molecule). Molecules **2** and **4** exhibited identical PL and CPL profiles. The cyclic trimer has chiral triangle-shaped chirality, i.e., a chiral second-ordered structure, in addition to the planar chirality of the [2.2]paracyclophane skeleton. Considering the identical CPL profiles between the linear and cyclic trimers, linear trimer **2** should form the chiral second-ordered structure as the additional chirality in addition to the planar chirality. It is suggested that a one-handed helical structure (left-handed helix for $(R_p)$-isomer) is formed in the excited stated of linear trimer **2**, as shown in Fig. 3.12.

4,7,12,15-Tetraethynyl[2.2]paracyclophane was prepared from 4,7,12-Tribromo-15-trifluoremethanesulfonyl[2.2]paracyclophane (Fig. 3.13), followed by the synthesis of the optically active cyclic molecule **5** [35]. The optically active molecule **5** formed the chiral second-ordered structure such as one-handed propeller-shaped

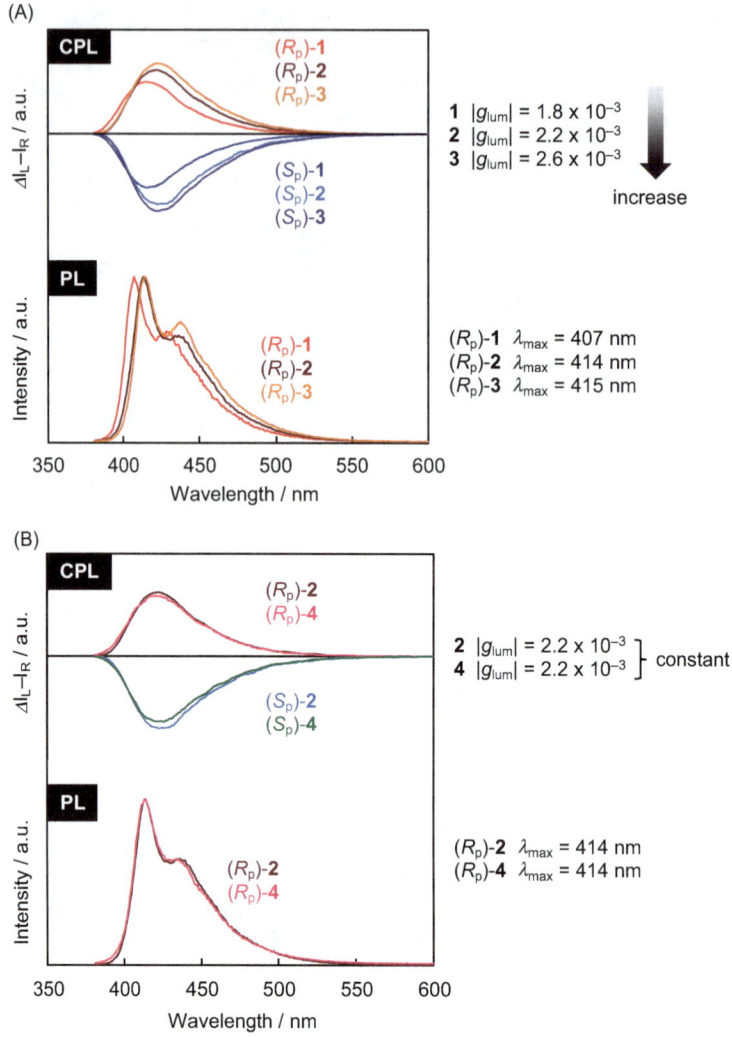

**Fig. 3.11** (**a**) CPL and PL spectra of **1–3** in $CHCl_3$ ($10 \times 10^{-5}$M), (**b**) CPL and PL spectra of **2** and **4** in $CHCl_3$ ($10 \times 10^{-5}$M)

structure, which induced large chirality in both the ground state and the excited state. The specific rotation reached 1500 (c 0.5, $CHCl_3$), and the molar ellipticity was approximately 3,000,000 deg $cm^2$ $dmol^{-1}$. Molecule **5** emitted intense CPL, and the spectra of the enantiomers are shown in Fig. 3.14. The $|g_{lum}|$ value at $\lambda_{PL,max}$ was estimated to be in the order of $10^{-2}$, with $|g_{lum}| = 1.3 \times 10^{-2}$. This $g_{lum}$ value is large for a monodispersed organic molecule in solution. The optically active second-ordered structure, i.e., optically active propeller-shaped structure, contributes greatly to the induction of chirality in the excited state. Although the $\Phi_{PL}$ was 0.45, the

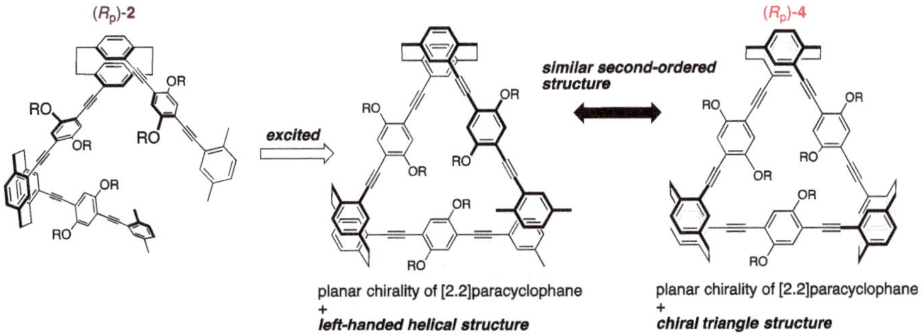

**Fig. 3.12** Plausible conformations of linear trimer **2** and cyclic trimer **4**

molar absorption coefficient ($\varepsilon$) of 130,000M$^{-1}$ cm$^{-1}$ indicates that this molecule emits bright PL. In addition, optically active propeller-shaped molecules **6** [45] and **7** [46] were prepared (Fig. 3.13). Both the compounds exhibited high molar absorption coefficients, good $\Phi_{PL}$, and large CPL $g_{lum}$ values.

Simple X-shaped molecules have been synthesized from 4,7,12,15-tetraethynyl [2.2]paracyclophane [47], and the structures and other data are briefly shown in Fig. 3.15. The phenyl- and naphthyl-containing molecules, **8** and **9**, emitted intense CPL, while a weak CPL signal was observed from the anthracene-containing molecule **10**. In particular, the naphthyl-containing molecule **9** was an excellent CPL emitter, with a large $g_{lum}$ in the order of $10^{-3}$, large $\varepsilon$ of $0.79 \times 10^5$ cm$^{-1}$ M$^{-1}$, and good $\Phi_{lum}$ of 0.78.

Figure 3.15 shows an optically active X-shaped molecule ($R_p$)-**11** that emits intense CPL, with a $\Phi_{PL}$ of 0.87 in a dilute solution and the $g_{lum}$ value of $-1.2 \times 10^{-3}$ [48]. The $g_{lum}$ value of the spin-coated film was positive with $g_{lum} = +2.1 \times 10^{-2}$, which was larger by an order of magnitude. Furthermore, the $g_{lum}$ value of the annealed thin film was negative and in the order of $10^{-1}$ ($-0.25$). The $g_{lum}$ value of the thin film obtained by the casting method was negative and still larger by an order of magnitude ($-3.0 \times 10^{-2}$). When it was annealed, the sign was reversed, and the same $g_{lum}$ value ($-0.25$) was observed. An optically active higher-ordered structure was formed in the thin film because of the van der Waals force of the long alkyl chains and $\pi$-$\pi$ interactions. Thus, a thermodynamically stable higher-ordered structure was formed by annealing.

Not only arylene-ethynylenes but also arylene-vinylenes were chosen as $\pi$-electron systems. Figure 3.16 shows the optically active X-shaped molecules consisting of arylene-vinylenes prepared from 4,7,12,15-tetraethynyl[2.2] paracyclophane [49]. Molecule **12** [40, 49] exhibited aggregation-caused quenching of PL ($\Phi_{PL} = 0.03$) because of the high planarity of the stacked $\pi$-electron system, while **13** exhibited moderate PL properties both in the dilute solution ($\Phi_{PL} = 0.58$) and in the aggregated state ($\Phi_{PL} = 0.24$). In the dilute solution, **12** showed good CPL properties ($\Phi_{PL} = 0.78$ and $|g_{lum}| = 3.7 \times 10^{-3}$), and in the aggregated state, **13** showed good CPL properties ($\Phi_{PL}$ of 0.24 and $|g_{lum}| = +0.90 \times 10^{-3}$) which were observed.

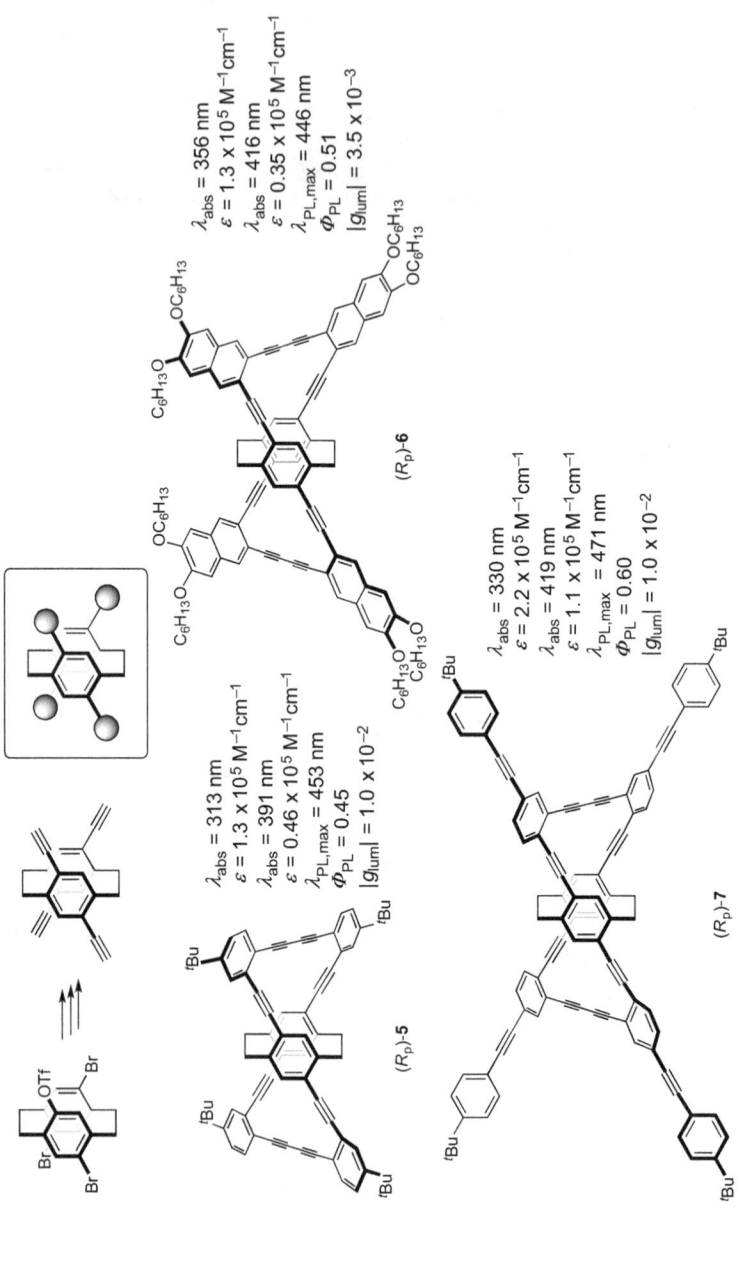

$\lambda_{abs}$ = 356 nm
$\varepsilon$ = 1.3 x 10$^5$ M$^{-1}$cm$^{-1}$
$\lambda_{abs}$ = 416 nm
$\varepsilon$ = 0.35 x 10$^5$ M$^{-1}$cm$^{-1}$
$\lambda_{PL,max}$ = 446 nm
$\Phi_{PL}$ = 0.51
$|g_{lum}|$ = 3.5 x 10$^{-3}$

$(R_p)$-6

$\lambda_{abs}$ = 330 nm
$\varepsilon$ = 2.2 x 10$^5$ M$^{-1}$cm$^{-1}$
$\lambda_{abs}$ = 419 nm
$\varepsilon$ = 1.1 x 10$^5$ M$^{-1}$cm$^{-1}$
$\lambda_{PL,max}$ = 471 nm
$\Phi_{PL}$ = 0.60
$|g_{lum}|$ = 1.0 x 10$^{-2}$

$(R_p)$-7

$\lambda_{abs}$ = 313 nm
$\varepsilon$ = 1.3 x 10$^5$ M$^{-1}$cm$^{-1}$
$\lambda_{abs}$ = 391 nm
$\varepsilon$ = 0.46 x 10$^5$ M$^{-1}$cm$^{-1}$
$\lambda_{PL,max}$ = 453 nm
$\Phi_{PL}$ = 0.45
$|g_{lum}|$ = 1.0 x 10$^{-2}$

$(R_p)$-5

**Fig. 3.13** Structures and optical data of propeller-shaped cyclic molecules **5–7**

**Fig. 3.14** CPL and PL
spectra of ($R_p$)- and ($S_p$)-**5** in
CHCl$_3$ (10 × 10$^{-6}$M)

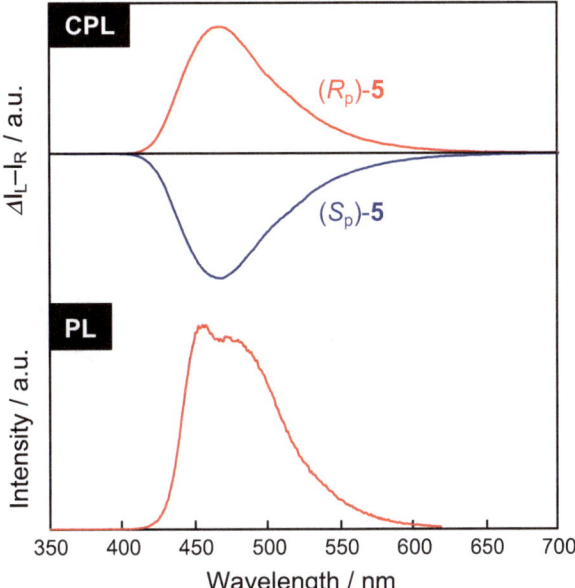

Although the planar chiral molecules discussed above emit intense CPL in a
dilute solution, their fluorescence quantum yield is basically lowered in the solid
state due to the general aggregation-caused quenching. The problem of quenching
can be overcome by introducing Fréchet-type dendrons [50, 51] into the X-shaped
molecules [52]. Figure 3.17 shows the third-generation dendrimer **14**, and the CPL
spectrum of the thin film. The $\Phi_{PL}$ of **14** was estimated to be 0.65, which is almost
the same as the value of 0.66 observed in the dilute solution. This was because of the
X-shaped core being isolated by the dendrons. The CPL |$g_{lum}$| value of **14** was
1.8 × 10$^{-3}$. A thin film emitting CPL with high intensity, high efficiency, and high
dissymmetry factor could be obtained because of the light-harvesting effect of the
benzene rings of the dendrimer.

Recently, catalytic system that enables chemoselective Sonogashira-Hagihara
coupling was developed [35]. The combination of Pd$_2$(dba)$_3$/P$^t$Bu$_3$ reacted predom-
inantly    with    Ar-Br    instead    of    Ar-OTf;    thus,    4,7,12-tribromo-15-
trifluoromethanesulfonyl[2.2]paracyclophane (Fig. 3.6) was converted to the
corresponding triyne (Fig. 3.18). This triyne could be used as a chiral building
block to obtain the X-shaped molecules **15**, which consisted of heterogeneous
π-electron systems could be obtained. The optical properties were almost identical
to those of the X-shaped molecule **16**, and they were excellent CPL emitters [53]. It
is possible to layer heterogeneous π-electron systems possessing various electron-
accepting and electron-donating groups.

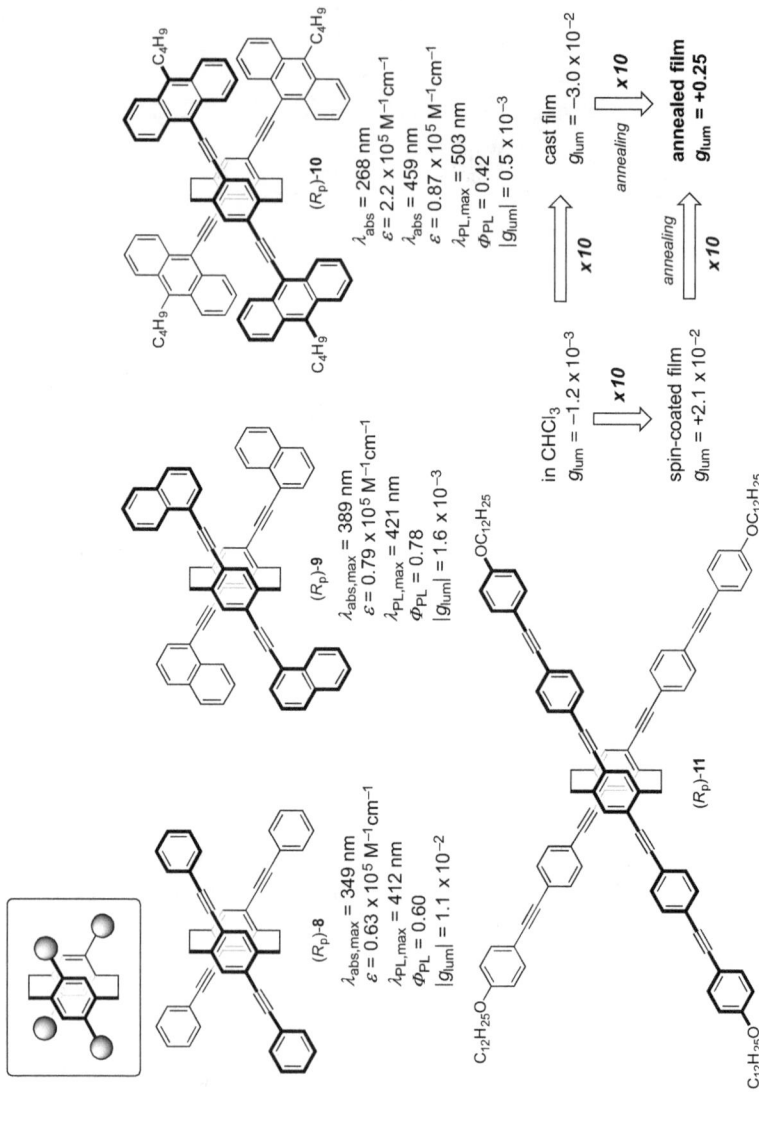

$(R_p)$-**10**

$\lambda_{abs}$ = 268 nm
$\varepsilon$ = 2.2 x 10$^5$ M$^{-1}$cm$^{-1}$
$\lambda_{abs}$ = 459 nm
$\varepsilon$ = 0.87 x 10$^5$ M$^{-1}$cm$^{-1}$
$\lambda_{PL,max}$ = 503 nm
$\Phi_{PL}$ = 0.42
$|g_{lum}|$ = 0.5 x 10$^{-3}$

cast film
$g_{lum}$ = −3.0 x 10$^{-2}$

*annealing* → **x10**

**annealed film**
$g_{lum}$ = **+0.25**

$(R_p)$-**9**

$\lambda_{abs,max}$ = 389 nm
$\varepsilon$ = 0.79 x 10$^5$ M$^{-1}$cm$^{-1}$
$\lambda_{PL,max}$ = 421 nm
$\Phi_{PL}$ = 0.78
$|g_{lum}|$ = 1.6 x 10$^{-3}$

in CHCl$_3$
$g_{lum}$ = −1.2 x 10$^{-3}$

**x10**

spin-coated film
$g_{lum}$ = +2.1 x 10$^{-2}$

*annealing* → **x10**

$(R_p)$-**8**

$\lambda_{abs,max}$ = 349 nm
$\varepsilon$ = 0.63 x 10$^5$ M$^{-1}$cm$^{-1}$
$\lambda_{PL,max}$ = 412 nm
$\Phi_{PL}$ = 0.60
$|g_{lum}|$ = 1.1 x 10$^{-2}$

$(R_p)$-**11**

**Fig. 3.15** Structures and optical data of X-shaped molecules **8–11**

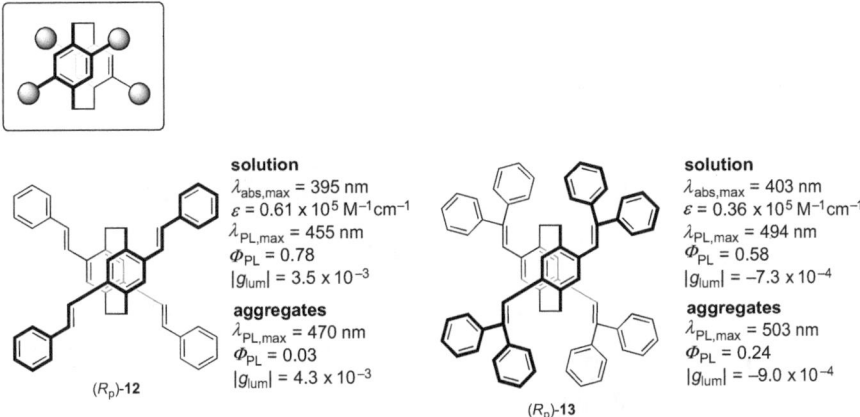

**Fig. 3.16** Structures and optical data of X-shaped molecules **12** and **13**

Bis-(*para*)-pseudo-*ortho*-4,7,12,15-tetrasubstituted [2.2]paracyclophanes (Fig. 3.7) were applied in the chemoselective Sonogashira-Hagihara coupling to afford the corresponding diyne, which could be employed as a chiral building block [36]. Different π-conjugated substituents could be introduced at the 4,12- and 7,15-positions. As shown in Fig. 3.19, one-handed double helical structure was constructed; for example, boomerang-shaped π-electron systems were stacked at the second and fourth phenylene rings to form a left-handed helix from the $(R_p)$-isomer. The double helical compounds $(R_p, R_p)$- and $(S_p, S_p)$-**17** were highly emissive with excellent chiroptical properties in the ground state and, in particular, in the excited state ($|g_{lum}| = 1.6 \times 10^{-3}$); that is to say, they were also excellent organic CPL emitters [36].

Optically active V-shaped molecule **18** (Fig. 3.20) could be prepared from MeO-substituted bis-(*para*)-pseudo-*ortho*-4,7,12,15-tetrasubstituted [2.2] paracyclophane (Fig. 3.7) [37]. π-Electron systems were stacked at the terminal benzene rings. The properties were compared with the corresponding X-shaped molecule **16**, in which the π-electron systems were stacked at the central benzene rings. The CPL signs of the V- and X-shaped molecules were positive and negative, respectively. Positive and negative CPL signs appeared from V-shaped molecule $(R_p)$-**18** and X-shaped molecule $(R_p)$-**16**, respectively. The stacking positions of the two π-electron systems leads to the different CPL signs despite the same absolute configuration.

The electronic transition dipole moments and magnetic transition dipole moments from S1 to S0 of molecules **18** and **16** were simulated (Fig. 3.21); the simulation was carried out for $(S_p)$-isomers. The $g_{lum}$ value is defined by $4|\boldsymbol{\mu}||\boldsymbol{m}|\cos\theta/(|\boldsymbol{\mu}|^2+|\boldsymbol{m}|^2)$, where $\boldsymbol{\mu}$ and $\boldsymbol{m}$ represent electric and magnetic transition dipole moments, respectively, and the $\theta$ represents the angle between the $\boldsymbol{\mu}$ and $\boldsymbol{m}$ [54–56]. The sign of a $g_{lum}$ value is decided by this angle. The angle $\theta$ between $\boldsymbol{\mu}$ and $\boldsymbol{m}$ of $(S_p)$-**18** was estimated to be 144°, while that of $(S_p)$-**16** was estimated to be 87°. Theoretical results

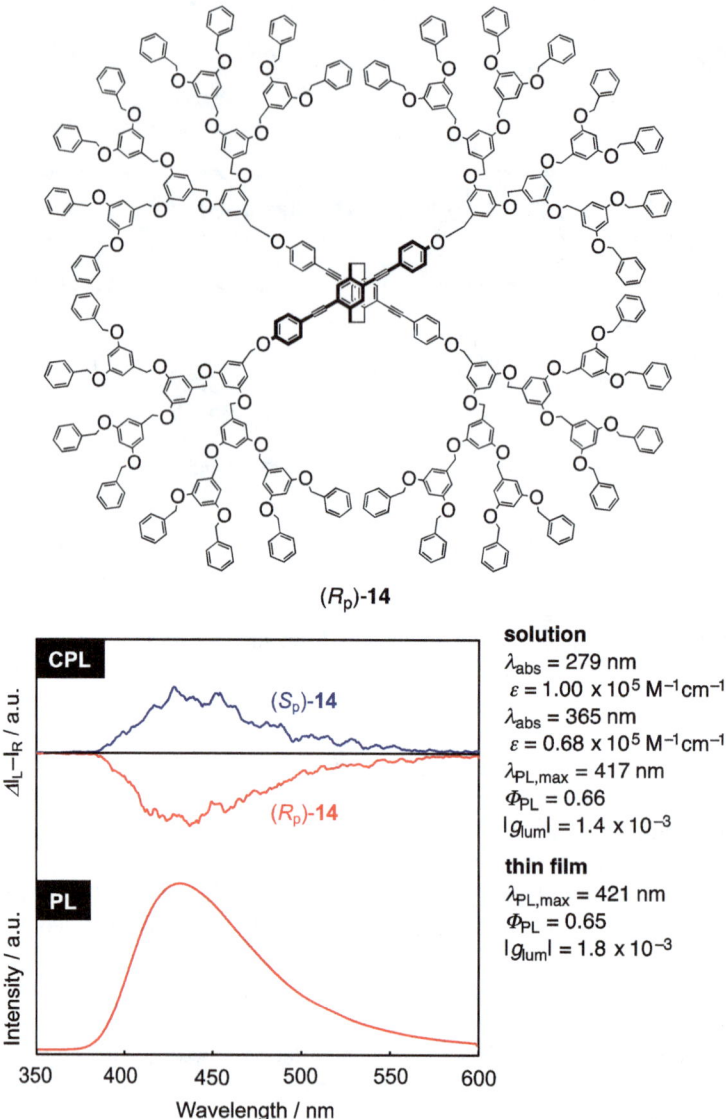

$(R_p)$-**14**

**Fig. 3.17** Structure of third-generation dendrimer **14**. The CPL and PL spectra of the dendrimer film excited at 279 nm are shown

predicted opposite CPL signs for **18** and **16**, and this was also supported by the experimental results.

TMS
Br
OTf
Br
Br
TMS—≡
$Pd_2(dba)_3$
$P^tBu_3$
CuI
in THF, $NEt_3$
OTf
TMS
TMS
MeO
MeO
OMe
($R_p$)-**15**
$\lambda_{abs,max}$ = 357 nm
$\varepsilon$ = 0.56 x $10^5$ $M^{-1}cm^{-1}$
$\lambda_{PL,max}$ = 424 nm
$\Phi_{PL}$ = 0.73
$|g_{lum}|$ = 1.7 x $10^{-3}$
MeO
OMe
MeO
OMe
($R_p$)-**16**
$\lambda_{abs,max}$ = 361 nm
$\varepsilon$ = 0.68 x $10^5$ $M^{-1}cm^{-1}$
$\lambda_{PL,max}$ = 427 nm
$\Phi_{PL}$ = 0.75
$|g_{lum}|$ = 1.7 x $10^{-3}$

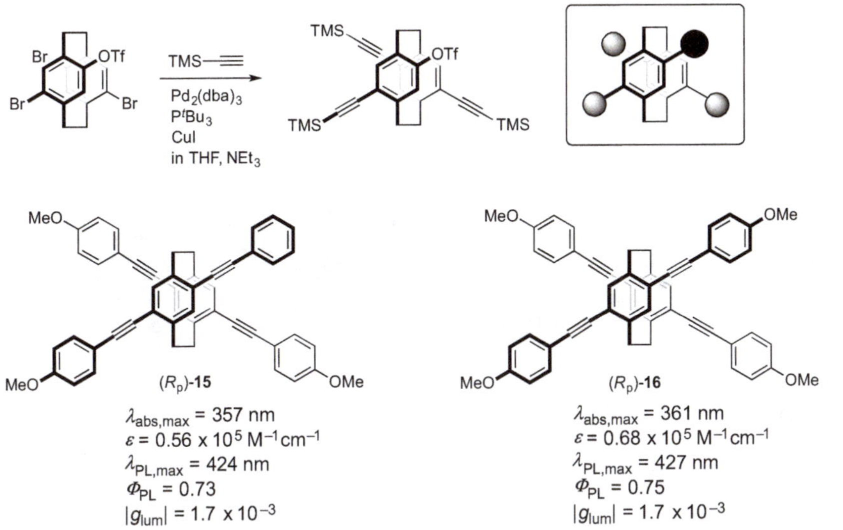

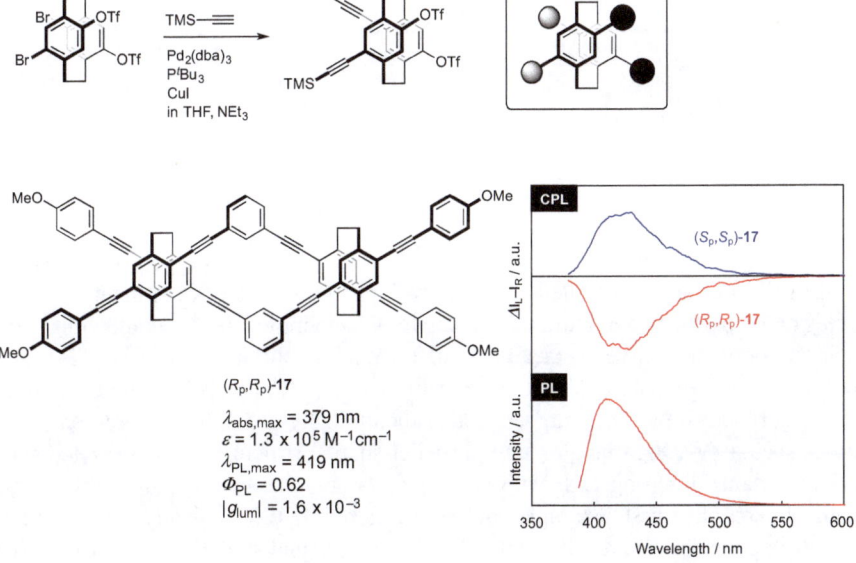

**Fig. 3.18** Structures and optical data of X-shaped molecules **15** and **16** in which different π-electron systems are stacked

Br
OTf
Br
OTf
TMS—≡
$Pd_2(dba)_3$
$P^tBu_3$
CuI
in THF, $NEt_3$
TMS
OTf
TMS
OTf
MeO
OMe
MeO
OMe
($R_p$,$R_p$)-**17**
$\lambda_{abs,max}$ = 379 nm
$\varepsilon$ = 1.3 x $10^5$ $M^{-1}cm^{-1}$
$\lambda_{PL,max}$ = 419 nm
$\Phi_{PL}$ = 0.62
$|g_{lum}|$ = 1.6 x $10^{-3}$
CPL
($S_p$,$S_p$)-**17**
($R_p$,$R_p$)-**17**
$\Delta I_L - I_R$ / a.u.
PL
Intensity / a.u.
350
400
450
500
550
600
Wavelength / nm

**Fig. 3.19** One-handed double helical molecule **17** and the CPL and PL spectra in $CHCl_3$ ($10 \times 10^{-5}$M)

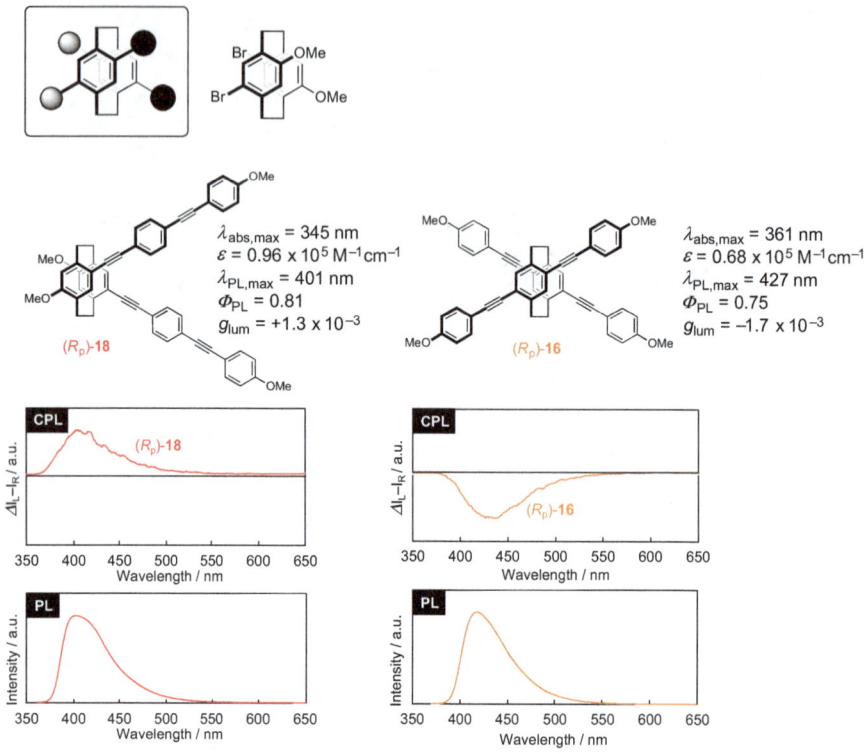

**Fig. 3.20** CPL and PL spectra of $(R_p)$-**18** and **16** in CHCl$_3$ ($10 \times 10^{-5}$M)

## 3.4 Conclusion

In summary, the author described the recent synthetic routes for the optical resolution of di- and tetrasubstituted [2.2]paracyclophane for the development of molecules based on the planar chiral [2.2]paracyclophane molecule that could emit CPL. Various optically active π-stacked small molecules, oligomers, macrocycles, and polymers were prepared using the optically active [2.2]paracyclophane as a chiral building block. It has been suggested that the optically active higher-ordered structures, such as V-, X-, triangle-, and propeller-shaped structures, in the excited state are important for strong CPL with large $g_{lum}$ values. The results introduced in this chapter are the first example of application of the optically active [2.2] paracyclophane molecule in the fields of polymer and materials chemistry. The π-conjugated molecules based on [2.2]paracyclophane emit luminescence basically with high $\Phi_{PL}$. Their $\varepsilon$ values are large due to the extended π-electron systems, which lead to the excellent CPL emission. It is difficult to obtain materials that emit

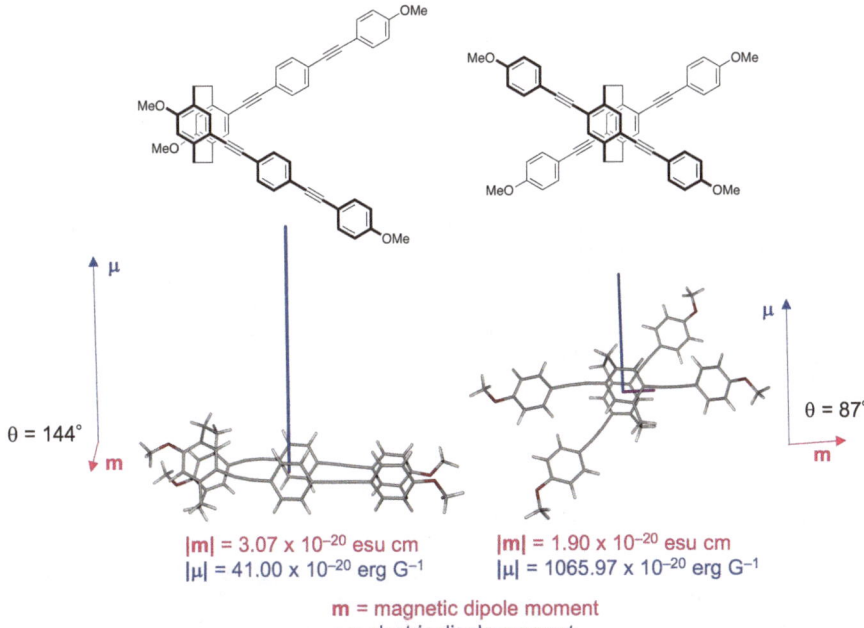

$|\mathbf{m}| = 3.07 \times 10^{-20}$ esu cm
$|\mu| = 41.00 \times 10^{-20}$ erg G$^{-1}$

$|\mathbf{m}| = 1.90 \times 10^{-20}$ esu cm
$|\mu| = 1065.97 \times 10^{-20}$ erg G$^{-1}$

$\theta = 144°$      $\theta = 87°$

m = magnetic dipole moment
μ = electric dipole moment

**Fig. 3.21** Simulations of the transition dipole moments of ($S_p$)-**18** and ($S_p$)-**16** in the excited states by using the TD-DFT calculations at the ωB97XD/6-31G(d,p) level of theory with following options: TD = ($N_{States} = 10$, Root = 1), SCRF = (SOLVENT = Chloroform). Lengths of the dipole moments on the structures are based on the atomic units

high intensity CPL, with high quantum efficiency, and have high anisotropy as well in other chiral scaffolds. The author believes that the planar chiral [2.2] paracyclophane skeleton is an ideal scaffold for developing materials exhibiting strong CPL.

# References

1. Vögtle F (1993) Cyclophane chemistry: synthesis, structures and reactions. Wiley, Chichester
2. Gleiter R (2004) In: Hopf H (ed) Modern cyclophane chemistry. Wiley, Weinheim
3. Brown CJ, Farthing AC (1949) Preparation and Structure of Di-p-xylylene. Nature 164:915–916
4. Cram DJ, Steinberg H (1951) Macro rings. I. Preparation and spectra of the paracyclophanes. J Am Chem Soc 73:5691–5704
5. Morisaki Y, Chujo Y (2006) Through-space conjugated polymers based on cyclophanes. Angew Chem Int Ed 45:6430–6437
6. Morisaki Y, Chujo Y (2008) Cyclophane-containing polymers. Prog Polym Sci 33:346–364
7. Morisaki Y, Chujo Y (2009) Synthesis of π-stacked polymers on the basis of [2.2] paracyclophane. Bull Chem Soc Jpn 8:1070–1082

8. Morisaki Y, Chujo Y (2011) Through-space conjugated polymers consisting of [2.2] paracyclophane. Polym Chem 2:1249–1257
9. Morisaki Y, Chujo Y (2012) π-electron-system-layered polymers based on [2.2] paracyclophane. Chem Lett 41:840–846
10. Morisaki Y, Chujo Y (2019) Planar chiral [2.2]paracyclophanes: optical resolution and transformation to optically active π-stacked molecules. Bull Chem Soc Jpn 92:265–274
11. Mizogami S, Yoshimura S (1985) Synthesis of a new crystalline polymer: polymetacyclophane. J Chem Soc Chem Commun 1985:1736–1738
12. Guyard L, Audebert P (2001) Synthesis and electrochemical polymerization of bis-dithienyl cyclophane. Electrochem Commun 3:164–167
13. Guyard L, Audebert P, Dolbier WR Jr, Duan JX (2002) Synthesis and electrochemical polymerization of new oligothiophene functionalized fluorocyclophanes. J Electroanal Chem 537:189–193
14. Salhi F, Lee B, Metz C, Bottomley LA, Collard DM (2002) Influence of π-stacking. On the redox properties of oligothiophenes: (α-alkyloligo-thienyl)para[2.2]cyclophane. Org Lett 4:3195–3198
15. Salhi F, Collard DM (2003) π-Stacked conjugated polymers: the influence of paracyclophane π-stacks on the redox and optical properties of a new class of broken conjugated polythiophenes. Adv Mater 15:81–85
16. Jagtap SP, Collard DM (2010) Multitiered 2D π-stacked conjugated polymers based on pseudogeminal disubstituted [2.2]paracyclophane. J Am Chem Soc 132:12208–12209
17. Cram DJ, Allinger NL (1955) Macro rings. XII. Stereochemical consequences of steric compression in the smallest paracyclophane. J Am Chem Soc 77:6289–6294
18. Rozenberg V, Sergeeva E, Hopf H (2004) Modern cyclophane chemistry. Wiley, Weinheim, pp 435–462
19. Rowlands GJ (2008) The synthesis of enantiomerically pure [2.2]paracyclophane derivatives. Org Biomol Chem 6:1527–1534
20. Gibson SE, Knight JD (2003) [2.2]Paracyclophane derivatives in asymmetric catalysis. Org Biomol Chem 1:1256–1269
21. Aly AA, Brown AB (2009) Asymmetric and fused heterocycles based on [2.2]paracyclophane. Tetrahedron 65:8055–8089
22. Paradies J (2011) [2.2]Paracyclophane derivatives: synthesis and application in catalysis. Synthesis 23:3749–3766
23. Pye PJ, Rossen K, Reamer RA, Tsou NN, Volante RP, Reider PJ (1997) A new planar chiral bisphosphine ligand for asymmetric catalysis: highly enantioselective hydrogenations under mild conditions. J Am Chem Soc 119:6207–6208
24. Rossen K, Pye PJ, Maliakal A, Volante RP (1997) Kinetic resolution of rac-4,12-dibromo[2.2] paracyclophane in a palladium [2.2]PHANEPHOS catalyzed amination. J Org Chem 62:6462–6463
25. Zhuravsky R, Starikova Z, Vorontsov E, Rozenberg V (2008) Novel strategy for the synthesis of chiral pseudo-ortho-substituted Hydrxy[2.2]paracyclophane-based ligands from the resolved 4-bromo-12-hydroxy[2.2]paracyclophane as a parent compound. Tetrahedron-Asymmetry 19:216–222
26. Jiang B, Zhao XL (2004) A simple and efficient resolution of (±)-4,12-dihydroxy[2.2] paracyclophane. Tetrahedron-Asymmetry 15:1141–1143
27. Jones PG, Hillmer J, Hopf H (2003) (S)-4,16-Dihydroxymethyl-[2.2]paracyclophane bis-(1S)-camphanoate. Acta Cryst E 59:o24–o25
28. Pamperin D, Hopf H, Syldatk C, Pietzsch M (1997) Synthesis of planar chiral [2.2] paracyclophanes by biotransformations: kinetic resolution of 4-formyl[2.2]paracyclophane by asymmetric reduction. Tetrahedron-Asymmetry 8:319–325
29. Pamperin D, Ohse B, Hopf H, Pietzsch M (1998) Synthesis of planar-chiral [2.2] paracyclophanes by biotransformations: screening for hydrolase activity for the kinetic resolution of 4-acetoxy-[2.2]paracyclophane. J Mol Cat B Enzym 5:317–319

30. Braddock DC, MacGilp ID, Perry BG (2002) Improved synthesis of ($\pm$) -4,12-dohydroxy[2.2] paracyclophane and its enantiomeric resolution by enzymatic methods: plana chiral (R)- and (S)-phanol. J Org Chem 67:8679–8681
31. Morisaki Y, Hifumi R, Lin L, Inoshita K, Chujo Y (2012) Through-space conjugated polymers consisting of planar chiral pseudo-*ortho*-linked [2.2]paracyclophane. Chem Lett 3:2727–2730
32. Meyer-Eppler G, Vogelsang E, Benkhäuser C, Schneider A, Schnakenburg G, Lützen A (2013) Synthesis, chiral resolution, and absolute configuration of dissymmetric 4,12-difunctionalized [2.2]paracyclophane. Eur J Org Chem 21:4523–4532
33. Meyer-Eppler G, Sure R, Schneider A, Schnakenburg G, Grimme S, Lützen A (2014) Synthesis, chiral resolution, and absolute configuration of dissymmetric 4,15-difunctionalized [2.2] paracyclophanes. J Org Chem 79:6679–6687
34. Vorontsova NV, Rozenberg VI, Sergeeva EV, Vorontsov EV, Starikova ZA, Lyssenko KA, Hopf H (2008) Symmetrically tetrasubstituted [2.2]paracyclophanes: their systematization and regioselective synthesis of several types of bis-bifunctional derivatives by double electrophilic substitution. Chem Eur J 14:4600–4617
35. Morisaki Y, Gon M, Sasamori T, Tokitoh N, Chujo Y (2014) Planar chiral tetrasubstituted [2.2] paracyclophane: optical resolution and functionalization. J Am Chem Soc 136:3350–3353
36. Morisaki Y, Sawada R, Gon M, Chujo Y (2016) New type of planar chiral [2.2]paracyclophanes and construction of one-handed double helices. Chem Asian J 11:2524–2527
37. Kikuchi K, Nakamura J, Nagata Y, Tsuchida H, Kakuta T, Ogoshi T, Morisaki Y (2019) Control of circularly polarized luminescence by orientation of stacked π-electron systems. Chem Asian J 14:1681–1685
38. Sawada R, Gon M, Nakamura J, Morisaki Y, Chujo Y (2018) Synthesis of enantiopure planar chiral bis-(*para*)-pseudo-*meta*-type [2.2]paracyclophanes. Chirality 30:1109–1114
39. Morisaki Y, Inoshita K, Chujo Y (2014) Chem Eur J 20:8386–8390
40. Wang S, Bazan GC, Tretiak S, Mukamel S (2000) J Am Chem Soc 122:1289
41. Bartholomew GP, Bazan GC (2001) Bichromophoric paracyclophanes: models for interchromophore delocalization. Acc Chem Res 34:30–39
42. Bartholomew GP, Bazan GC (2002) Strategies for the synthesis of 'through-space' chromophore dimers based on [2.2]paracyclophane. Synthesis 9:1245–1255
43. Hong JW, Woo HY, Bazan GC (2005) Solvatochromism of distyrylbenzene pairs bound together by [2.2]paracyclophane: evidence for a polarizable "through-space" delocalized state. J Am Chem Soc 127:7435–7443
44. Bazan GC (2007) Novel organic materials through control of multichromophore interactions. J Org Chem 72:8615–8635
45. Gon M, Kozuka H, Morisaki Y, Chujo Y (2016) Optically active cyclic compounds based on planar chiral [2.2]paracyclophane: extension of the π-surface with naphthalene units. Asian J Org Chem 5:353–359
46. Gon M, Morisaki Y, Chujo Y (2015) Optically active cyclic compounds based on planar chiral [2.2]paracyclophane: extension of the conjugated systems and chiroptical properties. J Mater Chem C 3:521–529
47. Gon M, Morisaki Y, Chujo Y (2015) Highly emissive circularly polarized luminescence from optically active conjugated dimers consisting of planar chiral [2.2]paracyclophane. Eur J Org Chem 2015:7756–7762
48. Gon M, Sawada R, Morisaki Y (2017) Enhancement and controlling the signal of circularly polarized luminescence based on a planar chiral tetrasubstituted [2.2]paracyclophane framework in aggregation system. Macromolecules 50:1790–1802
49. Gon M, Morisaki Y (2017) Optically active phenylethene dimers based on planar chiral tetrasubstituted [2.2]paracyclophane. Chem Eur J 23:6323–6329
50. Hawker CJ, Fréchet JMJ (1990) Preparation of polymers with controlled molecular architecture. A new convergent approach to dendritic macromolecules. J Am Chem Soc 112:7638–7647
51. Fréchet JMJ (1994) Functional polymers and dendrimers: reactivity, molecular architecture, and interfacial energy. Science 263:1710–1715

52. Gon M, Morisaki Y, Sawada R (2016) Synthesis of optically active X-shaped conjugated compounds and dendrimers based on planar chiral [2.2]paracyclophane, leading to highly emissive circularly polarized luminescence materials. Chem Eur J 22:2291–2298

53. Sasai Y, Tsuchida H, Kakuta T, Ogoshi T, Morisaki Y (2018) Synthesis of optically active π-stacked compounds based on planar chiral tetrasubstituted [2.2]paracyclophane. Mater Chem Front 2:791–795

54. Riehl JP, Richardson FS (1986) Circularly polarized luminescence spectroscopy. Chem Rev 86:1–16

55. Riehl JP, Muller F (2012) Comprehensive chiroptical spectroscopy. Wiley, New York

56. Dekkers HPJM (2000) Circular dichroism: principles and applications, 2nd edn. Wiley, New York, pp 185–215

# Chapter 4
# Circularly Polarized Luminescence in Helicene and Helicenoid Derivatives

**Jeanne Crassous**

**Abstract** In this chapter, we discuss the circularly polarized luminescence (CPL) of helicene and helicenoid derivatives. The organic helicenic derivatives are classified according to the type of atom (heteroatom or carbon) incorporated within the helical backbone. Transition-metal complexes and chiroptical devices incorporating helicene derivatives and exhibiting CPL activity are also presented.

## 4.1 Introduction

Circularly polarized luminescence (CPL) is a fascinating property of many classes of chiral emissive molecules. The well-known luminescence dissymmetry factor ($g_{lum}$) as a measure of the degree of circular polarization of emitted light is defined as $g_{lum} = 2\Delta I/I = 2(I_L - I_R)/(I_L + I_R)$, where $I_L$ and $I_R$ denote the left- and right-handed circularly polarized emission intensities, respectively. Various types of chiral fluorescent conjugated organic molecules exhibiting CPL are known and generally possess $g_{lum}$ values on the order of $10^{-4}$–$10^{-2}$ [1–6]. Among them, helicenes are a special class of chiral molecules which consist of n *ortho*-fused aromatic rings combining a helical topology with an extended π-conjugation [7–9]. Since the first preparation of enantioenriched helicenes in 1956 [10], their chirality has essentially been characterized through their very large optical rotation (OR) values, and their intense and characteristic electronic circular dichroism (ECD). In more rare cases, vibrational circular dichroism (VCD) and Raman optical activity (ROA) of helicenes have been measured. More recently, strong interest has grown on the elucidation of the chirality-related emission properties of helicenes. Indeed, helicene derivatives can be regarded as helically shaped polycyclic aromatic hydrocarbons (PAHs); they thus usually display organic semiconducting behavior and are efficient chiral emitters. With the development of circularly polarized emission techniques, the CPL

J. Crassous (✉)
Univ Rennes, CNRS, ISCR (Institut des Sciences Chimiques de Rennes) - UMR6226, Rennes, France
e-mail: jeanne.crassous@univ-rennes1.fr

© Springer Nature Singapore Pte Ltd. 2020

T. Mori (ed.), *Circularly Polarized Luminescence of Isolated Small Organic Molecules*, https://doi.org/10.1007/978-981-15-2309-0_4

53

activity of helicenes and helicenoids is therefore being more and more investigated. In this chapter, we present the helicene derivatives that have shown to display CPL. Helically twisted acenes will not be considered here.

Theoretically, the luminescence dissymmetry factor $g_{lum}$ for the electronic transition $i \rightarrow j$ can be expressed by the following equation: $g_{lum} = 4\dfrac{|\mathbf{\mu}_{ij}| \cdot |\mathbf{m}_{ji}| \cdot \cos\theta\mathbf{\mu},\mathbf{m}}{|\mathbf{\mu}_{ij}|^2 + |\mathbf{m}_{ji}|^2}$, where $\mathbf{\mu}_{ij}$ and $\mathbf{m}_{ij}$ are, respectively, the electric and magnetic transition dipole moment vectors, while $\theta\mathbf{\mu},\mathbf{m}$ is the angle between them. In the case of an electronic transition, the $|\mathbf{m}_{ji}|$ term is usually small with respect to $|\mathbf{\mu}_{ij}|$ so the equation becomes $g_{lum} = 4\dfrac{|\mathbf{m}_{ji}|}{|\mathbf{\mu}_{ij}|} \cos \theta\mathbf{\mu},\mathbf{m}$. High $g_{lum}$ values are therefore usually obtained for magnetic dipole-allowed transitions. For this reason, CPL spectroscopy is widely applied to chiral lanthanide(III) complexes: the magnetically allowed intraconfigurational $f \rightarrow f$ transitions of lanthanide metal ions often provide extraordinary $g_{lum}$ values [11]. In general, chiral organic π-conjugated systems display luminescence dissymmetry factors significantly lower than lanthanide(III) compounds, due to their electronic transitions with strong electric dipole character. However, their easy processability, the wide range of emission wavelengths accessible, and good quantum yields of fluorescence, together with their propensity to self-assemble into chiral supramolecules or aggregates makes chiral π-conjugated molecules appealing systems for improved materials with CPL activity [4]. Therefore, there is a growing interest in the investigation of the CPL properties of chiral π-conjugated systems.

In 2018, Mori et al. tried to see whether there was a correlation between excitation and emission dissymmetry factors; they examined the experimental ratio $g_{lum}/g_{abs}$ (where $g_{abs} = \Delta\varepsilon/\varepsilon$) for a series of chiral organic emissive molecules among which helicenes. They found that this ratio significantly depended on the structure of the helicenic molecule and varied between 0.16 and 28 [12].

This chapter is structured via the different types of CPL-active helicene derivatives, i.e., the N-, O-, S-, B-, Si-, P-, and C-based helicenes and helicenoids, together with transition metal complexes of helicenes exhibiting CPL.

## 4.2 CPL-Active N-Containing Helicenes

### 4.2.1 Helicenic Bridged Triarylamines

In 2003, Venkataraman et al. described the preparation of helical triarylamines. These compounds were among the first helicenic structures displaying clear CPL-activity [13]. The two diastereomers of (P,S)-1a and (M,S)-1b displayed identical absorption spectra (Fig. 4.1a, b and Table 4.1) in the UV-vis region with maximum absorption and emission occurring at 434 nm and 453 nm, respectively. These pseudoenantiomeric compounds showed mirror-image ECD and CPL spectra, revealing that the (1S)-camphanate substituent had no influence on the

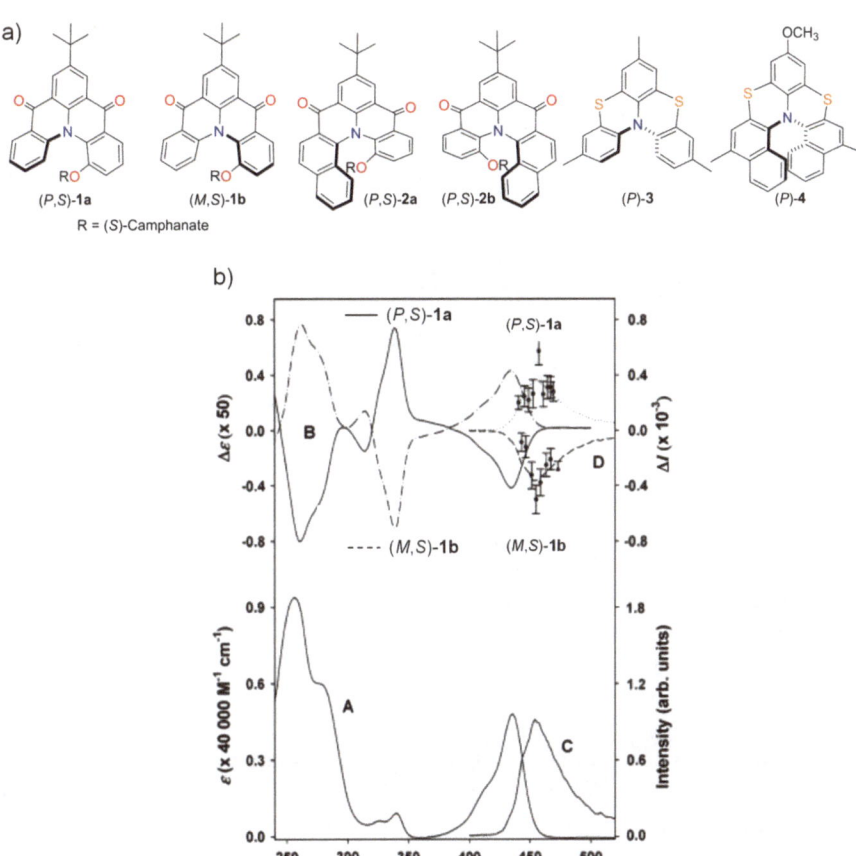

**Fig. 4.1** (a) Chemical structures of helical bridged triarylamines **1a,b**; **2a,b**; and **3–4**. (b) UV-vis (A), ECD (B), fluorescence (C), and CPL (D) spectra of (*P,S*)-**1a** and (*M,S*)-**1b** in CHCl₃. Reproduced with permission [13]. Copyright 2003, American Chemical Society

chiroptical properties. Fluorescence dissymmetry factors of ±0.001 were obtained in chloroform. For the longer and more π-conjugated helicenes (*P,S*)-**2a** and (*M,S*)-**2b**, the maximum emission occurred at longer wavelength, i.e., 478 nm, and slightly smaller fluorescence dissymmetry factors (±0.0008) were obtained. For both compounds **1** and **2**, and for the same transition, $g_{abs}$ and $g_{lum}$ have essentially the same value showing no significant geometry change upon population of the emitting state. This is corroborated by the small Stokes shifts of the emission maxima. In 2016, the CPL activity of sulfurated systems **3** and **4** was measured by Longhi et al. [14] who found a $g_{lum}$ as high $0.9 \times 10^{-2}$ at ~510 nm (positive for (*M*) and vice versa, Table 4.1) for **4** while racemization was observed upon excitation of **3**. Post-functionalization of heterohelicene **5** with formyl (**6**) and diacyanovinyl (**7**) groups enabled to extend the π-conjugation [15]. The tails of UV-vis and ECD spectra together with emission and CPL bands in CH₂Cl₂ are highlighted in Fig. 4.2.

**Table 4.1** Photophysical data of helical triarylamines and of azahelicenes

| Compound | $\lambda_{Abs}^{max\ a}$ (nm) | $\lambda_{Em}$ (nm) | $\Phi_F$ (%) | Solvent (CPL) | $10^3$ $g_{abs}$ | $10^3$ $g_{lum}$ | Ref. |
|---|---|---|---|---|---|---|---|
| (P,S)-**1a** | 434 | 453 | – | CHCl$_3$ | +1.1 | +0.9 | [13] |
| (M,S)-**1b** | 434 | 453 | – | CHCl$_3$ | −1.1 | −1.1 | [13] |
| (P,S)-**2a** | 442 | 478 | – | CHCl$_3$ | +1.0 | +0.8 | [13] |
| (M,S)-**2b** | 442 | 478 | – | CHCl$_3$ | −1.0 | −0.7 | [13] |
| (M)-**4** | 400 | 510 | 1.6 | CHCl$_3$ | ~7.3[b] | 9 | [14] |
| (M)-**5** | 395 | 488 | 86 | CH$_2$Cl$_2$ | +5.6 | +4.7 | [15] |
| (M)-**6** | 461 | 584 | 44 | CH$_2$Cl$_2$ | +2.1 | +1.4 | [15] |
| (M)-**7** | 552 | 685 | 9 | CH$_2$Cl$_2$ | +0.9 | +0.9 | [15] |
| (M)-**8** | 411 | 444 | – | CHCl$_3$ | +5.16 | +5.9 | [16] |
| (M)-**12** | 381 | 467 | 5.1 | CHCl$_3$ | 0.09[b] | <1 | [17] |
| (M)-**13** | 325 | 467 | 2.1 | CHCl$_3$ | 0.02[b] | <1 | [17] |
| (P)-**14** | 442 | 477,509 | 1.4 | CHCl$_3$ | 4.5 | ~0 | [18] |
| (P,P)-**15** | 448 | 471,492 | 19 | CHCl$_3$ | ~1[b] | 28 ± 2 | [17] |
| (P,P)-**16** | 445 | 454,480 | 9.4 | CHCl$_3$ | 6.5 | 11 ± 2 | [17] |
| (+)-**17** | 416 | 473 | 39 | CH$_2$Cl$_2$ | ~13[b] | 9 | [19] |
| (+)-**17**+TFA (200 eq.) | 416 | 514 | 80 | CH$_2$Cl$_2$ | – | 8 | [19] |

[a]Lowest-energy absorption band
[b]Taken from [12]

These compounds are strongly emissive, with quantum yields up to 0.86 for **5** in CH$_2$Cl$_2$ (Table 4.1). A clear red shift of absorption and emission was observed upon increasing the conjugation. The strong charge transfer and high polarity of these molecules were evidenced experimentally through the high solvent polarity dependence of the emission properties (emission wavelength and quantum yield). Regarding the chiroptics, similar $g_{abs}$ and $g_{lum}$ magnitudes were found for each compound with $g_{abs}/g_{lum}$ of +5.6 × 10$^{-3}$/+4.7 × 10$^{-3}$ for (M)-**5**, +2.1 × 10$^{-3}$/+1.4 × 10$^{-3}$ for (M)-**6**, and + 0.9 × 10$^{-3}$/+0.9 × 10$^{-3}$ for (M)-**7**. These properties were also compared in liquid state/solid state and in nanoparticles obtained by rapid precipitation in water. Respective $g_{lum}$ values of 4.5 × 10$^{-3}$, 1.5 × 10$^{-3}$, and 2.8 × 10$^{-3}$ were measured for nanoparticles of **5**, **6**, and **7** dispersed in water, showing that CPL activity was conserved. Note that the influence of solvent polarity was studied on the quantum yields but not on the absolute CPL values.

## 4.2.2 Emission Properties and CPL Activity of Azahelicenes

In 2014, Abbate et al. reported the CPL activity of blue-fluorescent 5-aza[6]helicene (**8**) enantiomers in relation to their ECD spectrum (Figs. 4.3 and 4.4) [20]. The sign of the CPL signal was controlled by the sign of the lower energy ECD named S-type

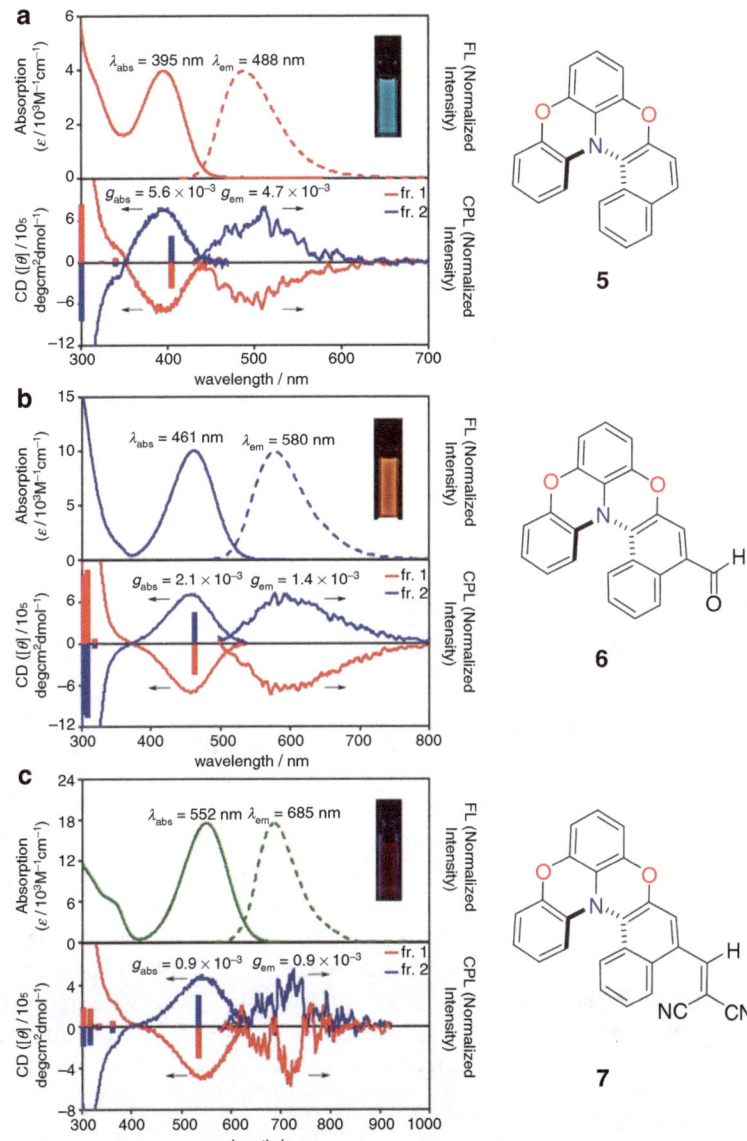

**Fig. 4.2** UV-vis absorption (solid) and fluorescence (dashed) spectra (top) and CD and CPL spectra (bottom) for **5**, **6**, and **7** in $CH_2Cl_2$. The red and blue bars show the calculated ECD bands (CAM-B3LYP/6-31G(d)) for the (*P*)- and (*M*)-helices, respectively. The transition energies have been calibrated using a factor of 0.88. Photographs show the emission of **5-7**. Reproduced with permission [15]. Copyright 2017, American Chemical Society

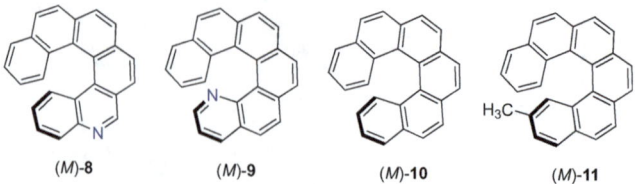

**Fig. 4.3** Structures of aza[6]helicenes **8,9** and carbo[6]helicenes **10,11** [20, 21]

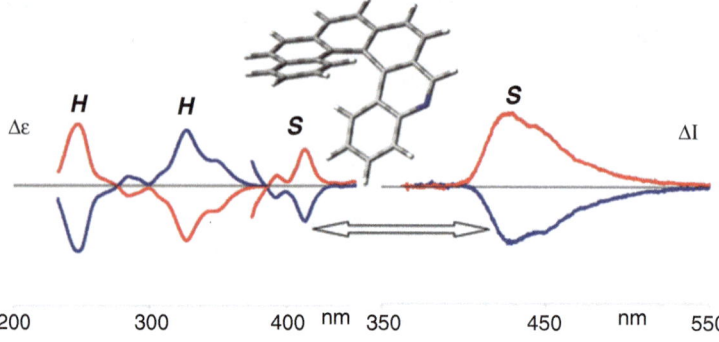

**Fig. 4.4** ECD and CPL spectra of **8** in CHCl₃. Reproduced with permission [20]. Copyright 2014, American Chemical Society

band in relation with Inoue and Mori's nomenclature [22]. This effect was also shown in carbo[6]helicenes (*M*)-**10** and (*M*)-**11** which exhibited positive CPL signals at ~410 and 415 nm, respectively, in CHCl₃ (vide infra) [20]. In 2016, Longhi and Santoro reported the vibronically resolved calculated UV-vis, ECD, emission, and CPL spectra of **8**. A CPL dissymmetry factor $g_{lum}$ of $+0.59 \times 10^{-3}$ at 444 nm was experimentally measured for (*M*)-**8** in CHCl₃ (Table 4.1) and used as a helically shaped chiral model to test the validity of advanced theoretical calculations of chiroptical techniques [16]. Note that the $g_{lum}$ value of corresponding 1-aza[6]helicene **9** was evaluated to be between $10^{-4}$ and $10^{-3}$ by Fuchter, Campbell, and coworkers who used this helicene as a chiral inducer in organic light-emitting diodes (OLEDs, see Sect. 4.10) [21]

### 4.2.3 Double Vs. Single Azahelicenes

In 2014 Tanaka and coworkers reported the enantioselective synthesis of azahelicenes **12** and **13** and of S-shaped double azahelicenes **15** and **16** (Fig. 4.5) [17, 18]. Their photophysical properties are summarized in Table 4.1. Double azahelicenes **15** and **16** showed red shifts of absorption and emission maxima as compared with their corresponding single azahelicenes **12** and **13**. They also showed higher quantum yields in CHCl₃ solution. Interestingly, the CPL activity of

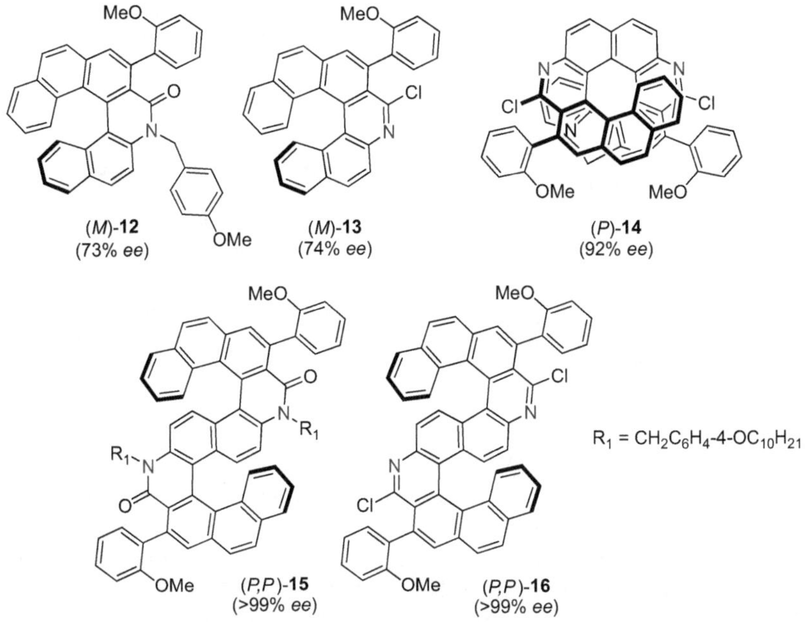

**Fig. 4.5** Chemical structures of enantioenriched single aza[6]helicenes **12–14** and S-shaped double aza[6]helicenes **15,16** [17, 18]

the S-shaped double azahelicenes was significantly higher than that of the single azahelicenes. Indeed, CPL measurements showed intensities for azahelicenes **12** and **13** were below their measurable limit ($g_{lum} < 0.001$), whereas double azahelicenes **15** and **16** exhibited strong CPL activities, with $g_{lum} = 0.028$ at 492 nm for (+)-**15** and $g_{lum} = +0.011$ at 454 nm for (+)-**16** in CHCl$_3$ [17]. In 2016, (−) and (+)-aza[10] helicenes **14** were found to display $|g_{abs}| = 4.5 \times 10^{-3}$ at 303 nm which correspond to a smaller value than for $S$-shaped **16** ($|g_{abs}| = 6.5 \times 10^{-3}$ at 331 nm) but no CPL activity could be measured for **14** [18].

To explain the enhancement of CPL in S-shaped double helicenes, Mori et al. proposed in 2018 a protocol for rationally aligning multiple chiral units to boost the chiroptical responses, using hexahelicene **10** as a prototype [23]. To do so, they aligned two hexahelicenes in various orientations and examined by theoretical calculations which orientation resulted in the highest chiroptical performance from X-shaped or S-shaped double hexahelicenes (see Sect. 4.8.2).

### 4.2.4   Polyaza[7]helicenes

In 2017, Shibata et al. reported the synthesis of enantiopure polyaza[7]helicenes such as **17** (Fig. 4.6) possessing a 6-5-6-6-6-5-6 skeleton [19] which showed high fluorescence quantum yields under both neutral ($\Phi_F = 0.39$) and acidic conditions

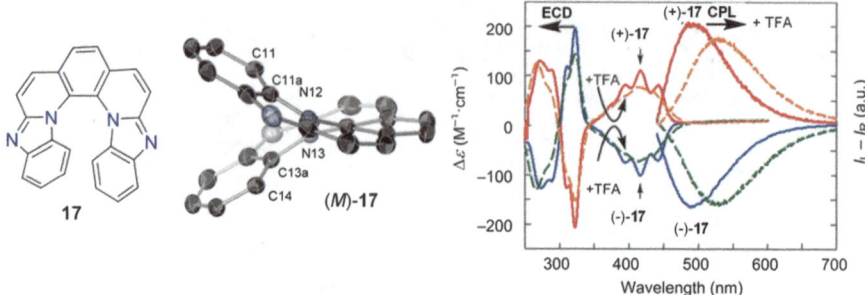

**Fig. 4.6** Chemical structure, X-ray structure, ECD, and CPL spectra of neutral and acidic form of **17** enantiomers. Adapted with permission [19]. Copyright 2017, Wiley

($\Phi_F = 0.80$). The ECD spectrum of (+)-**17** showed several positive Cotton bands in the longer wavelength region with very similar shapes for both neutral and acidic forms. These observations imply that the electronic transitions have similar features in the ground state. Under both conditions, helicene **17** also shows strong CPL activity with a $g_{lum}$ value under neutral conditions of 0.009 at 473 nm which is quite large for a heptahelicene derivative. Upon addition of 200 equivalent amounts of TFA, the $g_{lum}$ value remained high (0.008 at 514 nm). Overall, these systems combine both high $g_{lum}$ value with high quantum yields.

## 4.2.5 Azahelicenes with Fused Carbazole Cycles (Pyrrolohelicenes)

Pyrrole-incorporating PHAs have been shown to possess remarkable physical properties such as effective hole-transporting ability and bright emission. In 2012, Hiroto et al. reported pyrrole-fused system **18** displaying an aza[5]helicenic structure with a stable helical conformation, thanks to the presence of bulky ethynylsilylated groups (Fig. 4.7) [24]. Good fluorescence quantum yields were obtained for **18** ($\Phi_F = 0.36$) with a Stokes shift of 2220 cm$^{-1}$ reflecting the distorted conformation of **18**; a CPL anisotropy factor $|g_{lum}|$ of $3 \times 10^{-3}$ was measured for both enantiomers. The corresponding bis-butadiyne bridged azahelicene dimer (M,M)- and (P,P)-**20** with a figure-eight shape was prepared and exhibited red-shifted absorption and emission, with a higher fluorescence quantum yield ($\Phi_F = 0.55$) and higher $g_{abs}$ and $g_{lum}$ values as compared to (M)- and (P)-**18** (with opposite signs, see Table 4.2). This enhancement was attributed to the rigid conformation of the dimer [25]. See Sect. 4.3 for the corresponding CPL-active oxygen-containing helicene derivative **19**.

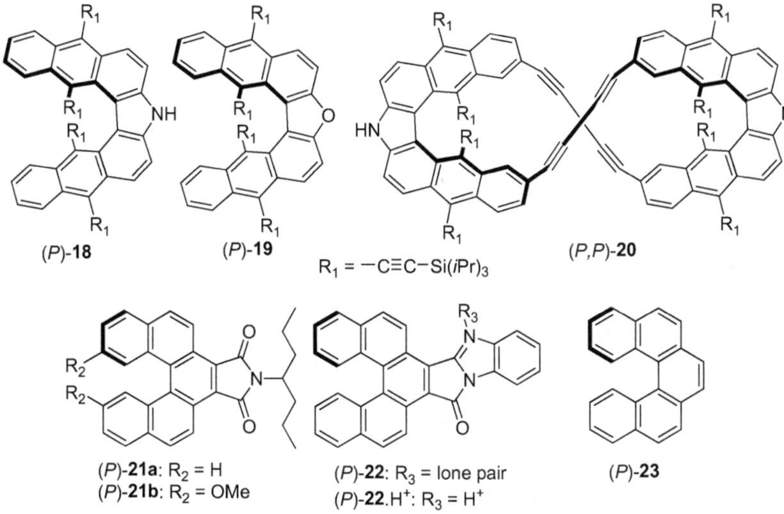

(P)-18    (P)-19    (P,P)-20

$R_1 = -C \equiv C - Si(iPr)_3$

(P)-21a: $R_2$ = H
(P)-21b: $R_2$ = OMe

(P)-22: $R_3$ = lone pair
(P)-22.H$^+$: $R_3$ = H$^+$

(P)-23

**Fig. 4.7** Chemical structures of pentahelicenes incorporating a carbazole (**18**) together with a dimer (**20**) and a dibenzofurane analogue (**19**); pentahelicenes including imide functions (**21a,b** and **22, 22.H$^+$**) together with pentacarbohelicene **23**

**Table 4.2** Photophysical data of helicenes fused with pyrroles or imide cycles

| Compound | $\lambda_{Abs}^{max\ a}$ (nm) | $\lambda_{Em}$ (nm) | $\Phi_F$ (%) | Solvent (CPL) | $10^3$ $g_{abs}$ | $10^3$ $g_{lum}$ | Ref. |
|---|---|---|---|---|---|---|---|
| (P)-(+)-**18** | 503 | 560 | 36 | CH$_2$Cl$_2$ | +1.4$^b$ | +3 | [24] |
| (P,P)-(+)-**20** | 526 | 588 | 55 | CH$_2$Cl$_2$ | −8.5 | −8.5 | [25] |
| (P)-(+)-**21a** | 456 | ~480 | 37 | THF | −4.8 | −2.4 | [26] |
| (P)-(+)-**21b** | 475 | ~525 | 22 | THF | −5.7 | −2.3 | [26] |
| (P)-(+)-**22** | 450 | 575 | 6 | CH$_2$Cl$_2$ | – | −9.45 | [27] |
| (P)-(+)-**22.**H$^+$ | 570 | 650 | 6 | CH$_2$Cl$_2$ | – | −5.92 | [27] |
| (P)-(−)-**24a** | 366 | 445 | 12.8 | THF | −1.55 | −1.2 | [28] |
| (P)-(−)-**24b** | 366 | 457 | 19.2 | THF | −1.33 | −1.5 | [28] |
| (P)-(−)-**24c** | 371 | 482 | 64.8 | THF | −1.35 | −0.8 | [28] |
| (P)-(−)-**24d** | 387 | 556 | 40.3 | THF | −1.31 | −0.8 | [28] |
| (P)-(−)-**24e** | 415 | 617 | 7.4 | THF | −3.53 | −0.2 | [28] |

$^a$Lowest-energy absorption band
$^b$Taken from [12]

## 4.2.6 Helicene Imide Derivatives

Aromatic diimides are known to display bright emission properties. In 2016, Hasobe and coworkers reported [5]carbohelicene derivatives **21a,b** fused with an electron-withdrawing maleimide and substituted with electron-donating methoxy groups [26]. Compared to pristine [5]carbohelicene **23**, the introduction of an electron-withdrawing maleimide group onto a [5]carbohelicene core contributes to the stabilization of the LUMO level in **21a,** whereas the energy level of HOMO level in MeO-substituted **21b** increases due to the electron donation. As a result, the HOMO-LUMO gap of **21b** is smaller than that of **21a** and of carbo[5]helicene **23**, giving bathochromic shift of absorption and emission bands. The absolute fluorescence quantum yield of **21a** was found higher (0.37) as compared to [5] carbohelicene **23** (0.04), whereas $\Phi_F$ of **21b** was slightly smaller (0.22). The chirality of these [5]carbohelicene derivatives **21a,b** was evidenced by their ECD and CPL activities. In particular, **21a** and **21b** gave good CPL activity with anisotropy factors $g_{lum}$ estimated to be $\pm 2.4 \times 10^{-3}$ and $\pm 2.3 \times 10^{-3}$ in THF. This example was the first observation of CPL in [5]carbohelicene derivatives which are usually thought to be configurationally unstable. Here, the authors verified that the CPL signal was stable with time in solution. Note that this negative CPL signal for the $(P)$-(+) isomer is a S-type one (vide supra and [22]), i.e. corresponding to the small negative ECD signal at 435 nm with $g_{abs}$ values of $-4.8/-5.7 \times 10^{-3}$ for $(P)$-(+)-**21a/b**. The same authors also reported a carbo[5]helicene **22** bearing a fused benzimidazole and its protonated form **22**.H$^+$ (see Fig. 4.7 and Table 4.2) [27].

In 2016, Chen and coworkers reported the preparation of configurationally stable helical aromatic imides displaying full-color CPL responses [28]. For this purpose, they prepared enantiomers $(P)$-(−)- and $(M)$-(+)-**24a-e** with high *ee*'s (98.4–99%). Each pair of enantiomers displayed mirror-image ECD spectra of moderate magnitude and absolute configurations opposite to those of classical heterohelicenes. $(P)$-(−)- and $(M)$-(+)-**24a-e** exhibited full color fluorescence emission (from 445 to 617 nm) and mirror-image CPL signals in THF (Fig. 4.8). The $g_{abs}$ values of the enantiomers fell within the range of $\pm 1.5 \times 10^{-3}$ to $\pm 3.5 \times 10^{-4}$ and the $g_{lum}$ values between $\pm 0.2 \times 10^{-3}$ and $\pm 1.5 \times 10^{-3}$ (Table 4.2).

### 4.2.7 Carbocationic Azahelicenes

Lacour and coworkers prepared functionalized carbocationic [4]helicene (**25–30**) and [6]helicene (**31–33**) derivatives and studied their photophysical and chiroptical properties (Figs. 4.9 and 4.10 and Table 4.3) [29–31]. Interestingly, these compounds exhibit fluorescence emissions in the red to near infrared region (with $\Phi_F$ ranging from 0.01 to 0.445), which corresponds to an unusual spectral range for helicene-based chromophores, especially for fully organic ones. As a result, these helical derivatives may be interesting for chiral bioimaging. The same authors investigated different diaza[4]helicenium chiral dyes functionalized with different donor and acceptor groups in order to tune their chiroptical properties in terms of ECD and CPL responses (Fig. 4.9) [29]. These helical derivatives present ECD signatures up to 750 nm with moderate intensity in the visible-red region ($\Delta\varepsilon \sim 10$ M$^{-1}$ cm$^{-1}$), resulting from partial charge-transfer transitions involving the nitrogen atoms and the central carbocation. Furthermore, CPL

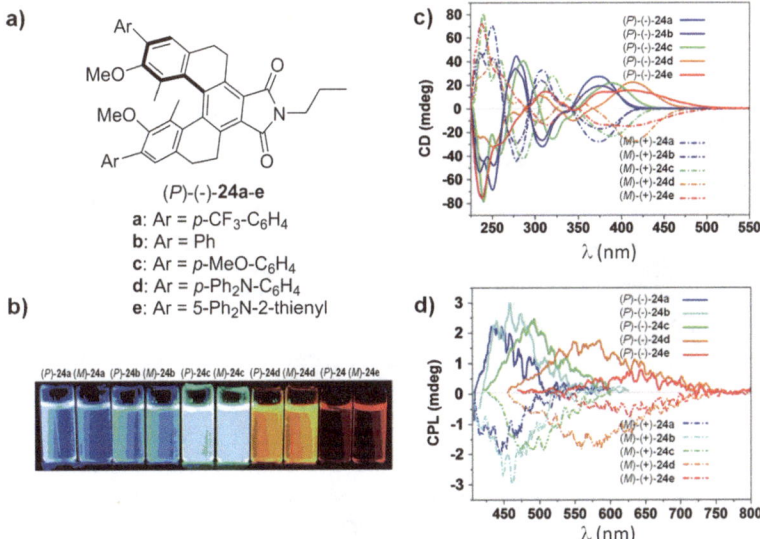

**Fig. 4.8** (**a**) Chemical structure of **24a-e** ((*P*)-(−) enantiomers); (**b**) emission color panel of **24a-e**. (**c**) ECD spectra in THF of pure enantiomers. (**d**) CPL spectra in THF of pure enantiomers. Adapted with permission [28]. Copyright 2016, Royal Society of Chemistry

emissions were recorded between 650 and 700 nm and were characterized by a $g_{lum}$ of $\sim 10^{-4}$ to $10^{-3}$. Such cationic chiral dyes were also used as pH-triggered ECD and CPL chiroptical switches when they possessed pH-sensitive group such as carboxylic acid. For instance, zwitterionic [4]helicene **29** was reported in 2016 as a reversible pH-triggered ECD/CPL chiroptical switch (Fig. 4.10) [30]. Protonated **30** displayed $g_{lum}$ of around $5 \times 10^{-4}$ and of similar order as $g_{abs}$ ($4 \times 10^{-4}$), while the carboxylate derivative **29** displayed no CPL, probably due to very low emission. Overall it represented an on-off CPL switch. Similarly, longer O-containing and N-containing [6]helicenium derivatives **31–33** displayed CPL signals (with opposite signs compared to their optical rotation values) with $|g_{lum}|$ between $0.32 \times 10^{-3}$ and $2.1 \times 10^{-3}$ in the infrared region [31].

## 4.3 CPL-Active Oxygen-Containing Helicene Derivatives

In 2011, Tanaka et al. reported a phthalhydrazide-functionalized [7]oxahelicene derivative **34** (Fig. 4.11), displaying a strong increase of $g_{lum}$, i.e., one order of magnitude, as compared to other helicenic derivatives [32]. This strong CPL enhancement was attributed to the presence of multiple-hydrogen-bonding sites enabling the formation of a trimeric structure which further organizes into chiral fibers (Fig. 4.12). These chiral fibers were 200 nm wide and 3–4 μm long in chloroform solutions, as characterized by SEM and AFM images. While UV-vis

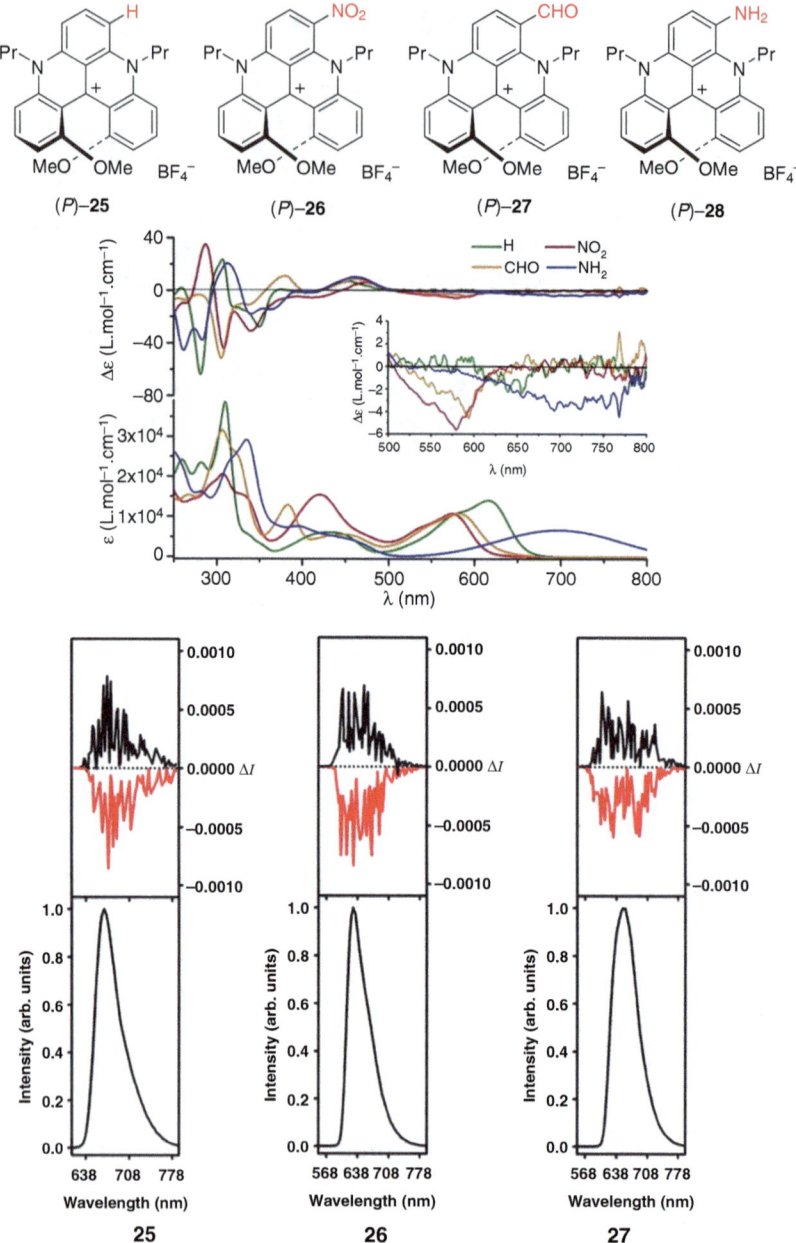

**Fig. 4.9** Chemical structures of carbocationic diaza[4]helicenes **25–28** and their UV-vis, ECD and CPL activities. Adapted with permission [29]. Copyright 2016, Royal Society of Chemistry

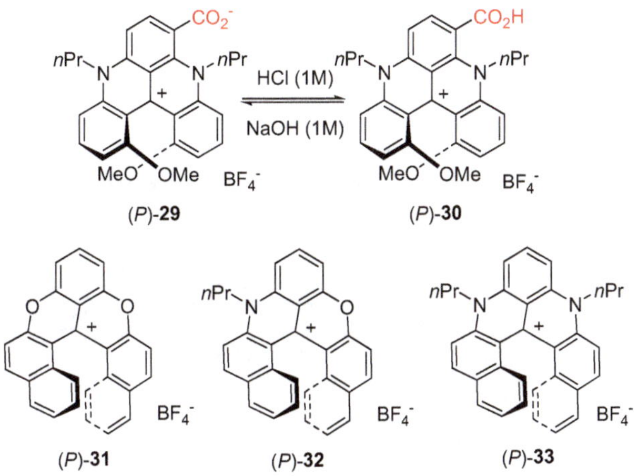

**Fig. 4.10** Chemical structures of carbocationic diaza[4]helicenes **29–30** (pH-triggered switch) and O,N-containing carbocationic [6]helicenes [30, 31]

**Table 4.3** Photophysical data of carbocationic azahelicenes

| Compound | $\lambda_{Abs}^{max\ a}$ (nm) | $\lambda_{Em}$ (nm) | $\Phi_F$ (%) | Solvent (CPL) | $10^3$ $g_{abs}$ | $10^3$ $g_{lum}$ | Ref. |
|---|---|---|---|---|---|---|---|
| (P)-(+)-**25** | 616 | 667 | 13 (CH$_3$CN) | CH$_2$Cl$_2$ | +0.15 | +1.3 | [29] |
| (P)-(+)-**26** | 575 | 624 | 35 (CH$_3$CN) | CH$_2$Cl$_2$ | +0.45 | +1.6 | [29] |
| (P)-(+)-**27** | 582 | 640 | 37 (CH$_3$CN) | CH$_2$Cl$_2$ | +0.36 | +0.9 | [29] |
| (P)-(+)-**29** | 626 | 709 | 1 | CH$_3$CN | ~0.4 | 0 | [30] |
| (P)-(+)-**30** | 590 | 654 | 29 | CH$_3$CN | 0.4 | 0.5 | [30] |
| (P)-(+)-**31** | 562 | 595 | 44.5 | CH$_2$Cl$_2$ | −0.86[b] | −0.32 | [31] |
| (P)-(+)-**32** | 562 | 614 | 28 | CH$_2$Cl$_2$ | −1.3[b] | −2.1 | [31] |
| (P)-(+)-**33** | 614 | 658 | 28.4 | CH$_2$Cl$_2$ | −0.57[b] | −1.2 | [31] |

[a]Lowest-energy absorption band
[b]Taken from [12]

and ECD spectra seemed hardly sensitive to the formation of such aggregates, CPL measurements of **34** afforded $g_{lum}$ of −0.035 for the (M) enantiomer in chloroform solutions, which was larger than in methanol solutions (−0.021). Similar systems, namely, [7]oxahelicene **35** (71% ee) and [9]oxahelicene derivative **36** (88% ee) were reported in 2017 [33]. Comparison of the photophysical data of these 9-oxahelicene **36** compared to **35** shows a redshift of both the absorption and luminescence spectra by ~20 to 50 nm, along with a decrease in fluorescence quantum yield (0.23–0.18). ECD and CPL spectra followed the same trend as for unpolarized UV-vis and fluorescence measurements, but with more intense ECD and CPL signals for **36**. Here the $g_{lum}$ values around $10^{-3}$ in chloroform are classical for helicenic solutions

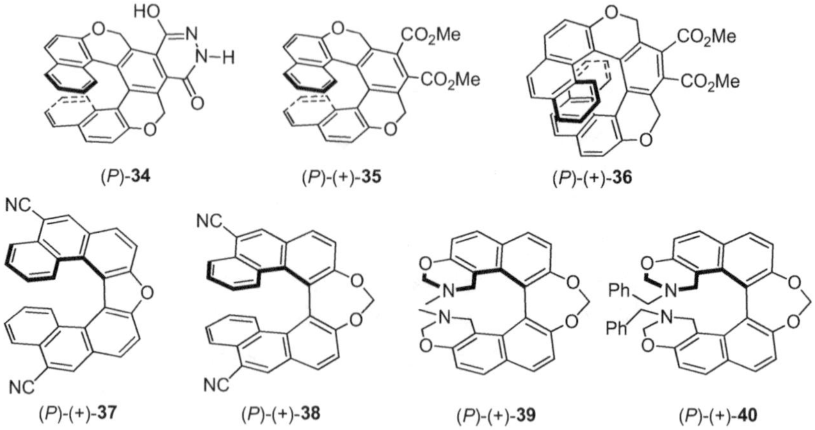

**Fig. 4.11** Chemical structures of oxahelicenic derivatives

(see Table 4.4). Similarly to **18** (see Fig. 4.7), Hiroto and coworkers prepared $P$-(+) and $M$-(−) enantiomers of **19** which displayed strong orange fluorescence ($\Phi_F = 0.66$ in dichloromethane) and moderate CPL activity ($\pm 1.2 \times 10^{-3}$) [34]. Note that different emission properties between racemic and pure enantiomers were obtained for this compound in the solid state but no CPL in the solid state was reported.

In 2018, Bedekar and coworkers reported the preparation of $(P)$-(+) and $(M)$-(−) enantiomers of two 5,13-dicyano-9-oxa[7]helicene derivatives **37** and **38** and reported their photophysical and chiroptical properties [35]. These helical compounds are also active in CPL and mirror-like CPL spectra were measured with positive sign for $(P)$ enantiomers, and $g_{lum}$ values of 3–5 × $10^{-3}$ in DMSO solutions, i.e., falling within the classical range of helicene compounds. In 2014, Bedekar and coworkers also reported helicene-like bis-oxazines **39** and **40** from atropisomeric 7,7′-dihydroxy BINOL derivatives displaying good CPL activity, with $g_{lum}$ of +0.0015/−0.0009 and + 0.0014/−0.0013 for $(P)$−/$(M)$-**39** and $(P)$−/$(M)$-**40**, respectively [36].

## 4.4   CPL-Active Sulfur-Containing Helicene Derivatives

There are few examples of CPL-active sulfur containing helicenes and helicenoids reported in the literature. First of all, it is worth to compare results obtained for **34** with those reported by Katz and coworkers in 2001 on thia[7]helicene-bisquinone derivative **41** decorated with four dodecyloxy groups [37]. Indeed, enantiopure [7] helicene **41** aggregated into columnar structures depending on the solvent type. The aggregation occurred in dodecane and in pure materials, but not in chloroform. The specific rotation showed strong enhancement with increasing concentration, i.e., $[\alpha]_D = 2800$ and 10,400 at $2 \times 10^{-5}$ M and $2 \times 10^{-2}$ M, respectively, in

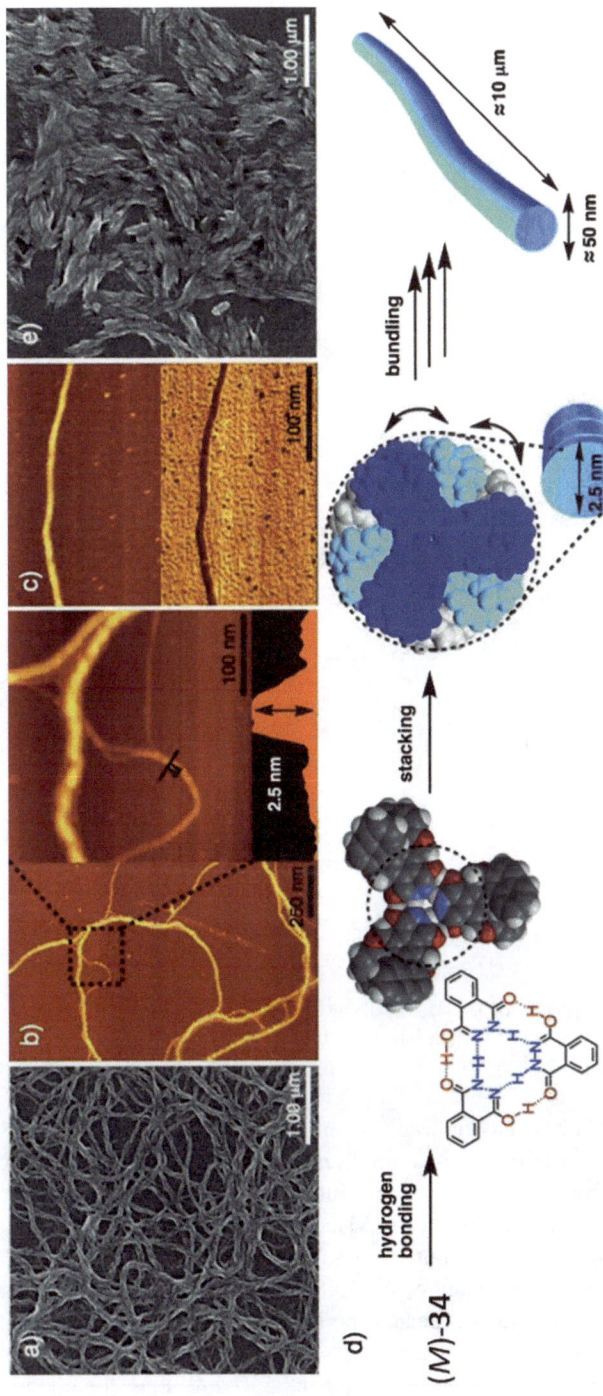

**Fig. 4.12** AFM (**b, c**) and SEM images of (*M*)-**34** (**a**) and (*rac*)-**34** (**e**) prepared in toluene solutions. (**d**) Postulated mechanism for the formation of supramolecular chiral aggregates from a trimeric association. Reproduced with permission [32]. Copyright 2011, Wiley

**Table 4.4** Photophysical data of oxahelicenic derivatives

| Compound | $\lambda_{Abs}^{\ max\ a}$ (nm) | $\lambda_{Em}$ (nm) | $\Phi_F$ (%) | Solvent (CPL) | $10^3\ g_{abs}$ | $10^3\ g_{lum}$ | Ref. |
|---|---|---|---|---|---|---|---|
| (M)-34 | 400 | 476 | – | CHCl₃ | −0.22[b] | −35 | [32] |
|  |  |  |  | MeOH | – | −21 | [32] |
| (P)-(+)-35 | 380 | 479 | 23 | CHCl₃ | +0.56[b] | +0.95 | [33] |
| (P)-(+)-36 | 388 | 514 | 18 | CHCl₃ | +1.3[b] | +1.1 | [33] |
| (P)-19 | 500 | 550 | 66 | CH₂Cl₂ | -1.1[b] | −1.2 | [34] |
| (P)-(+)-37 | 323 | 421 | – | DMSO | – | 3 | [35] |
| (P)-(+)-38 | 310 | 439 | – | DMSO | – | 5 | [35] |
| (P)-(+)-39 | ~360 | 430 | – | CH₃CN | ~+1.2[b] | +1.5 | [36] |
| (P)-(+)-40 | ~360 | 430 | – | CH₃CN | ~+1.2[b] | +1.4 | [36] |

[a]Lowest-energy absorption band
[b]Taken form [12]

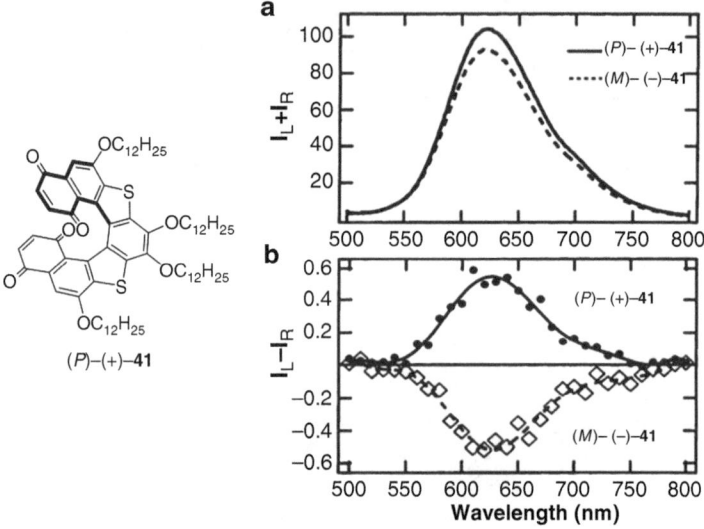

**Fig. 4.13** Chemical structure of (P)-**41**. (**a**) Total luminescence and (**b**) CPL from solutions at 23 °C of (P)-(+) and (M)-(−)-**41** in dodecane ($1 \times 10^{-3}$ M) after excitation with unpolarized light ($\lambda_{ex} = 325$ nm). Adapted with permission [37]. Copyright 2001, American Chemical Society

dodecane. The same behavior was observed by ECD spectroscopy showing significant change for concentrated solutions in dodecane and for films obtained by drop casting from nonane solutions. Interestingly, for a concentration higher than $2 \times 10^{-3}$ M, the solution was viscous and at 0.05 M it became a gel. CPL measurement was carried out for a solution with $1 \times 10^{-3}$ M in dodecane (Fig. 4.13). Emission from enantiomer aggregates between 600 and 700 nm was found to be mirror image for enantiomeric systems. Interestingly, aggregation offered a strong dissymmetry factor ($g_{lum} = 0.01$ at around 630 nm). Likewise, the $g_{abs}$ was large ($g_{abs} = 0.01$ between 500 and 550 nm); thus, the good similarity suggests that the aggregates adopt the same chiral geometry in the ground and the excited states. However, the increased ordering of these aggregates also resulted in a large degree of linear polarization ($P = 0.39$) which can greatly affect the CPL measurement [38].

In 2016, Yamamoto et al. reported the synthesis of tetrasulfone[9]helicene **43** via the oxidation of tetrathia[9]helicene **42** [39]. Remarkably, it was found that the quantum yield of fluorescence for sulfone[9]helicene **43** ($\Phi_F = 0.27$) was ten times higher than tetrathia[9]helicene precursor **42** ($\Phi_F = 0.03$). The author explain this strong enhancement by a significant increase of the energy gap between the lowest singlet ($S_1$) and the triplet ($T_1$) excited states $\Delta E_{ST}$ in the case of **43** ($\Delta E_{ST} = 1.02$ eV) compared to ($\Delta E_{ST} = 0.60$ eV) for **42**, which may efficiently decrease the intersystem crossing (ISC) rate. Additionally, ECD spectra for **43** were recorded and an anisotropy factor $g_{abs} = -4.7 \times 10^{-3}$ was measured for the ($P$) enantiomer. Likewise, plotting the fluorescence CPL mirror-image spectra **43** enantiomers gave an estimated anisotropy factor value of $g_{lum} = -8.3 \times 10^{-4}$ for the ($P$)-**43** (Fig. 4.14).

A series of fluorescent "push-pull" tetrathia[9]helicenes based on quinoxaline (acceptor) fused with tetrathia[9]helicene (donor) derivatives was synthesized for control of the excited-state dynamics and circularly polarized luminescence (CPL) properties [40]. Introduction of a quinoxaline onto the tetrathia[9]helicene skeleton induced a "push-pull" character, which was enhanced by further introduction of electron-releasing or electron-withdrawing groups onto the quinoxaline unit (Fig. 4.15). Significant enhancement in the fluorescence quantum yields ($\Phi_F$) was, for instance, obtained for **44**: ($\Phi_F = 0.30$, Table 4.5), which is more than 20 times

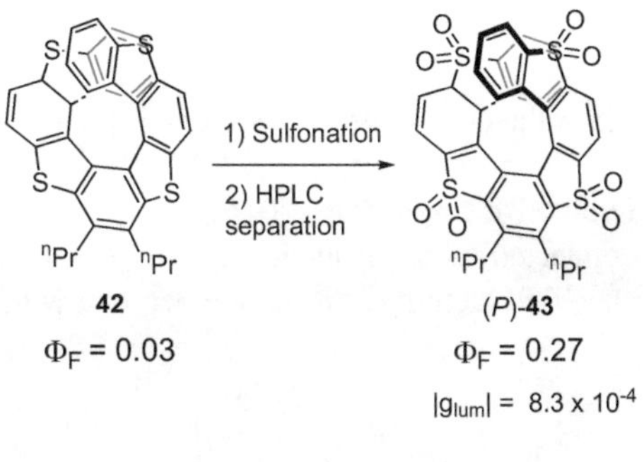

**Fig. 4.14** Chemical structures of **42** and **43** and their emission data

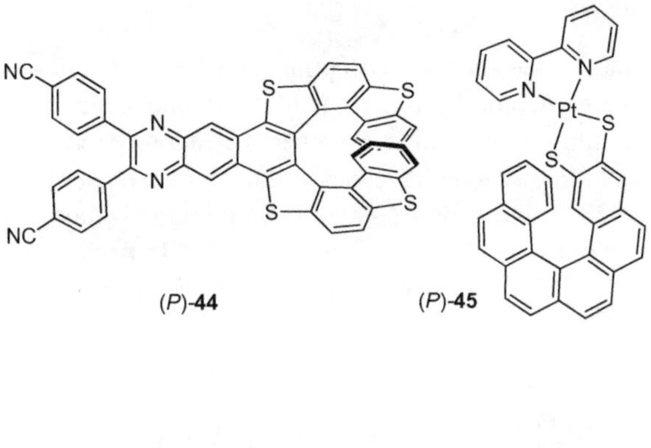

**42**
$\Phi_F = 0.03$

1) Sulfonation
2) HPLC separation

($P$)-**43**
$\Phi_F = 0.27$
$|g_{lum}| = 8.3 \times 10^{-4}$

**Fig. 4.15** Chemical structures of push-pull systems with improved luminescence and CPL emission and of a [6] helicene Pt(diimine) (dithiolene) complex

($P$)-**44**

($P$)-**45**

**Table 4.5** Photophysical data of thiahelicenic derivatives

| Compound | $\lambda_{Abs}^{max\ a}$ (nm) | $\lambda_{Em}$ (nm) | $\Phi_F$ (%) | Solvent (CPL) | $10^3\ g_{abs}$ | $10^3$ $g_{lum}$ | Ref. |
|---|---|---|---|---|---|---|---|
| (P)-(+)-**41** | 470 | 630 | – | Dodecane | +10 | +10 | [37] |
| (P)-**43** | 390 | 450 | 27 | THF | −4.7 | −0.83 | [39] |
| (P)-**44** | 475 | 600 | 30 | THF | +6.8 | +3.0 | [40] |
| (P)-**45** | 562 | 715 | 15 | CH$_3$CN | ~+0.91[b] | +0.3 | [41] |

[a]Lowest-energy absorption band
[b]Taken form [12]

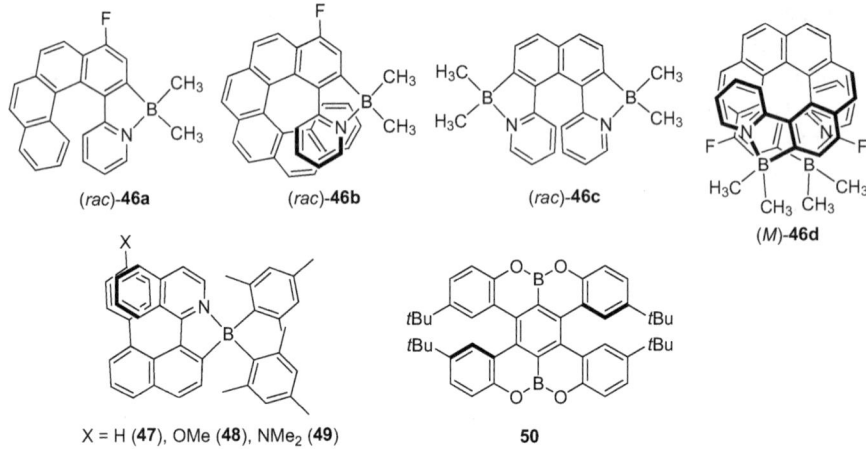

$X = H$ (**47**), OMe (**48**), NMe$_2$ (**49**)                **50**

**Fig. 4.16** Chemical structures of azabora[$n$]helicenes **46a–d** and **47–49** and oxaborahelicene **50**

larger than that of pristine tetrathia[9]helicene ($\Phi_F = 0.02$). Good CPL properties, with an anisotropy factor $g_{lum}$ of $3.0 \times 10^{-3}$ were measured for **44**. In 2017, Avarvari et al. reported the synthesis of (P) and (M)-**45**, corresponding to Pt(diimine)(dithiolene) connected with a [6]helicene unit [41]. Interestingly, (P) and (M)-**45** showed triplet state CPL activity at room temperature with low $g_{lum}$ values of $3 \times 10^{-4}$.

## 4.5   CPL-Active Borahelicenes

Due to the electron-accepting and Lewis acidic character of boron, introducing one or several boron atoms into carbohelicenes generally results in strongly blue-emitting fluorophores. For chemical stability reasons, helicenes incorporating boron atoms are azaborahelicenes and oxaborahelicenes, i.e., also including N or O atoms. In 2017, our group prepared enantiopure azabora[n]helicenes **46a-d** incorporating one or two boron atoms and with 6,8 or 10 *ortho*-fused rings ($n = 6, 8, 10$, Fig. 4.16) [42]. These compounds displayed strong absorption between 250 and 450 nm and blue fluorescence ($\lambda_{Em} \sim$ 420–450 nm) with rather strong quantum yields (0.21–0.49) for azaborahexahelicenes **46a,c** and more modest ones (~0.07) for the octa- and decahelicenes **46b,d**. The introduction of one additional boron atom on **46c** strongly increased the emission efficiency compared to **46a**, but at the same time strongly decreased the configuration stability (enantiomerization barrier $\Delta G^{\neq}$ of 27.5 kcal mol$^{-1}$ at 78 °C, in ethanol) due to the presence of two azaborapentacycles. From the UV-vis spectra, the longer the helicene, the stronger were the absorption coefficients and the more red-shifted the absorption wavelengths. Similarly, the ECD spectra were more red-shifted and more intense for azaboraoctahelicene **46b** and azaboradecahelicene **46d** as compared to azaborahexahelicenes **46a,c**. Note that, except for **46c**, the overall ECD signature appeared typical of helicene derivatives and that the (P)-enantiomers display positive optical rotation values. Regarding the CPL responses, $g_{lum}$ values were found negative for (P)-**46a-c** and positive for (P)-**46d** (see Table 4.6). As mentioned above, the sign of CPL greatly varies with the substituents grafted onto the helicenic core and generally follows the sign of the lower energy ECD-active band. The absolute values of $g_{lum}$ (between $7 \times 10^{-4}$ and $10^{-3}$) for **46a-d** are typical of enantiopure organic helicenes.

Enantiopure azabora[5]helicenes **47–49** were also prepared; they displayed different charge transfer characters and fluorescence quantum yields ranging from 0.13 to 0.30 in toluene, governed by the electron-donor substitution (p-MeO-phenyl, p-Me$_2$N-phenyl) at the helicene [43]. The dimethylamino-substituted derivative emitted at the most red-shifted wavelength and showed the highest Stokes shift in toluene. These helicenes also showed CPL activity with dissymmetry factors $g_{lum}$ between $2.5 \times 10^{-4}$ and $3.5 \times 10^{-3}$. Their ECD spectra and optical rotation values of **47–49** were very different from azaborahelicenes **46a–d**, and it was shown that the sign of the ECD band corresponding to the first transition and the CPL spectrum depended on the electron-donor substitution.

**Table 4.6** Photophysical data of borahelicenes

| Compound | $\lambda_{Abs}^{max\ a}$ (nm) | $\lambda_{Em}$ (nm) | $\Phi_F$ (%) | Solvent (CPL) | $10^3$ $g_{abs}$ | $10^3$ $g_{lum}$ | Ref. |
|---|---|---|---|---|---|---|---|
| (P)-**46a** | 398 | 404 | 21 | CH$_2$Cl$_2$ | ~−1.9[b] | −0.9 | [42] |
| (P)-**46b** | 429 | 435 | 6.9 | CH$_2$Cl$_2$ | ~−2.7[b] | −0.7 | [42] |
| (P)-**46c** | 391 | 427 | 49 | CH$_2$Cl$_2$ | ~−0.8[b] | −2.3 | [42] |
| (P)-**46d** | 440 | 471 | 7.4 | CH$_2$Cl$_2$ | ~+2.6[b] | +1 | [42] |
| (P)-**47** | 414 | 495 | 29/toluene | CHCl$_3$ | −0.7 | −0.25 | [43] |
| (P)-**48** | 420 | 502 | 30/toluene | CHCl$_3$ | +1.1 | +0.95 | [43] |
| (P)-**49** | 433 | 586 | 13/toluene | CHCl$_3$ | +2 | +3.5 | [43] |
| (P,P)-**50** | 411 | 436 | 65/CH$_2$Cl$_2$ (26)[b] | CH$_2$Cl$_2$ | ~−1.1[c] | −1.7 | [44] |

[a]Lowest-energy absorption band
[b]Fluorescence quantum yield measured in the solid state
[c]Taken from [12]

Oxabora[6]helicene **50** was prepared in 2016 by Hatakeyama and coworkers and revealed deep and almost pure blue fluorescence with Commission Internationale de l'Eclairage coordinates of (0.15, 0.08) [44]. Its enantiomers showed CPL high fluorescence quantum yields of 0.65 at 436 nm activity with $g_{lum}$ of $1.7 \times 10^{-3}$. Achiral structural analogues of **50** have proven efficient B-containing PAH dopants in organic OLEDs and in field-effect transistors [45]. These compounds are indeed known to display good carrier mobilities. Note also that such BN and BO aromatic compounds display increasing interest in the domain of thermally activated delayed fluorescence (TADF) [46].

## 4.6 CPL-Active Silahelicenes

Silylated π-conjugated molecules also display strong blue emission. In 2013, Nozaki and coworkers reported the synthesis of enantiopure sila[7]helicene **51**, bearing a silole as the central cycle (Fig. 4.17) [47]. The UV-vis absorption spectrum of (*rac*)-**51** showed longest absorption at 412 nm, which is much longer than pristine phenanthrene (293 nm) and dibenzosilole (286 nm), due to extended delocalization of the π-electrons over the molecule. The absorption edge of (*rac*)-**51** at 431 nm is similar to that of $\lambda^5$-phospha[7]helicene (432 nm) and red-shifted compared to the related aza- and oxa-[7]helicenes (425 nm for aza[7]helicene and 409 nm for oxa[7] helicene). Upon excitation at 320 nm, compound (*rac*)-**51** exhibited a strong blue fluorescence with $\lambda_{max}$ at 450 nm and good quantum yields in solution and

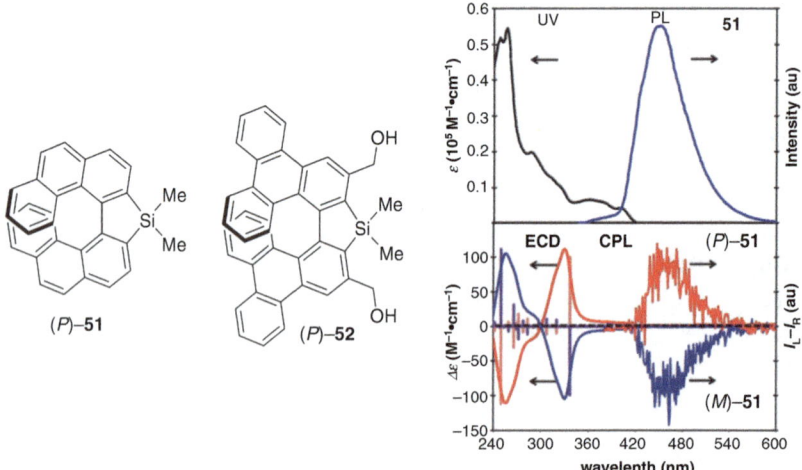

**Fig. 4.17** Chemical structures of silahelicenes (*P*)-**51** and (*P*)-**52**. UV-vis/fluorescence spectra and ECD/CPL spectra of sila[7]helicene **51** in $CH_2Cl_2$. Blue lines in ECD and CPL spectra: (*M*)-isomer. Red lines: (*P*)-isomer. The blue and red bars show the calculated ECD spectra. Reproduced with permission [47]. Copyright 2013, American Chemical Society

**Table 4.7**  Photophysical data of silahelicenes

| Compound | $\lambda_{Abs}^{max\ a}$ (nm) | $\lambda_{Em}$ (nm) | $\Phi_F$ (%) | Solvent (CPL) | $10^3$ $g_{abs}$ | $10^3$ $g_{lum}$ | Ref. |
|---|---|---|---|---|---|---|---|
| (P)-**51** | 412 | 450 | 23/CH$_2$Cl$_2$ (26)$^b$ | CH$_2$Cl$_2$ | ~1.4$^c$ | −3.5 | [47] |
| (P)-**52** | 400 | 482 | 6.9 | CHCl$_3$ | ~4.2$^c$ | +16 | [48] |

$^a$Lowest-energy absorption band
$^b$Fluorescence quantum yield measured in the solid state
$^c$Taken from [12]

in the solid state (see Fig. 4.17 and Table 4.7). The CPL spectra of enantiopure sila[7]helicene (P)- and (M)-**32**, which are mirror image of positive and negative sign, respectively; dissymmetry factors of 3.5 × 10$^{-4}$ at 470 nm were measured. The authors concluded that the $g_{lum}$ derives mainly from the helical biphenanthryl moiety, while the heterole moiety plays essential roles in the luminescent properties. In 2015, Tanaka et al. prepared enantioenriched 1,1′-bis-triphenylene-based sila[7] helicenes (P)-**52** with 91% ee [48]. Compared to sila[7]helicene **51**, **52** displayed red-shifted absorption and fluorescence responses explained by the presence of fused 1,1′-bistriphenylenes resulting in more extended π-conjugation. Probably for the same reason, enantiopure **52** show high $g_{lum}$ values, i.e., 1.6 × 10$^{-2}$, which is uncommonly high for an organic helicene. These values appear larger than that for the 3,3-biphenanthrene-based sila[7]helicene **51** ($g_{lum}$ = −0.0035 at 470 nm) but smaller than that for the 1,1-bitriphenyl based carbo[7]helicene **61** ($g_{lum}$ = −0.030 at 428 nm for the (M)-(−) enantiomer; see Sect. 4.8.3) [49].

## 4.7  CPL-Active Phosphahelicenes

It is now well recognized that phosphorus-containing π-conjugated small molecules, oligomers, polymers, and supramolecular assemblies are important classes of heteroatomic molecular materials for many applications in optoelectronics including OLEDs [50]. P-containing building blocks can indeed lead to materials with unique properties (emission, charge transport, coordination, (anti)aromaticity, etc.). So far, most phosphorus derivatives having helical chirality have displayed polyaromatic (or heteroaromatic) helical scaffolds with pendant phosphorus functions (phosphites, trivalent phosphines and phosphine oxides, helicene-phospholes derivatives) but a few classes of P-containing heterohelicenes have appeared in the literature in the last years. Although several phospholes and helicenes are known to exhibit efficient fluorescence properties, the only example of CPL-active phosphane-containing helicenes has been reported recently [51]. Compounds **53a** and **53b** are benzooxophosphole derivatives that display a carbo[6] helicene unit that is *meta*-fused with one terminal oxophosphole ring containing a pendant phenyl ring at position 5 (Fig. 4.18). In both systems a *l*-menthyl group at the P atom is directed toward the inner groove of the helix (*endo* isomer).

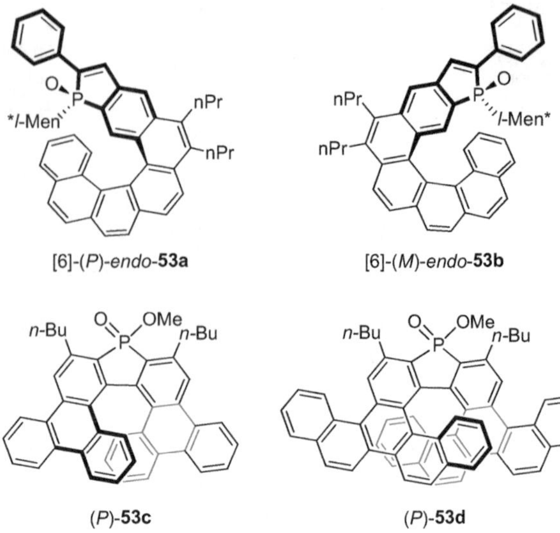

**Fig. 4.18** CPL-active
phosphahelicenes

[6]-(*P*)-*endo*-**53a**                    [6]-(*M*)-*endo*-**53b**

(*P*)-**53c**                             (*P*)-**53d**

**Table 4.8** Photophysical data of phosphahelicenes

| Compound | $\lambda_{Abs}$ max a (nm) | $\lambda_{Em}$ (nm) | $\Phi_F$ (%) | Solvent | $10^3 g_{abs}$ | $10^3 g_{lum}$ | Ref. |
|---|---|---|---|---|---|---|---|
| (*P*)-(+)-**53a** | 436 | 449 | 10 | CH$_2$Cl$_2$ | – | +0.8 | [51] |
| (*M*)-(−)-**53b** | 438 | 455 | 7 | CH$_2$Cl$_2$ | – | −0.7 | [51] |
| (+)-**53c**[b] | 388 | 487 | 22 | CHCl$_3$ | – | +0.81 | [52] |
| (+)-**53d**[b] | 315 | 502 | 8.5 | CHCl$_3$ | – | +0.48 | [52] |

[a]Lowest energy UV-vis band
[b]Absolute configuration not determined

These epimeric compounds can be considered as pseudo-enantiomers since they demonstrate reverse stereochemistry of the helix (*P*/*M*) and of the P atom ($R_P$/$S_P$) but unchanged stereochemistry of the chiral *l*-menthyl group. The epimeric helicenes [6]-(*P*)-*endo*-**53a** and [6]-(*M*)-*endo*-**53b** displayed almost identical UV-vis spectra, blue fluorescence with moderate quantum yield (0.07–0.10), and mirror-image ECD spectra. Similarly, they exhibited mirror-image CPL spectra with luminescence anisotropy factor $g_{lum} = +8 \times 10^{-4}$ and $- 7 \times 10^{-4}$ at 452 nm (excitation at 404–416 nm) for (*P*)-**53a** and (*M*)-**53b**, respectively (Table 4.8). Note that the other phosphahelicenes tested underwent photodegradation under the conditions of the CPL measurement.

In 2018, Tanaka and coworkers reported the enantioselective synthesis of [7]- and [9]phosphahelicenes **53c** and **53d** (Fig. 4.18) by [2+2+2] cycloaddition. These heterohelicenes displayed modest quantum yields (0.22 and 0.085) and $g_{lum}$ values (+8.1 × 10$^{-4}$ and + 3.8 × 10$^{-4}$) [52].

## 4.8  CPL-Active Carbohelicenic Derivatives

### 4.8.1  Pentahelicenic Structures

In 2018, Mori and coworkers reported a combined experimental and theoretical study to elucidate the ECD and CPL behaviors of parent pentahelicene (P)-**23** and of $D_3$-symmetric triple pentahelicene (P,P,P)-**54** [53]. They showed that the pentahelicene unit exhibits absorption and luminescence, with dissymmetry factors $g_{abs}$ and $g_{lum}$ that are intrinsically larger than those of higher homologues. Thus, (P)-**23** emitted strong CPL with $g_{lum}$ of $-2.7 \times 10^{-3}$ at low temperature (see Table 4.9 and Fig. 4.19), which is about a half value of the $g_{abs}$. Due to its photolabile nature, **23** is not suitable to be incorporated in chiroptical materials. However, such undesirable reactivities can be excluded by merging three pentahelicenic units into **54**, for which the $g_{lum}$ and $g_{abs}$ factors were found to be as high as $-1.3 \times 10^{-3}$ and $-1.8 \times 10^{-3}$, respectively, for the (P) enantiomer, corresponding to a high $g_{lum}/g_{abs}$ ratio of 0.72, indicating moderate excited-state relaxation. Theoretical calculations provided further insights into the improved chiroptical responses at the main band ($^1B_b$ transition) in the triple pentahelicene **23**, which was ascribed to its symmetric nature.

In 2017, Lu and coworkers prepared enantiopure tetrahydrocarbo[5]helicenic derivatives (Fig. 4.20) with high configurational stability, thanks to the presence of phenyl groups placed in the inner groove of the helix [54]. These compounds displayed strong blue fluorescence with quantum yield of up to 0.59. Compounds (P)-(+)-**55a–e,56** and (M)-(−)-**55a–e,56** also exhibited mirror-imaged ECD and

**Table 4.9**  Photophysical data of pentahelicenic derivatives

| Compound | $\lambda_{Abs}^{max\ a}$ (nm) | $\lambda_{Em}$ (nm) | $\Phi_F$ (%) | Solvent (CPL) | $10^3\ g_{abs}$ | $10^3\ g_{lum}$ | Ref. |
|---|---|---|---|---|---|---|---|
| (P)-(+)-**23** | 398 | 406 | 5.7 | CH$_2$Cl$_2$ | −7.6 | −2.7 | [53] |
| (M)-(−)-**54** | 460 | 483 | 1.8 | CH$_2$Cl$_2$ | −1.8 | −1.3 | [53] |
| (P)-(+)-**55a** | 347 | 440 | 48 | THF | 2.63 | 5.63 | [54] |
| (P)-(+)-**55b** | 349 | 442 | 59 | THF | 3.55 | 3.66 | [54] |
| (P)-(+)-**55c** | 344 | 439 | 40 | THF | 4.34 | 6.32 | [54] |
| (P)-(+)-**55d** | 344 | 438 | 41 | THF | 3.82 | 6.41 | [54] |
| (P)-(+)-**55e** | 345 | 440 | 42 | THF | 3.21 | 4.58 | [54] |
| (P)-(+)-**56** | 338 | 465 | 3 | THF | 7.87 | 28.2 | [54] |
| (P)-(−)-**57a** | 305 | 424 | 32.6 | CH$_2$Cl$_2$ | −0.47 | −0.25 | [55] |
| (P)-(−)-**57b** | 310 | 425 | 41.0 | CH$_2$Cl$_2$ | −0.50 | −0.28 | [55] |
| (P)-(−)-**57c** | 301 | 425 | 39.1 | CH$_2$Cl$_2$ | −0.83 | −0.41 | [55] |
| (P)-(+)-**57d** | 330 | 455 | 18.7 | CH$_2$Cl$_2$ | −0.328 | −4.52 | [55] |

[a]Lowest-energy UV-vis band

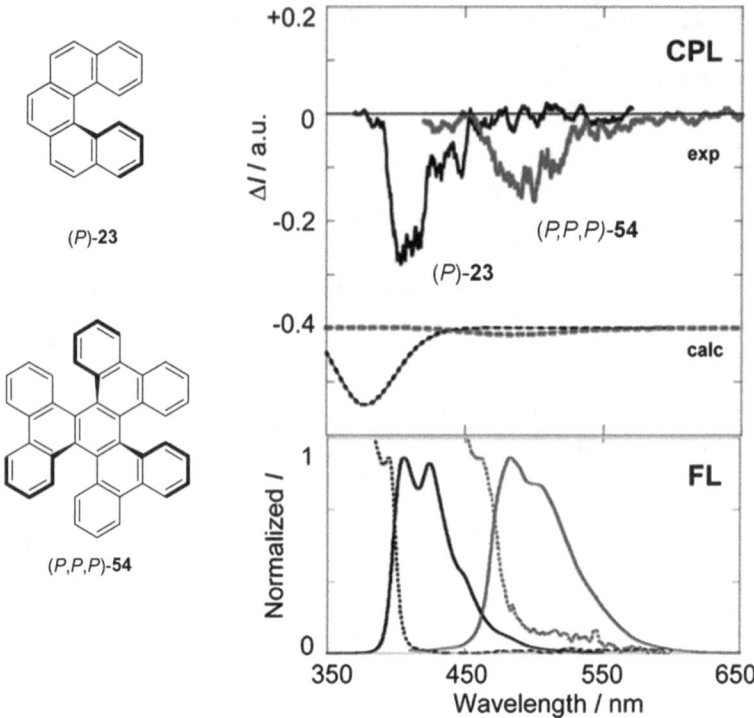

**Fig. 4.19** Chemical structures, fluorescence, and CPL activity of pentahelicenic $(P)$-**23** and of $D_3$-symmetric triple pentahelicene $(P,P,P)$-**54**. Adapted with permission [53]. Copyright 2018, American Chemical Society

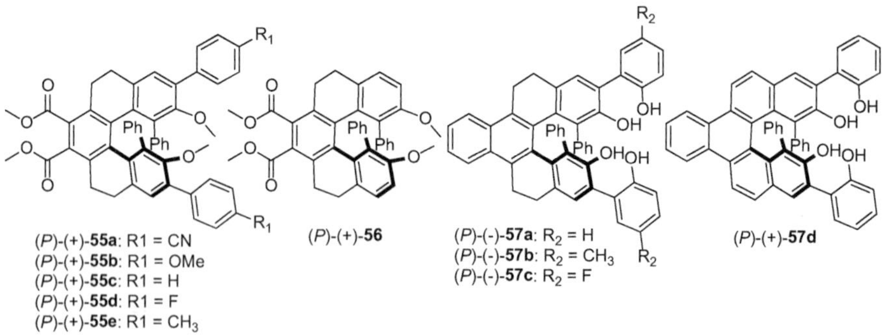

**Fig. 4.20** Enantioenriched tetrahydrocarbo[5]helicenic derivatives prepared by Chen et al

CPL spectra in THF. Due to similar helical backbones, all systems showed similar Cotton effects with strong negative Cotton effects at ~290 nm and positive ones at 315–320 nm for the $(P)$-(+)-**55a-e**,**56** enantiomers. The $g_{abs}$ values were found within the range of $+2.63 \times 10^{-3}$ to $+4.77 \times 10^{-3}$ for the $(P)$ enantiomers and $-3.40 \times 10^{-3}$ to $-2.73 \times 10^{-3}$ for the ($M$) enantiomers, respectively. These compounds also showed intense CPL signals, matching with the region of emission spectra and corresponding ECD signals at the longest wavelength (305–400 nm). The pure enantiomers all exhibited relatively high $g_{lum}$ with values between $+3.66 \times 10^{-3}$ and $+ 6.41 \times 10^{-3}$ for the ($P$) configurations and $-3.40 \times 10^{-3}$ to $-6.59 \times 10^{-3}$ for the ($M$) configurations (Table 4.9) thus showing that chirality existed in both the ground and excited states and could be attributed to their rigid helical structures. Interestingly, absorption dissymmetry factors ($g_{abs}$) of ($P$)-(+)-**56** and ($M$)-(−)-**56** were found stronger (+7.9 × $10^{-3}$ and $-7.7 \times 10^{-3}$, respectively), as well as the emission dissymmetry factors ($g_{lum}$) as large as $2.8 \times 10^{-2}$ and $-3.1 \times 10^{-2}$ at 466 nm, respectively, considerably higher than values reported for other CPL-active single organic molecules. Similar tetrahydropentahelicenic structures, i.e., ($P$)-(−)-**57a-c** and ($M$)-(+)-**57a-c**, were prepared in 2018 by Chen et al. [55]. They exhibited mirror-image ECD and CPL spectra in dichloromethane, with $g_{abs}$ of $-4.7 \times 10^{-4}$ to $-8.3 \times 10^{-4}$ for the ($P$) configuration, and $+4.9 \times 10^{-4}$ to $+8.7 \times 10^{-4}$ for the ($M$) configuration (Table 4.10 and [55]). As for fully conjugated derivative pentahelicene ($P$)-(+) and (M)-(−)-**57d**, they display $g_{abs}$ values of similar magnitude (+3.28 × $10^{-4}$ and $-3.57 \times 10^{-4}$, respectively). Furthermore, the enantiomers of **57a–c** showed CPL signals with $g_{lum}$ values between $-2.5 \times 10^{-4}$ and $-4.1 \times 10^{-4}$ for the ($P$) enantiomers and $+2.6 \times 10^{-4}$ and $+4.2 \times 10^{-4}$ for the ($M$) ones. ($P$)-(+) and ($M$)-(−)-**57d** in dichloromethane exhibited stronger CPL activity, with $g_{lum}$ values of $-4.52 \times 10^{-3}$ and $+4.43 \times 10^{-3}$, respectively. In 2018, Chen et al. also reported the one-pot oxidative aromatization and dearomatization (OADA) reactions of similar tetrahydro[5]helicene diols, with DDQ as the oxidant, which provided a new method for the synthesis of novel CPL-active chiral π-extended diones [58].

**Table 4.10** Photophysical data of carbon-based helicenes

| Compound | $\lambda_{Abs}$ max a (nm) | $\lambda_{Em}$ (nm) | $\Phi$ (%) | Solvent (CPL) | $10^3$ $g_{abs}$ | $10^3$ $g_{lum}$ | Ref. |
|---|---|---|---|---|---|---|---|
| ($P$)-**58a** | 582 | 610 | 39 | CH$_2$Cl$_2$ | +0.15 | +0.1 | [56] |
| ($P$)-**58b** | 582 | 610 | 41 | CH$_2$Cl$_2$ | +0.9 | +0.6 | [56] |
| ($P,P$)**58c** | 622 | 650 | 35 | CH$_2$Cl$_2$ | +1.3 | +0.9 | [56] |
| ($P$)-(+)-**59a** | 326 | 442 | 14 | CHCl$_3$ | −0.56 | ~0 | [57] |
| ($P$)-(+)-**59b** | 327 | 452 | 29 | CHCl$_3$ | −3.1 | ~0 | [57] |
| ($P,P$)-(+)-**59c** | 291 | 464 | 11 | CHCl$_3$ | −0.68 | −2.7 | [57] |
| ($P,P$)-(+)-**59d** | 292 | 466 | 7.6 | CHCl$_3$ | −4.5 | −1.5 | [57] |
| ($P$)-**10** | 410 | 413 | 14 | CHCl$_3$ | −1.2[b] | −0.9 | [23] |
| ($P$)-**11** | 410 | 416 | 5 | CHCl$_3$ | −2.5[b] | −0.02[b] | [20] |
| ($P,P$)-**60** | 471 | 494 | 1.8 | CH$_2$Cl$_2$ | −2.6 | −2.5 | [23] |
| ($P,P$)-**61** | 430 | 434 | 4.1 | CH$_2$Cl$_2$ | −2.8 | −2.1 | [23] |

[a]Lowest energy UV-vis band
[b]Taken from [12]

### 4.8.2 CPL-Active Hexahelicenic Structures

Hexahelicenic derivatives are prototypic helicenes exhibiting CPL activity. In 2018, our group showed that grafting diketopyrrolopyrrole (dpp) dyes onto a carbo[6]helicene structure through ethynyl bridges (see **58a–c** in Fig. 4.21) leads to exciton coupling circular dichroism in the red region arising from the achiral - red-absorbing DPP units in the helical environment [56]. Furthermore, red to near-infrared circularly CPL was obtained. Indeed, the association of enantiopure [6] helicene and dpp units provided helical π-conjugated molecules with strong ECD signal in the visible region (~600 nm), intense red and near-infrared fluorescence ($\phi_F \sim 0.4$), and CPL activity up to 650 nm with $g_{lum}$ found to increase from $1 \times 10^{-4}$ to $6 \times 10^{-4}$ then $9 \times 10^{-4}$ with the increase of exciton coupling (i.e., through the series **58a** → **58b** → **58c**). The $g_{abs}$ values were also found to follow the same increasing trend with the increasing exciton coupling. These results highlighted the synergy between the chiral hexahelicene structure and the organic dye. Thus, decorating carbohelicenes with dyes constitutes an appealing strategy of chemical engineering of a π-helical platform to further improve the chiroptical responses.

In 2018, Tanaka and coworkers reported the enantioselective synthesis of fully benzenoid single (**59a,b**) and double (**59c,d**) carbo[6]helicenes via efficient gold-catalyzed intramolecular hydroarylation (Fig. 4.22) [57]. Similar to the single (**12–14**) and double azahelicenes (**15–16**) described in Sect. 4.2.3, the double carbo[6]helicenes **59c,d** exhibited relatively large CPL activities (up to $2.7 \times 10^{-3}$, see Table 4.10), as compared to the single carbo[6]helicenes **59a,b** whose CPL was below the limit of the apparatus.

In 2018, X-shaped and S-shaped pristine double hexahelicenes (**60** and **61**, Fig. 4.23) were prepared and used as representative molecular models, and a theory-guided, symmetry-based protocol was developed [23]. Compound **60** and **61** exhibited a strong increase in intensity of ECD and CPL. The enhanced chiroptical responses were theoretically assigned to the electric ($\mu_e$) and magnetic ($\mu_m$) transition dipole moments of component hexahelicenes aligned in the correct symmetry. Indeed, **60** and **61**, constructed by merging two hexahelicenes in $D_2$ and $C_2$ symmetry, respectively, showed absorption dissymmetry factors per benzene unit ($g_{abs}/n$) for the $^1B_b$ band that are larger by a factor of up to 1.5 than that of parent **10**. This enhancement was well rationalized by $\mu_e$ and $\mu_m$ and their relative angle ($\theta$) evaluated theoretically. In the double helicenes, $\mu_e$ and $\mu_m$ were parallel-aligned ($\theta = 0$) to maximize the orientation factor ($\cos \theta$) up to unity, which was mere 0.24 ($\cos 76°$) in **10**, while $|\mu_e|$ and $|\mu_m|$ were comparable or only slightly improved. Similarly, the luminescence dissymmetry factor per benzene unit ($g_{lum}/n$) was up to 1.7-fold larger for the double helicenes than for **10**, for which the increased $|\mu_e|$ and $\theta$ are responsible. The enhanced $g_{abs}/n$ and $g_{lum}/n$ values for double helicenes mean that merging two helicenes is 50–70% more resource efficient than simply assembling them, in favor of the molecular, rather than supramolecular strategy for constructing advanced chiroptical devices.

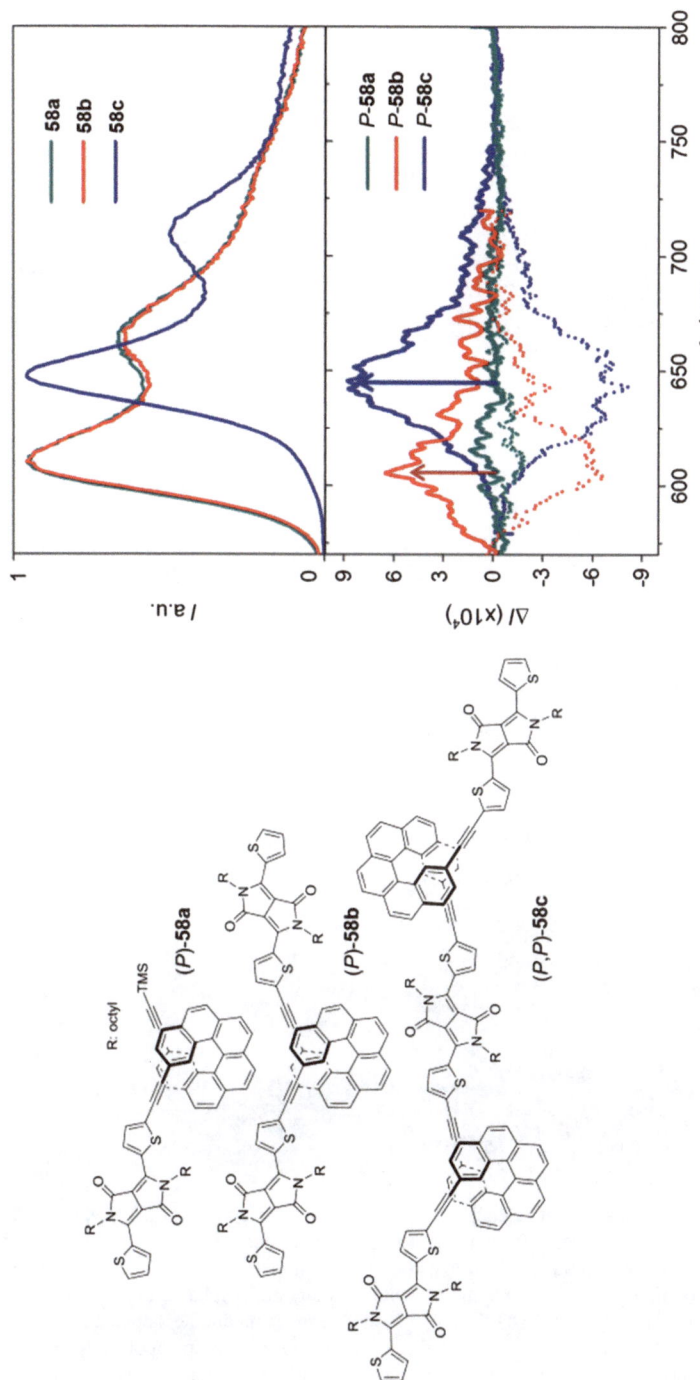

**Fig. 4.21** Chemical structures, fluorescence, and CPL spectra in CH₂Cl₂ of hexahelicene-dpp derivatives (*P*)-**58a–c**. Adapted with permission [56]. Copyright 2018, Royal Society of Chemistry

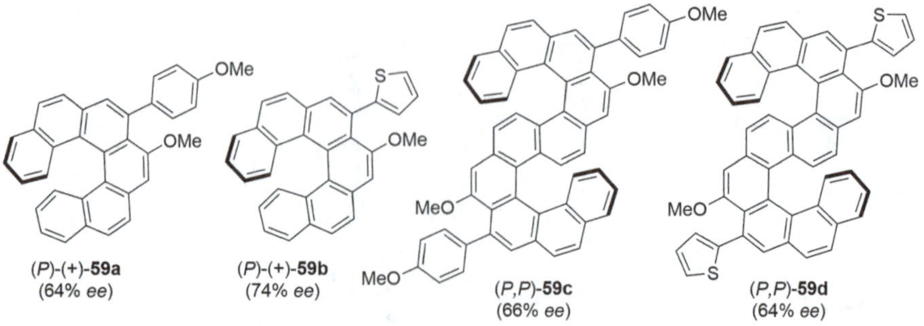

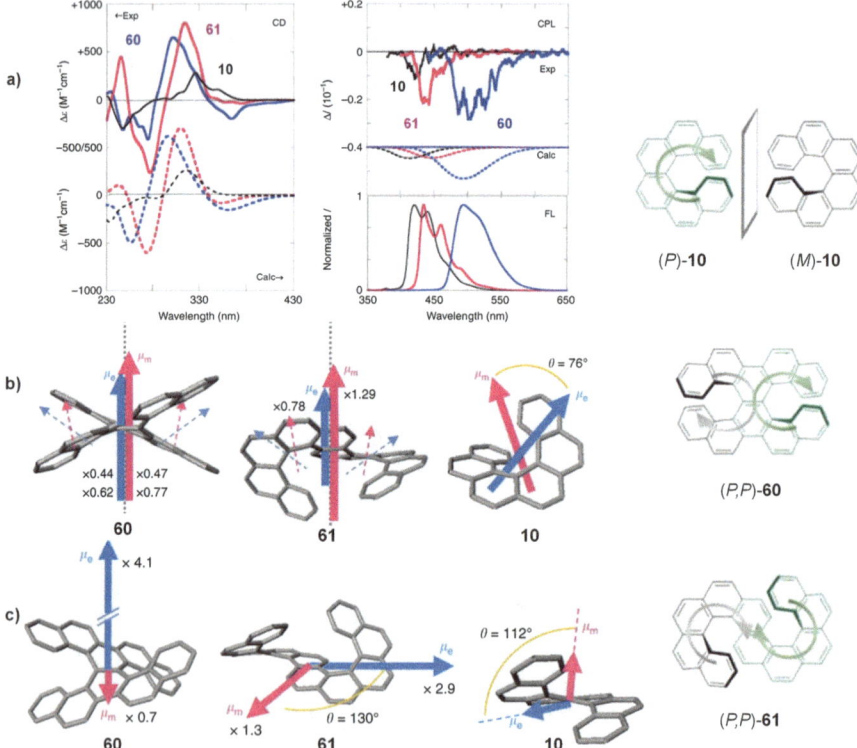

**Fig. 4.22** Chemical structures of enantioenriched single carbo[6]helicenes **59a,b** and S-shaped double carbo[6]helicenes **59c,d** [57]

**Fig. 4.23** (**a**) Experimental and calculated ECD and CPL responses of **10**, **60**, and **61** ((*P*) enantiomers). (**b**) Transition dipole moments in the ground state. Schematic representations of electric ($\mu_e$, blue) and magnetic ($\mu_m$, red) transition dipole moments of the $^1B_b$ band for X-shaped and S-shaped double hexahelicenes **60** and **61**, with the magnitudes relative to parent helicene **10**, calculated at the RI-CC2/def2-TZVPP level. Dashed arrows in double helicenes indicate the transition dipole moments of component helicene units. (**c**) Transition dipole moments in the excited state. Schematic representations of the electric ($\mu_e$, blue) and magnetic ($\mu_m$, red) transition dipole moments of the $^1L_b$ band of **60** and **61** in the excited states, with the magnitudes relative to those for parent helicene **10**, calculated at the RI-CC2/def2-TZVPP level. Adapted with permission [23]. Copyright 2018, Nature Publishers

### 4.8.3   CPL-Active Heptahelicenic Structures

In 2012, Tanaka et al. reported the preparation of helically chiral 1,1′-bitriphenylenes displaying a central fluorenyl cycle. These compounds correspond to heptahelicenic structures and exhibit among the strongest CPL activity [59]. Indeed, compounds **62** and **63** (Fig. 4.24) obtained with high enantiomeric excesses (93% and 91% *ee*, respectively) exhibited mirror-imaged CPL spectra with particularly high fluorescence dissymmetry factors ($g_{lum} = -0.030$ at 428 nm for (*M*)-(−)-**62** and $g_{lum} = -0.032$ at 449 nm for (*M*)-(−)-**63** in chloroform, Table 4.11), which are comparable to phthalhydrazide-functionalized [7]helicene-like molecule **34** ($g_{lum} = -0.035$ at 476 nm for the assembly state and −0.021 for the molecularly dispersed state) and significantly larger than those for helically chiral molecules reported to date. However, the $g_{lum}$ values are found very high compared to the $g_{abs}$ values, which are around $2.5$–$2.9 \times 10^{-3}$, i.e., one order of magnitude lower. This is clearly seen in the ECD shape, which displays the typically large bands of (*P*)-helicenes (a large positive band at ~320 nm, and two large negative bands at ~260 and 285 nm, Fig. 4.25) but small bands between 340 and 400 nm which correspond to the lowest energy transitions. Later on in 2016, Nozaki and coworkers reported the synthesis and photophysical properties of enantiopure [7]helicene-like fluorenyl systems **64a,b** [60]. These compounds showed similar absorption spectra with the longest absorption maximum at ~400 nm slightly blue-shifted compared to that of carbo[7]helicene, while they are significantly red-shifted compared to those of fluorene and phenanthrene. Indeed, the π-conjugation is well extended over the whole molecule despite their helically twisted structures. In addition, the longest absorption maximum is slightly red-shifted compared to that of compound **62** despite the fact that **62** possesses triphenylene units [57]. This phenomenon may be attributed to a smaller twist of compounds **64** than compound **62**, resulting in more effective π-conjugation along the helical structure. [7]Helicene-like compounds **64a,b** exhibited an emission maximum at around 420 nm with a very small Stokes shift and high fluorescence quantum yields (up to 40%) among the highest reported helicenes at that time and slightly higher than those of [7]helicene-like compounds [59]. This makes helicenes incorporating a fluorene unit very appealing for highly emissive chiral molecular materials. ECD spectra of **64a,b** enantiomers display a similar shape as for **62** and

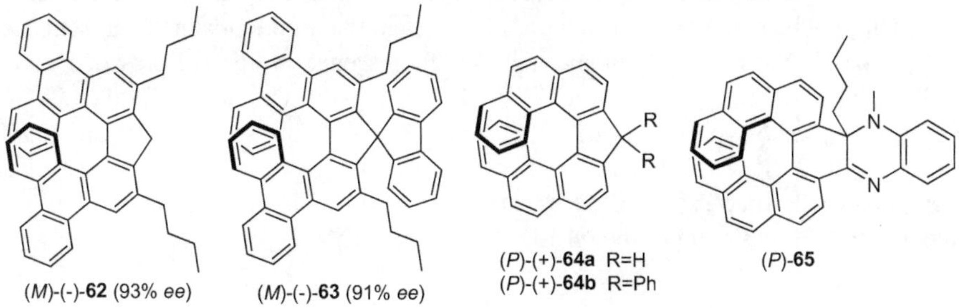

(*M*)-(−)-**62** (93% *ee*)        (*M*)-(−)-**63** (91% *ee*)        (*P*)-(+)-**64a** R=H        (*P*)-**65**
                                                                        (*P*)-(+)-**64b** R=Ph

**Fig. 4.24** Chemical structures of CPL-active heptahelicenic derivatives

**Table 4.11** Photophysical data of heptahelicenic derivatives

| Compound | $\lambda_{Abs}^{max\ a}$ (nm) | $\lambda_{Em}$ (nm) | $\Phi$ (%) | Solvent (CPL) | $10^3\ g_{abs}$ | $10^3\ g_{lum}$ | Ref. |
|---|---|---|---|---|---|---|---|
| $(M)$-$(-)$-**62** | 388 | 428 | 32 | CHCl$_3$ | $-2.5^b$ | $-30$ | [59] |
| $(M)$-$(-)$-**63** | 400 | 449 | 29.6 | CHCl$_3$ | $-2.9^b$ | $-32$ | [59] |
| $(P)$-$(+)$-**64a** | 400 | 417 | 39 | CH$_2$Cl$_2$ | $1.3^b$ | 3 | [60] |
| $(P)$-$(+)$-**64b** | 408 | 421 | 40 | CH$_2$Cl$_2$ | – | 2.5 | [60] |
| $(P)$-$(+)$-**65** | 435 | 550 | 25 | THF | $-1.3$ | $-4$ | [61] |

[a]Lowest energy UV-vis band
[b]Taken from [12]

**Fig. 4.25** (a) ECD/CPL spectra of heptahelicenic fluorene **63** in CH$_2$Cl$_2$. Blue lines in ECD and CPL spectra: $(P)$-isomer. Red lines: $(M)$-isomer and (b) UV-vis/fluorescence spectra. Reproduced with permission [59]. Copyright 2012, American Chemical Society

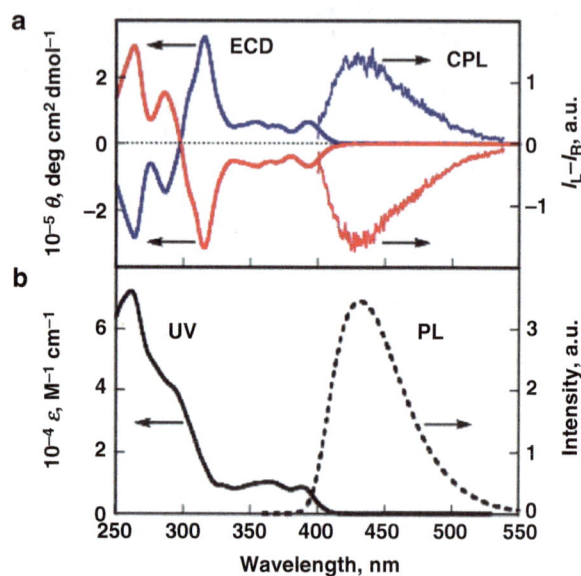

**63** but their CPL activities were smaller, with dissymmetry factors of $3.0 \times 10^{-3}$ and $2.5 \times 10^{-3}$, respectively. These values are comparable to that of silole-fused compound **51** (vide supra). In 2015, Hasobe and coworkers reported a highly yellow fluorescent [7]carbohelicene fused by asymmetric 1,2-dialkyl-substituted quinoxaline (**65**) [61]. It displayed a fluorescence quantum yield of 0.25 at 550 nm emission wavelength which is more than 10 times larger than that of the pristine heptahelicene ($\Phi_F = 0.02$). Such a large enhancement of fluorescence in **65** also provided good CPL activity, with $g_{lum}$ values of $\pm 4.0 \times 10^{-3}$. A negative CPL signal for the $(P)$-$(+)$ isomer, i.e., an S-type one (vide supra and [20, 22]), corresponding to the small negative ECD signal at 435 nm with a $g_{abs}$ of $-1.3 \times 10^{-3}$. Note also that this compound was successfully used as an emitter in OLEDs but the authors did not report any CPL emission of the OLED.

## 4.9   CPL-Active Transition Metal Complexes of Helicenes

### 4.9.1   Cycloplatinated Helicenes

Coordination chemistry offers a simple way to tune the optical and electronic properties of the $\pi$-ligands since both the coordination sphere geometry and the nature of the metal-ligand interaction can be readily modified by varying the metal center. This will produce a great impact on the properties of the molecule [62]. Recent studies have demonstrated many potential applications of N-containing helicenes in coordination chemistry and in materials science [63]. Indeed, their transition metal complexes may show interesting properties in harvesting (visible) light and re-emitting it at a wavelength that depends on the metallic ion used, thus allowing the development of light-emitting devices, chemosensors, photovoltaic dye-sensitized devices, etc. In 2010, our group prepared the first class of organometallic helicenes incorporating a metallic ion, i.e., Pt, within their helical backbone, named platinahelicenes [64, 65]. Enantiopure platina[6] helicene **66a**, platina[8]helicene **66b**, bisplatina[6]helicene **66c**, and bisplatina[10] helicene **66d** (Fig. 4.26), displayed absorption spectra that were strongly red-shifted compared to the starting ligands, with longer absorption wavelengths above 450 nm. Furthermore, platinahelicenes **66a-d** are efficient deep-red phosphors, with emission maxima between 630 and 700 nm, quantum yields around 0.05–0.10 in deoxygenated solution at room temperature and luminescence lifetimes of 10–20 μs. Interestingly, platinahelicenes **66a–c** displayed circularly polarized phosphorescence with dissymmetry factors as high as $10^{-2}$, which is one order of magnitude bigger than for most of organic helicenes. These $g_{lum}$ values appeared positive for the (*P*) enantiomers and negative for the (*M*), which was not always the case in azaborahelicenes analogues **46a–d** [45]. Note that bis-platina[10]helicene **66d** also exhibited red

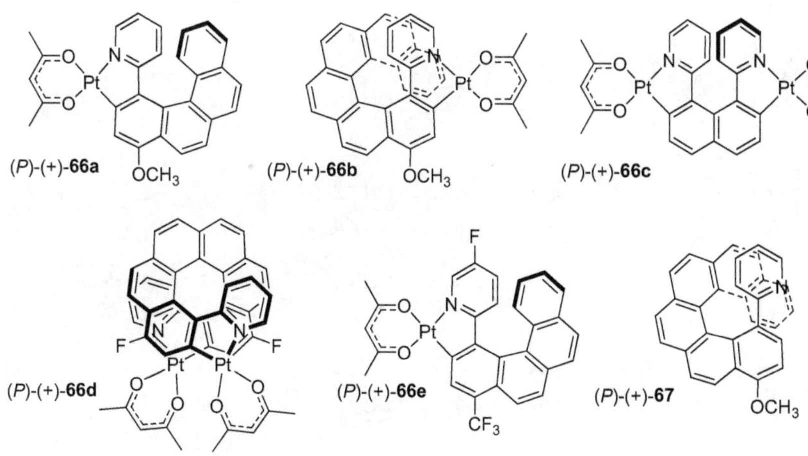

**Fig. 4.26**   Chemical structures of platinahelicenes **66a–e** and of precursor **67**

phosphorescence at room temperature, but no CPL activity was detected. This can be explained by the weakly chiral environment around the two Pt centers and the high sensitivity to oxygen. Note also that the precursor 1-(2-pyridyl)-hexahelicene **67** displayed fluorescence emission and CPL activity (see Table 4.12). Recently, polyfluorinated platina[6]helicene **66e** was prepared by Zheng et al. [66] It displayed similar molecular behavior as **66a–c**, namely red phosphorescence, and CPL activity with a $g_{lum}$ of $-3.7 \times 10^{-3}$ in dichloromethane solution for the (*P*)-(+) enantiomer. The same compound displayed a $g_{lum}$ of $-4.1 \times 10^{-3}$ when incorporated in a DCzppy film (DCzppy: 2,6-bis(3-(9H-carbazol-9-yl)phenyl)pyridine). Interestingly, compounds **66a** and **66e** were used as chiral dopants in OLED devices to conceive efficient circularly polarized OLEDs (CP-OLEDs; see Sect. 4.10).

## 4.9.2  Coordination of Helicene-Bipyridine Ligands

Our group also prepared complexes bearing a helicene-bipyridine-type ligand. Bipyridine ligands are classical N^N chelate ligands but can also act as C^N ones toward different transition metal ions such as platinum. In 2015, we reported the preparation of enantiopure helical cycloplatinated complexes (*P*)- and (*M*)-**69** from a [6]helicene-bipyridine-type ligand, namely, 3-(2-pyridyl)-4-aza[6]helicene ((*P*)- and (*M*)-**68** in Fig. 4.27) [67]. Due to the presence of an additional N atom in organometallic species (*P*)- and (*M*)-**69**, the acid-base triggering of UV-vis, ECD,

**Table 4.12** Photophysical data of platinahelicenes and helicene-bipy rhenium complexes

| Compound | $\lambda_{Abs}^{max\ a}$ (nm) | $\lambda_{Em}$ (nm) | $\Phi$ (%) | Solvent (CPL) | $10^3$ $g_{abs}$ | $10^3$ $g_{lum}$ | Ref. |
|---|---|---|---|---|---|---|---|
| (*P*)-(+)-**66a** | 452 | 644 | 10 | CH$_2$Cl$_2$ | ~3.5[b] | +13 | [65] |
| (*P*)-(+)-**67** | 422 | 430 | – | CH$_2$Cl$_2$ | ~7.3[b] | +0.8 | [65] |
| (*P*)-(+)-**66b** | 467 | 648 | 5.6 | CH$_2$Cl$_2$ | ~12[b] | +0.5 | [65] |
| (*P*)-(+)-**66c** | 471 | 633 | 13 | CH$_2$Cl$_2$ | ~3.2[b] | +0.4 | [65] |
| (*P*)-(+)-**66d** | 479 | 639 | 6.6 | CH$_2$Cl$_2$ | – | ~0 | [65] |
| (*P*)-(+)-**66e** | 463 | 612 | 27 | CH$_2$Cl$_2$ | – | −3.7 (−4.1)[c] | [66] |
| (*M*)-(−)-**68** | 417 | 421 | 8.4 | CH$_2$Cl$_2$ | ~−4.1[b] | −3.2 | [67] |
| (*M*)-(−)-**68.2H⁺** | 418 | 590 | 8.2 | CH$_2$Cl$_2$ | ~−5.3[b] | −2.9 | [67] |
| (*M*)-(−)-**69** | 430 | 547 | 0.38 | CH$_2$Cl$_2$ | ~−5.2 | −1.1 | [67] |
| (*M*)-(−)-**69.H⁺** | 450 | 555 | 2.7 | CH$_2$Cl$_2$ | ~−10 | −2 | [67] |
| ((*M*,$C_{Re}$)-**70a**[2] | 445 | 673 | 0.16 | CH$_2$Cl$_2$ | ~−2 | −3 | [68] |
| (*M*)-**71a**[1,2] | 444 | 598 | 6 | CH$_2$Cl$_2$ | ~−0.4 | −1.5 | [68] |

[a]Lowest-energy UV-vis band
[b]Taken from [12]
[c]Fluorescence quantum yield measured in a DCzppy film

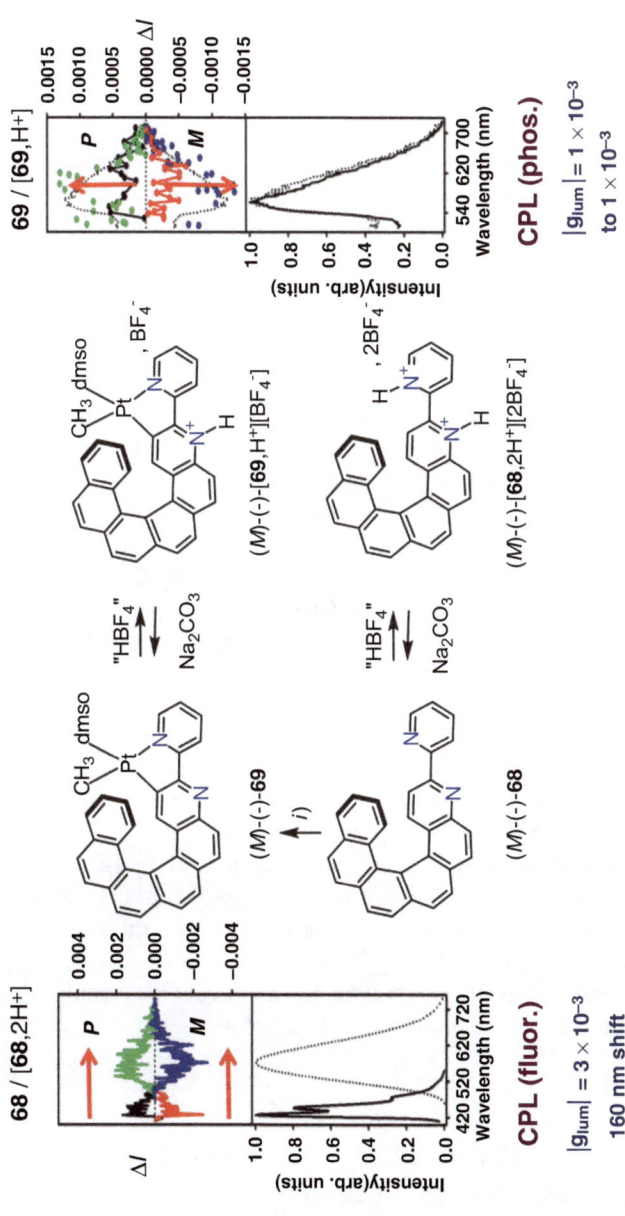

**Fig. 4.27** Synthesis of cycloplatinated helicene (*M*)-**69** from (*M*)-**68** and reversible protonation and deprotonation process of organic and organometallic systems, observed by emission and CPL spectroscopies. (*i*) Pt(dmso)$_2$(CH$_3$)$_2$, acetone, 50 °C, 5 h, 90%. Variation of emission and CPL responses upon protonation [67]

phosphorescence, and CPL were achieved, thus yielding the first acid-based CPL switch (see the increase of $g_{lum}$ upon protonation in Fig. 4.27). Furthermore, we showed that organic helicene ligand (P)- and (M)-**68** was also an efficient chiroptical switch since, after double protonation, it displayed a strong bathochromic shift in emission wavelength while keeping strong CPL fluorescence signal ($g_{lum} = \pm 2 \times 10^{-3}$ in $CH_2Cl_2$). TDDFT calculations showed that, upon protonation, the HOMO-to-LUMO transition changed from a $\pi$–$\pi*$-type to a charge transfer-type transition.

Rhenium(I)-chloro-tricarbonyl complexes bearing a bipy ligand are known to display efficient luminescence, usually a $^3$CT emission from an excited state based on the bis-imine ligand. In this context, organic helicene-bipy ligand (P)- and (M)-**68** was used as N^N chelate to prepare enantioenriched CPL-active helicene-bipyridine-rhenium complexes **70** (Fig. 4.28) [68]. Starting from (M)-**68** ligand, two diastereomeric complexes, i.e., (M,$A_{Re}$)-**70a**$^1$ and (P,$C_{Re}$)-**70a**$^2$, were formed, since the Re(I) atom is also a stereogenic center. These stereoisomers were separated by regular silica gel column chromatography and their chiroptical and emissive properties were studied. They revealed strong ECD spectra in $CH_2Cl_2$ (whose intensity depends on the rhenium stereochemistry; see Fig. 4.28), accompanied by substantial phosphorescence and CPL activity. Indeed (M,$A_{Re}$)-**70a**$^1$ and (M,$C_{Re}$)-**70a**$^2$ displayed phosphorescence emission ($\lambda_{max}^{phos} = 673$–$680$ nm, $\phi = 0.13$–$0.16\%$, $\tau = 27$–$33$ ns) and good $g_{lum}$ values ((M,$C_{Re}$)-**70a**$^2$: $g_{lum} \sim -3 \times 10^{-3}$ around 670 nm). Upon reaction with AgOTf and 2,6-dimethylphenyl isocyanide in the presence of $NH_4PF_6$, (M,$C_{Re}$) and (P,$A_{Re}$)-**71a**$^2$ were transformed to cationic complexes (P)- and (M)-**71a**$^{1,2}$, respectively (see Fig. 4.28). The latter displayed stronger phosphorescence ($\lambda_{max}^{phos} = 598$ nm, $\phi = 6\%$, $\tau = 79$ μs) and still good CPL activity ($g_{lum} \sim \pm 1.5 \times 10^{-3}$). However, the stereochemical information at the Re(I) center was lost (epimerization to 50:50 mixture). Nevertheless, the ECD spectrum of (P)-**71a**$^{1,2}$ displayed an additional positive ECD-active band around 450 nm as compared to (P,$C_{Re}$)-**114a**$^1$ and (P,$A_{Re}$)-**70a**$^2$. According to TDDFT calculations, this band does not involve the Re center but corresponds to the HOMO-to-LUMO transition with strong intra-ligand charge transfer from the $\pi$-helicene to the bipy moiety [68]. We have thus shown that the incorporation of a rhenium atom within an extended helical $\pi$-conjugated bipyridine system can impact the chiroptical and photophysical properties of the resulting neutral or cationic complexes, leading to the first rhenium-based circularly polarized phosphors.

### 4.9.3 Coordination Chemistry of Bis-Helicene-Terpyridine and Bis-Helicene-Bipy Ligands

In 2016, our group also prepared the bis-helical terpyridine (terpy) ligand **72** which acted as a chiroptical switch upon reversible coordination-decoordination to zinc(II). The strong conformational changes induced led to a multi-responsive chiroptical

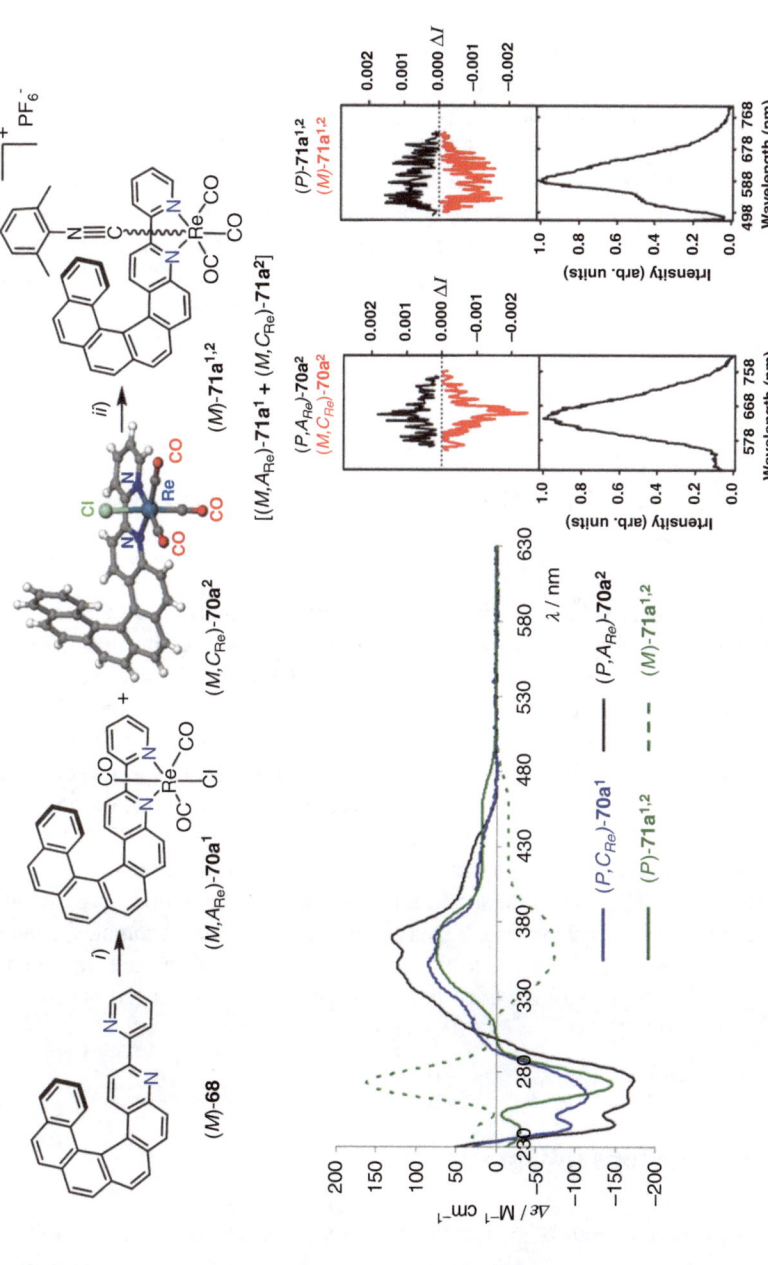

**Fig. 4.28** Synthesis of enantioenriched rhenium complexes ($M,A_{Re}$)-**70a**$^1$, ($M,C_{Re}$)-**70a**$^2$, and ($M$)-**71a**$^{1,2}$ (mixture of two diastereomers). (*i*) Re(CO)$_5$Cl, toluene, reflux; (*ii*) AgOTf, EtOH/THF, then 2,6-dimethylphenyl isocyanide, THF, NH$_4$PF$_6$. X-ray crystallographic structure of **70a**$^2$. ECD spectra of ($P,C_{Re}$)-**70a**$^1$, ($P,A_{Re}$)-**70a**$^2$ and ($M$)/($P$)-**71a**$^{1,2}$ isomers and ($M$)/($P$)-**71a**$^{1,2}$ isomers. Phosphorescence and CPL spectra of ($P,C_{Re}$)-**70a**$^1$, ($P,A_{Re}$)-**70a**$^2$ and ($M$)/($P$)-**71a**$^{1,2}$ isomers in CH$_2$Cl$_2$ [68]

switch (Fig. 4.29) [69]. The interconversion between the ligand and zinc-complexed states was analyzed via first-principles calculations, which highlighted the change from π-π* transitions in the organic ligand to charge transfer transitions in the Zn complex. Overall, this system behaved as a chiroptical switch offering multi-output readout (UV-vis, ECD, luminescence, and CPL). Furthermore, the switching process triggered conformational changes and molecular motion around the Zn center, from a clear *trans* (W-shape) conformation in the free ligand to a *cis* (U-shape) one in the Zn-complex **73** (Fig. 4.29, Table 4.13). Recently, we have prepared a novel enantiopure bis-helicenic 2,2′-bipyridine system ligand, **74** [70]. Thanks to the bipyridine unit, the coordination to **75** with $Zn^{II}$ and protonation processes to **74.2H$^+$** were studied revealing efficient tuning of photophysical (UV-vis and emission) and chiroptical properties (ECD and CPL) of the system (Fig. 4.29, Table 4.13). The coordination-decoordination and protonation/deprotonation processes appeared reversible thus constituting novel chiroptical switches.

### 4.9.4   CPL-Active Iridium Complexes from Helicene-NHC Ligands

In the last two decades, octahedral cyclometalated iridium(III) complexes have attracted attention due to their appealing properties as phosphors in high-efficiency organic light-emitting devices (OLEDs) [71]. In 2017, the first fused π-helical NHC system was prepared and examined through its diastereoisomerically pure cyclometalated complexes *mer*-($P,\Lambda_{Ir}$)-**77a$^1$** and *mer*-($P,\Delta_{Ir}$)-**77a$^2$** from pentahelical imidazolium **76** (Fig. 4.30) [49]. These chiral organometallic species displayed light-green phosphorescence with (*i*) circular polarization that depends on both the -helical-NHC (*P*)/(*M*) stereochemistry and the iridium ($\Delta$)/($\Lambda$) one (Fig. 4.30 and Table 4.14) and (*ii*) unusually long lifetimes (up to 250 μs as compared to 530 ns for model *mer*-**78**). The unprecedented features of **77a$^{1,2}$** can be attributed to extended π-conjugation within helical carbenic ligand. A similar cycloiridiated complex, namely, *mer*-($\Delta_{Ir}/\Lambda_{Ir}$)-**79**, bearing a NHC N-substituted with a carbo[4]helicene unit was recently obtained and also displayed long-lived mirror-image circularly polarized phosphorescence (Table 4.14) [72].

## 4.10   Applications in Optoelectronics

There is an obvious interest to develop the use of helicenes and helicenoids as chiral molecular materials in chiral OLEDs, in chiral sensors, in chiral bioimaging agents, chiroptical switching activity [73, 74] applications that directly take benefit from circularly polarized emission. For instance, there is a strong potential of CP

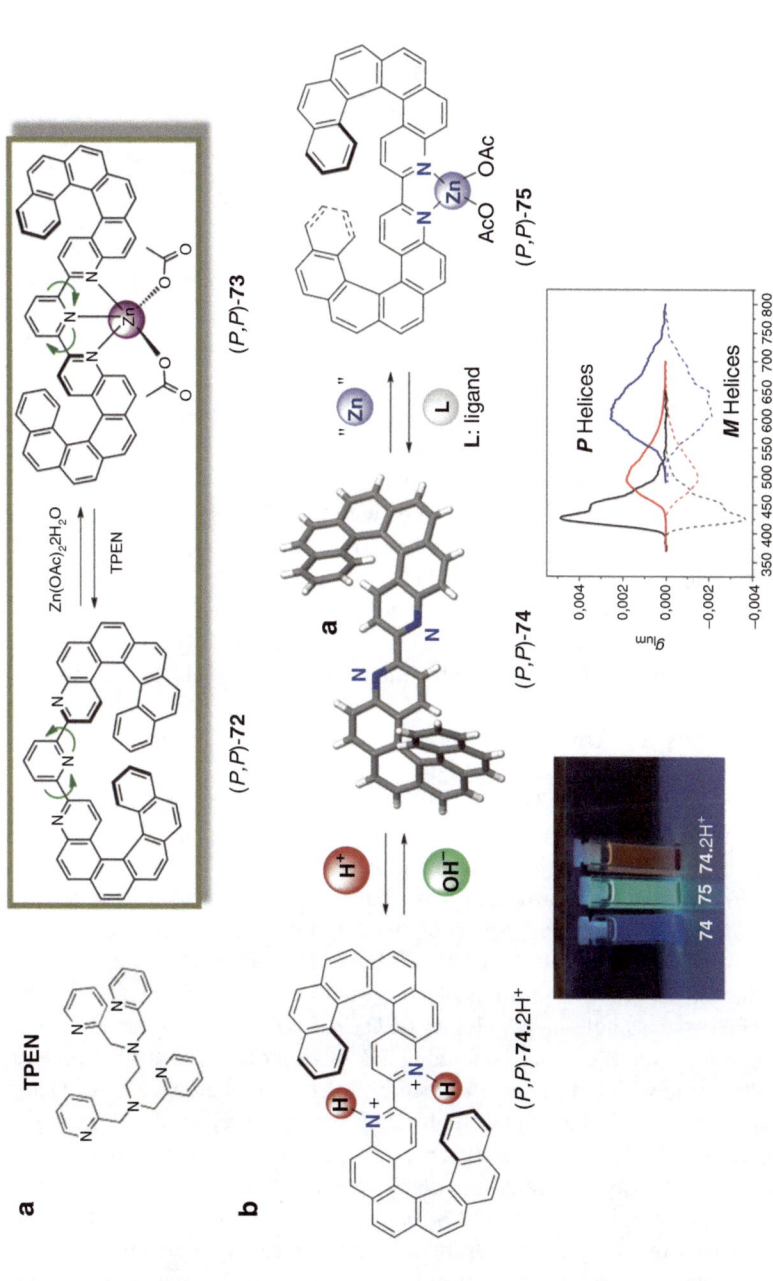

**Fig. 4.29** (**a**) Reversible Zn(II) complexation-decomplexation process of (*P*,*P*)-**72** to (*P*,*P*)-**73** using Zn(OAc)₂ and TPEN as the chemical stimuli [69]. (**b**) Reversible Zn(II) complexation-decomplexation process of (*P*,*P*)-**74** to (*P*,*P*)-**75** using Zn(OAc)₂ and TPEN as the chemical stimuli and acid-base triggered switch between (*P*,*P*)-**74** and (*P*,*P*)-**74.2H⁺**. Emission colors and CPL activity of **72** (black), **75** (red) and **74.2H⁺** (blue). Plain lines are for the (*P*) helicenes and dotted lines for the (*M*) ones. Adapted with permission [70]. Copyright 2019, American Chemical Society

**Table 4.13** Photophysical data of helicene-bipy and terpy ligands together with heir Zn and proton complexes

| Compound | $\lambda_{Abs}^{max~a}$ (nm) | $\lambda_{Em}$ (nm) | $\Phi$ (%) | Solvent (CPL) | $10^3$ $g_{abs}$ | $10^3$ $g_{lum}$ | Ref. |
|---|---|---|---|---|---|---|---|
| (P,P)-(+)-**72** | 416 | 421 | 8.4 | CH$_2$Cl$_2$ | ~6.5[b] | +8.6 | [69] |
| (P,P)-(+)-**73** | 430 | 480 | 19 | CH$_2$Cl$_2$ | ~4.3[b] | +1.2 | [69] |
| (P,P)-(+)-**74** | 420 | 421 | 22 | CH$_2$Cl$_2$ | – | +4.8 | [70] |
| (P,P)-(+)-**75** | 453 | 520 | 44 | CH$_2$Cl$_2$ | – | +1.8 | [70] |
| (P,P)-(+)-**74.2H$^+$** | 507 | 600 | 28 | CH$_2$Cl$_2$ | – | +2.5 | [70] |

[a]Lowest-energy UV-vis band
[b]Taken from 12

light technologies in the development of OLEDs in which the electroluminescence is directly circularly polarized thus giving CP-OLEDs. Indeed, antiglare filters commonly used for OLED displays exploit the physics of CP light to eliminate glare from external light sources (e.g., the sunlight). Unfortunately, this technology removes approximately 50% of the non-polarized light emitted from the OLED pixels. If the non-polarized OLEDs are replaced with CP-OLEDs (with a comparable device performance), an improved amount of CP light component of the correct handedness would pass through the antiglare filter with less loss, thus increasing the energy efficiency of the display in proportion to the increasing dissymmetry of the light. In addition, the use of CP-OLED will enable to simplify the architecture of the device by avoiding the use of extra filter components, which will directly impact the overall cost of the device.

In 2013, Fuchter and coworkers, reported the use of 1-aza[6]helicene **9** as a chiral dopant in light-emitting polymer, i.e., poly[9,9-dioctylfluorene-co-benzothiadiazole] **80** (Fig. 4.31a) [21]. It was found that blends consisting of a small amount (7%) of enantiopure 1-aza[6]helicene dopant gave a strong CP-photoluminescence response of the **80** films. Increasing the 1-aza[6]helicene blending ratio resulted in improvements of the $g_{PL}$ factor, up to a significantly high value of 0.5 for the 53% helicene blend (while the starting azahelicene displayed only modest $g_{lum}$ ~$10^{-4}$ to $10^{-3}$). To explain this behavior, the authors suggested the formation of a chiroptical co-crystalline phase. The authors were then able to fabricate a single-layer polymer LED (PLED) device emitting circularly polarized light from the **80** blends containing 7% of either (−)-1-aza[6]helicene or (+)-1-aza[6]helicene with a dissymmetry factor of electroluminescence ($g_{EL}$) factor as high as 0.2. In 2016, Fuchter and Campbell succeeded in preparing a single layer CP-phosphorescent OLEDs (CP-PHOLEDs), using **66a** as a chiral emissive dopant; these PHOLEDs displayed strong circularly polarized electrophosphorescence (CPEL), with $g_{EL}$ reaching −0.38 and + 0.22 at 615 nm for (−)- and (+)-**66a**, respectively (see Fig. 4.31b) [75]. Although not yet clearly demonstrated, the increase of $g_{EL}$ as compared to the molecular $g_{lum}$ value ($10^{-2}$) may be explained by a supramolecular organization of **66a** in the solid state. Recently, by decorating the pyridyl-helicene ligands with –CF$_3$ and –F groups [66], the platinahelicene enantiomers **66e** featured good configurational stability as well a s high sublimation

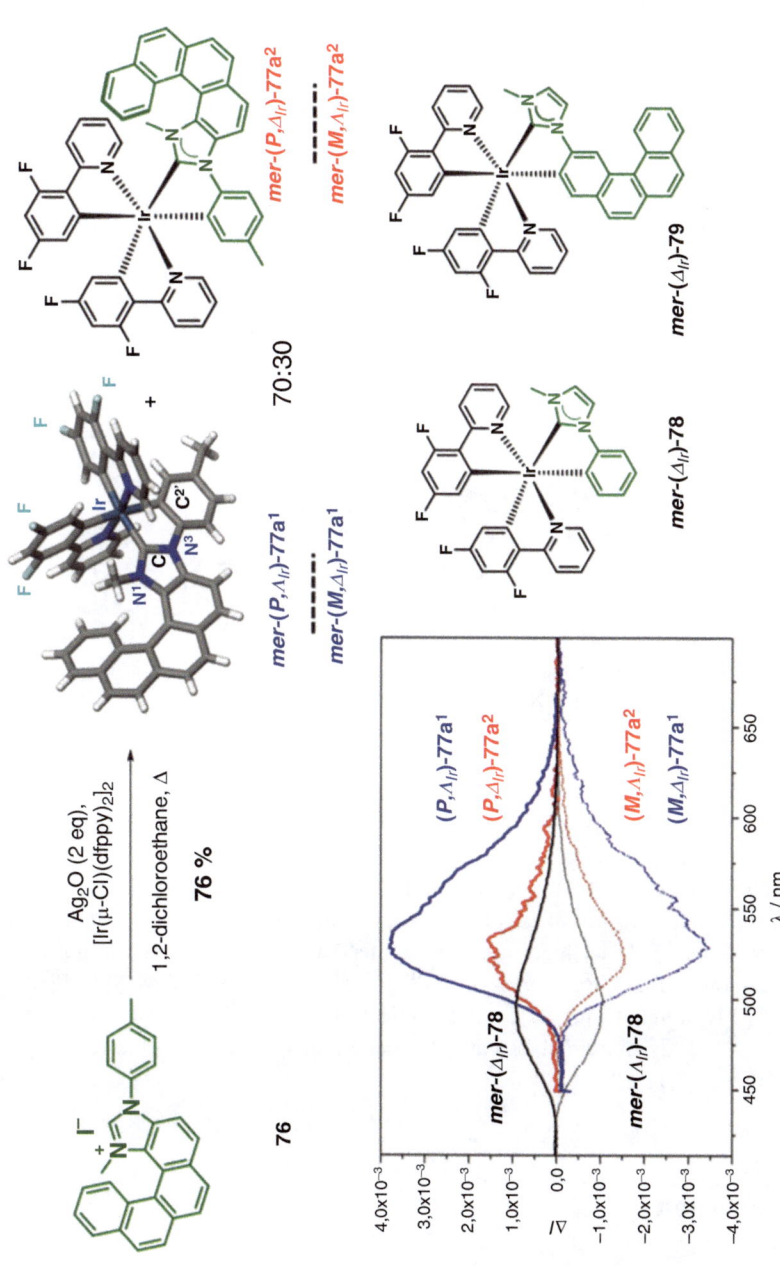

**Fig. 4.30** Preparation of cycloiridiated complexes **77**[1,2]. Chemical structures and CPL spectra of iridium(III) complexes $(P,\Lambda_{Ir})$-**77a**[1]/$(M,\Delta_{Ir})$-**77a**[1], $(P,\Delta_{Ir})$-**77a**[2]/$(M,\Lambda_{Ir})$-**77a**[2], and $(\Delta_{Ir})$-**78**/$(\Lambda_{Ir})$-**78**. Chemical structure of $(\Delta_{Ir})$-**79**. X-ray structure of stereoisomer$(P,\Lambda_{Ir})$-**77a**[1]. All complexes have the *mer* geometry [49, 72]

**Table 4.14** Photophysical data of helicene-NHC iridium complexes

| Compound | $\lambda_{Abs}^{max\ a}$ (nm) | $\lambda_{Em}$ (nm) | $\Phi$ (%) | Solvent (CPL) | $10^3\ g_{lum}$ | Ref. |
|---|---|---|---|---|---|---|
| $(P,\Delta_{Ir})$-(+)-**77a**[1] | 402 | 525 | 9 | $CH_2Cl_2$ | +3.7 | [49] |
| $(P,\Lambda_{Ir})$-(+)-**77a**[2] | 403 | 526 | 13 | $CH_2Cl_2$ | +1.5 | [49] |
| $(\Delta_{Ir})$-(+)-**78** | 390 | 498 | 29 | $CH_2Cl_2$ | +0.9 | [49] |
| $(\Delta_{Ir})$-(+)-**79** | 394 | 510 | 5 | $CH_2Cl_2$ | +3.1 | [72] |

[a]Lowest-energy UV-vis band

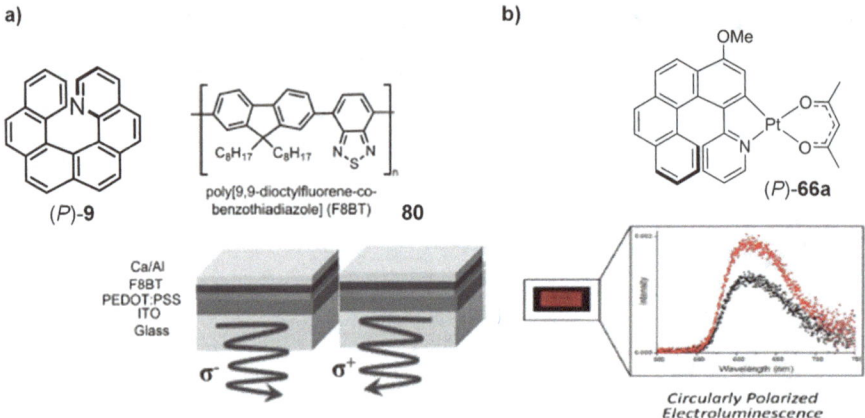

**Fig. 4.31** (a) CP-PLED based on blends between enantiopure **9** and **80**. Adapted with permission 19 Copyright 2013, Wiley. (b) CP-PHOLED based on pure enantiopure cycloplatinahelicene **66a**. Adapted with permission [75]. Copyright 2016, American Chemical Society

yield (>90%) and clear circularly polarized phosphorescence, with dissymmetry factors ($|g_{PL}|$) of approximately $3.7 \times 10^{-3}$ in solution and about $4.1 \times 10^{-3}$ in doped films. The CP-PHOLEDs with two enantiomers as emitters exhibited symmetric CPEL signals with $|g_{EL}|$ of $(1.1–1.6) \times 10^{-3}$ and good device performances, achieving a maximum brightness of 11,590 cd m$^{-2}$, a maximum external quantum efficiency up to 18.81%, which are the highest values among the reported devices based on chiral phosphorescent Pt$^{II}$ complexes. To suppress the effect of reverse CPEL signal from the cathode reflection, the further implementation of semi-transparent aluminum/silver cathode successfully boosted up the $|g_{EL}|$ by over three times to $5.1 \times 10^{-3}$.

## 4.11 Conclusion

Since the first examples of helicenes displaying circularly polarized luminescence described in the literature [13, 37], there has been a growing interest in CPL-active helicenes and helicenoids, and the number of reported examples is growing very fast. This is concomitant to the development of helicenes chemistry and the creation of structural diversity of helical structures, either fully carbonated or containing main-group elements, and to the development of CPL instrumentations and CPL technology. However, a much higher degree of circular polarization is still highly desirable for applications, for example, in chiroptical devices. For this purpose, future work will be probably dedicated to the development of supramolecular architectures, sophisticated homogeneous/heterogeneous mixtures, aggregation induced emission (AIE) materials, or chiral TADF systems, to obtain good CPL properties together with overall good quantum yield of photoluminescence. In these domains, there is still plenty of room for fundamental discoveries and for further applications. Finally, although not detailed in the present chapter, theoretical calculations of CPL activity may be of great help to anticipate highly efficient CPL-active systems. In this context, CPL-active helicenes are good models for testing and improving theoretical tools.

# References

1. Dekkers HPJM (2000) Circularly polarized luminescence: a probe for chirality in the excited state. In: Berova N, Nakanishi K, Woody RW (eds) Circular dichroism: principles and applications. Wiley-VCH, New York, pp 185–215
2. Riehl JP, Muller G (2012) Circularly polarized luminescence spectroscopy and emission-detected circular dichroism. In: Berova N, Polavarapu PL, Nakanishi K, Woody RW (eds) Comprehensive chiroptical spectroscopy, vol 1. Wiley, Hoboken
3. Maeda H, Bando Y (2013) Recent progress in research on stimuli responsive circularly polarized luminescence based on π-conjugated molecules. Pure Appl Chem 85:1967–1978
4. Kumar J, Nakashima T, Kawai T (2015) Circularly polarized luminescence in chiral molecules and supramolecular assemblies. J Phys Chem Lett 6:3445–3452
5. Sánchez-Carnerero EM, Agarrabeitia AR, Moreno F, Maroto BL, Muller G, Ortiz MJ, de la Moya S (2015) Circularly polarized luminescence from simple organic molecules. Chem Eur J 21:13488–13500
6. Chen N, Yan B (2018) Recent theoretical and experimental progress in molecules circularly polarized luminescence of small organic molecules. Molecules 23:3376
7. Shen Y, Chen C-F (2012) Helicenes: synthesis and applications. Chem Rev 112:1463–1535
8. Chen C-F, Shen Y (2017) Helicene chemistry: from synthesis to applications. Springer, Berlin
9. Gingras M (2012) One hundred years of Helicene chemistry. Part 3: applications and properties of carbohelicenes. Chem Soc Rev 42:1051–1095
10. Newman MS, Lednicer D (1956) The synthesis and resolution of hexahelicene. J Am Chem Soc 78:4765–4770
11. Zinna F, Di Bari L (2015) Lanthanide circularly polarized luminescence: bases and applications. Chirality 27:1–13
12. Tanaka H, Inoue Y, Mori T (2018) Circularly polarized luminescence and circular dichroisms in small organic molecules: correlation between excitation and emission dissymmetry factors. ChemPhotoChem 2:386–402
13. Field JE, Muller G, Riehl JP, Venkataraman D (2003) Circularly polarized luminescence from bridged triarylamine helicenes. J Am Chem Soc 125:11808–11809
14. Longhi G, Castiglioni E, Villani C, Sabia R, Menichetti S, Viglianisi C, Devlin F, Abbate S (2016) Chiroptical properties of the ground and excited states of two thia-bridged triarylamine heterohelicenes. J Photochem Photobiol A 331:138–145

15. Nishimura H, Tanaka K, Morisaki Y, Chujo Y, Wakamiya A, Murata Y (2017) Oxygen-bridged diphenylnaphthylamine as a scaffold for full-color circularly polarized luminescent materials. J Org Chem 82:5242–5249
16. Liu Y, Cerezo J, Mazzeo G, Lin N, Zhao X, Longhi G, Abbate S, Santoro F (2016) Vibronic coupling explains the different shape of electronic circular dichroism and of circularly polarized luminescence spectra of hexahelicenes. J Chem Theory Comput 12:2799–2819
17. Nakamura K, Furumi S, Takeuchi M, Shibuya T, Tanaka K (2014) Enantioselective synthesis and enhanced circularly polarized luminescence of S-shaped double azahelicenes. J Am Chem Soc 136:5555–5558
18. Tanaka M, Shibata Y, Nakamura K, Teraoka K, Uekusa H, Nakazono K, Takata T, Tanaka K (2016) Gold-catalyzed enantioselective synthesis, crystal structure, and photophysical/chiroptical properties of aza[10]helicenes. Chem Eur J 22:9537–9541
19. Otani T, Tsuyuki A, Iwachi T, Someya S, Tateno K, Kawai H, Saito T, Kanyiva KS, Shibata T (2017) Facile two-step synthesis of 1,10-phenanthroline-derived polyaza[7]helicenes with high fluorescence and CPL efficiency. Angew Chem Int Ed 56:3906–3910
20. Abbate S, Longhi G, Lebon F, Castiglioni E, Superchi S, Pisani L, Fontana F, Torricelli F, Caronna T, Villani C, Sabia R, Tommasini M, Lucotti A, Mendola D, Mele A, Lightner DA (2014) Helical sense-responsive and substituent-sensitive features in vibrational and electronic circular dichroism, in circularly polarized luminescence, and in Raman spectra of some simple optically active hexahelicenes. J Phys Chem C 118:1682–1695
21. Yang Y, da Costa RC, Smilgies D-M, Campbell AJ, Fuchter MJ (2013) Induction of circularly polarized electroluminescence from an achiral light-emitting polymer via a chiral small-molecule dopant. Adv Mater 25:2624–2628
22. Nakai Y, Mori T, Sato K, Inoue Y (2013) Theoretical and experimental studies of circular dichroism of mono and diazonia[6]helicenes. J Phys Chem A 117:5082–5092
23. Tanaka H, Ikenosako M, Kato Y, Fujiki M, Inoue Y, Mori T (2018) Symmetry-based rational design for boosting chiroptical responses. Commun Chem 1:38
24. Goto K, Yamaguchi R, Hiroto S, Ueno H, Kawai T, Shinokubo H (2012) Intermolecular oxidative annulation of 2-aminoanthracenes to diazaacenes and aza[7]helicenes. Angew Chem Int Ed 51:10333–10336
25. Ushiyama A, Hiroto S, Yuasa J, Kawai T, Shinokubo H (2017) Synthesis of a figure-eight azahelicene dimer with high emission and CPL properties. Org Chem Front 4:664–667
26. Sakai H, Kubota T, Yuasa J, Araki Y, Sakanoue T, Takenobu T, Wada T, Kawai T, Hasobe T (2016) Synthetic control of photophysical process and circularly polarized luminescence of [5]carbohelicene derivatives substituted by maleimide units. J Phys Chem C 120:7860–7869
27. Sakai H, Kubota T, Yuasa J, Araki Y, Sakanoue T, Takenobu T, Wada T, Kawai T, Hasobe T (2016) Protonation-induced red-coloured circularly polarized luminescence of [5]carbohelicene fused by benzimidazole. Org Biomol Chem 14:6738–6743
28. Li M, Lu H-Y, Zhang C, Shi L, Tang Z, Chen C-F (2016) Helical aromatic imide based enantiomers with full color circularly polarized luminescence. Chem Commun 52:9921–9924
29. Delgado IH, Pascal S, Wallabregue A, Duwald R, Besnard C, Guenee L, Nancoz C, Vauthey E, Tovar RC, Lunkley JL, Muller G, Lacour J (2016) Functionalized cationic [4]helicenes with unique tuning of absorption, fluorescence and chiroptical properties up to the far-red range. Chem Sci 7:4685–4693
30. Pascal S, Besnard C, Zinna F, Di Bari L, Le Guennic B, Jacquemin D, Lacour J (2016) Zwitterionic [4]helicene: a water-soluble and reversible PH-triggered ECD/CPL chiroptical switch in the UV and red spectral regions. Org Biomol Chem 14:4590–4594
31. Bosson J, Labrador GM, Pascal S, Miannay FA, Yushchenko O, Li H, Bouffier L, Sojic N, Tovar RC, Muller G, Jacquemin D, Laurent AD, Le Guennic B, Vauthey E, Lacour J (2016) Physicochemical and electronic properties of cationic [6]helicenes: from chemical and electrochemical stabilities to far-red (polarized) luminescence. Chem Eur J 22:18394–18403
32. Kaseyama T, Furumi S, Zhang X, Tanaka K, Takeuchi M (2011) Hierarchical assembly of a phthalhydrazide-functionalized helicene. Angew Chem Int Ed 50:3684–3687

33. Yamano R, Hara J, Murayama K, Sugiyama H, Teraoka K, Uekusa H, Kawauchi S, Shibata Y, Tanaka K (2017) Rh-mediated enantioselective synthesis, crystal structures, and photophysical/ chiroptical properties of phenanthrenol-based[9]helicene-like molecules. Org Lett 19:42–45

34. Matsuno T, Koyama Y, Hiroto S, Kumar J, Kawai T, Shinokubo H (2015) Isolation of a 1, 4-diketone intermediate in oxidative dimerization of 2-hydroxyanthracene and its conversion to oxahelicene. Chem Commun 51:4607–4610

35. Gupta R, Cabreros TA, Muller G, Bedekar AV (2018) Enantiomerically pure 5,13-dicyano-9-oxa[7]helicene: synthesis and study. Eur J Org Chem 2018:5397–5405

36. Sundar MS, Talele HR, Mande HM, Bedekar AV, Tovar RC, Muller G (2014) Synthesis of enantiomerically pure helicene like bis-oxazines from atropisomeric 7,7′-dihydroxy BINOL: preliminary measurements of the circularly polarized luminescence. Tetrahedron Lett 55:1760–1764

37. Phillips KES, Katz TJ, Jockusch S, Lovinger AJ, Turro NJ (2001) Synthesis and properties of an aggregating heterocyclic helicene. J Am Chem Soc 123:11899–11907

38. Riehl JP, Richardson FS (1986) Circularly polarized luminescence spectroscopy. Chem Rev 86:1–16

39. Yamamoto Y, Sakai H, Yuasa J, Araki Y, Wada T, Sakanoue T, Takenobu T, Kawai T, Hasobe T (2016) Controlled excited-state dynamics and enhanced fluorescence property of tetrasulfone[9]helicene by a simple synthetic process. J Phys Chem C 120:7421–7427

40. Yamamoto Y, Sakai H, Yuasa J, Araki Y, Wada T, Sakanoue T, Takenobu T, Kawai T, Hasobe T (2016) Synthetic control of the excited-state dynamics and circularly polarized luminescence of fluorescent "push–pull" tetrathia[9]helicenes. Chem Eur J 22:4263–4273

41. Biet T, Cauchy T, Sun Q, Ding J, Hauser A, Oulevey P, Bürgi T, Jacquemin D, Vanthuyne N, Crassous J, Avarvari N (2017) Triplet state CPL active helicene–dithiolene platinum bipyridine complexes. Chem Commun 53:9210–9213

42. Shen C, Srebro-Hooper M, Jean M, Vanthuyne N, Toupet L, Williams JAG, Torres AR, Riives AJ, Muller G, Autschbach J, Crassous J (2017) Synthesis and chiroptical properties of hexa-, octa-, and decaazaborahelicenes: influence of helicene size and of the number of boron atoms. Chem Eur J 23:407–418

43. Dominguez Z, Lopez-Rodriguez R, Alvarez E, Abbate S, Longhi G, Pischel U, Ros A (2018) Azabora[5]helicene charge-transfer dyes show efficient and spectrally variable circularly polarized luminescence. Chem Eur J 24:12660–12668

44. Katayama T, Nakatsuka S, Hirai H, Yasuda N, Kumar J, Kawai T, Hatakeyama T (2016) Two-step synthesis of boron-fused double helicenes. J Am Chem Soc 138:5210–5213

45. Miyamoto F, Nakatsuka S, Yamada K, Nakayama K, Hatakeyama T (2015) Synthesis of boron-doped polycyclic aromatic hydrocarbons by tandem intramolecular electrophilic arene borylation. Org Lett 17:6158–6161

46. Hatakeyama T, Shiren K, Nakajima K, Nomura S, Nakatsuka S, Kinoshita K, Ni J, Ono Y, Ikuta T (2016) Ultrapure blue thermally activated delayed fluorescence molecules: efficient HOMO-LUMO separation by the multiple resonance effect. Adv Mater 28:2777–2781

47. Oyama H, Nakano K, Harada T, Kuroda R, Naito M, Nobusawa K, Nozaki K (2013) Facile synthetic route to highly luminescent Sila[7]helicene. Org Lett 15:2104–2107

48. Murayama K, Oike Y, Furumi S, Takeuchi M, Noguchi K, Tanaka K (2015) Enantioselective synthesis, crystal structure, and photophysical properties of a 1,1′-bitriphenylene-based sila[7] helicene. Eur J Org Chem 2015:1409–1414

49. Hellou N, Srebro-Hooper M, Favereau L, Zinna F, Caytan E, Toupet L, Dorcet V, Jean M, Vanthuyne N, Williams JAG, Di Bari L, Autschbach J, Crassous J (2017) Enantiopure cycloiridiated complexes bearing a pentahelicenic N-heterocyclic carbene and displaying long-lived circularly-polarized phosphorescence. Angew Chem Int Ed 56:8236–8239

50. Duffy MP, Delaunay W, Bouit P-A, Hissler M (2016) π-Conjugated phospholes and their incorporation into devices: components with a great deal of potential. Chem Soc Rev 45:5296–5310

51. Yavari K, Delaunay W, De Rycke N, Reynaldo T, Aillard P, Srebro-Hooper M, Chang VY, Muller G, Tondelier D, Geffroy B, Voituriez A, Marinetti A, Hissler M, Crassous J (2019) Phosphahelicenes: from chiroptical and photophysical properties to OLED applications. Chem Eur J 25:5303–5310

52. Nishigaki S, Murayama K, Shibata Y, Tanaka K (2018) Rhodium-mediated enantioselective synthesis of a benzopicene-based phospha[9]helicene: the structure–property relationship of triphenylene- and benzopicene-based carbo- and phosphahelicenes. Mater Chem Front 2:585–590

53. Tanaka H, Kato Y, Fujiki M, Inoue Y, Mori T (2018) Combined experimental and theoretical study on circular dichroism and circularly polarized luminescence of configurationally robust $D_3$-symmetric triple pentahelicene. J Phys Chem A 122:7378–7384

54. He D-Q, Lu H-Y, Li M, Chen C-F (2017) Intense blue circularly polarized luminescence from helical aromatic esters. Chem Commun 53:6093–6096

55. Fang L, Li M, Lin W-B, Chen C-F (2018) Enantiopure (*P*)- and (*M*)-3,14-bis(*O*-hydroxyaryl) tetrahydrobenzo[5]helicenediols and their helicene analogues: synthesis, amplified circularly polarized luminescence and catalytic activity in asymmetric hetero-Diels–Alder reactions. Tetrahedron 74:7164–7172

56. Dhbaibi K, Favereau L, Srebro-Hooper M, Jean M, Vanthuyne N, Zinna F, Jamoussi B, Di Bari L, Autschbach J, Crassous J (2018) Exciton coupling in diketopyrrolopyrrole-helicene derivatives leads to red and near-infrared circularly polarized luminescence. Chem Sci 9:735–742

57. Satoh M, Shibata Y, Tanaka K (2018) Enantioselective synthesis of fully benzenoid single and double carbohelicenes via gold-catalyzed intramolecular hydroarylation. Chem Eur J 24:5434–5438

58. Fang L, Li M, Lin W-B, Shen Y, Chen C-F (2018) One-pot oxidative aromatization and dearomatization of tetrahydro[5]helicene diols: synthesis, structure, photophysical and chiroptical properties of chiral π-extended diones. Asian J OrgChem 12:2518–2526

59. Sawada Y, Furumi S, Takai A, Takeuchi M, Noguchi K, Tanaka K (2012) Rhodium-catalyzed enantioselective synthesis, crystal structures, and photophysical properties of helically chiral 1,1-bitriphenylenes. J Am Chem Soc 134:4080–4083

60. Oyama H, Akiyama M, Nakano K, Naito M, Nobusawa K, Nozaki K (2016) Synthesis and properties of [7]helicene-like compounds fused with a fluorene unit. Org Lett 18:3654–3657

61. Sakai H, Shinto S, Kumar J, Araki Y, Sakanoue T, Takenobu T, Wada T, Kawai T, Hasobe T (2015) Highly fluorescent [7]carbohelicene fused by asymmetric 1,2-dialkyl-substituted quinoxaline for circularly polarized luminescence and electroluminescence. J Phys Chem C 119:13937–13947

62. Saleh N, Shen C, Crassous J (2014) Helicene-based transition metal complexes: synthesis, properties and applications. Chem Sci 5:3680–3694

63. Ou-Yang J-K, Crassous J (2018) Chiral multifunctional molecules based on organometallic helicenes: recent advances. Coord Chem Rev 376:533–547

64. Norel L, Rudolph M, Vanthuyne N, Williams JAG, Lescop C, Roussel C, Autschbach J, Crassous J, Réau R (2010) Metallahelicenes: easily accessible helicene derivatives with large and tunable chiroptical properties. Angew Chem Int Ed 49:99–102

65. Shen C, Anger E, Srebro M, Vanthuyne N, Deol KK, Jefferson TD, Muller G, Williams JAG, Toupet L, Roussel C, Autschbach J, Réau R, Crassous J (2014) Straightforward access to mono- and bis- cycloplatinated helicenes displaying circularly polarized phosphorescence by using crystallization resolution methods. Chem Sci 5:1915–1927

66. Yan Z-P, Luo X-F, Liu W-Q, Wu Z-G, Liang X, Liao K, Wang Y, Zheng Y-X, Zhou L, Zuo J-L, Pan Y, Zhang H (2019) Configurationally stable platinahelicene enantiomers for efficient circularly polarized phosphorescent organic light-emitting diodes. Chem Eur J 25:5672–5676

67. Saleh N, Moore B, Srebro M, Vanthuyne N, Toupet L, Williams JAG, Roussel C, Deol KK, Muller G, Autschbach J, Crassous J (2015) Acid/base-triggered switching of circularly

polarized luminescence and electronic circular dichroism in organic and organometallic helicenes. Chem Eur J 21:1673–1681

68. Saleh N, Srebro M, Reynaldo T, Vanthuyne N, Toupet L, Chang VY, Muller G, Williams JAG, Roussel C, Autschbach J, Crassous J (2015) Enantio-enriched CPL-active helicene-bipyridine-rhenium complexes. Chem Commun 51:3754–3757

69. Isla H, Srebro-Hooper M, Jean M, Vanthuyne N, Roisnel T, Lunkley JL, Muller G, Williams JAG, Autschbach J, Crassous J (2016) Conformational changes and chiroptical switching of enantiopure bis-helicenic terpyridine upon $Zn^{2+}$ binding. Chem Commun 52:5932–5935

70. Isla H, Saleh N, Ou-Yang J-K, Dhbaibi K, Jean M, Dziurka M, Favereau L, Vanthuyne N, Toupet L, Jamoussi B, Srebro-Hooper M, Crassous J (2019) Bis-4-aza[6]helicene: a bis-helicenic 2,2′-bipyridine with chemically-triggered chiroptical switching activity. J Org Chem 84:5383–5393

71. Lee J, Chen H-F, Batagoda T, Coburn C, Djurovich PI, Thompson ME, Forrest SR (2016) Deep blue phosphorescent organic light-emitting diodes with very high brightness and efficiency. Nat Mater 15:92–98

72. Macé A, Hellou N, Hammoud J, Martin C, Gauthier ES, Favereau L, Roisnel T, Caytan E, Nasser G, Vanthuyne N, Williams JAG, Berrée F, Carboni B, Crassous J (2019) An enantiopure cyclometallated iridium complex displaying long-lived phosphorescence both in solution and in the solid state. Helv Chim Acta 102:e1900044

73. Isla H, Crassous J (2016) Helicene-based chiroptical switches. C R Chim 19:39–49

74. Brandt JR, Salerno F, Fuchter MJ (2017) The added value of small-molecule chirality in technological applications. Nat Rev Chem 1:0045

75. Brandt JR, Wang X, Yang Y, Campbell AJ, Fuchter MJ (2016) Circularly polarized phosphorescent electroluminescence with a high dissymmetry factor from PHOLEDs based on a platinahelicene. J Am Chem Soc 138:9743–9746

# Chapter 5
# Structural Control of Fluorescent Helicates for Improved Circularly Polarized Luminescence Properties

**Taku Hasobe**

**Abstract** This chapter focuses on the structural and photophysical control of fluorescent helicene derivatives and related homoleptic zinc(II) helicates for improved circularly polarized luminescence properties. Our synthetic strategy enables to significantly enhance the fluorescence quantum yield ($\Phi_{FL}$) of helicene derivatives for observation of circularly polarized luminescence (CPL). Moreover, the zinc(II) helicate also demonstrated unusually strong chiroptical responses with absorption and luminescence dissymmetry factors such as $|g_{abs}| = 0.20$ at 615 nm and $|g_{lum}| = 0.022$ at 660 nm, respectively. The $g_{lum}$ profiles significantly expanded up to ca. 850 nm.

## 5.1 Introduction

$\pi$-Conjugated molecules generally possess the characteristic optical and electronic behaviors concerning the fundamental properties and advanced applications [1, 2]. These are dependent on the electronic structures of molecules. Especially, the conformational distortion of $\pi$-electron units enforced by an intramolecular and/or intermolecular steric hindrance and electronic interaction trigger the chirality, which is seen in biological molecular assemblies. Such helical topology of these molecules with distinctive photophysical properties provide good possibilities for various functionalities such as separation [3], chiroptical, and electronic properties [4, 5].

Helicene derivatives are among the unusual organic dye molecules composed of *ortho*-condensed aromatic rings that form the peculiar screwed $\pi$-conjugated structures [6]. These molecular systems have been investigated in various research fields such as supramolecular assemblies [7–9], fluorescent sensors [10], and liquid crystals [11–14] because of the nonplanar conformations. As is well known, these

T. Hasobe (✉)
Department of Chemistry, Faculty of Science and Technology, Keio University, Yokohama, Japan
e-mail: hasobe@chem.keio.ac.jp

© Springer Nature Singapore Pte Ltd. 2020           99
T. Mori (ed.), *Circularly Polarized Luminescence of Isolated Small Organic Molecules*, https://doi.org/10.1007/978-981-15-2309-0_5

molecules show the peculiar chiroptical properties such as circular dichroism (CD) with large anisotropy factor ($g$-value).

In contrast, the chirality derived from luminescence compounds and materials induce circularly polarized luminescence (CPL), i.e., the differential emission intensity of right- and left-circularly polarized light [15–24]. However, the fluorescence quantum yields ($\Phi_{FL}$) of helicenes are generally extremely small because of the acceleration of the intersystem crossing (ISC) processes (the ISC quantum yield $\Phi$ISC: ~0.9). Therefore, synthetic strategies for improvements of $\Phi_{FL}$ are definitely required [21, 25–27]. To optimize the excited-state dynamics of π-conjugated molecules, one of the useful ways is substitution of electron-accepting groups (e.g., maleimide groups) on the peripheral positions. Moreover, the supramolecular dimer with the helical chirality is also interesting for improvement of CPL properties [24].

Based on the above concepts, we discuss the structural and photophysical control of fluorescent helicene derivatives and related homoleptic zinc(II) helicates for improved circularly polarized luminescence properties. The details on the structural and photophysical control as well as CPL properties are presented.

## 5.2  Results and Discussion

### 5.2.1  *[7]Carbohelicene Derivatives Fused by Asymmetric 1,2-Dialkyl-Substituted Quinoxaline*

Kang et al. previously demonstrated that the imine unit of quinoxalines reduced by alkyl/aryllithiums results in a significant change of steric hindrance on the quinoxaline unit [28]. This is derived from the presence of two dissimilar and asymmetric conditions of nitrogen environments in a quinoxaline unit (one is an electron-donating sp$^3$ site and the other is an electron-accepting sp$^2$ site). Such a structural change is expected to control the optical and electronic properties. Based on the above points, a new compound, 1,2-dialkylquinoxaline-substituted [7] carbohelicene (denoted as [7]Hl-NAIQx) by introduction of two alkyl chain groups onto a quinoxaline unit, was designed and synthesized (Fig. 5.1) [21].

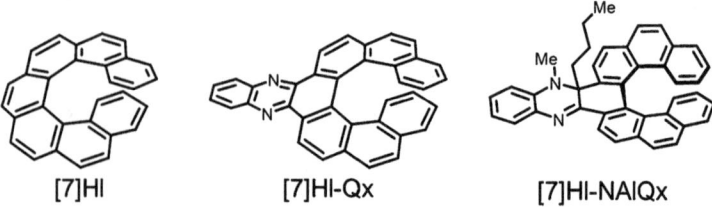

[7]Hl                    [7]Hl-Qx                    [7]Hl-NAlQx

**Fig. 5.1**  Structures of [7]carbohelicene in this work

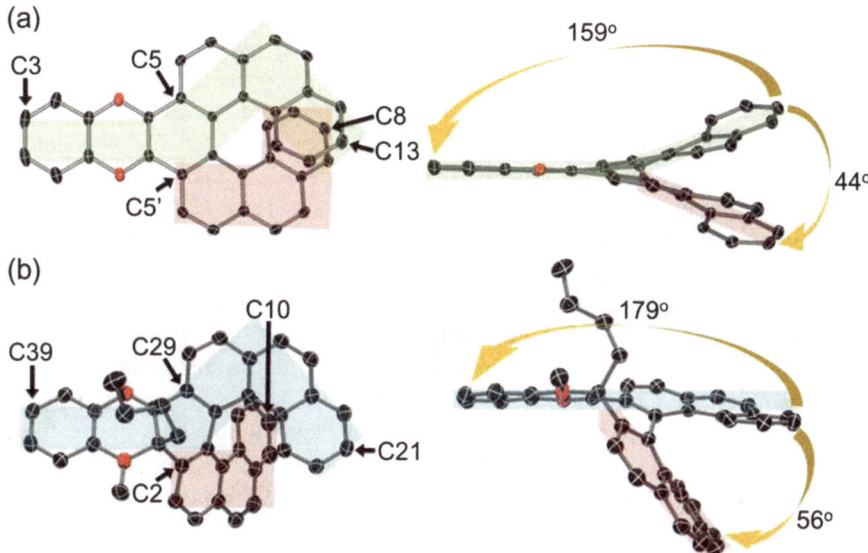

**Fig. 5.2** ORTEP diagrams. (**a**) rac-[7]Hl-Qx (proton units were omitted for clarity, ellipsoids set at 50% probability) and (**b**) rac-[7]Hl-NAIQx (proton units were omitted for clarity, ellipsoids set at 50% probability) (reprinted with permission from Ref. [21] Copyright 2015 American Chemical Society)

The single crystals of helicene derivatives were arranged using vapor diffusion method. The crystal structures of rac-[7]Hl-Qx and rac-[7]Hl-NAIQx are shown in Fig. 5.2. [7]Hl-Qx possesses a nonplanar formation between the two phenanthrenes (Fig. 5.2a). The C3–C5–C5'–C13 torsion angle between the alkyl-quinoxaline and phenanthrene is calculated to be 159°. On the other hand, [7]Hl-NAIQx has a helical conformation, which is formed by the two phenanthrenes and a planar structure between the alkyl-quinoxaline and phenanthrene units (Fig. 5.2b). The torsion angle between the alkyl-quinoxaline and phenanthrene units (C39–C29–C2–C21) is determined to be 179°. This trend is significantly different from that of [7]Hl-Qx as discussed above. In contrast, the torsion angle of the two phenanthrenes (C21–C29–C2–C10) in [7]Hl-NAIQx (56°) is larger than that of C13–C5–C5'–C8 torsion in [7]Hl-Qx (44°). The differential angles may directly have an effect on the delocalized π-electrons in the entire molecules, which results in the enhancement of fluorescence emission properties (vide infra) [21].

To quantitatively discuss the photophysical properties of these carbohelicene derivatives, fluorescence lifetimes were measured. Fluorescence lifetimes ($\tau_{FL}$) were successfully estimated from monoexponential fitting analysis. The $\tau_{FL}$ values of [7]Hl-Qx (1.66 ns) and [7]Hl-NAIQx (4.00 ns) are calculated, respectively. These are shorter as compared to the reference molecule: [7]Hl (13.8 ns). Then, fluorescence quantum yields ($\Phi_{FL}$) of these [7]carbohelicene derivatives were evaluated. As compared to the $\Phi_{FL}$ value of [7]Hl ($\Phi_{FL} = 0.02$), those of [7]Hl-Qx: $\Phi_{FL} = 0.05$ and [7]Hl-NAIQx: $\Phi_{FL} = 0.25$ largely increased. In particular, the

$\Phi_{FL}$ of [7]Hl-NAIQx is an order of magnitude greater that of pristine [7] carbohelicene, i.e., [7]Hl, which is quite comparable to the previous related systems of helical molecules [25, 32, 33]. Then, the quantum yields of ISC ($\Phi$ISC) of these compounds were estimated by $^1O_2$ phosphorescence utilizing triplet-triplet energy transfer from these helicene derivatives to $O_2$. Noted that we assume that the quenching process from the singlet-excited state was negligible [34–37]. Singlet oxygen generated from triplet states of these helicene derivatives in oxygen-saturated toluene was confirmed by $^1O_2$ phosphorescence at ~1270 nm. Following the reported data of $\Phi_{ISC}$ of fullerene ($C_{60}$) in solution [$\Phi_{ISC} = 0.96$] [38], $C_{60}$ was employed as the standard molecule. The $\Phi_{ISC}$ of pristine [7] carbohelicene (i.e., [7]Hl) was not reported, so far. Regarding the quantum yields of internal conversion ($\Phi_{IC}$), we can estimate the yields by subtracting $\Phi_{FL}$ and $\Phi_{ISC}$ from the unity. The summarized quantum yields including $\Phi_{ISC}$ and $\Phi_{IC}$ are shown in Table 5.1. It should be noted that the total sum of the above three quantum yields is close to unity within the experimental error although measurement instruments were totally different. The $\Phi_{FL}$ of [7]Hl-NAIQx (0.25) is much larger than those of [7]Hl-Qx ($\Phi_{FL} = 0.05$) and [7]Hl ($\Phi_{FL} = 0.02$) regardless of the corresponding low $\Phi_{ISC}$ values.

Finally, the respective rate constants of these three processes such as fluorescence emission ($k_{FL}$), intersystem crossing ($k_{ISC}$), and internal conversion ($k_{IC}$) were estimated as shown in Table 5.1. It should be emphasized that the $k_{FL}$ of [7]Hl-NAIQx is much larger than that of [7]Hl. In particular, the $k_{FL}$ values such as $6.25 \times 10^7$ $s^{-1}$ for [7]Hl-NAIQx and $3.01 \times 10^7$ $s^{-1}$ for [7]Hl-Qx are one order of magnitude greater than $1.45 \times 10^6$ $s^{-1}$ for [7]Hl. According to the basic theory of photochemistry [39], $k_{FL}$ is largely related to the molar extinction coefficients. The $\varepsilon$ values are therefore estimated from the 0–0 absorption bands ($\varepsilon_{0-0}$), as shown in Table 5.1. The evaluated $\varepsilon_{0-0}$ values of [7]Hl-NAIQx and [7]Hl-Qx are significantly larger as compared to [7]Hl, which coincides with the photophysical trends of the $k_{FL}$. Then, we also discussed the ISC processes such as the rate constant: $k_{ISC}$ and quantum yield: $\Phi_{ISC}$. The estimated $k_{ISC}$ constants of [7]Hl-Qx and [7]Hl-NAIQx are $5.78 \times 10^8$ and $1.55 \times 10^8$ $s^{-1}$, respectively. In the same manner as $k_{FL}$, these rate constants are significantly larger than that of pristine [7]carbohelicene i.e., [7]Hl ($k_{ISC} \approx 6.5 \times 10^7$ $s^{-1}$). Generally, $k_{ISC}$ is highly associated with the energy gaps between singlet- and triplet-excited states. To experimentally estimate the energy differences of these derivatives, phosphorescence spectra were measured at low temperature, as shown in Table 5.1. The obtained similar gaps of singlet- and triplet-excited states in the range from ca. 0.7 to 0.8 eV presumably do not have an effect on the $k_{ISC}$ rate constants. Regarding the enhanced fluorescence behaviors of [7]Hl-NAIQx, we need to consider the following two contents from the above discussions. The first content is the internal charge-transfer (ICT) emission associated with the asymmetrical dialkylquinoxaline (i.e., electron donor-electron acceptor pair unit) [28]. The other point includes the planar aromatic structure between the quinoxaline and phenanthrene units originated from alkylation. This enables to decelerate the ISC pathway in comparison with the nonplanar structure. The excited state of [7]Hl-NAIQx largely depends on the planar steric configuration but not the

**Table 5.1** Summarized photophysical parameters of [7]carbohelicene derivatives (reprinted with permission from Ref. [21]. Copyright 2015 American Chemical Society)

| Helicene | $\tau_{FL}$[a] (ns) | $\Phi_{FL}$[b] | $\Phi_{ISC}$[c] | $\Phi_{IC}$ | $k_{FL} \times 10^{-7}$ ($s^{-1}$) | $k_{ISC} \times 10^{-7}$ ($s^{-1}$) | $k_{IC} \times 10^{-7}$ ($s^{-1}$) | $S_1-S_0$ (eV) | $T_1-S_0$ (eV) | $S_1-T_1$ (eV) | $\varepsilon_{0-0}$ ($M^{-1}$ $cm^{-1}$) |
|---|---|---|---|---|---|---|---|---|---|---|---|
| [7]HI-Qx | 1.66 | 0.05 | 0.96 | 0.00 | 3.01 | 57.8 | – | 2.62 | 1.93 | 0.69 | 14,500 |
| [7]HI-NAIQx | 4.00 | 0.25 | 0.62 | 0.13 | 6.25 | 15.5 | 3.25 | 2.42 | 1.56 | 0.86 | 4700 |
| [7]HI | 13.8[d] | 0.02[d] | (~0.9)[e] | – | 0.145[d] | (~6.5)[f] | – | 2.79[g] | 2.12[h] | 0.67[h] | 474[i] |

$\tau_{FL}$ fluorescence lifetime, $\Phi_{FL}$ fluorescence quantum yield, $\Phi_{ISC}$ quantum yield of intersystem crossing, $\Phi_{IC}$ quantum yield of internal conversion, $k_{FL}$ fluorescence rate constant, $k_{ISC}$ rate constant of intersystem crossing, $k_{IC}$ rate constant of internal conversion. $S_1-S_0$: determined by steady-state absorption and fluorescence spectra. $T_1-S_0$: determined by phosphorescence spectra at 77 K. $\varepsilon_{0-0}$: molar absorption coefficients of 0–0 absorption bands. $\Phi_{IC} = 1-\Phi_{FL}-\Phi_{ISC}$, $k_{FL} = \Phi_{FL} \cdot \tau_{FL}^{-1}$, $k_{ISC} = \Phi_{ISC} \cdot \tau_{FL}^{-1}$, $k_{IC} = \Phi_{IC} \cdot \tau_{FL}^{-1}$
[a]Excitation wavelength: 404 nm
[b]Excited at 330 nm
[c]Excited at 450 nm
[d]Reported values in 1,4-dioxane [29–31]
[e]The value was estimated using the reported results such as pristine [6]carbohelicene ($\Phi_{ISC} = 0.91$) [30], pristine [9]carbohelicene ($\Phi_{ISC} = 0.91$) [30] and pristine [8]carbohelicene ($\Phi_{ISC} = 0.92$) [30]
[f]Estimated value from $\Phi_{ISC}$ of [7]HI: 0.9
[g]Reported data [29]
[h]Reported data [30]
[i]Reported data [9]

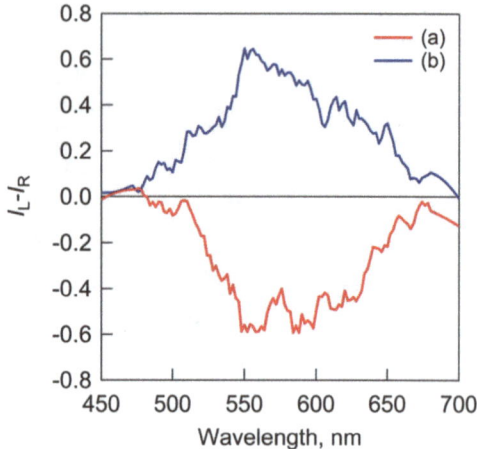

**Fig. 5.3** CPL profiles of **(a)** (+)-(*P*)-[7]Hl-NAIQx and **(b)** (−)-(*M*)-[7]Hl-NAIQx in THF (reprinted with permission from Ref. [21]. Copyright 2015 American Chemical Society)

nonplanar helicene configuration according to the molecular orbital structure of the LUMO in the original report [21]. The spin–orbit coupling is highly associated with the degree of nonplanar configuration as discussed by Schmidt and coworkers [40]. Thus, our synthetic strategies for [7]carbohelicene derivatives successfully resulted in the improved $\Phi_{FL}$ of [7]Hl-NAIQx.

In order to discuss chiral nature in fluorescence emission, CPL spectra of [7] carbohelicene derivatives were measured in THF (Fig. 5.3). No CPL signals were seen in [7]Hl-Qx. In contrast with [7]Hl-Qx, (+)-(*P*)-[7]Hl-NAIQx (spectrum *a*) and (−)-(*M*)-[7]Hl-NAIQx (spectrum *b*) successfully demonstrate CPL spectra with negative and positive signals, respectively as shown in Fig. 5.3. The signals of CPL are identical with those of the corresponding CD profiles in the spectral range of longest wavelength (i.e., around 400–500 nm) [41]. The degree of CPL, i.e., the luminescence dissymmetric factor ($g_{lum}$), is calculated by the following equation: $g_{lum} = 2(I_L - I_R)/(I_L + I_R)$, where $I_L$ and $I_R$ are the luminescence intensities of left and right circularly polarized light, respectively. By contrast with the negligible $g_{lum}$ of [7]Hl-Qx, the $g_{lum}$ of [7]Hl-NAIQx is calculated to be $4.0 \times 10^{-3}$. This result is equivalent to the $g_{lum}$ values observed in the monomeric form of small molecules [25, 27, 33, 42].

To investigate the electroluminescence behaviors of [7]Hl-NAIQx, an OLED doped with [7]Hl-NAIQx was constructed. As previously discussed [21], absorption spectral range of [7]Hl-NAIQx extended from UV to ca. 500 nm. Fluorescence spectrum of poly(9,9-dioctylfluorene) (PFO) range from ca. 400 to 500 nm. PFO was therefore chosen as the host material for [7]Hl-NAIQx because of the sufficient spectral overlap [43]. The OLED was composed of a bilayer structure with an indium tin oxide (ITO) anode, a hole transport layer of poly(3,4-ethylene dioxythiophene): poly-(styrenesulfonate) (PEDOT:PSS), an electron transport and emissive layer of PFO/[7]Hl-NAIQx, and a calcium/aluminum cathode [denoted as ITO/PEDOT: PSS/-PFO:[7]Hl-NAIQx (95:5)/Ca/Al]. Then, we obtained the *J–V–L* characteristics and electroluminescence spectrum as shown in Fig. 5.4a, b. Although we observed

**Fig. 5.4** (**a**) Current–luminance–voltage (*J–L–V*) characteristics and (**b**) an electroluminescence (EL) spectrum of ITO/PEDOT:PSS/PFO: [7] Hl-NAIQx-(95:5)/Ca/Al) (Reprinted with permission from Ref. [21]. Copyright 2015 American Chemical Society)

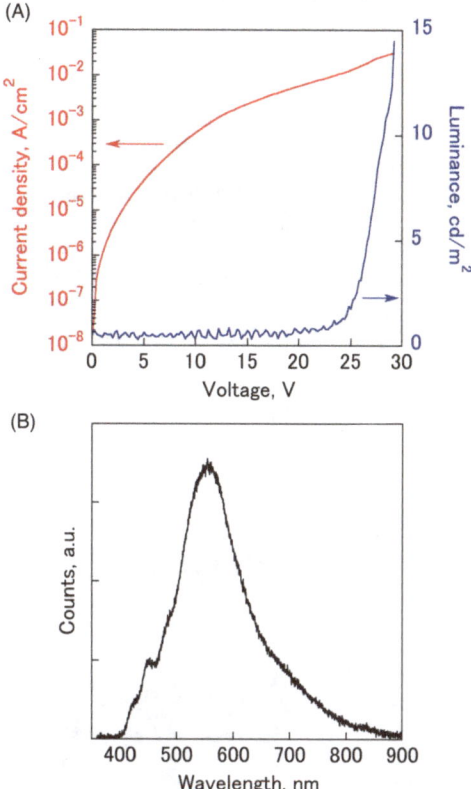

the relative high device operation voltage, a stable electroluminescence spectrum was successfully confirmed (Fig. 5.4b). This coincides with the observed steady-state fluorescence spectrum [21]. However, the gradual decrease of fluorescence intensity was seen with increasing the operation time due to the oxidation of Ca electrode. Anyway, we can summarize that [7]Hl-NAIQx is stable enough for OLED operation because of no spectral change during operation.

## 5.2.2 [5]Carbohelicene Derivatives for CPL

In association with the above section, [5]carbohelicene (i.e., a shorter carbohelicene relative to [7]carbohelicene) derivatives were similarly examined. These brief summaries are discussed here. A [5]carbohelicene derivative having electron-withdrawing maleimide, i.e., maleimide-substituted [5]carbohelicene (HeliIm), was designed and synthesized to investigate the photophysical behaviors and circularly polarized luminescence (CPL) as shown in Fig. 5.5a [44]. The fluorescence quantum yield of HeliIm ($\Phi_{FL} = 0.37$) significantly improved as compared to the reference

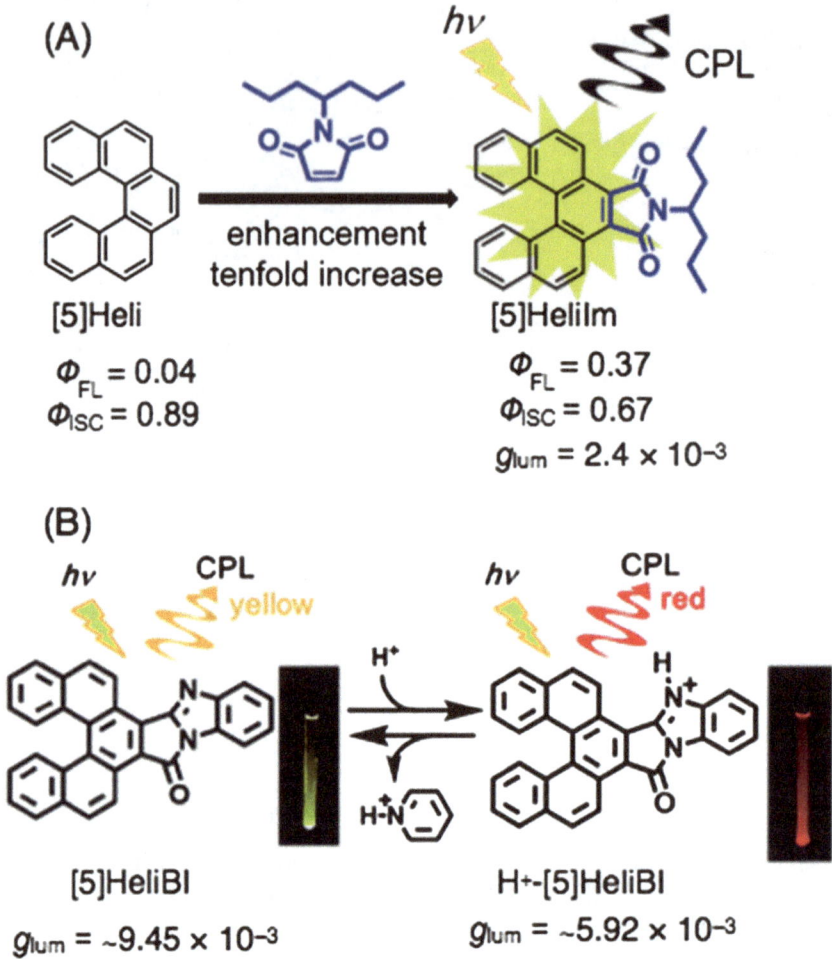

**Fig. 5.5** Schematic illustrations of (**a**) maleimide-substituted [5]carbohelicene (HeliIm) (reprinted with permission from Ref. [44] Copyright 2015 American Chemical Society) and (**b**) benzimidazole-fused [5]carbohelicene (reprinted with permission from Ref. [45] Copyright 2016 Royal Society of Chemistry)

compound: Heli ($\Phi_{FL} = 0.04$). These photophysical processes including intersystem crossing are successfully discussed by the kinetic discussions [44]. Circular dichroism (CD) and circularly polarized luminescence (CPL) of [5]carbohelicene derivatives were measured because they show the chirality. In particular, HeliIm provides excellent CPL and the anisotropy factor $g_{lum}$ was calculated to be $2.4 \times 10^{-3}$. Note that this result is the first example of observation of CPL for [5]carbohelicene derivatives (Fig. 5.5a) [44].

Then, [5]carbohelicene fused by benzimidazole unit ([5]HeliBI) was synthesized to discuss the photophysical and chiroptical behaviors [45]. The reversible photophysical switching between protonation and deprotonation of [5]HeliBI was investigated by $^1$H NMR and steady-state spectroscopy (i.e., absorption and fluorescence spectra). We also confirmed the CPL signals of protonated [5]HeliBI (H$^+$-[5]HeliBI). This is the first example regarding red-colored CPL behaviors of helicene derivatives (Fig. 5.5b) [45].

### 5.2.3  Tetrathia[9]helicene Derivatives for CPL

In this section, we briefly focus on the longer helicene derives. A series of tetrathia [9]helicene derivatives such as quinoxaline (acceptor)-fused tetrathia[9]helicene (donor) were synthesized to optimize the photophysical behaviors and CPL properties (Fig. 5.6a) [46]. In this work, the "push-pull" character was induced by introducing a quinoxaline onto the tetrathia[9]helicene unit, which was enhanced by further introduction of an electron-donating Me$_2$N unit or an electron-accepting NC unit onto the quinoxaline unit (i.e., Me$_2$N-QTTH and NC-QTTH in Fig. 5.6a). These properties were successfully evaluated by electrochemical methods and DFT calculations [46]. Large enhancements in the fluorescence quantum yields ($\Phi_{FL}$) were accordingly obtained. Particularly, the maximum $\Phi_{FL}$ of Me$_2$N-QTTH attains 0.43 in benzene (NC-QTTH: $\Phi_{FL} = 0.30$), which is more than 20 times higher than that of a pristine tetrathia[9]helicene, i.e., TTH ($\Phi_{FL} = 0.02$). These enhanced trends can be quantitatively explained by kinetic parameters such as the rate constants of fluorescence and intersystem crossing (ISC) pathways. This result enabled to demonstrate the excellent CPL behaviors. The anisotropy factor, $g_{lum}$ of NC-QTTH, was estimated to be $3.0 \times 10^{-3}$ (Fig. 5.6a) [46].

On the other hand, tetrasulfone[9]helicene (PTSH) was newly synthesized by a single-step oxidation reaction of tetrathia[9]helicene (PTTH) as shown in Fig. 5.6b [47]. In electrochemical evaluation, the first reduction potential of PTSH was positively shifted (ca. 1.0 V) as compared to that of PTTH due to the electron-accepting group: sulfone units. The electrochemical trends are identical with the energy levels estimated by DFT methods and steady-state spectroscopy. Furthermore, a significant improvement of the $\Phi_{FL}$ was confirmed. The $\Phi_{FL}$ of PTSH is 0.27, which is approximately an order of magnitude greater than that of PTTH ($\Phi_{FL} = 0.03$). The improved $\Phi_{FL}$ can be successfully discussed by the corresponding kinetic comparison. The plausible reason is due to the increased energy difference between the lowest singlet ($S_1$) and triplet ($T_1$) excited states ($\Delta E_{ST}$). Finally, efficient anisotropy factor ($g_{lum}$) of PTSH was also confirmed. The obtained $g_{lum}$ value was calculated to be $8.3 \times 10^{-4}$ (Fig. 5.6b) [47].

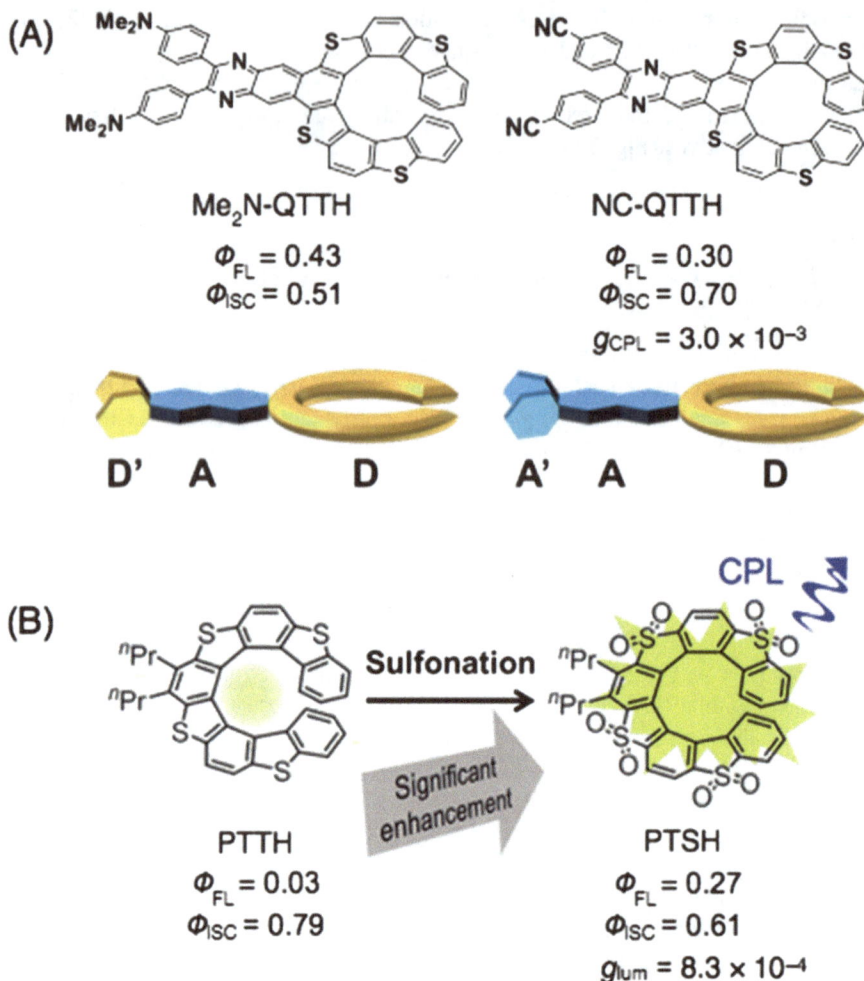

**Fig. 5.6** Schematic illustrations of (**a**) quinoxaline fused tetrathia[9]helicene derivatives (reprinted with permission from Ref. [46] Copyright 2016 Wiley-VCH) and (**b**) tetrasulfone[9]helicene in this study (reprinted with permission from Ref. [47] Copyright 2016 American Chemical Society)

### 5.2.4  A Zinc(II) Homoleptic Helicate for Largely Enhanced Absorption and Luminescence Dissymmetry Factors in the Near-Infrared Region

Although the extended π-conjugated structure is a straightforward method for enhanced absorption and fluorescence properties in the near-infrared (NIR) region, the degree of the redshift-trend by such an approach is quite limited in the case of carbohelicenes. For example, [16]carbohelicene [48] demonstrates an absorption maximum at ca. 430 nm (cf. 324 nm for [6]-helicene and 390 nm for [9]-helicenes)

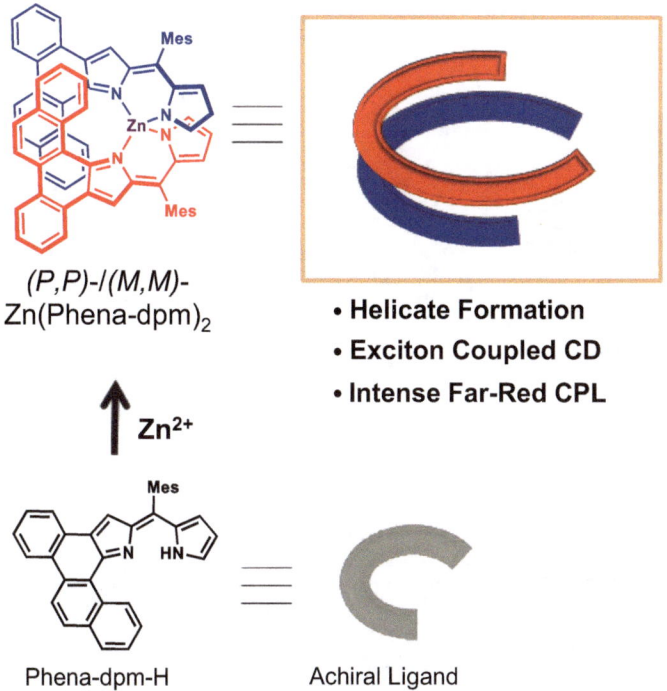

**(P,P)-/(M,M)-**
**Zn(Phena-dpm)₂**

- **Helicate Formation**
- **Exciton Coupled CD**
- **Intense Far-Red CPL**

Zn²⁺

Phena-dpm-H                    Achiral Ligand

**Fig. 5.7** A proposed chiral formation of a homoleptic zinc(II) helical complex composed by a pair of achiral dipyrromethene ligands: Phena-dpm-H (reprinted with permission from Ref. [52] Copyright 2018 Wiley-VCH)

[49]. Additionally, the $\Phi_{FL}$ of [$n$]carbohelicenes largely decreased with increasing the number ($n$) of the benzene units ($\Phi_{FL} \leq \sim 0.02$ for $n \geq 7$) [31]. Thus, this hampers the utilization of simple carbohelicene derivatives for CPL evaluation in the NIR region. In contrast, chiroptical properties of dipyrromethenes have improved recently, and the dimeric formations using supramolecular method were highly promising [50, 51]. Accordingly, we can expect that coordination-driven chiral structure composed of a pair of dipyrromethene chromophores are suitable for enhanced CD and CPL behaviors.

Therefore, in this section, we discuss the synthesis and enhanced chiroptical properties of a rare-earth- and precious-metal-free homoleptic zinc(II) helicate arranged by a pair of achiral ligands (benzo[$a$]phenanthrene-fused dipyrromethene) (Fig. 5.7) [52]. Selective configuration of the chiral helical structures is obtained by the possibility of tetrahedral conformation composed of two unsymmetrical bidentate ligands.

First, Zn(Phena-dpm)₂ was synthesized by the reported method [52]. In the steady-state spectroscopic measurements, the absorption bands of Zn(Phena-dpm)₂ (Fig. 5.8a, spectrum $a$) were red-shifted and broadened as compared to that of the reference compound: Phena-dpm-H (Fig. 5.8a, spectrum $b$). More importantly, the

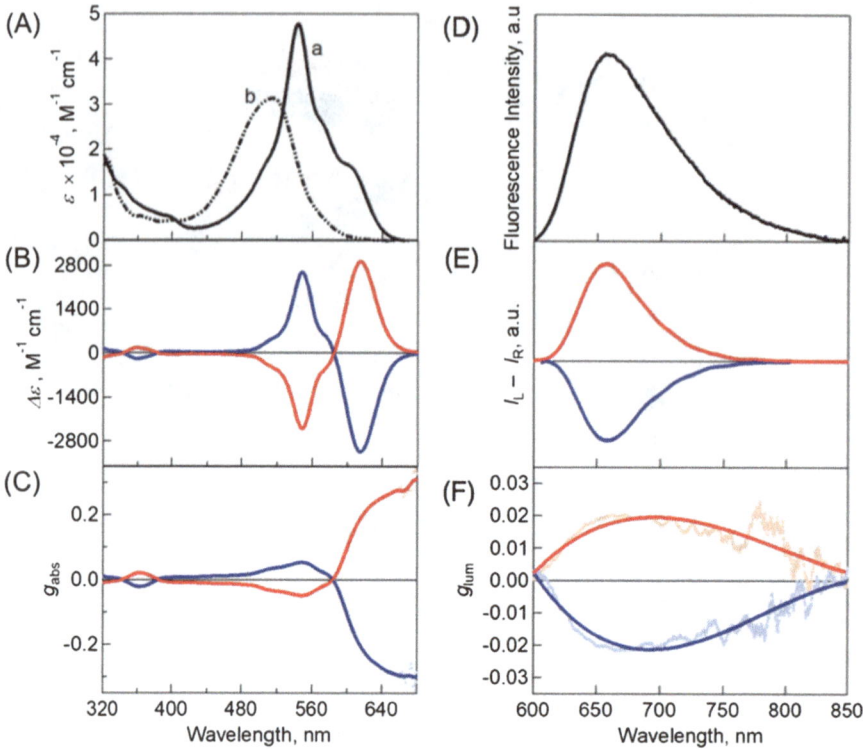

**Fig. 5.8** (**a**) Steady-state absorption spectra of (*a*) Zn(Phena-dpm)$_2$ and (*b*) free Phena-dpm-H. (**b**) CD spectra and (**c**) corresponding dissymmetry factor $g_{abs}$ profiles in toluene. (**d**) Fluorescence and (**e**) CPL spectra of Zn(Phena-dpm)$_2$ and (**f**) corresponding dissymmetry factor $g_{lum}$ profiles in toluene ($\lambda_{ex} = 580$ nm). Red and blue lines shows *M,M* and *P,P* forms of Zn(Phena-dpm)$_2$, respectively. The spectra were corrected by a smoothing method (Reprinted with permission from Ref. [52] Copyright 2018 Wiley-VCH)

absorption maximum of the pristine ligands at 522 nm was found to be split peaks at 548 and 615 nm, due to the exciton coupling [53]. A forbidden peak of Zn(Phena-dpm)$_2$ at 575 nm is attributable to the distinct transition derived from the distorted formation. To perform chiroptical measurements, chiral HPLC analysis using a DAICEL CHIRALPAK IE column was performed to purify enantiopure compounds. Figure 5.8b demonstrates the CD spectra of (*P,P*)-Zn(Phena-dpm)$_2$ and (*M,M*)-Zn(Phena-dpm)$_2$ in toluene solution. The enantiomers displayed mirror-image Cotton effects in the UV and visible regions. The value of molar CD $\Delta\varepsilon$ in the visible region (~3000 M$^{-1}$ cm$^{-1}$ at ca. 500–650 nm) was significantly larger as compared to than in the UV region (~160 M$^{-1}$ cm$^{-1}$ at ca. 360 nm). The dissymmetry factors $g_{abs}$ of the *P,P* form were calculated to be +0.022 (at 360 nm), +0.054 (at 540 nm), and −0.20 (at 615 nm), respectively (Fig. 5.8c) [52].

Then, stable fluorescence spectrum of Zn(Phena-dpm)$_2$ in the NIR region was observed (toluene) at 25 °C (Fig. 5.8d), together with a good fluorescence quantum yield ($\Phi_{FL} = 0.23$) and lifetime ($\tau_{FL} = 3.0$ ns) [52]. Although a mirror-image relation between absorption and fluorescence emission spectra was noted, a Stokes shift of $1100$ cm$^{-1}$ indicated appropriate the occurrence of excited-state relaxation regarding the coordination-driven helicate (vide infra). Strong and exact mirror image CPL signals were observed for (P,P)- and (M,M)-forms of Zn(Phena-dpm)$_2$ in the corresponding fluorescence spectrum (Fig. 5.8e). The signs of CPL agree with those of CD at the lowest-energy band, which is in consistent with the general trends for the ECD and CPL spectra, including those reported for exciton-coupled systems. The $g_{lum}$ of the P,P form was assigned to be $-0.022$ at 660 nm (Fig. 5.8f). Moreover, the $g_{lum}$ profiles expanded up to ca. 850 nm (i.e., far-red region), as shown in Fig. 5.8f.

To further discuss the origin of the chiroptical response in Zn(Phena-dpm)$_2$, the analysis of relevant electronic transitions were performed by DFT calculations. A similar way has been successfully employed for a few simple organic compounds [41, 54]. Therefore, the optimized structures of Zn(Phena-dpm)$_2$ in the ground and excited states were firstly optimized at the (TD)DFT-M062X/def2-QZV(GD3) level of theory [55]. Based on these calculated geometries in the $S_0$ and $S_1$ states, the relevant electronic ($\mu$) and magnetic ($m$) transition dipole moments were evaluated. Noted that these values are highly associated with the rotational strength $R$ for the corresponding CD and CPL spectra. In contrast with small structural change of single-crystal and calculated structures in the ground state, the structural relaxation should be substantial on photoexcitation, which subsequently have an effect on chiroptical response. Thus, the entire structural change and distortion around the zinc (II) ion was somewhat decreased in the excited state because the dihedral and torsion angles of the two dipyrromethenes were smaller (53.7 and 39.2°, respectively). More importantly, the overlapped trend between the two phenanthrene ligands was considerably changed in the excited state (Fig. 5.9), whereas the interplanar distances are approximately similar each other (3.40 vs. 3.25 Å). The structural change is attributable to the exciplex formation because of sufficient π-overlap between two aromatic groups, whereas this is generally hampered in the ground state considering the electrostatic and/or Pauli repulsion [56]. Thus, a slightly smaller chiroptical response was probably due to the $S_1$-to-$S_0$ transition ($R = 770$, for CPL) in contrast with the reverse $S_0$-to-$S_1$ transition ($R = 1013$, for CD).

Associated with the Rosenfeld equation, $R$ is the imaginary part in the scalar product of $\mu$ and $m$ of the relevant electronic transition (i.e., $R = |\mu||m| \cos\theta$). Among the possible factors contributing to the large $R$ values for the dissymmetry factors $g_{abs}$ and $g_{lum}$, the small angle $\theta$ between these two moments was found to play an important role. Indeed, these factors for the $S_0$-to-$S_1$ and $S_1$-to-$S_0$ transitions in the complex were estimated to be 0.94 (i.e., $\theta = 20.6°$) and 0.96 (i.e., $\theta = 15.7°$), respectively, which is totally different from the much smaller value of 0.027 ($\theta = 91.6°$) for free Phena-dpm-H. Such a drastic improvement is established by the C2-symmetrical alignment of two identical phenanthrene-fused helical units in Zn(Phena-dpm)$_2$. In addition, the increase trend in absolute $|m|$ value was also attributable to the dimeric structure of Zn(Phena-dpm)$_2$. Briefly, a proper alignment of two identical helical components via coordination-organized contributes to

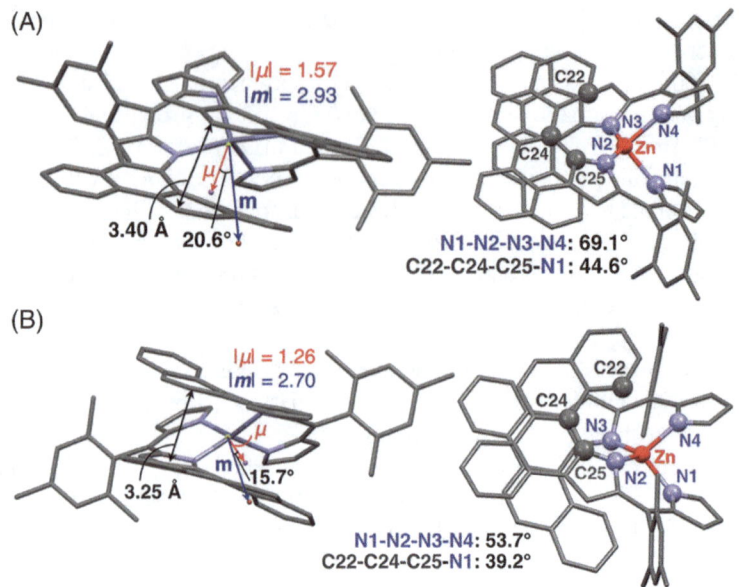

**Fig. 5.9** Optimized structures of $(P,P)$-Zn(Phena-dpm)$_2$. (**a**) ground state and (**b**) excited state. Hydrogen atoms are omitted for clarity. Electric transition dipole moment $\boldsymbol{\mu}$ and magnetic transition dipole moment $\boldsymbol{m}$ demonstrate the $S_0$-to-$S_1$ or $S_1$-to-$S_0$, respectively (Reprinted with permission from Ref. [52] Copyright 2018 Wiley-VCH)

achieve strong chiroptical responses. This may have been apparent in CD, where an effective exciton coupling should be observed, but now turned out to be also applicable to the materials that emit strong CPL in the NIR region. Recently, a related approach has been reported [57].

## 5.3 Summary

In conclusion, we successfully demonstrated the structural and photophysical control of fluorescent helicene and related zinc(II) helicates for improved CPL properties. The absolute fluorescence quantum yields ($\Phi_{FL}$) of helicene derivatives are generally extremely low because the intersystem crossing (ISC) pathways are highly accelerated ($\Phi$ISC: ~0.9). However, our synthetic strategy (e.g., "push-pull" method) enables to significantly enhance the $\Phi_{FL}$ values, which resulted in the observation of CPL signals. Then, a homoleptic chiral zinc(II) helicate selectively formed by a pair of achiral benzo[$a$]phenanthrene-fused dipyrromethene ligands afforded unusually strong chiroptical responses with absorption and luminescence dissymmetry factors such as $|g_{abs}| = 0.20$ (at 615 nm) and $|g_{lum}| = 0.022$ (at 660 nm), respectively. More interestingly, the $g_{lum}$ profiles also extended up to ca. 850 nm. They are highest values among the rare earth- and precious metal-free small molecules that emit in the NIR region. We believe that our synthetic strategies exemplified herein for enhanced chiroptical properties will provide a new perspective for future development of advanced luminescent materials.

# References

1. Wu J, Pisula W, Müllen K (2007) Graphenes as potential material for electronics. Chem Rev 107(3):718–747
2. Hammer BAG, Müllen K (2016) Dimensional evolution of polyphenylenes: expanding in all directions. Chem Rev 116(4):2103–2140
3. Ho R-M, Li M-C, Lin S-C, Wang H-F, Lee Y-D, Hasegawa H, Thomas EL (2012) Transfer of chirality from molecule to phase in self-assembled chiral block copolymers. J Am Chem Soc 134(26):10974–10986
4. Watanabe K, Iida H, Akagi K (2012) Circularly polarized blue luminescent spherulites consisting of hierarchically assembled ionic conjugated polymers with a helically π-stacked structure. Adv Mater 24(48):6451–6456
5. Gingras M (2013) One hundred years of helicene chemistry. Part 3: applications and properties of carbohelicenes. Chem Soc Rev 42(3):1051–1095
6. Gingras M (2013) One hundred years of helicene chemistry. Part 1: non-stereoselective syntheses of carbohelicenes. Chem Soc Rev 42(3):968–1006
7. Verbiest T, Elshocht SV, Kauranen M, Hellemans L, Snauwaert J, Nuckolls C, Katz TJ, Persoons A (1998) Strong enhancement of nonlinear optical properties through supramolecular chirality. Science 282(5390):913–915
8. Phillips KES, Katz TJ, Jockusch S, Lovinger AJ, Turro NJ (2001) Synthesis and properties of an aggregating heterocyclic helicene. J Am Chem Soc 123(48):11899–11907
9. Sakai H, Shinto S, Araki Y, Wada T, Sakanoue T, Takenobu T, Hasobe T (2014) One-dimensional helical columnar formation and excimer-like excited-state of racemic quinoxaline-fused [7]carbohelicenes in the crystal state. Chem Eur J 20(32):10099–10109
10. Reetz MT, Sostmann S (2001) 2,15-dihydroxy-hexahelicene (helixol): synthesis and use as an enantioselective fluorescent sensor. Tetrahedron 57(13):2515–2520
11. Kaseyama T, Furumi S, Zhang X, Tanaka K, Takeuchi M (2011) Hierarchical assembly of a phthalhydrazide-functionalized helicene. Angew Chem Int Ed 50(16):3684–3687
12. Shcherbina MÄ, Zeng X-B, Tadjiev T, Ungar G, Eichhorn SÄ, Phillips KÄÄ, Katz TÄ (2009) Hollow six-stranded helical columns of a helicene. Angew Chem Int Ed 48(42):7837–7840
13. Verbiest T, Sioncke S, Persoons A, Vyklický L, Katz TJ (2002) Electric-field-modulated circular-difference effects in second-harmonic generation from a chiral liquid crystal. Angew Chem Int Ed 41(20):3882–3884
14. Nuckolls C, Shao R, Jang W-G, Clark NA, Walba DM, Katz TJ (2002) Electro-optic switching by helicene liquid crystals. Chem Mater 14(2):773–776
15. Yuasa J, Ohno T, Tsumatori H, Shiba R, Kamikubo H, Kataoka M, Hasegawa Y, Kawai T (2013) Fingerprint signatures of lanthanide circularly polarized luminescence from proteins covalently labeled with a [small beta]-diketonate europium (iii) chelate. Chem Commun 49 (41):4604–4606
16. Riehl JP, Richardson FS (1986) Circularly polarized luminescence spectroscopy. Chem Rev 86 (1):1–16
17. Kumar J, Nakashima T, Tsumatori H, Kawai T (2014) Circularly polarized luminescence in chiral aggregates: dependence of morphology on luminescence dissymmetry. J Phys Chem Lett 5(2):316–321
18. Kumar J, Nakashima T, Kawai T (2015) Circularly polarized luminescence in chiral molecules and supramolecular assemblies. J Phys Chem Lett 6(17):3445–3452

19. Kumar J, Nakashima T, Kawai T (2014) Inversion of supramolecular chirality in bichromophoric perylene bisimides: influence of temperature and ultrasound. Langmuir 30 (21):6030–6037
20. Carr R, Evans NH, Parker D (2012) Lanthanide complexes as chiral probes exploiting circularly polarized luminescence. Chem Soc Rev 41(23):7673–7686
21. Sakai H, Shinto S, Kumar J, Araki Y, Sakanoue T, Takenobu T, Wada T, Kawai T, Hasobe T (2015) Highly fluorescent [7]carbohelicene fused by asymmetric 1,2-dialkyl-substituted quinoxaline for circularly polarized luminescence and electroluminescence. J Phys Chem C 119(24):13937–13947
22. Morisaki Y, Gon M, Sasamori T, Tokitoh N, Chujo Y (2014) Planar chiral tetrasubstituted [2.2] paracyclophane: optical resolution and functionalization. J Am Chem Soc 136(9):3350–3353
23. Nakamura K, Furumi S, Takeuchi M, Shibuya T, Tanaka K (2014) Enantioselective synthesis and enhanced circularly polarized luminescence of s-shaped double azahelicenes. J Am Chem Soc 136(15):5555–5558
24. Kögel JF, Kusaka S, Sakamoto R, Iwashima T, Tsuchiya M, Toyoda R, Matsuoka R, Tsukamoto T, Yuasa J, Kitagawa Y, Kawai T, Nishihara H (2016) Heteroleptic [bis (oxazoline)](dipyrrinato)zinc(ii) complexes: bright and circularly polarized luminescence from an originally achiral dipyrrinato ligand. Angew Chem Int Ed 55(4):1377–1381
25. Oyama H, Nakano K, Harada T, Kuroda R, Naito M, Nobusawa K, Nozaki K (2013) Facile synthetic route to highly luminescent sila[7]helicene. Org Lett 15(9):2104–2107
26. Li M, Niu Y, Zhu X, Peng Q, Lu H-Y, Xia A, Chen C-F (2014) Tetrahydro[5]helicene-based imide dyes with intense fluorescence in both solution and solid state. Chem Commun 50 (23):2993–2995
27. Goto K, Yamaguchi R, Hiroto S, Ueno H, Kawai T, Shinokubo H (2012) Intermolecular oxidative annulation of 2-aminoanthracenes to diazaacenes and aza[7]helicenes. Angew Chem Int Ed 51(41):10333–10336
28. Son H-J, Han W-S, Wee K-R, Yoo D-H, Lee J-H, Kwon S-N, Ko J, Kang SO (2008) Turning on fluorescent emission from c-alkylation on quinoxaline derivatives. Org Lett 10(23):5401–5404
29. Birks JB, Birch DJS, Cordemans E, Vander Donckt E (1976) Fluorescence of the higher helicenes. Chem Phys Lett 43(1):33–36
30. Sapir M, Donckt EV (1975) Intersystem crossing in the helicenes. Chem Phys Lett 36 (1):108–110
31. Vander Donckt E, Nasielski J, Greenleaf JR, Birks JB (1968) Fluorescence of the helicenes. Chem Phys Lett 2(6):409–410
32. Tsumatori H, Nakashima T, Kawai T (2010) Observation of chiral aggregate growth of perylene derivative in opaque solution by circularly polarized luminescence. Org Lett 12(10):2362–2365
33. Sawada Y, Furumi S, Takai A, Takeuchi M, Noguchi K, Tanaka K (2012) Rhodium-catalyzed enantioselective synthesis, crystal structures, and photophysical properties of helically chiral 1,1′-bitriphenylenes. J Am Chem Soc 134(9):4080–4083
34. Pineiro M, Carvalho AL, Pereira MM, Gonsalves AM, Arnaut LG, Formosinho SJ (1998) Photoacoustic measurements of porphyrin triplet-state quantum yields and singlet-oxygen efficiencies. Chem Eur J 4(11):2299–2307
35. Bonnett R, McGarvey DJ, Harriman A, Land EJ, Truscott TG, Winfield UJ (1988) Photophysical properties of meso-tetraphenylporphyrin and some meso-tetra(hydroxyphenyl) porphyrins. Photochem Photobiol 48(3):271–276
36. Hirayama S, Sakai H, Araki Y, Tanaka M, Imakawa M, Wada T, Takenobu T, Hasobe T (2014) Systematic control of the excited-state dynamics and carrier-transport properties of functionalized benzo[ghi]perylene and coronene derivatives. Chem Eur J 20(29):9081–9093
37. Ida K, Sakai H, Ohkubo K, Araki Y, Wada T, Sakanoue T, Takenobu T, Fukuzumi S, Hasobe T (2014) Electron-transfer reduction properties and excited-state dynamics of benzo[ghi] peryleneimide and coroneneimide derivatives. J Phys Chem C 118(14):7710–7720

38. Arbogast JW, Darmanyan AP, Foote CS, Diederich FN, Whetten RL, Rubin Y, Alvarez MM, Anz SJ (1991) Photophysical properties of sixty atom carbon molecule (C60). J Phys Chem 95 (1):11–12
39. Turro NJ (1991) Modern molecular photochemistry. University Science Books, Sausalito
40. Schmidt K, Brovelli S, Coropceanu V, Beljonne D, Cornil J, Bazzini C, Caronna T, Tubino R, Meinardi F, Shuai Z, Brédas J-L (2007) Intersystem crossing processes in nonplanar aromatic heterocyclic molecules. J Phys Chem A 111(42):10490–10499
41. Abbate S, Longhi G, Lebon F, Castiglioni E, Superchi S, Pisani L, Fontana F, Torricelli F, Caronna T, Villani C, Sabia R, Tommasini M, Lucotti A, Mendola D, Mele A, Lightner DA (2014) Helical sense-responsive and substituent-sensitive features in vibrational and electronic circular dichroism, in circularly polarized luminescence, and in raman spectra of some simple optically active hexahelicenes. J Phys Chem C 118(3):1682–1695
42. Field JE, Muller G, Riehl JP, Venkataraman D (2003) Circularly polarized luminescence from bridged triarylamine helicenes. J Am Chem Soc 125(39):11808–11809
43. Lane P, Palilis L, O'Brien D, Giebeler C, Cadby A, Lidzey D, Campbell A, Blau W, Bradley D (2001) Origin of electrophosphorescence from a doped polymer light emitting diode. Phys Rev B 63(23):235206
44. Sakai H, Kubota T, Yuasa J, Araki Y, Sakanoue T, Takenobu T, Wada T, Kawai T, Hasobe T (2016) Synthetic control of photophysical process and circularly polarized luminescence of 5 carbohelicene derivatives substituted by maleimide units. J Phys Chem C 120(14):7860–7869
45. Sakai H, Kubota T, Yuasa J, Araki Y, Sakanoue T, Takenobu T, Wada T, Kawai T, Hasobe T (2016) Protonation-induced red-coloured circularly polarized luminescence of [5]carbohelicene fused by benzimidazole. Org Biomol Chem 14(28):6738–6743
46. Yamamoto Y, Sakai H, Yuasa J, Araki Y, Wada T, Sakanoue T, Takenobu T, Kawai T, Hasobe T (2016) Synthetic control of the excited-state dynamics and circularly polarized luminescence of fluorescent "push-pull" tetrathia[9]helicenes. Chem Eur J 22(12):4263–4273
47. Yamamoto Y, Sakai H, Yuasa J, Araki Y, Wada T, Sakanoue T, Takenobu T, Kawai T, Hasobe T (2016) Controlled excited-state dynamics and enhanced fluorescence property of tetrasulfone[9]helicene by a simple synthetic process. J Phys Chem C 120(13):7421–7427
48. Mori K, Murase T, Fujita M (2015) One-step synthesis of [16]helicene. Angew Chem Int Ed 54 (23):6847–6851
49. Nakai Y, Mori T, Inoue Y (2012) Theoretical and experimental studies on circular dichroism of carbo[n]helicenes. J Phys Chem A 116(27):7372–7385
50. Toyoda M, Imai Y, Mori T (2017) Propeller chirality of boron heptaaryldipyrromethene: unprecedented supramolecular dimerization and chiroptical properties. J Phys Chem Lett 8 (1):42–48
51. Maeda H, Bando Y, Shimomura K, Yamada I, Naito M, Nobusawa K, Tsumatori H, Kawai T (2011) Chemical-stimuli-controllable circularly polarized luminescence from anion-responsive π-conjugated molecules. J Am Chem Soc 133(24):9266–9269
52. Ito H, Sakai H, Okayasu Y, Yuasa J, Mori T, Hasobe T (2018) Significant enhancement of absorption and luminescence dissymmetry factors in the far-red region: a zinc(II) homoleptic helicate formed by a pair of achiral dipyrromethene ligands. Chem Eur J 24(63):16889–16894
53. Zinna F, Bruhn T, Guido CA, Ahrens J, Bröring M, Bari LD, Pescitelli G (2016) Circularly polarized luminescence from axially chiral bodipy dyemers: an experimental and computational study. Chem Eur J 22(45):16089–16098
54. Longhi G, Castiglioni E, Koshoubu J, Mazzeo G, Abbate S (2016) Circularly polarized luminescence: a review of experimental and theoretical aspects. Chirality 28(10):696–707
55. Zhao Y, Truhlar DG (2008) The M06 suite of density functionals for main group thermochemistry, thermochemical kinetics, noncovalent interactions, excited states, and transition elements: two new functionals and systematic testing of four M06-class functionals and 12 other functionals. Theor Chem Accounts 120(1):215–241

56. Mori T, Inoue Y (2013) Charge-transfer excitation: unconventional yet practical means for controlling stereoselectivity in asymmetric photoreactions. Chem Soc Rev 42(20):8122–8133
57. Dhbaibi K, Favereau L, Srebro-Hooper M, Jean M, Vanthuyne N, Zinna F, Jamoussi B, Di Bari L, Autschbach J, Crassous J (2018) Exciton coupling in diketopyrrolopyrrole-helicene derivatives leads to red and near-infrared circularly polarized luminescence. Chem Sci 9 (3):735–742

# Chapter 6
# BODIPY Based Emitters of Circularly Polarized Luminescence

**Michael John Hall and Santiago de la Moya**

**Abstract** The boron dipyrromethenes (BODIPYs) are a ubiquitous class of fluorescent dyes which have found utility in a wide range of photonics applications. More recently the BODIPYs have been investigated for their use as small organic molecule (SOM) emitters of circularly polarized luminescence (CPL). Herein, we will discuss recent developments in the field of BODIPY-based CPL-SOMs. In particular, we will review the current design strategies for the induction of chirality in the otherwise planar BODIPY fluorophore, for both mono- and multi-fluorophore containing systems, and examine the impact of these designs on the resulting chiroptical properties, including CPL emission.

## 6.1 Introduction

It is obvious that CPL-enabling organic systems have great potential for the development of optical materials, and photonic tools based on such materials, because they combine the advantages offered by their ability to emit circularly polarized light with those derived from their organic nature [1, 2]. Circularly polarized light endows the optical tools based on it with a higher spatio-temporal resolution owing to the additional signal parameters introduced by the circular polarization, both the level and handedness of the circular polarization [3–8]. Moreover circularly polarized light can interact selectively with chiral matter, making it indispensable for the study and discernment of key phenomena based on molecular chirality, omnipresent in the world around us, Life included [9–11]. In addition, circularly polarized light can be

M. J. Hall (✉)
Chemistry, School of Natural and Environmental Sciences, Newcastle University, Newcastle upon Tyne, UK
e-mail: michael.hall@newcastle.ac.uk

S. de la Moya (✉)
Departamento de Química Orgánica, Facultad de Ciencias Químicas, Universidad Complutense de Madrid, Madrid, Spain
e-mail: santmoya@ucm.es

© Springer Nature Singapore Pte Ltd. 2020
T. Mori (ed.), *Circularly Polarized Luminescence of Isolated Small Organic Molecules*, https://doi.org/10.1007/978-981-15-2309-0_6

used to conduct light-induced asymmetric reactions [12–14] or to control chiral morphologies in nanostructures [15, 16]. Organic systems present significant advantages when it comes the modulation of key physical and chemical properties through workable structural modifications which are available via the rich variety of chemical transformations offered by modern organic chemistry [1, 2]. Furthermore, organic systems offer additional advantages in other fundamental aspects related to the development of specific materials, for example, in materials processing or in gaining biocompatibility, facilitating the preparation of ultra-fine materials or biomaterials, respectively. Additionally, the characteristic broadband emission of the organic luminescent systems allows the selection of multiple wavelengths from the same photonic material, which is interesting for the development of applications requiring such a tunability [1, 2].

Among the CPL-enabling organic systems (molecules, polymers, supramolecular aggregates, etc.), those based on simple and non-aggregated small organic molecules (SOMs) have aroused great interest in recent years. The interest in CPL-enabling SOMs (CPL-SOMs) is due to (1) their high potential for the development of specific CPL materials due to properties associated with low molecular weight (e.g. in the development of certain biomaterials beyond biosensors) and organic-solvent solubility (e.g. in the development of CPL-active dye-doped inclusion materials); (2) their capability to achieve high emission quantum yields ($\phi$); and (3) the ease in which emission signatures can be modulated by accessible structural modification of the chromophoric scaffold [1, 2]. However, examples of CPL-SOMs are scarce, are based on a small set of chromophoric molecular scaffolds (biaryls, helicenes, perylenes, BODIPYs) and usually exhibit low levels of circular polarization, as measured by the luminescent dissymmetry factor ($|g_{lum}|$ typically in the $10^{-5}$–$10^{-2}$ range) [1, 2]. The development of chiral SOMs that able to exhibit CPL, with both a large luminescent dissymmetry factor and a high emission quantum yield, is thus an exciting challenge in the field of molecular photonics, due to the difficulty of combining both key properties in the same SOM.

Boron dipyrromethenes (4-bora-3a,4a-diaza-$s$-indacenes or BODIPYs; see Fig. 6.1) are valuable fluorescent dyes, which have been successfully used in the development of many photonic tools [17–23]. BODIPYs are normally defined as SOMs, despite the presence of a chelated boron atom in the core structure. This is due to the metalloid nature of boron allowing properties closer to those of organic molecules than to inorganic metal complexes. On the other hand, the BODIPY chromophore itself is purely organic, as it is located in the aromatic-like π-conjugated system of its organic dipyrromethene ligand (dipyrrin; highlighted in

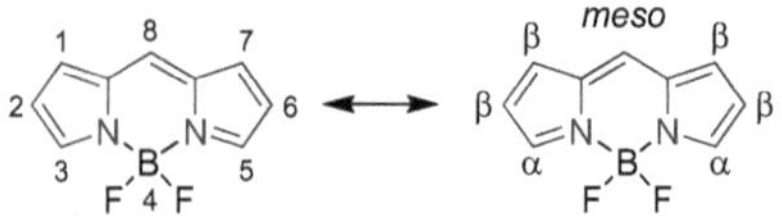

Fig. 6.1 Boron dipyrromethene scaffold for the simplest $F$-BODIPY, including both commonly used numbering schemes. Dipyrrin chromophore highlighted in blue

blue in Fig. 6.1), the boron acting as a mere linking atom maintaining chromophoric planarity and rigidity, thus minimizing the rate of non-radiative relaxation of the excited state and favouring radiative emission (i.e. fluorescence) [24].

Among the different organic chromophoric scaffolds, BODIPYs, and mainly *F*-BODIPYs (4,4-difluoro-4-bora-3a,4a-diaza-*s*-indacenes; see Fig. 6.1), present significant advantages for the development of CPL-SOMs. This is due to their outstanding physical, photophysical and chemical properties [25–30], including:

- High solubility in a wide range of solvents.
- Low aggregation capability.
- High chemical robustness.
- High light-absorption coefficients ($\varepsilon_{max}$).
- High fluorescent quantum yields ($\phi$).
- High photostability.
- Fluorescence signatures which are mostly independent of solvent polarity.
- Ample structural diversity.
- Potential for modulation of physical, chemical, and biological properties through molecular modification.
- Potential for modulation of photophysical properties, including CPL signatures (e.g. $g_{lum}$) and CPL efficiency.

In other words, high fluorescence efficiency combined with a rich, well-known, and workable synthetic chemistry, the latter allowing synthetic access and fine modulation of key properties, makes the BODIPYs one of the most interesting chromophoric scaffolds with which to develop novel photonic materials, including CPL-SOMs [31].

Although the π-conjugated BODIPY chromophore is inherently achiral, it can be chirally perturbed as to make it efficiently exhibit chiroptical properties, including CPL [1, 2, 31]. This perturbation is generally achieved by embedding the BODIPY chromophore within a chirally resolved molecular design, which is crucial to gain chiroptical efficiency (i.e. to maximize $|g_{lum}|$ or ideally to maximize the product of $|g_{lum}|$ and $\phi$) [1, 2]. On the other hand, such a design should be as simple as possible to maximize synthetically accessibility, thus allowing easy access to low-cost chiroptical materials, including molecular CPL materials [1, 2].

To date, a number of preferred chiral designs have been reported as successful approaches with which to endow the BODIPY chromophore with CPL activity. These designs which can be initially separated into those based on (a) monomeric chiral BODIPYs (mono(BODIPY)s) and (b) those involving two or more BODIPY units (poly(BODIPY)s). Mono(BODIPY)-based CPL-SOMs can be further subdivided into those in which (1) chiral moieties (one or more) are covalently linked to the BODIPY core to obtain a chiral $C_2$-symmetric mono(BODIPY) or (2) the BODIPY core is embedded into a molecular architecture displaying helical chirality. Whilst poly(BODIPY)-based CPL-SOMs can also be subdivided into those in which (1) two (or more) BODIPY cores are directly linked via a covalent bond to obtain an axially-chiral atropoisomeric bis(BODIPY) structure, (2) two BODIPY moieties are covalently linked through a flexible chiral bridge to obtain a configurationally-labile helically-chiral bis(BODIPY), and (3) two BODIPY moieties with intrinsic helical chirality are combined into a figure-of-eight motif.

Thus, this chapter aims to review the current state of the art with respect to the design and synthesis of CPL-SOMs based on the BODIPY scaffold, including an evaluation of how recent findings could inform chiral design strategies for the development of future BODIPY based CPL-SOM with highly efficient CPL emission.

## 6.2   Chirality in BODIPYs

Prior to the examination of BODIPYs as potential CPL-SOMs, a small number of chiral BODIPYs had been described in the literature, either through their synthesis as single enantiomers or through chiral resolution, thus allowing preliminary evaluation of the chiroptical properties of BODIPYs, albeit not with a focus on CPL. We will therefore discuss a number of these chiral BODIPYs to provide both context for the discussion of current BODIPY CPL-SOMs and to underpin the conceptual basis for future developments in the field [31].

### 6.2.1   BODIPYs with Covalently Attached Chiral Moieties

Early examples of enantiomerically pure BODIPYs were typically obtained through the decoration of the central chromophore with asymmetric substituents. One of the first examples of such systems was published by Gossauer et al. and involved the preparation of two homochiral mono(BODIPY) architectures bearing asymmetric phenyl substituents at either the *meso*- (**1**) or $\alpha$-positions (**2**) [32]. Measurement of the circular dichroism (CD) spectra of both **1** and **2** showed Cotton effects corresponding to the $S_0$–$S_1$ transition of the BODIPY fluorophores, with $\alpha$-functionalized BODIPY **2** showing a significantly stronger Cotton effect ($\lambda = 520$ nm, $\Delta\varepsilon_{max} = +12.5$ Lmol$^{-1}$ cm$^{-1}$) than that of the *meso*-functionalized BODIPY **1** ($\lambda = 538$ nm, $\Delta\varepsilon_{max} = +2$ Lmol$^{-1}$ cm$^{-1}$). Although no CPL were reported for these compounds, these early observations led Gossauer et al. to speculate that "...both a 'chiral perturbation' of an inherently planar dipyrrin chromophore and a twisting deformation of the latter may give rise to high optical activity of the corresponding derivatives...". As we will see in later examples, the requirement for a "twisted" fluorophore core remains a key design component in the development of chiroptically active BODIPY CPL-SOMs (Fig. 6.2).

An alternate strategy for the synthesis of chirally substituted BODIPYs was developed by Grimme et al. through the introduction of binaphthyl moieties at the *meso*-position [33]. Mono(BODIPY) **3** and bis(BODIPY) **4** were prepared through condensation of the appropriate enantiomerically pure 1,1'-binaphthalene carbaldehydes with 2,4-dimethylpyrrole. The CD spectra of these two compounds

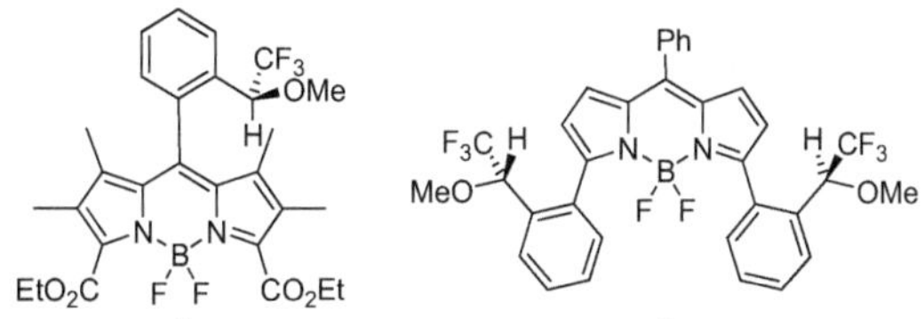

**Fig. 6.2** Mono(BODIPY)s bearing chiral-phenyl substituents at the *meso-* (**1**) or α-positions (**2**)

showed a number of short-wavelength transitions corresponding to those arising from the binaphthalene moiety, along with longer wavelength absorptions corresponding to those of the BODIPY fluorophores. Binaphthalene-substituted mono(BODIPY) **3** showed a positive long-wavelength transition ($\lambda = 495$ nm, $\Delta\varepsilon_{max} = +10$ Lmol$^{-1}$ cm$^{-1}$) related to the $S_0$–$S_1$ absorption band ($\lambda = 507$ nm) of the BODIPY fluorophore. More interestingly binaphthalene-substituted bis (BODIPY) **4** showed a positive, strong asymmetric couplet ($\lambda = 501$ nm, $\Delta\varepsilon_{max} = +53$ Lmol$^{-1}$ cm$^{-1}$; $\lambda = 485$ nm, $\Delta\varepsilon_{max} = -5$ Lmol$^{-1}$ cm$^{-1}$). Quantum-chemical calculations of the excited states of both mono(BODIPY) **3** and bis (BODIPY) **4** suggest that the observed long-wavelength CD transition of binaphthalene-substituted bis(BODIPY) **4** corresponds to an exciton-coupled transition arising from the interaction between the two BODIPY fluorophores (Fig. 6.3).

Due to a combination of long-wavelength CD transitions and the redox behaviour of the BODIPY fluorophores, binaphthalene-substituted bis(BODIPY) **4** was subsequently shown to act as an electrochemically controlled chiroptical switch. Thus CD-spectroelectrochemical measurements showed a decrease in the intensity of the Cotton effect observed at 501 nm when reductive potential was increased, attributed to the reversible reduction of BODIPY **4** to the corresponding dianion ($E_{1/2} = -1510$ mV) in which one electron is transferred to each chromophore [33].

In related system by Beer, Rurack, and Daub, an enantiopure binaphthalene-substituted bis(BODIPY) **5** was constructed containing free hydroxyl groups on the chiral binaphthalene ((*R*)-BINOL) moiety, amine-mediated deprotonation of which led to static quenching of the BODIPY fluorescence [34]. Interestingly in the presence of optically active bases (i.e. (*R*)-(+)- or (*S*)-(−)-1-phenylethylamine),

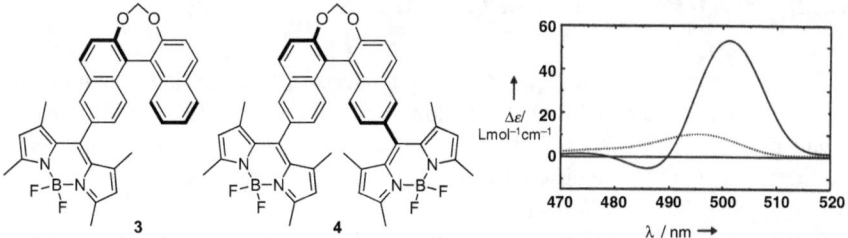

**Fig. 6.3** 1,1′-Binaphthalene-based mono(BODIPY) **3** and bis(BODIPY) **4**, including CD spectra (**3** (broken line) and **4** (solid line)) [adapted with permission from [33], copyright 2000 WILEY-VCH Verlag GmbH]

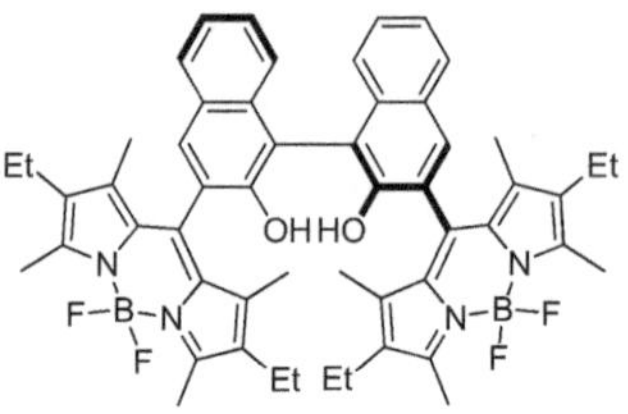

**Fig. 6.4** BINOL-based bis
(BODIPY) **5**

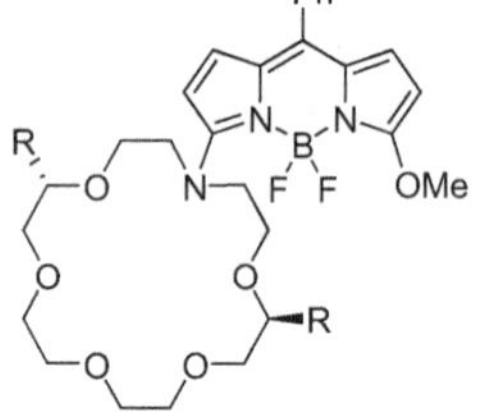

**Fig. 6.5** Aza-crown-ether-
based mono(BODIPY)s
**6** and **7**

R = Me; **6**

R = *i*-Bu; **7**

differences in static quenching constants $K_S$ could be observed, likely due to the formation of a diastereotopic tight ion pair between deprotonated **5** and the chiral base (i.e. $RO^- \cdots HNR_3^+$), the ratio of $K_S(R–S)/K_S(R–R) = 1.40$ indicating that bis (BODIPY) **5** may have potential for use as an enantioselective molecular sensor (Fig. 6.4).

A final example of a BODIPY covalently attached to a chiral moiety was disclosed by Huszthy et al. in which two mono(BODIPY)-linked azacrown ethers **6** and **7** were prepared through the reaction of 3-chloro-5-methoxyBODIPY with the corresponding enantiopure 1-aza-18-crown-6 ethers [35]. Although **6** and **7** performed admirably as fluorescent sensors of divalent metal cations, attempts to employ either as enantioselective molecular sensors of chiral ammonium cations (e.g. (*R*)- or (*S*)-α-phenylethylammonium perchlorate) proved unsuccessful (Fig. 6.5).

## *6.2.2 BODIPYs Containing a Chiral Boron Centre*

In 2010 the group of Ulrich and Ziessel published one of the first examples of a BODIPY based around a chiral boron atom, termed a $B^*$-BODIPY [36]. In order to obtain a stable chiral $B^*$-BODIPY, Ziessel et al. proposed the following design requirements: (1) lateral differentiation of the dipyrromethene core; (2) a polar group attached via the 3/5-position capable of intramolecular association with the central boron atom; and (3) a polyaromatic residue causing moderate steric conges-tion around the boron atom (also aiding in chromatographic resolution on a chiral solid phase). Therefore racemic $B^*$-BODIPY **8** (Fig. 6.6) was prepared starting from the parent BODIPY through displacement of single fluoride with

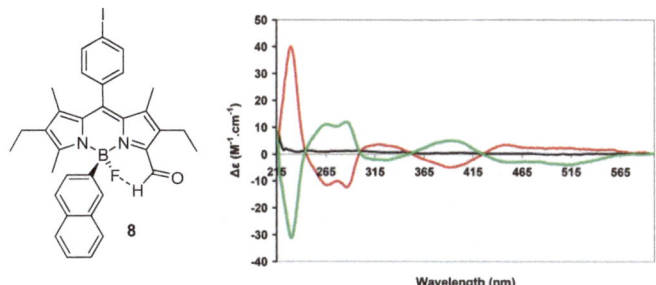

**Fig. 6.6** $B^*$-BODIPY **8** (one enantiomer shown) and its associated CD spectra (racemic **8** (black); firstly eluted (red) and secondly eluted (green) enantiomers by HPLC (Chiralcel-OD, hexane/2-propanol)) [adapted with permission from [36], copyright 2010 American Chemical Society]

naphthylmagnesium bromide and subsequent selective oxidation of the 3-methyl group with 2,3-dichloro-5,6-dicyanoquinone (DDQ) to form the corresponding aldehyde. The loss of lateral symmetry on formation of $B^*$-BODIPY **8** was evident through signal doubling in the $^1$H NMR and further supported by the observation of through space coupling of the formyl proton to the adjacent fluorine atom, arising from the formation of an intramolecular NH···F H-bond. Resolution of racemic $B^*$-BODIPY **8** was achieved by chiral HPLC (Chiralcel-OD, hexane/2-propanol) to give the corresponding enantiomers (absolute configurations were not determined). The enantiomeric nature of the thus resolved $B^*$-BODIPYs was confirmed by the presence of mirror image CD spectra. A strong Cotton effect was observed at the short-wavelength transitions associated with the carbonyl group; however only weak signals were obtained corresponding to the $S_0$-$S_1$ transition band of the BODIPY chromophore (Fig. 6.6).

## 6.2.3   BODIPYs Displaying Axial Chirality

The aromatic structure of the BODIPY chromophore lends itself well to the introduction of axial chirality, via the formation of atropisomers containing rotationally restricted aryl-aryl bonds. The formation of an atropisomeric BODIPY requires both a laterally differentiated dipyrromethene core and a conformationally restricted BODIPY-aryl bond. Hall et al. have demonstrated such an axially chiral BODIPY through the preparation of **9**. In this case lateral differentiation is achieved through introduction of a substituent at the 2-position (through Heck coupling of the parent 2-bromo-BODIPY with ethyl acrylate), whilst the required conformationally restricted BODIPY-aryl bond (methyl—methyl clash) is introduced via the an *ortho*-substituted aryl group in the *meso*-position (Fig. 6.7) [37]. The restricted rotation of the *meso*-aryl substituent could be observed via the imposition of a diastereotopic relationship on the two fluorine atoms of the chelating $BF_2$ moiety, observable in the $^{19}$F NMR spectrum (ABX coupling, each fluorine showing both

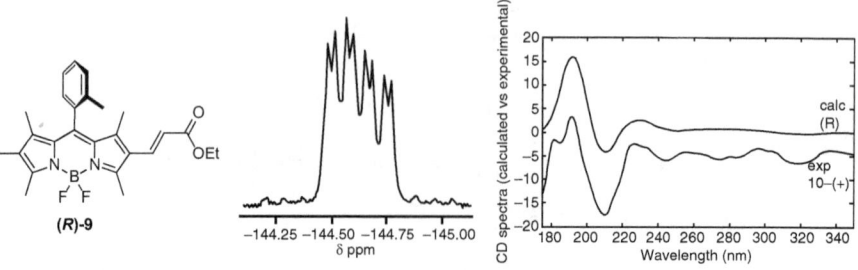

**Fig. 6.7** Axially chiral BODIPYs (*R*)-**9** (one enantiomer shown), $^{19}$F NMR spectrum of BODIPY **9** (showing ABX coupling pattern), and comparison of the experimental and calculated CD spectra for BODIPYs (*R*)-**9** [adapted with permission from [37], published by the Royal Society of Chemistry]

geminal $^{19}$F–$^{19}$F and $^{19}$F–$^{11}$B couplings, Fig. 6.7). Resolution of racemic BODIPY **9** into its two enantiomers was achieved through semi-preparative chiral HPLC (Chiralpak AD-H, heptane/2-propanol). CD spectroscopy of the resolved axially chiral enantiomers, (*R*)-**9** and (*S*)-**9**, did not however reveal any significant Cotton effect at long wavelengths (corresponding to the $S_0$–$S_1$ transitions of the BODIPY), only short-wavelength transitions were observed associated with the electronic transitions of the *meso*-aryl group (Fig. 6.7). However the CD spectra of each enantiomer was sufficient to allow absolute stereochemical assignment, through a comparison of the experimental and calculated CD spectra for the postulated (*R*)-**9** enantiomer (CD spectra were calculated via a Boltzmann-weighted averaging of the calculated CD spectra of each low energy conformer of (*R*)-**9**, obtained at via TD-DFT (cam-B3LYP/6-311++G(2d,p)).

Akkaya et al. employed a similar design principle in the construction of atropisomeric bis(BODIPY) **10** and tris(BODIPY) **11** (Fig. 6.8) [38]. In the case of axially chiral bis(BODIPY) **10**, one of the involved BODIPY chromophores is laterally differentiated through the introduction of a single 2-formyl group to the non-formylated parent bis(BODIPY) through a Vilsmeier-Haack formylation, whilst its *meso*-position is occupied by a second 2-linked BODIPY moiety. Rotation around the BODIPY-BODIPY bond is thus restricted by a double methyl−methyl clash, resulting in atropisomeric bis(BODIPY) **10**. Further elaboration, through the introduction of a third *meso*-linked BODIPY at the 2-position of the central

**Fig. 6.8** Atropisomeric bis (BODIPY) **10** and tris (BODIPY) **11** (one enantiomer shown for each)

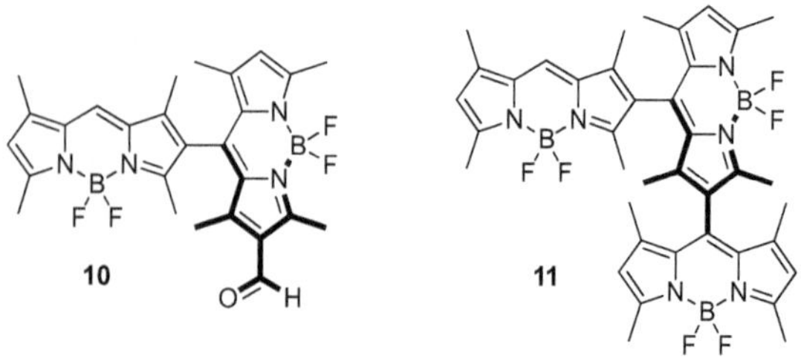

**10**

**11**

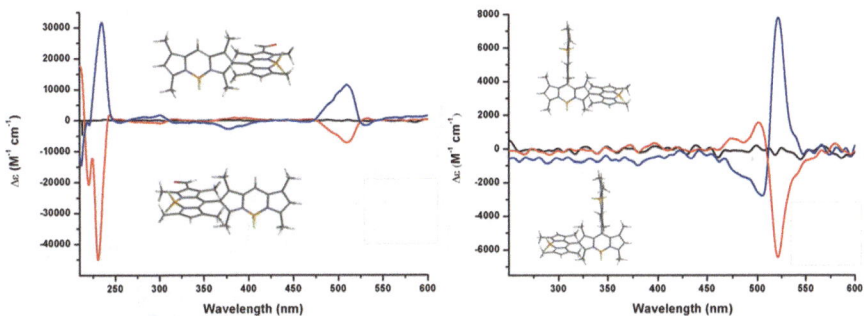

**Fig. 6.9**  CD spectra of resolved enantiomers (red and blue lines) of both bis(BODIPY) **10** and tris (BODIPY) **11** [adapted with permission from [38], copyright 2014 American Chemical Society]

chromophore, maintains the atropisomeric nature of the system resulting in tris (BODIPY) **11** (Fig. 6.8).

Resolution of the racemic mixtures of these axially chiral BODIPYs was again achieved through chromatography on a chiral solid support (Chiralcel-OD, heptane/ 2-propanol or ethanol) and their chiroptical properties studied. The enantiomers of bis(BODIPY) **10** and tris(BODIPY) **11** both gave mirror image CD spectra with large anisotropy factors (bis(BODIPY) **10**, $g_{abs} \approx 1 \cdot 10^{-1}$; tris(BODIPY) **11**, $g_{abs} \approx 5 \cdot 10^{-2}$ (estimated from the reported CD spectra and extinction coefficients) for the CD maxima relating to the $S_0$-$S_1$ transition of the two BODIPY systems. Bis (BODIPY) **10** gave a single long-wavelength signal corresponding to the $S_0$-$S_1$ transition, whilst tris(BODIPY) **11** gave a long-wavelength asymmetric couplet (Fig. 6.9).

A related set of atropisomeric bis(BODIPY)s (**12** and **13**), alongside an atropisomeric bis(aza-BODIPY) (**14**), were prepared by Bruhn et al. based around a direct 2,2′-linkage (Fig. 6.10) [39]. In these cases the lateral differentiation is achieved for both BODIPY/aza-BODIPY chromophores via the 2,2′-linkage, whilst rotational restriction is achieved through either a mutual double methyl−methyl clash (**12** and **13**) or mutual double phenyl−phenyl clash (**14**). Atropisomeric bis

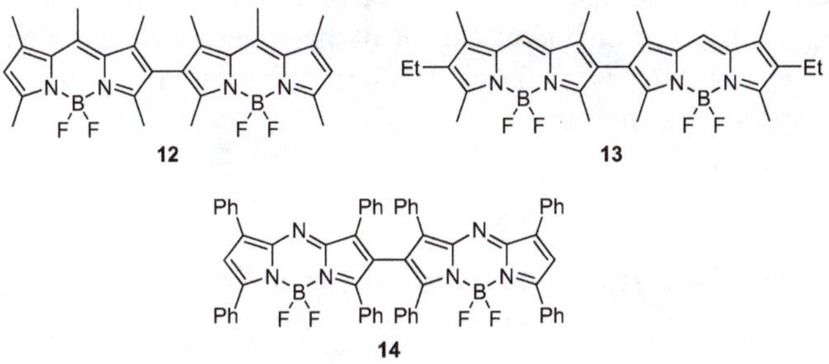

**Fig. 6.10**  Cryptochiral atropisomeric bis(BODIPY)s **12** and **13** and cryptochiral atropisomeric bis (azaBODIPY) **14**

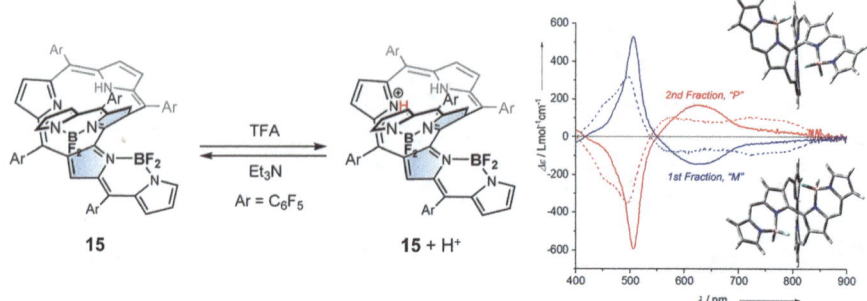

**Fig. 6.11** Atropisomeric bis(BODIPY) **15**, including acid/base induced interconversion (left). CD spectra of atropisomeric bis(BODIPY) **15** in $CH_2Cl_2$ in the absence (solid line) and presence (broken line) of TFA (right) [adapted with permission from [40], copyright 2018 WILEY-VCH Verlag GmbH]

(BODIPY)s **12** and **13** and bis(azaBODIPY) **14** were synthesized as racemates through oxidative coupling of the corresponding monomers, with subsequent resolution achieved via chiral HPLC (Chirex, dichloromethane/*n*-hexane (**12** and **13**) or dichloromethane/2-propanol/*n*-hexane (**14**)). Interestingly, despite the axial chirality imbued to these systems, the CD spectra of the resolved enantiomers of **12–14** showed only very low intensity signals between 300 and 800 nm, an interesting example of cryptochirality (i.e. two enantiomers for which the specific rotation is non-measurable). This cryptochirality is thought to arise in these systems due to the existence of two different conformations, for each compound **12–14**, around the biaryl axis which exhibit nearly mirror-image CD spectra, resulting in weak experimental CD spectra dominated by vibronic coupling effects.

An alternative approach to the construction of atropisomeric bis(BODIPY) **15** was subsequently demonstrated by Ishida et al. in which they constructed a 3,3'-linked bis (BODIPY) unit within the architecture of a π-extended corrorin (Fig. 6.11) [40]. This atropisomeric bis(BODIPY) **15** could be resolved by chiral HPLC to give the corresponding *P* and *M* isomers, the CD spectra of each enantiomer showing a strong Cotton couplet ($g_{abs} = 0.012$ at $\lambda = 501$ nm). Interestingly the CD spectra showed a reversible change in peak shape upon addition of a strong protic acid (trifluoroacetic acid), arising from a conformational change of the molecule upon protonation of the inner dipyrrin ring, suggesting possible future applications of these systems in chiroptical sensing (Fig. 6.11).

### 6.2.4 BODIPYs Displaying Helical Chirality

During their studies into new red-shifted BODIPYs, Burgess et al. investigated the synthesis and optical properties of 3,5-diaryl-substituted BODIPY **16**, in which the oxygen atoms of the 3,5-*ortho*-phenolic groups chelate the central boron atom

**Fig. 6.12** Helically chiral
*N,N,O,O*-boron-chelated
BODIPY **16** (one
enantiomer shown) [adapted
with permission from [41],
published by the Royal
Society of Chemistry

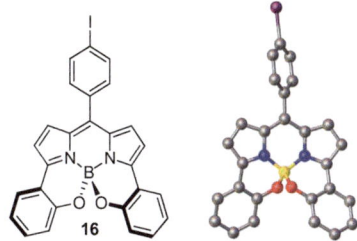

(Fig. 6.12) [41]. The required tetrahedral geometry of the central boron atom ensures that the two aryl groups must twist to direct the two B-O bonds to opposite faces of the BODIPY chromophore, inducing an overall helical chirality to the molecule (Fig. 6.12). Although helically chiral BODIPY **16** was synthesized as a racemate, Burgess et al. showed that the involved enantiomers can be detected by analytical chiral HPLC (Pirkle covalent (*S,S*) whelk-01; 2-propanol/hexane). Unfortunately, single enantiomers of BODIPY **16** could not be isolated and no further analysis of their chiroptical properties was undertaken.

Several related helical *N,N,O,O*-boron-chelated BODIPYs and azaBODIPYs were subsequently reported by a number of groups; however in each case researchers focused on the red-shifting effect of such a substitution pattern rather than their potential as CPL-SOMs [42–44].

Interestingly, only a single chelating *ortho*-phenolic moiety in the 3-position is required in order to induce helicity in such BODIPY architectures. Thus, helically chiral *N,N,O*-chelated BODIPYs such as **17** were first synthesized by Nabeshima et al. through the chelation of the corresponding 3,5-diarylBODIPY with an appropriate arylboronic acid (Fig. 6.13) [45]. [1]H NMR analysis of *N,N,O*-boron-chelated BODIPY **17** in the presence of a chiral shift reagent (europium tris [3-(heptafluoropropylhydroxymethylene)-(+)-camphorate]) showed splitting of the methyl signal at 2.14 ppm into two, due to the formation of transient diastereotopic complexes in solution, indicating stable helical chirality on the NMR timescale.

Related helically chiral *N,N,O*-boron-chelated BODIPYs have been reported by the groups of both Harriman and Jiao, albeit with limited discussion of the potential for CPL [46, 47]. Whilst in a more recent example, Nabeshima et al. reported the

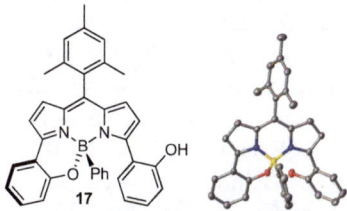

**Fig. 6.13** Helically chiral *N,N,O*-boron-chelated BODIPY **17** (one enantiomer shown) [adapted with permission from [45], copyright 2009 Elsevier]

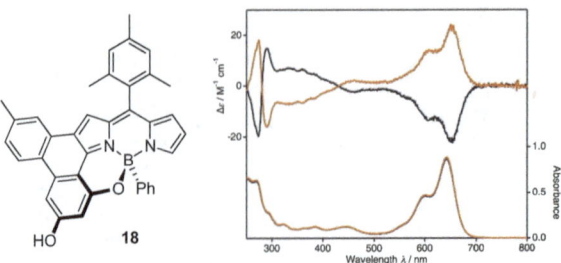

**Fig. 6.14** Helically chiral *N,N,O*-boron-chelated BODIPY (one enantiomer shown). CD spectra (top) and UV-Vis spectra (bottom) of (*S*)-**18** (black) and (*R*)-**18** (orange) [adapted with permission from [48], copyright 2018 Elsevier]

synthesis of helically chiral *N,N,O*-boron-chelated BODIPY **18** and subsequent resolution by chiral HPLC (Daicel Chiralpak IA; CHCl$_3$/hexane 1:1). The resolved enantiomers showed mirror image CD spectra ($\lambda_{max} = 655$ nm); however no CPL studies on these compounds have been disclosed to date (Fig. 6.14) [48].

## 6.3    CPL BODIPYs

### 6.3.1    Mono(BODIPY)S

#### 6.3.1.1    $C_2$-Symmetric Mono(BODIPY)S

The first example of a CPL-SOM based on BODIPY was a chiral $C_2$-symmetric mono(BODIPY) published by the group of Gossauer et al. in their study of the urobilins, an important family of natural-product chromophores in algal and bacterial photosynthesis [49]. As part of their work examining the conformation of synthetic urobilin analogues in solution, a number of urobilin-based difluoroboron chelates were prepared. In the case of the non-racemisable, difluoroboron-chelated urobilin derivative **19** (Fig. 6.15), study of the CD spectra suggests that the, albeit peripheral, asymmetric carbon centres can effectively induce chirality to the chromophore (the observed long-wavelength CD transition correlating with that of the visible absorption maxima at approximately 535 nm). Interestingly, comparison of the CD spectra of difluoroboron-chelated urobilin derivative **19** in either CH$_2$Cl$_2$ or DMF shows an inversion of sign of the long-wavelength CD transition. This suggests an inversion of conformation of **19** in solution, presumably as a consequence of competition between intramolecular NH···F bonding in CH$_2$Cl$_2$ and intermolecular NH···F bonding in DMF (Fig. 6.15). Difluoroboron-chelated urobilin derivative **19** in CH$_2$Cl$_2$ solution was also reported to exhibit CPL at 546 nm with a $g_{lum}$ of $+0.94 \cdot 10^{-3}$, following indirect excitation (366 nm) of the pendant 1,5-dihydro-2*H*-pyrrol-2-one groups and subsequent energy transfer to the BODIPY chromophore [32].

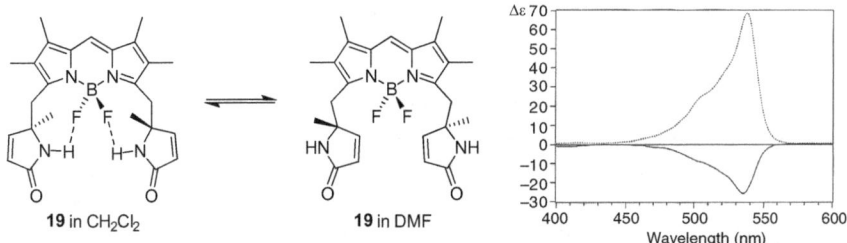

**Fig. 6.15** Molecular structure of difluoroboron-chelated urobilin derivative **19** showing postulated conformational change on solvation (left $CH_2Cl_2$; right DMF) and CD spectra (in $CH_2Cl_2$ (broken line) and DMF (solid line)) [adapted with permission from [49], copyright 1997 American Chemical Society]

### Spiranic *O*-BODIPYs

The group of de la Moya demonstrated that simple fluorine-substitution reactions of accessible *F*-BODIPYs with enantiopure 1,1′-binapth-2-ol (BINOL) allowed the easy preparation of spiranic $C_2$-symmetric mono(BODIPY)s endowed with CPL activity, as exemplified by the straightforward preparation (ca. 60% chemical yield) and CPL behaviour of enantiomers **20** (Fig. 6.16) [50]. Thus, (*R*)-**20** and (*S*)-**20** exhibit visible spectrum CPL (BODIPY emission) upon irradiation with visible light (BODIPY excitation) in CHCl$_3$ solution, the recorded CPL spectra appearing as mirror images with a maximum |$g_{lum}$| value ca. $10^{-3}$ (positive $g_{lum}$ for the *R* enantiomer). The CPL maxima observed match those of the visible light emission maxima of the involved BODIPY chromophore ($\lambda_{max}$ ca. 570 nm; $\phi$ ca. 0.45), clearly indicating that the observed CPL arises from BODIPY not BINOL emission. Thus spiranic $C_2$-symmetric mono(BODIPY)s (*R*)-**20** and (*S*)-**20** represent the first reported examples of CPL from a BODIPY chromophore following direct excitation.

The CPL-enabling design of **20** is interesting since it involves the use of a single chiral moiety (the 1,1′-biphen-2-ol-based BINOL unit) to chirally perturb the

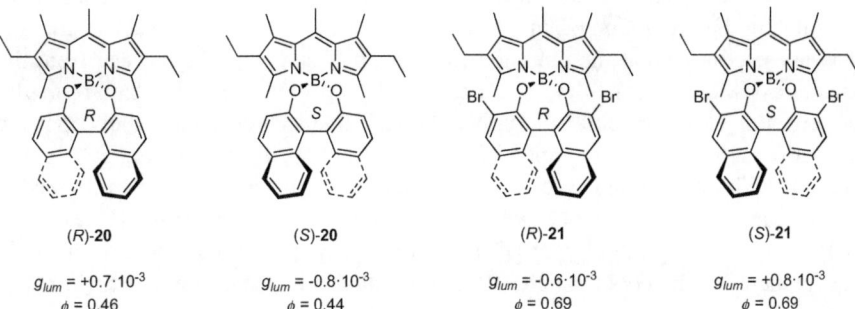

| (*R*)-**20** | (*S*)-**20** | (*R*)-**21** | (*S*)-**21** |
|---|---|---|---|
| $g_{lum}$ = +0.7·10$^{-3}$ | $g_{lum}$ = -0.8·10$^{-3}$ | $g_{lum}$ = -0.6·10$^{-3}$ | $g_{lum}$ = +0.8·10$^{-3}$ |
| $\phi$ = 0.46 | $\phi$ = 0.44 | $\phi$ = 0.69 | $\phi$ = 0.69 |

**Fig. 6.16** Molecular structures and visible CPL signatures (upon visible light irradiation in solution) of the first CPL-SOMs based on spiranic $C_2$-symmetric mono(BODIPY)s **20** and **21**

inherently achiral BODIPY chromophore. The success of such a perturbation is attributed to the following key factors: (1) axial chirality and $C_2$ symmetry provided by such a chiral moiety; (2) an orthogonally fixed arrangement of the perturbing moiety with respect to the perturbed one, which is imposed by the involved spiranic geometry; and (3) electronic isolation of both chromophores, due to the aforementioned orthogonal arrangement and the involvement of boron at the spiranic junction [25]. Additionally, synthetic access to this class of BODIPY based CPL-SOMs is straightforward, allowing the direct preparation of enantiopure CPL-enabling BODIPYs from two readily accessible and oft commercial available components: (1) achiral $F$-BODIPY dyes and (2) enantiopure 1,1′-biphen-2-ols such as BINOL and its derivatives.

The fluorescence behaviour of these 1,1′-biphen-2-ol-based spiranic $O$-BODIPYs can be easily tuned by properly modulating electronic factors in both the BODIPY moiety and the oxygenated biphen-2-ol connected to it, as de la Moya et al. have demonstrated recently [51]. This approach adds significant value to the use of such a $C_2$-symmetric mono(BODIPY) design when developing CPL-SOMs. As an example, the said photophysical tuning has allowed the successful preparation of **21** (Fig. 6.16) exhibiting enhanced CPL activity ($\phi \cdot |g_{lum}|$) when compared to its analogue **20** [52]. This enhancement is achieved by simply introducing 3,3′-dibromoBINOL instead of BINOL in the $C_2$-symmetric mono(BODIPY) structure. Strikingly, the sign of the $g_{lum}$ values exhibited by **20** and **21**, having identical chiral absolute configuration, are opposite (see Fig. 6.16). Besides, each individual enantiomer of **20** shows opposite signs for its maximum $g_{abs}$ and $g_{lum}$ values [50], whereas the enantiomers of **21** do not show such a striking $g_{abs}$-versus-$g_{lum}$ sign reversal [52]. Interestingly, the significant CPL activity of mono(BODIPY) **21** joined to its lasing capability under laser pumping has also served to explore factors affecting CPL activity in dye lasers [52].

The aforementioned differential behaviour of **20** and **21** in relation to the sign of the corresponding $g_{lum}$ values [50, 52], joined to the fact that both mono(BODIPY)s involve highly similar BODIPY chromophores, but with very different capability to populate emissive intramolecular charge transfer (ICT) states (higher for BINOL-based **20**; undetectable for 3,3′-dibromoBINOL-based **21**) [51], served de la Moya's group to establish a new strategy to manipulate the CPL sign in chiral emitters (i.e. to change the handedness of the observed circularly polarized emission) by means of modulating the promotion of luminescent ICT states [53]. To apply such a strategy, it is not necessary that the emission from the ICT state surpass the one from the locally excited state, since the ICT emission is expected to involve larger circular polarization. However, both circularly polarized emissions do need to show opposite handedness [53]. This is an unprecedented way to manipulate the CPL sign in chiral emitters without changing absolute configurations, being specifically useful for rigid emitters and emitters without capability to promote excimer emission [53]. The new strategy was exemplified by de la Moya et al. by means of a set of CPL-enabling 1,1′-biphen-2-ol-based spiranic $O$-BODIPYs **20–25**, where the BODIPY (in red in Fig. 6.17) provides the functional visible light chromophore and the 1,1′-biphen-2-ol moiety (in blue in Fig. 6.17) acts as the ICT-switching moiety depending on the electronic nature of the BODIPY chromophore connected to it. Thus, in dyes with

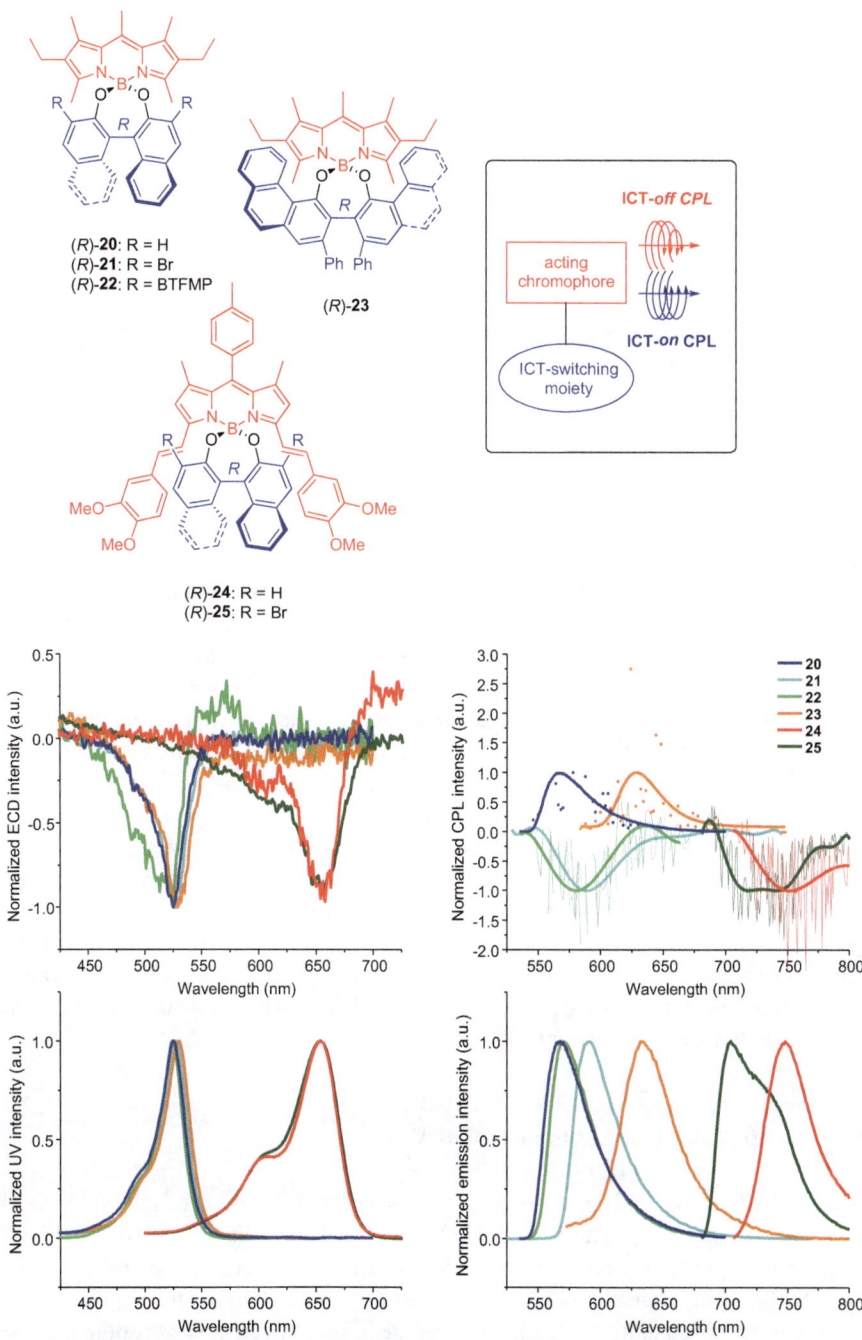

**Fig. 6.17** Manipulation of the CPL sign in a set of chiral 1,1′-biphen-2-ol-based spiranic *O*-BODIPYs by modulating ICT emission by means of selecting electronic factors in both the BODIPY core (highlighted in red in the molecular structures) and the chelating 1,1′-biphen-2-ol (highlighted in blue in the molecular structures). BTFM: bis(3,5-trifluoromethyl)phenyl [adapted with permission from [52], published by the Royal Society of Chemistry]

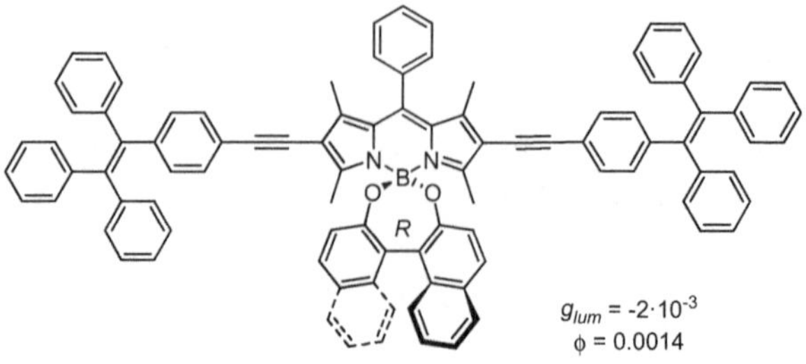

$g_{lum} = -2 \cdot 10^{-3}$
$\phi = 0.0014$

(R)-26

**Fig. 6.18** Energy transfer in a CPL mono(BODIPY) based on the spiranic *O*-BODIPY design. Visible CPL is achieved by UV irradiation in dichloromethane

enabled ICT (**20** and **23**), the CPL spectrum is inverted with respect to the CD one (i.e. sign reversal for the corresponding visible light maximum-$g_{abs}$ and -$g_{lum}$ values). However, in the cases for which ICT is not significantly promoted (**21–22, 24,** and **25**), the signs of the corresponding $g_{abs}$ and $g_{lum}$ values are equal (Fig. 6.17).

Although the maximum |$g_{lum}$| values reached for these spiranic $C_2$-symmetric mono(BODIPYs) are of the order $10^{-3}$, the aforementioned possibility of modulating their fluorescent signatures, including their CPL sign, has made this design approach an exciting avenue for the development of highly functional CPL-SOMs. Thus Cheng et al. have taken advantage of this in the creation of an interesting enantiopure mono(BODIPY) dye capable of visible light CPL emission following ultraviolet (UV) irradiation (340 nm) (**26** in Fig. 6.18; $\lambda_{max}$ ca. 630 nm; maximum |$g_{lum}$| = $2 \cdot 10^{-3}$, positive $g_{lum}$ for the *S* enantiomer) due to an efficient donor-acceptor energy transfer resulting in a large pseudo-Stokes shifted CPL emission [54]. Besides which CPL-enabling mono(BODIPY) **26** has been shown to aggregate in specific solvent mixtures (dichloromethane and hexane) of low polarity, resulting in aggregation induced fluorescence (AIE), from $\phi$ = 0.014 in pure dichloromethane (emission from monomeric **26**) to $\phi = 0.582$ in a less polar 9:1 mixture of hexane and dichloromethane (emission from aggregated **26**), without disturbing the CPL activity (in terms of $g_{lum}$ value) [54].

### 6.3.1.2  Helically Chiral Mono(BODIPY)S

Inspired by the work of Burgess et al. [41], the group of Knight and Hall synthesized a series of helically chiral *N,N,O,O*-boron-chelated dipyrromethenes **27–30** (Fig. 6.19), with a range of *meso*-substituents, for evaluation as potential CPL-SOMs [30]. Helical chirality is induced in these BODIPYs through chelation of the central boron atom by the oxygens of the 3,5-*ortho*-phenolic substituents. The *meso*-substituents are electronically discrete from the BODIPY chromophore itself (Fig. 6.19), and as such their direct electronic influence on the chromophore is typically minimal; however they can sterically influence the conformational

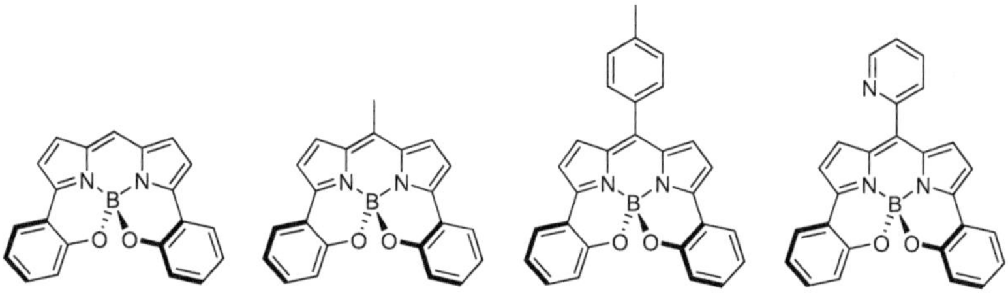

**27** $|g_{lum}| = 4.7{\cdot}10^{-3}$ $\phi = 0.65$

**28** $|g_{lum}| = 3.3{\cdot}10^{-3}$ $\phi = 0.73$

**29** $|g_{lum}| = 4.3{\cdot}10^{-3}$ $\phi = 0.52$

**30** $|g_{lum}| = 4.2{\cdot}10^{-3}$ $\phi = 0.28$

**Fig. 6.19** Helically chiral *N,N,O,O*-boron-chelated dipyrromethenes **27–30** and their visible CPL signatures upon visible light irradiation in solution (one enantiomer shown in each case)

flexibility of the BODIPY core resulting in subtitle changes to the observed photophysical properties [55].

Helically chiral mono(BODIPY)s **27–30** were synthesized as racemates and were subsequently resolved by semi-preparative chiral HPLC (Chiralpak OB, toluene/*n*-hexane 9:1 (**27**); Chiralcel OD-H, *i*-PrOH/*n*-hexane 1:4 (**28–30**)). Measurement of the CD spectra of **27–30** gave mirror image spectra for the resolved *P* and *M* enantiomers, with a strong Cotton effect apparent, corresponding to the $S_0$–$S_1$ BODIPY transition. Assignment of absolute stereochemistry was performed through comparison of the experimental and calculated CD spectra (TD-DFT). *P* and *M* helically-chiral mono(BODIPY)s **27–30** showed red-shifted CPL upon irradiation, the CPL spectra giving maxima ($|g_{\mathrm{lum}}| = 4.7{\cdot}10^{-3}$, $3.3{\cdot}10^{-3}$, $4.3{\cdot}10^{-3}$ and $4.2{\cdot}10^{-3}$ for **27–30,** respectively) corresponding to the BODIPY emission maxima of each fluorophore. Helically chiral mono(BODIPY) **27** gave a marginally larger $|g_{\mathrm{lum}}|$ than **28–30**, suggesting a subtle influence of the *meso*-substituent on the conformation and thus chiroptical properties of these systems. X-ray crystal structures were obtained for helically-chiral mono(BODIPY)s **27–30** (albeit via crystals containing a racemic mixture of the corresponding mono(BODIPY)). In all cases, a significant twisting of the fluorophore was observed; the twist angle between the planes as defined by the two pyrrolic rings being 11.2°, 9.0° and 9.8° for **27**, **28** and **29,** respectively. This provides some support for Gossauer's proposition that for a BODIPY to be CPL active, a twisting deformation of the planar fluorophore is required, although crystallographically determined twist angles can be influenced by crystal packing effects and would require confirmation through computational prediction of solution conformations.

Subsequently, Nabeshima et al. examined the synthesis of a π-skewed helically chiral mono(BODIPY) **31** (Fig. 6.20), formed through an oxidative annulation of the corresponding 2,6-bis(biphenyl)BODIPY precursor [56]. Following separation from the *R,S meso* form, the remaining *R,R* and *S,S* enantiomers of mono(BODIPY) **31** were separated by chiral HPLC (Chiralpak IA, chloroform/*n*-hexane 1:1). As previously, the twisting deformation of the fluorophore core was evaluated through examination of the X-ray crystal structure, which displayed a twist angle between

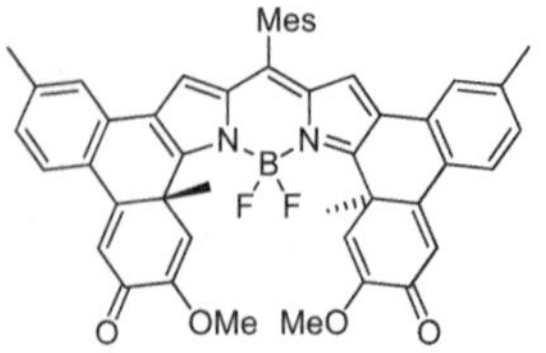

**Fig. 6.20** Helically chiral ring-fused skewed mono(BODIPY) **31** and visible CPL signatures upon visible irradiation in solution (one enantiomer shown)

the planes of the two pyrrole rings of 7.3°. The CD spectra of the (R,R)- and (S,S)-enantiomers of **31** showed opposing Cotton effects with maxima in the red region of the visible spectra corresponding to the $S_0$–$S_1$ transition of the BODIPY fluorophore. The absolute stereochemistry of the (R,R)-enantiomer of **31**, which showed a negative Cotton effect in CD, was assigned by comparison of the experimental with the corresponding TD-DFT calculated CD spectra. The CPL spectra for the resolved enantiomers of **31** also showed mirror image signals with $|g_{lum}| = 6 \cdot 10^{-4}$ (Fig. 6.20).

More recently, Hall et al. described the serendipitous discovery of a new class of helically chiral mono(BODIPY) [57]. Whilst exploring improved synthetic procedures for the synthesis of N,N,O,O-boron-chelated dipyrromethenes, the unexpected formation of red-shifted unsymmetrical helically chiral mono(BODIPY) **32** was observed, thought to arise from a boron metathesis, nucleophilic aromatic substitution ($S_NAr$), Suzuki coupling and boron-chelation reaction sequence (Fig. 6.21).

Resolution by semi-preparative chiral HPLC (Chiralpak IB, ethyl acetate/n-hexane 15:85) gave both the dextro- and levorotatory enantiomers of the unsymmetrical helically chiral mono(BODIPY) **32**, which in turn gave mirror-image CD spectra with good correlation with the corresponding absorption spectra (Fig. 6.22). Again absolute stereochemistry was assigned through comparison of the calculated (TD-DFT at the cam-B3LYP/6-311++G(3df,2pd) level) and the experimental

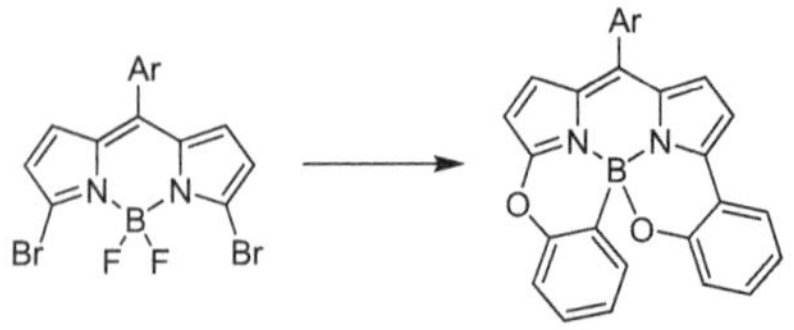

**Fig. 6.21** Synthesis of helically chiral mono(BODIPY) **32**. Reaction conditions: (2-hydroxyphenyl)boronic acid, $Na_2CO_3$, $Pd(PPh_3)$ (5 mol%), toluene/1,4-dioxane, 90 °C, 80 min. Chemical yield ca. 35%. Ar: $p$-$C_6H_4CO_2CH_3$

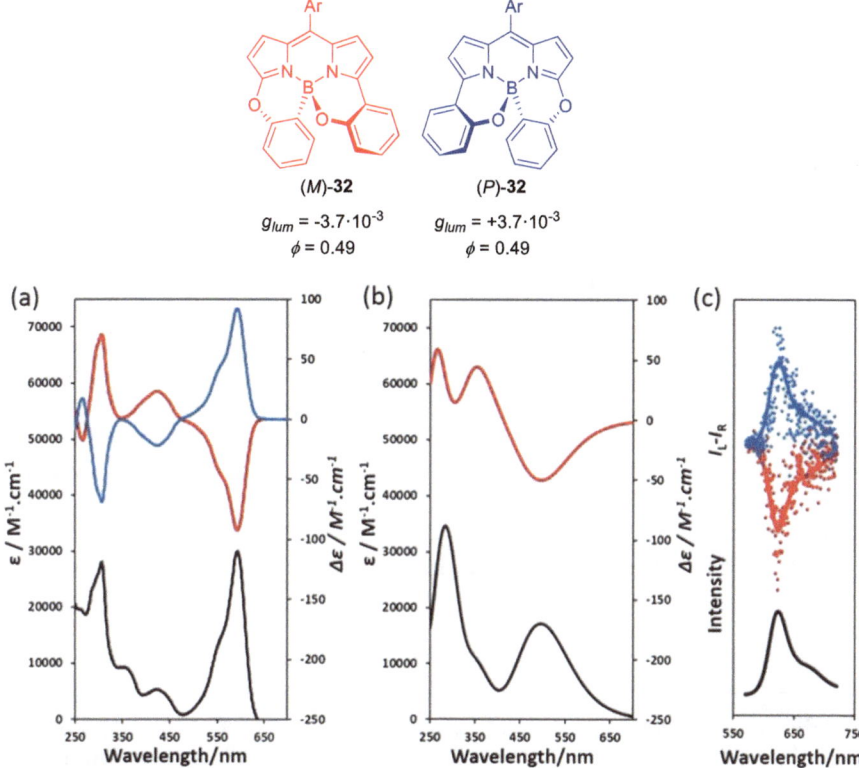

**Fig. 6.22** Unsymmetrical helically chiral mono(BODIPY) **32** and their visible CPL activity upon visible irradiation in solution. (**a**) Experimental CD (red = (*M*)-**32**, blue = (*P*)-**32**) and UV-Vis absorption spectra (black). (**b**) Calculated Boltzmann-weighted spectra, CD (red = (*M*)-**32**) and UV-Vis absorption spectra (black). (**c**) CPL (red = (*M*)-**32**, blue = (*P*)-**32**) and fluorescence spectra (black) [adapted with permission from [57], copyright 2017 WILEY-VCH Verlag GmbH]

CD spectra. Both enantiomers of mono(BODIPY) **32** were also CPL active, with $|g_{lum}| = 3.7 \cdot 10^{-3}$ (Fig. 6.22) [57].

Cyclic voltammetry of racemic **32** suggested that the process of symmetry breaking had resulted in an increase in charge transfer character in the excited state [57]. This is of interest in the design of CPL-SOMs as the typically low $|g_{lum}|$ values observed arise as a consequence of the mismatch in the magnitude of the magnetic (***m***) and electric (***μ***) transition dipole moments, since $|g_{lum}| = 4(|\boldsymbol{\mu}| \cdot |\boldsymbol{m}| \cdot \cos\tau)/(|\boldsymbol{\mu}|^2 + |\boldsymbol{m}|^2)$ where $\tau$ is the angle between the two transition dipole moments. In CPL-SOMs the electric transition dipole moment is commonly several orders of magnitude greater than the corresponding magnetic transition dipole moment. However, an increase in charge transfer character in the excited state would be expected to decrease the electric transition dipole moment. Calculation of these parameters for both the unsymmetrical helically chiral mono(BODIPY) **32** and the comparable symmetrical helically chiral mono(BODIPY) **27** did indeed show a slight decrease in $\mu$ (from 7.6 to $6.0 \cdot 10^{-1}$); however a concurrent decrease in $m$ (from $1.4 \cdot 10^{-3}$ to $7.3 \cdot 10^{-4}$) coupled with an unfavourable shift in $\tau$ (from 65 to 70°) resulted in an overall decrease in $|g_{lum}|$ for **32** vs. **27**. However examination of this system may provide useful lessons for the rational control of $m$, $\mu$, and $\tau$ as an important design paradigm for future CPL-SOMs.

Propeller-like PolyarylBODIPYs

Propeller chirality has been also explored to gain CPL activity in BODIPYs. Thus, Mori et al. have recently reported a heptaarylBODIPY **33** having *quasi* propeller chirality (Fig. 6.23) [58]. Although the arylic propeller blades have pendant chiral groups, almost-complete one-directional propeller chirality is unfortunately not achieved, even by using non-polar solvents (e.g. cyclohexane) and low temperatures [58]. This fact is attributed to the computed small energy difference between the corresponding propeller diastereomers ($0.8$ kcal·mol$^{-1}$), in combination with a computed barrier for the *P*-to-*M* propeller flipping of 3.8 kcal·mol$^{-1}$ [58]. Nonetheless, decreasing the temperature below $-70$ °C results in the detection of a visible (above 600 nm) clearly bisignalized dichroic CD signal corresponding to the emission maxima of the BODIPY chromophore and attributed to the formation of head-to-tail propeller dimers which induce an efficient chiral perturbation of the involved BODIPY chromophores. Interestingly, such an efficient perturbation also allows the detection of a visible CPL signal from the said BODIPY dimers (maximum $g_{lum} = +2.0 \cdot 10^{-3}$ at $-120$ °C) combined with a high fluorescence quantum yield ($\phi = 0.45$) [58].

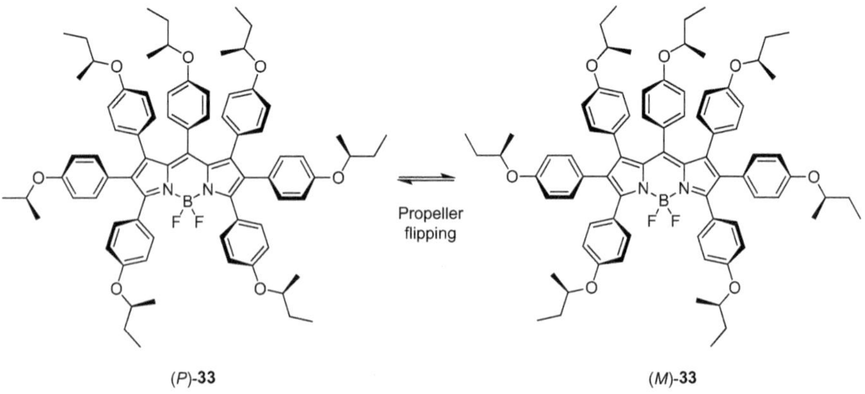

(*P*)-**33**                                                    (*M*)-**33**

**Fig. 6.23** BODIPY **33** with propeller-like chirality. At low temperature, efficient chiral perturbation of the BODIPY chromophore, detected by CD and CPL signalization, is attributed to the formation of head-to-tail dimers

## 6.3.2   Poly(BODIPY)S

### 6.3.2.1   Axially Chiral Bis(BODIPY)S

Bruhn et al. examined the synthesis of axially chiral 1,1′- and 3,3′-linked bis (BODIPY)s **34** and **35**, through the oxidative C–C coupling of the corresponding 1- and 3- unsubstituted mono(BODIPY)s [59, 60]. Steric crowding around the bis (BODIPY) axis of both **34** and **35** prevents free rotation resulting in configuration-ally stable atropisomers (Fig. 6.24).

Axially chiral 1,1′- and 3,3′-linked bis(BODIPY)s **34** and **35** were resolved by chiral HPLC (Chirex (S)-Val, CH$_2$Cl$_2$/n-hexane 3:7 (**34**); Chiralpak IA, CH$_2$Cl$_2$/n-hexane 8:2 (**35**)) and showed near mirror-image CD spectra. 3,3′-linked bis (BODIPY) **35** gave an intense CD couplet arising from a strong coupling of the transition dipole moments of the two BODIPY fluorophores, whilst also giving a maximum $g_{lum}$ of $3.8 \cdot 10^{-3}$ (Fig. 6.25). Interestingly, **34** is one of the rare examples in which exciton rule fails due to non-negligible **μ-m** coupling, resulting in an inverted and weaker CD couplet with respect to that of compound **35** (with same absolute configuration) as well as CPL. The demonstration of CPL from these directly linked atropisomeric BODIPYs provided some evidence that the coupling of the transition dipole moments of two (or more) BODIPY cores may be an interesting strategy for the creation of future efficient CPL-SOMs.

### 6.3.2.2   Helically Chiral Poly(BODIPYs)

Helically Labile Bis(BODIPYs)

The group of de la Moya has shown that nucleophilic aromatic substitution reaction of accessible 3,5-dichloroBODIPY with commercial 1,2-diphenyl-1,2-ethanediamine provides facile access to bis(haloBODIPY) **36** (Fig. 6.26) where two identical BODIPY cores are covalently linked by a conformationally flexible chiral centre containing bridge [61]. The synthesis of bis(haloBODIPY) **36** is straightforward (ca. 50% chemical yield) and allows the direct preparation of enantiopure chiral bis(BODIPY)s via the corresponding enantiopure diamines.

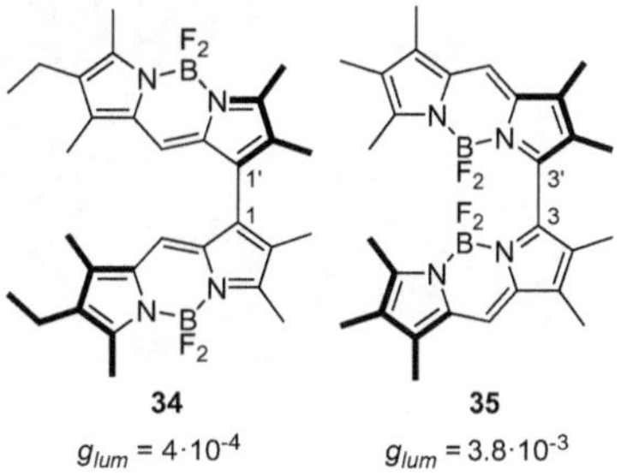

**Fig. 6.24** Axially chiral 1,1′-linked bis(BODIPY) **34** and axially chiral 3,3′-linked bis(BODIPY) **35** (one enantiomer shown in each case), as well as their visible CPL signatures upon visible light irradiation in solution (no data on fluorescence efficiency available)

34
$g_{lum} = 4 \cdot 10^{-4}$

35
$g_{lum} = 3.8 \cdot 10^{-3}$

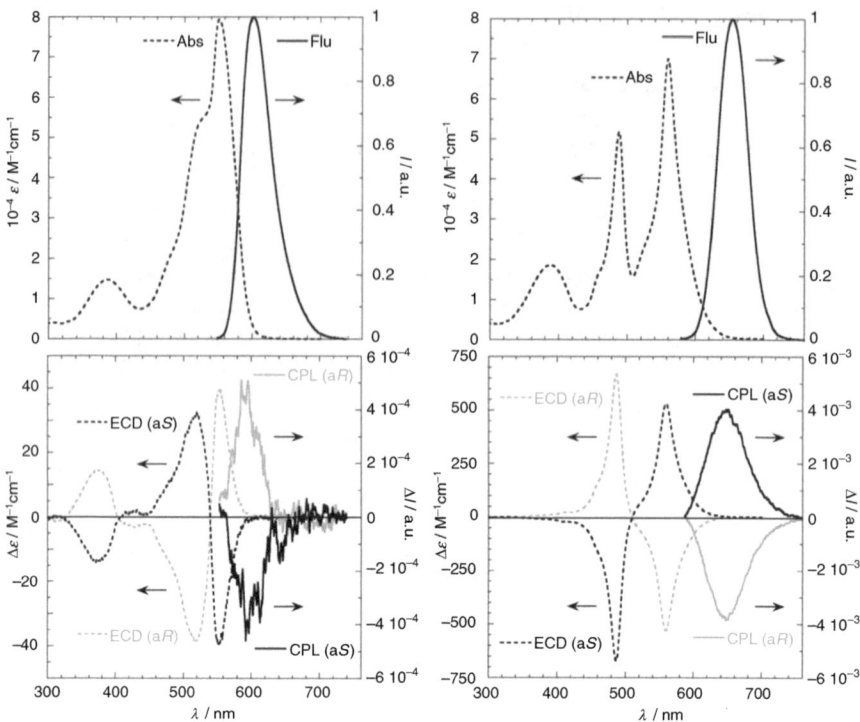

**Fig. 6.25** Experimental UV-Vis absorption and fluorescence spectra (top), CD and CPL spectra (bottom) for axially chiral 1,1′-linked bis(BODIPY) **34** (left), and axially chiral 3,3′-linked bis (BODIPY) **35** (right) [reproduced with permission from [60], copyright 2016 WILEY-VCH Verlag GmbH]

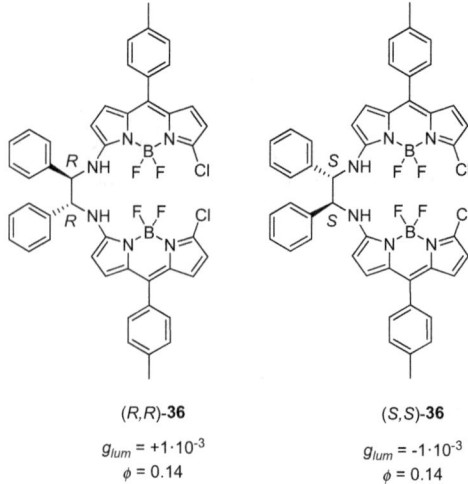

$$(R,R)\text{-}\mathbf{36}$$

$$g_{lum} = +1 \cdot 10^{-3}$$
$$\phi = 0.14$$

$$(S,S)\text{-}\mathbf{36}$$

$$g_{lum} = -1 \cdot 10^{-3}$$
$$\phi = 0.14$$

**Fig. 6.26** Helically labile bis(haloBODIPY)s (*R,R*)- and (*S,S*)-**36** and their visible CPL signatures upon visible light irradiation in solution

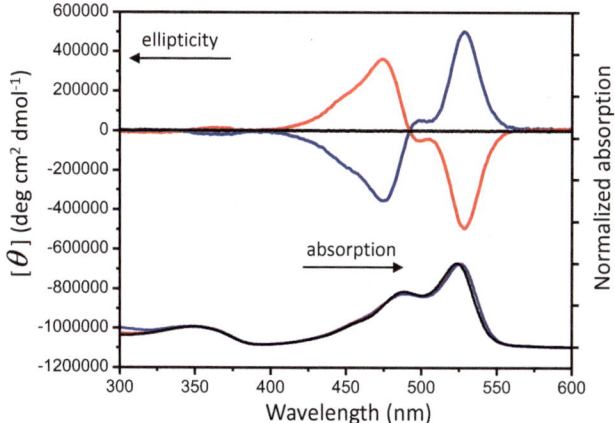

**Fig. 6.27**  UV-Vis absorption (down) and CD (up) spectra of bis(BODIPY)s (*R,R*)-**36** (red), (*S,S*)-**36** (blue), and *meso*-**36** (black) in CHCl$_3$. A visible dichroic signal with strongly bisignalized Cotton effect is observed for each enantiomer [adapted with permission from [61], published by the Royal Society of Chemistry]

Notably enantiopure bis(BODIPY)s of this type exhibit a strong and clearly bisignated visible wavelength CD signals in solution (Fig. 6.27) [61].

The observed bisignalized dichroism is attributed to the formation of a molecular helix, with a preferred helical conformation (Fig. 6.28), allowing near-degenerate visible wavelength absorption transitions, due to the localization of two near identical

**Fig. 6.28** Computed helical structure of (*R,R*)-**36**, showing preferred *P* configuration for the chiral helix and negative disposition for the involved exciton couplet

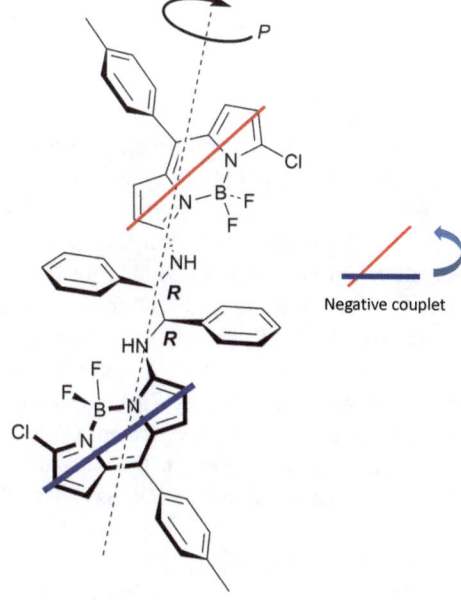

BODIPY chromophores at the either end of said chiral helix [61]. In fact, the observed bisignalization can be easily rationalized by the exciton coupling theory. Thus, if both BODIPY units are connected by the $R,R$ bridge, the preferred helical configuration is anticipated to be $P$, because such a configuration places the exciton partners (BODIPY chromophores) in an anticlockwise arrangement (negative exciton couplet; see Fig. 6.28), giving rise to the observed negative Cotton effect (plus-to-minus pattern in the bisignalized CD signal of the $R,R$ enantiomer with decreasing light energy; see Fig. 6.27). Furthermore $(R,R)$-**36** is computed to adopt a stable helical conformation with $P$ configuration (Fig. 6.28) [61], whereas $(S,S)$-**36** is computed to adopt preferably the helical $M$ configuration. However, when an $M$ configuration is computationally enforced for $(R,R)$-**36**, or a $(P)$ configuration enforced for $(S,S)$-**36**, the corresponding helical conformation becomes unstable and a pleated (CD-silent) conformation is shown to be the optimum geometry within these constraints [61].

The helically chiral geometry of $(R,R)$-**36** and $(S,S)$-**36** is not only able to perturb efficiently the BODIPY absorption transitions but also the BODIPY emission, giving place to CPL upon excitation with visible wavelength light in CHCl$_3$ solution [62]. The recorded CPL spectra for $(R,R)$-**36** and $(S,S)$-**36** were virtually mirror images, with maxima matching the maximum fluorescent emission of the involved BODIPY chromophores ($\lambda_{max}$ ca. 570 nm). Although the obtained maximum $g_{lum}$ values for each enantiomer are small ($|g_{lum}|$ ca. $1 \cdot 10^{-3}$, positive $g_{lum}$ for the $R,R$ enantiomer), they demonstrate that the involved helical architecture, despite its conformational lability, is able to endow the involved BODIPY chromophores with CPL activity in solution, since the emitted light is equally polarized with opposite handedness for each labile enantiomer [62].

The feasibility of this helically labile design to construct CPL-enabled bis(BODIPY)s was tested by the synthesis and study of $(R,R)$- and $(S,S)$-**37** (Fig. 6.29). These new bis(BODIPY)s are oxygenated analogues of $(R,R)$-**36** and $(S,S)$-**36**, respectively, in that **37** involve an oxygen-based bridge in place of the nitrogen-based bridge of **36**. These second-generation bis(BODIPY)s **37** were also obtained by nucleophilic aromatic substitution reaction of the same 3,3′-dichloroBODIPY precursor as used previously, but now using the corresponding commercially available enantiopure diol as the nucleophile (chemical yield ca. 60%) [62].

An experimental and computational study of bis(BODIPY)s **37** demonstrated that they too both adopt a preferred helical configuration in solution (similarly to **36**, $(P)$ for the $(R,R)$ enantiomer and $(M)$ for the $(S,S)$ enantiomer), giving rise to similar, opposite, and bisignalized CD spectra in CHCl$_3$ solution (again negative Cotton Effect for the $(R,R)$ enantiomer, positive for the $(S,S)$ enantiomer), as well as opposite CPL spectra under visible irradiation (maximum $|g_{lum}|$ ca. $1 \cdot 10^{-3}$). However, whilst nitrogenated $(R,R)$-**36** preferentially emits left circularly polarized light (positive $g_{lum}$ value), oxygenated $(R,R)$-**37** preferentially emits right circularly polarized light, even though their absolute configuration is the same (Fig. 6.30). The same reversal behaviour is also observed for the corresponding $(S,S)$ couple [62].

This striking behaviour, which is similar to that previously discussed for both **20** and **21** and related spiranic $O$-BODIPYs, was attributed to a differential ability for

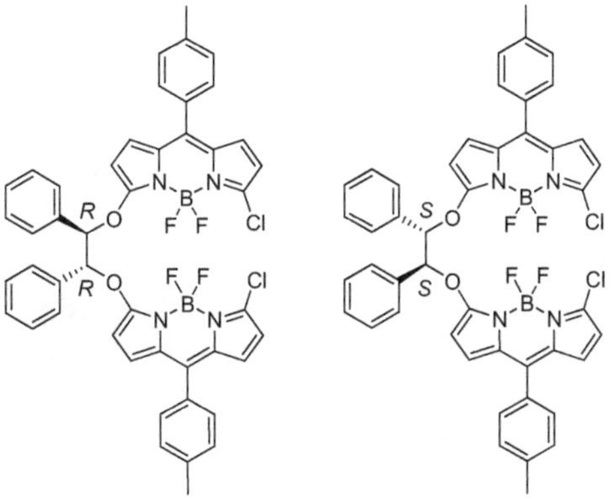

(R,R)-**37**                    (S,S)-**37**

$g_{lum} = -1 \cdot 10^{-3}$          $g_{lum} = +1 \cdot 10^{-3}$
$\phi = 0.17$                    $\phi = 0.17$

**Fig. 6.29** Helically labile bis(haloBODIPY)s (R,R)- and (S,S)-**37** and their visible CPL signatures upon visible wavelength irradiation in solution

promote ICT upon the excitation [62]. Thus, it is known that the fluorescent behaviour of BODIPY chromophores functionalized with both an efficient electron-donating group and an efficient electron-withdrawing group (push-pull effect) is ruled by the population of an ICT state upon excitation [63–65]. In this context, the differential ability of **36** and **37** to promote ICT (which is expected to be higher for nitrogenated **36**) could cause significant differences in the charge distribution upon excitation, which would explain the observed CPL reversal [62]. Indeed, the promotion of an ICT state upon excitation is demonstrated to be acting in these bis(BODIPY)s [66]. Other than by structural changes, ICT modulation can be achieved by alteration in solvent polarity, alongside the finely tuning other key fluorescent signatures such as the fluorescent quantum yield (e.g. $\phi$ up to 0.30 for **37** in less polar cyclohexane) [62]. The demonstrated capability for modulating ICT, together with synthetic accessibility, capability for reversing the polarization handedness, and configurational lability, is likely to make the helically labile bis (BODIPY) design a key element in the development of future tunable CPL-SOMs.

Figure-of-Eight Bis(BODIPY)S

The figure-of-eight geometry involves a peculiar helicity, which is expected to provide unique and interesting chiroptical properties. However, few examples of enantiopure figure-of-eight compounds have been described, and CPL-SOMs based on this geometry are limited [67–69]. Nonetheless, Nabeshima et al. have reported a CPL-enabling macrocyclic bis(BODIPY) **38** with a figure-of-eight design

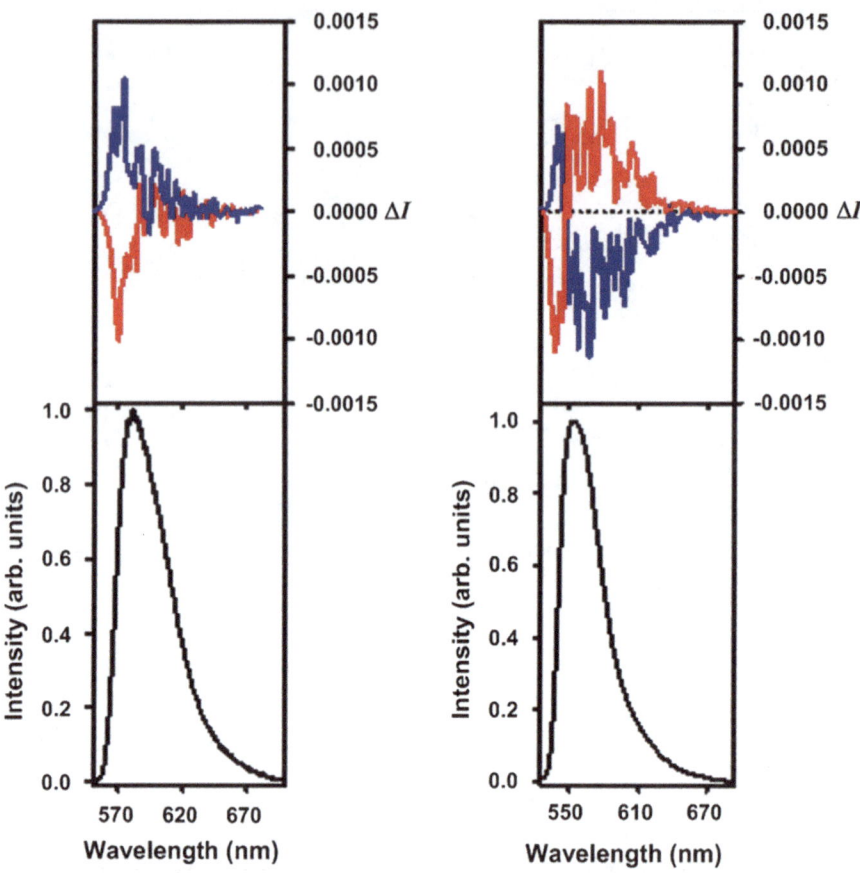

**Fig. 6.30** CPL (upper curves; *R,R* enantiomers in blue, *S,S* in red) and total luminescence (lower curves) spectra of **36** (left) and **37** (right) in CHCl₃ solution upon visible excitation [reproduced with permission from [62], copyright 2016 WILEY-VCH Verlag GmbH

(Fig. 6.31). This structure was obtained with an 88% chemical yield, as a racemic mixture, by reacting the corresponding bis(dipyrrin)-based macrocyclic precursor with phenylboronic acid (Fig. 6.31) [70]. In such a chelation process, the figure-of-eight helicity was achieved by the tetrahedral geometry imposed by the requirement for tetrahedral boron coordination. Optical resolution by chiral HPLC provided the corresponding *M,M* and *P,P* enantiomers, which exhibited an intense red fluorescence ($\lambda_{max} = 655$ nm, $\phi = 0.58$ in CHCl₃) and mirror-image CPL spectra with a maximum $|g_{lum}|$ value of $9 \cdot 10^{-3}$ (positive for the *P,P* isomer) in CHCl₃ solution upon visible wavelength excitation. This results in a chiroptical efficiency ($\phi \cdot |g_{lum}|$) value of $5.3 \cdot 10^{-3}$, which is one of the highest values reported for a CPL-SOMs to date [70]. Noticeably, the $|g_{lum}|$ value is three times higher than that exhibited by a

**Fig. 6.31** Synthesis, molecular structure, and visible CPL signatures (upon irradiation in solution) of figure-of-eight bis(BODIPY) **38**

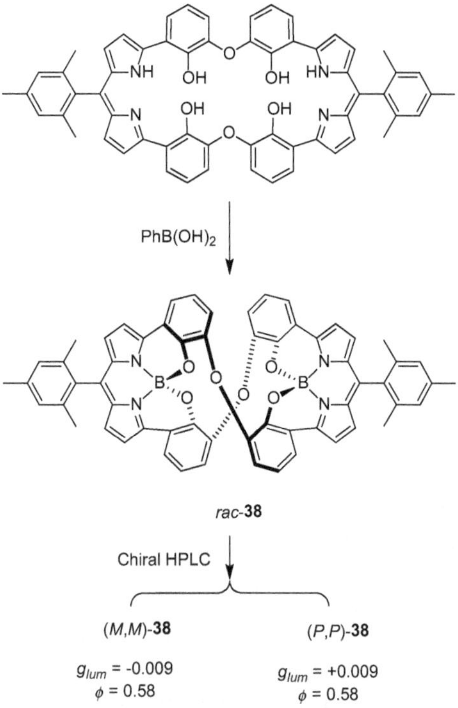

related mono(BODIPY) used as a reference [70], which demonstrates the advantageous synergistic effect achieved by the figure-of-eight geometry.

In an extension to this work, Nabeshima et al. have prepared several macrocycles exhibiting *pseudo*-figure-of-eight geometries, through the unsymmetrical chelation of the same bis(dipyrrin)-based macrocyclic precursor with one tetrahedral boron atom (forming a BODIPY moiety) and one octahedral group-13 element (aluminium, gallium, or indium) [71]. Single crystal X-ray diffraction analysis showed the presence of both enantiomers of the corresponding gallium and indium complexes in the crystals examined; however the resolution and examination of the chiroptical properties of these molecules is yet to be disclosed.

## Chiral BODIPY Polymers

Li et al. have developed the series of linear polymers based on BODIPY and chiral-BINOL **39** (Fig. 6.32) [72]. These chiral polymers, which were formed via a Sonogashira coupling of 2,2′-diethynylBODIPYs with the corresponding enantiopure 3,3′-diiodoBINOL derivatives, were soluble in a range of organic solvents and exhibited red fluorescence ($\phi$ up to 0.26 at 622 nm for **39b** in THF solution upon visible wavelength excitation) [72]. Interestingly, mirror-image CD spectra were obtained for the corresponding enantiomeric polymers, showing a

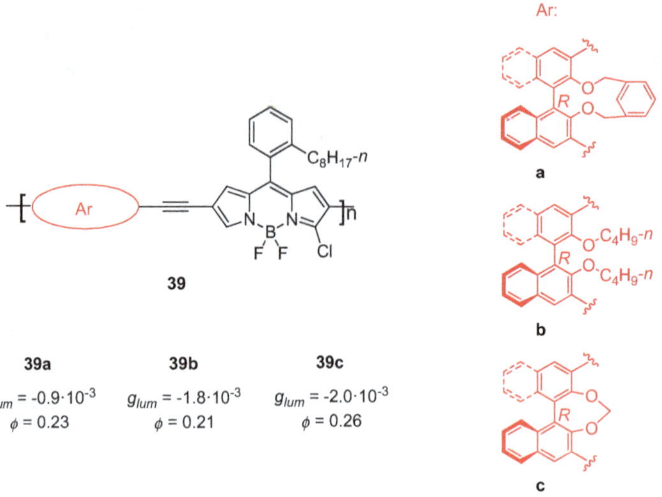

**Fig. 6.32** BINOL-based BODIPY polymers **39** and visible CPL signatures (upon visible light irradiation in THF solution)

bisignalizated visible wavelength dichroic signal (negative Cotton effect for polymers based on $(R)$-BINOL moiety) matching with the visible absorption spectra of the involved $\pi$-extended BODIPY chromophores (maximum $|g_{abs}|$ values up to ca. $3 \times 10^{-3}$ for **39c**; negative values for the $R$ polymers) [72]. The observed bisignalization suggests exciton coupling between the neighbouring BODIPY chromophores, disposed in a helically chiral architecture. Moreover, these BODIPY polymers show CPL activity in the red range of the spectra, upon excitation with visible light in THF solution, the CPL spectra for enantiomers being mirror images (maximum $|g_{lum}|$ values up to ca. $2 \cdot 10^{-3}$ at 611 nm for **39c**; negative values for the $R$ polymers) [72]. Interestingly, both the $|g_{abs}|$ and $|g_{lum}|$ values were demonstrated to depend on the dihedral angle of the involved BINOL-based moiety: as the angle increases, the corresponding size of CD or CPL signal decreases [72].

## 6.4 Summary and Outlook

Despite the inherently planar, achiral nature of the BODIPY fluorophore, a diverse array of strategies have been explored in an effort to induce efficient chiral perturbation directed towards the gain of CPL activity. The key twisting deformation postulated to be a requirement for an efficient CPL emission (in terms of $|g_{lum}|$ value) [32], has been demonstrated in mono(BODIPY)s through embedding the BODIPY core in an helicene-like architecture (e.g. by fusing $\pi$-conjugated rings to the BODIPY core [30], by intramolecular F$\cdots$H bonding in chiral-urobilin derivatives [49], or by constrained $N,N,O$,- and $N,N,O,O$-boron chelation [55, 57]). Beyond these helicenic-mono(BODIPY) designs, those based on $C_2$-symmetric BINOL-based $N,N,O,O$-boron-chelated dipyrromethane (spiranic mono($O$-BODIPY)) have proven interesting due to both synthetic accessibility [50, 52, 53] and their chiroptical versatility, notably their ability to modulate their fluorescence signatures, including the CPL sign, by simple manipulation of the ICT probability [54]. However, these simple chiral mono(BODIPY)s have yet to yield particularly high $|g_{lum}|$ values, with no CPL-SOM mono(BODIPY) to date exceeding $|g_{lum}| = 10^{-2}$. This is likely a consequence of the difference in magnitude of the magnetic ($\boldsymbol{m}$) and electric ($\boldsymbol{\mu}$) transition dipole moments in CPL-SOMs of this nature; thus future chiral mono (BODIPY) designs will have to carefully manage these key parameters. Interestingly Mori et al. have surveyed the chiroptical properties of the known CPL active mono (BODIPY)s and have shown a good linear correlation between $|g_{lum}|$ and $|g_{abs}|$ (absolute luminescent and absorption dissymmetry factors, respectively) [73], suggesting that rapid evaluation of CPL emission in chiral mono(BODIPY)s may be possible through simple CD measurement. However, it should be noted that correlation of $|g_{lum}|$ with $|g_{abs}|$ became less linear when more flexible bis(BODIPY)s were included in the analysis, suggesting more complex photophysical behaviours, including conformational changes between the $S_0$ and $S_1$ states, in these systems. In the related poly(BODIPY)s, several bis(BODIPY)-based designs have been explored in an attempt to introduce the required efficient chiral perturbation for CPL emission. The to-date published chiral bis(BODIPY) molecular architectures have included direct bonding of two BODIPY cores around a conformationally restricted axis (i.e. axially chiral bis(BODIPY)s) [34], indirect bonding via a flexible chiral linker to gain chiral helicity (i.e. helically labile bis(BODIPY)s) [62], and construction of macrocyclic figure-of-eight geometries involving $N,N,O,O$-boron-chelated dipyrromethane [70]. In particular, the figure-of-eight design has resulted in some of the highest reported values of $|g_{lum}| \approx 10^{-2}$, suggesting that multi-chromophoric systems may be required in the development of bright CPL-SOM BODIPYs. Usefully, some of these BODIPY-based chiral designs, aimed to chirally perturb the inherently achiral BODIPY fluorophore, have been shown to operate in other related achiral fluorophores to endow them with CPL activity [74–78], especially when such a fluorophore is closely related to BODIPY (e.g. azaBODIPYs or BOPHYs) [76–78].

# References

1. Sanchez-Carnerero EM, Agarrabeitia AR, Moreno F, Maroto BL, Muller G, Ortiz MJ, de la Moya S (2015) Circularly polarized luminescence from simple organic molecules. Chem Eur J 21:13488–13500
2. Kumar J, Nakashima T, Kawai T (2015) Circularly polarized luminescence in chiral molecules and supramolecular assemblies. Phys Chem Lett 6:3445–3452
3. Jan C-M, Lee Y-H, Wu K-C, Lee C-K (2011) Integrating fault tolerance algorithm and circularly polarized ellipsometer for point-of-care applications. Opt Express 19:5431–5441

4. Farshchi R, Ramsteiner M, Herfort J, Tahraoui A, Grahn HT (2011) Optical communication of spin information between light emitting diodes. Appl Phys Lett 98:162508. (3pp)
5. Wagenknecht C, Li C-M, Reingruber A, Bao X-H, Goebel A, Chen Y-A, Zhang Q, Chen K, Pan J-W (2010) Experimental demonstration of a heralded entanglement source. Nat Photonics 4:549–552
6. Sherson JF, Krauter H, Olsson RK, Julsgaard B, Hammerer K, Cirac I, Polzik ES (2006) Quantum teleportation between light and matter. Nature 443:557–560
7. Yu CJ, Lin CE, Yu LP, Chou C (2009) Paired circularly polarized heterodyne ellipsometer. Appl Opt 48:758–764
8. Schadt M (1997) Liquid crystal materials and liquid crystal displays. Annu Rev Mater Sci 27:305–379
9. Castiglioni E, Abbate S, Lebon F, Longhi G (2014) Chiroptical spectroscopic techniques based on fluorescence. Methods Appl Fluoresc 2:024006. (7pp)
10. Muller G (2014) In: de Bettencourt-Dias A (ed) Luminescence of lanthanide ions in coordination compounds and nanomaterials. Wiley, Chichester, pp 77–124
11. Riehl JP, Muller G (2012) In: Berova N, Polavarapu PL, Nakanishi K, Woody RW (eds) Comprehensive chiroptical spectroscopy vol. 1. Instrumentation, methodologies, and theoretical simulations. Wiley, Hoboken, pp 65–90
12. Meinert C, Hoffmann SV, Cassam-Chenaï P, Evans AC, Giri C, Nahon L, Meierhenrich UJ (2014) Photonenergy-controlled symmetry breaking with circularly polarized light. Angew Chem Int Ed 53:210–214
13. Cave RJ (2009) Inducing chirality with circularly polarized light. Science 323:1435–1436
14. Pagni RM, Compton RN (2005) Is circularly polarized light an effective reagent for asymmetric synthesis? Mini-Rev Org Chem 2:203–209
15. Bisoyi HK, Li Q (2016) Light-directed dynamic chirality inversion in functional self-organized helical superstructures. Angew Chem Int Ed 55:2994–3010
16. Yeom J, Yeom B, Chan H, Smith KW, Dominguez-Medina S, Bahng JH, Zhao G, Chang W-S, Chang S-J, Chuvilin A, Melnikau D, Rogach AL, Zhang P, Link S, Král P, Kotov NA (2015) Chiral templating of self-assembling nanostructures by circularly polarized light. Nat Mater 14:66–72
17. Wu D, Chen L, Lee W, Ko G, Yin J, Yoon J (2018) Recent progress in the development of organic dye based near-infrared fluorescence probes for metal ions. Coord Chem Rev 354:74–97
18. Kolemen S, Akkaya EU (2018) Reaction-based BODIPY probes for selective bio-imaging. Coord Chem Rev 354:121–134
19. Turskoy A, Yildiz D, Akkaya EU (2019) Photosensitization and controlled photosensitization with BODIPY dyes. Coord Chem Rev 379:47–64
20. Antina EV, Bumagina NA, V'yugin AI, Solomonov AM (2017) Fluorescent indicators of metal ions based on dipyrromethene platform. Dyes Pigments 136:368–381
21. Marfin YS, Solomonov AV, Timin AS, Rumyantsev EV (2017) Recent advances of individual bodipy and bodipy-based functional materials in medical diagnostics and treatment. Curr Med Chem 24:2745–2772
22. Kowada T, Maeda H, Kikuchi K (2015) BODIPY-based probes for the fluorescence imaging of biomolecules in living cells. Chem Soc Rev 44:4953–4972
23. Bessette A, Hanan G (2014) Design, synthesis and photophysical studies of dipyrromethene-based materials: insights into their applications in organic photovoltaic devices. Chem Soc Rev 43:3342–3405
24. López Arbeloa F, Bañuelos J, Martínez V, Arbeloa T, López Arbeloa I (2005) Structural, photophysical and lasing properties of pyrromethene dyes. Int Rev Phys Chem 24:339–374
25. Clarke RG, Hall MJ (2019) In: Scriven E, Ramsden C (eds) Advances in heterocyclic chemistry, vol 128. Academic Press, San Diego, pp 181–261
26. Bañuelos J (2016) BODIPY dye, the most versatile fluorophore ever? Chem Rec 16:335–348
27. Lakshmi V, Sharma R, Ravikanth M (2016) Functionalized boron-dipyrromethenes and their applications. Curr Org Chem 6:1–24

28. Boens N, Verbelen B, Dehaen W (2015) Postfunctionalization of the BODIPY core: synthesis and spectroscopy. Eur J Org Chem 2015:6577–6595
29. Lakshami V, Rao MR, Ravinkanth M (2015) Halogenated boron-dipyrromethenes: synthesis, properties and applications. Org Biomol Chem 13:2501–2517
30. Lu H, Mack J, Yang Y, Shen Z (2014) Structural modification strategies for the rational design of red/NIR region BODIPYs. Chem Soc Rev 43:4778–4823
31. Lu H, Mack J, Nyokong T, Kobayashi N, Shen Z (2016) Optically active BODIPYs. Coord Chem Rev 318:1–15
32. Gossauer A, Nydegger F, Kiss T, Sleziak R, Stoeckli-Evans H (2004) Synthesis, chiroptical properties, and solid-state structure determination of two new chiral dipyrrin difluoroboryl chelates. J Am Chem Soc 126:1772–1780
33. Beer G, Niederalt C, Grimme S, Daub J (2000) Redox switches with chiroptical signal expression based on binaphthyl boron dipyrromethene conjugates. Angew Chem Int Ed 39:3252–3255
34. Beer G, Rurack K, Daub J (2001) Chiral discrimination with a fluorescent boron–dipyrromethene. Chem Commun 2001:1138–1139
35. Móczár I, Huszthy P, Maidics Z, Kádár M, Tóth K (2009) Synthesis and optical characterization of novel enantiopure BODIPY linked azacrown ethers as potential fluorescent chemosensors. Tetrahedron 65:8250–8258
36. Haefele A, Zedde C, Retailleau P, Ulrich G, Ziessel R (2010) Boron asymmetry in a BODIPY derivative. Org Lett 12:1672–1675
37. Lerrick RI, Winstanley TPL, Haggerty K, Wills C, Clegg W, Harrington RW, Bultinck P, Herrebout W, Benniston AC, Hall MJ (2014) Axially chiral BODIPYs. Chem Commun 50:4714–4716
38. Kolemen S, Cakmak Y, Kostereli Z, Akkaya EU (2014) Atropisomeric dyes: axial chirality in orthogonal BODIPY oligomers. Org Lett 16:660–663
39. Bruhn T, Pescitelli G, Witterauf F, Ahrens J, Funk M, Wolfram B, Schneider H, Radius U, Bröring M (2016) Cryptochirality in 2,2′-coupled BODIPY DYEmers. Eur J Org Chem 2016:4236–4243
40. Hisamune Y, Kim T, Nishimura K, Ishida M, Toganoh M, Mori S, Kim D, Furuta H (2018) Switch-ON near IR fluorescent dye upon protonation: helically twisted bis(boron difluoride) complex of π-extended corrorin. Chem Eur J 24:4628–4634
41. Kim H, Burghart A, Welch MB, Reibenspies J, Burgess K (1999) Synthesis and spectroscopic properties of a new 4-bora-3a, 4a-diaza-s-indacene (BODIPY®) dye. Chem Commun 1999:1889–1890
42. Kubo Y, Minowa Y, Shoda T, Takeshita K (2010) Synthesis of a new type of dibenzopyrromethene-boron complex with near-infrared absorption property. Tetrahedron Lett 51:1600–1602
43. Tomimori Y, Okujima T, Yano T, Mori S, Ono N, Yamada H, Uno H (2011) Synthesis of π-expanded O-chelated boron-dipyrromethene as an NIR dye. Tetrahedron 67:3187–3193
44. Loudet A, Bandichhor R, Burgess K, Palma A, McDonnell SO, Hall MJ, O'Shea DF (2008) B,O-Chelated azadipyrromethenes as near-IR probes. Org Lett 10:4771–4774
45. Ikeda C, Maruyama T, Nabeshima T (2009) Convenient and highly efficient synthesis of boron-dipyrrins bearing an arylboronate center. Tetrahedron Lett 50:3349–3351
46. Sirbu D, Benniston AC, Harriman A (2017) One-pot synthesis of a mono-O, B, N-strapped BODIPY derivative displaying bright fluorescence in the solid state. Org Lett 19:1626–1629
47. Chen N, Zhang W, Chen S, Wu Q, Yu C, Wei Y, Xu Y, Hao E, Jiao L (2017) Sterically protected $N_2O$-type benzopyrromethene boron complexes from boronic acids with intense red/near-infrared fluorescence. Org Lett 19:2026–2029
48. Gobo Y, Matsuoka R, Chiba Y, Nakamura T, Nabeshima T (2018) Synthesis and chiroptical properties of phenanthrene-fused $N_2O$-type BODIPYs. Tetrahedron Lett 59:4149–4152

49. Gossauer A, Fehr F, Nydegger F, Stöckli-Evans H (1997) Synthesis and conformational studies of urobilin difluoroboron complexes. unprecedented solvent-dependent chiroptical properties of the BF$_2$ chelate of an urobilinoid analogue. J Am Chem Soc 119:1599–1608

50. Sánchez-Carnerero EM, Moreno F, Maroto BL, Agarrabeitia AR, Ortiz MJ, Vo BG, Muller G, de la Moya S (2014) Circularly polarized luminescence by visible-light absorption in a chiral *O*-BODIPY dye: unprecedented design of CPL organic molecules from achiral chromophores. J Am Chem Soc 136:3346–3349

51. Gartzia-Rivero L, Sánchez-Carnerero EM, Jiménez J, Bañuelos J, Moreno F, Maroto BL, López-Arbeloa I, de la Moya S (2017) Modulation of ICT probability in bi(polyarene)-based *O*-BODIPYs: towards the development of low-cost bright arene-BODIPY dyad. Dalton Trans 46:11830–11839

52. Jiménez J, Cerdán L, Moreno F, Maroto BL, García-Moreno I, Lunkley JL, Muller G, de la Moya S (2017) Chiral organic dyes endowed with circularly polarized laser emission. J Phys Chem C 121:5287–5292

53. Jiménez J, Moreno F, Maroto BL, Cabreros TA, Huy AS, Muller G, Bañuelos J, de la Moya S (2019) Modulating ICT emission: a new strategy to manipulate the CPL sign in chiral emitters. Chem Commun 55:1631–1634

54. Zhang S, Wang Y, Meng F, Dai C, Cheng Y, Zhu C (2015) Circularly polarized luminescence of AIE-active chiral *O*-BODIPYs induced via intramolecular energy transfer. Chem Commun 51:9014–9017

55. Alnoman RB, Rihn S, O'Connor DC, Black FA, Costello B, Waddell PG, Clegg W, Peacock RD, Herrebout W, Knight JG, Hall MJ (2016) Circularly polarized luminescence from helically chiral *N,N,O,O*-boron-chelated dipyrromethenes. Chem Eur J 22:93–96

56. Gobo Y, Yamamura M, Nakamura T, Nabeshima T (2016) Synthesis and chiroptical properties of a ring-fused BODIPY with a skewed chiral π skeleton. Org Lett 18:2719–2721

57. Clarke R, Ho KL, Alsimaree AA, Waddell PG, Bogaerts J, Herrebout W, Knight J, Pal R, Penfold T, Hall MJ (2017) Circularly polarised luminescence from helically chiral "confused" *N,N,O,C*-boron-chelated dipyrromethenes (BODIPYs). ChemPhotoChem 1:513–517

58. Toyoda M, Imai Y, Mori T (2017) Propeller chirality of boron heptaaryldipyrromethene: unprecedented supramolecular dimerization and chiroptical properties. Phys Chem Lett 8:42–48

59. Bruhn T, Pescitelli G, Jurinovich S, Schaumlöffel A, Witterauf F, Ahrens J, Bröring M, Bringmann G (2014) Axially chiral BODIPY DYEmers: an apparent exception to the exciton chirality rule. Angew Chem Int Ed Engl 53:14592–14595

60. Zinna F, Bruhn T, Guido CA, Ahrens J, Bröring M, Di Bari L, Pescitelli G (2016) Circularly polarized luminescence from axially chiral BODIPY DYEmers: an experimental and computational study. Chem Eur J 22:16089–16098

61. Sanchez-Carnerero EM, Moreno F, Maroto BL, Agarrabeitia AR, Bañuelos J, Arbeloa T, López-Arbeloa I, Ortiz MJ, de la Moya S (2013) Unprecedented induced axial chirality in a molecular BODIPY dye: strongly bisignated electronic circular dichroism in the visible region. Chem Commun 49:11641–11643

62. Ray C, Sanchez-Carnerero EM, Moreno F, Maroto BL, Agarrabeitia AR, Ortiz MJ, López-Arbeloa I, Bañuelos J, Cohovi KD, Lunkley JL, Muller G, de la Moya S (2016) Bis (haloBODIPYs) with labile helicity: valuable simple organic molecules that enable circularly polarized luminescence. Chem Eur J 22:8805–8808

63. Niu S, Ulrich G, Retailleau P, Ziessel R (2011) BODIPY-bridged push–pull chromophores: optical and electrochemical properties. Tetrahedron Lett 52:4848–4853

64. Boens N, Leen V, Dehaen W (2012) Fluorescent indicators based on BODIPY. Chem Soc Rev 41:1130–1172

65. Bonnier C, Machin DD, Abdi O, Koivisto BD (2013) Manipulating non-innocent π-spacers: the challenges of using 2,6-disubstituted BODIPY cores within donor-acceptor light-harvesting motifs. Org Biomol Chem 11:3756–3760

66. Ray C, Bañuenlos J, Arbeloa T, Maroto BL, Moreno F, Agarrabeitia AR, Ortiz MJ, López-Arbeloa I, de la Moya S (2016) Push-pull flexibly-bridged bis(haloBODIPYs): solvent and spacer switchable red emission. Dalton Trans 45:11839–11848
67. Morisaki Y, Gon M, Sasamori T, Tokitoh N, Chujo Y (2014) Planar chiral tetrasubstituted [2.2] paracyclophane: optical resolution and functionalization. J Am Chem Soc 136:3350–3353
68. Gon M, Morisaki Y, Chujo Y (2015) Optically active cyclic compounds based on planar chiral [2.2] paracyclophane: extension of the conjugated systems and chiroptical properties. J Mater Chem C 3:521–529
69. Gon M, Kozuka H, Morisaki Y, Chujo Y (2016) Optically active cyclic compounds based on planar chiral [2.2] paracyclophane with naphthalene units. Asian J Org Chem 5:353–359
70. Saikawa M, Nakamura T, Uchida J, Yamamura M, Nabeshima T (2016) Synthesis of figure-of-eight helical bisBODIPY macrocycles and their chiroptical propertie. Chem Commun 52:10727–10730
71. Saikawa M, Noda T, Matsuoka R, Nakamura T, Nabeshima T (2019) Heterodinuclear group 13 element complexes of $N_4O_6$-type dipyrrin with an unsymmetrical twisted structure. Eur J Inorg Chem 2019:766–769
72. Wang Y, Li Y, Liu S, Li F, Zhu C, Li S, Cheng Y (2016) Regulating circularly polarized luminescence signals of chiral binaphthyl-based conjugated polymers by tuning dihedral angles of binaphthyl moieties. Macromolecules 49:5444–5451
73. Tanaka H, Inoue Y, Mori T (2018) Circularly polarized luminescence and circular dichroisms in small organic molecules: correlation between excitation and emission dissymmetry factors. ChemPhotoChem 2:386–402
74. Feuillastre S, Pauton M, Gao L, Desmarchelier A, Riives AJ, Prim D, Tondelier D, Geffroy B, Muller G, Clavier G, Pieters G (2016) Design and synthesis of new circularly polarized thermally activated delayed fluorescence emitters. J Am Chem Soc 138:3990–3993
75. Li J, Peng X, Huang C, Qi Q, Lai W-Y, Huang W (2018) Control of circularly polarized luminescence from a boron ketoiminate-based π-conjugated polymer via conformational locks. Polym Chem 9:5278–5285
76. Wu Y, Wang S, Li Z, Shenb Z, Lu H (2016) Chiral binaphthyl-linked BODIPY analogues: synthesis and spectroscopic properties. J Mater Chem C 4:4668–4674
77. Meng F, Sheng Y, Li F, Zhu C, Quan Y, Cheng Y (2017) Reversal aggregation-induced circular dichroism from axial chirality transfer via self-assembled helical nanowires. RSC Adv 7:15851–15856
78. Maeda C, Nagahata K, Takaishi K, Ema T (2019) Synthesis of chiral carbazole-based BODIPYs showing circularly polarized luminescence. Chem Commun 55:3136–3139

# Chapter 7
# Propeller Chirality: Circular Dichroism and Circularly Polarized Luminescence

Tadashi Mori

**Abstract** Hexaarylbenzenes (HABs) and related highly substituted aromatic molecules possess unique propeller-shaped structure. The propeller chirality derived therefrom turned out not to be a simple multiple combination of axial chirality. A domino-like corporation between the radial aromatic blades is critical to produce the highly enhanced chiroptical responses. More interestingly, such chirality can be either static or dynamic depending on the sterical hindrance between the aromatic blades. The latter systems could be easily controlled by subtle environmental factors such as temperature, solvent, as well as pressure, and hence feasibly manipulatable. The molecular systems with propeller chirality thus provide new design principle for advanced and superior chiroptical materials, particularly those emit efficient circularly polarized light at desired wavelength and, in addition, are modulable and are switched on-off at will.

## 7.1 Introduction

A propeller with uni-directionally twisted blades generates a forward or backward thrust by inverting the rotating direction (Fig. 7.1, left pair). The same can be done by switching the twist direction of the blades (Fig. 7.1, right pair). Certainly, such a mechanical motif can be implemented in a molecule to afford a chiral molecular propeller. Indeed, several propeller chiral molecules have been prepared using benzene as a hub and aromatic rings as blades and their structures and inter-blade interactions have been elucidated in some depth [1–6]. However, the dynamic propeller chirality inversion and its chiroptical consequences have rarely befallen to be the targets of intensive experimental and theoretical studies until very recently [7, 8], making a keen contrast to molecular rotors and motors.

T. Mori (✉)
Department of Applied Chemistry, Graduate School of Engineering, Osaka University, Suita, Japan
e-mail: tmori@chem.eng.osaka-u.ac.jp

© Springer Nature Singapore Pte Ltd. 2020                                              151
T. Mori (ed.), *Circularly Polarized Luminescence of Isolated Small Organic Molecules*, https://doi.org/10.1007/978-981-15-2309-0_7

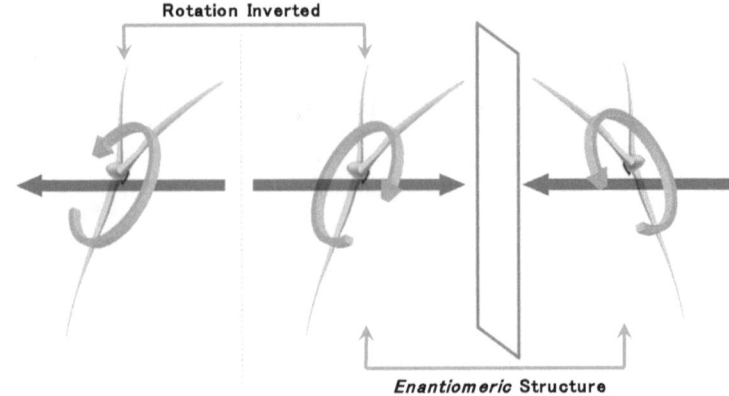

**Fig. 7.1** Schematic illustrations of the thrust direction changes by inverting the rotating direction (the left pair) or by using the enantiomeric propeller (the right pair)

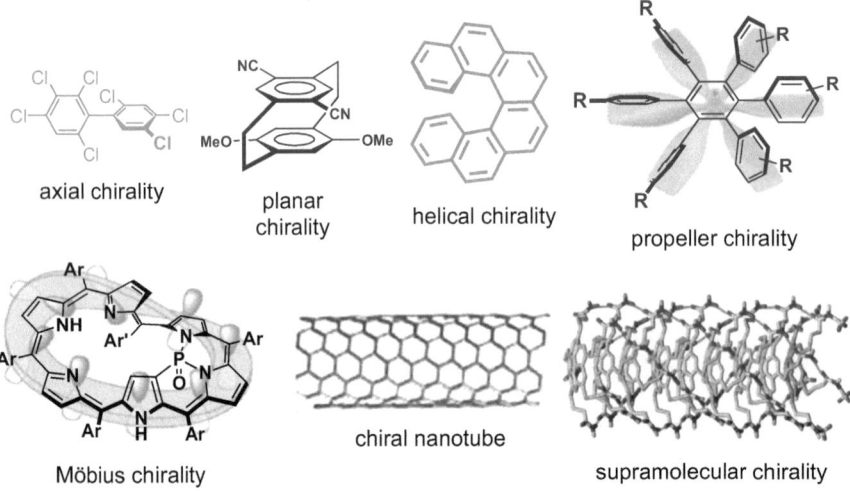

**Fig. 7.2** Exemplars of common and less common categories of (supra)molecular chirality

We have systematically investigated the chiroptical behaviors of chiral molecules of various categories, including point, axial [9–13], planar [14–17], and helical chirality [18–22]. Recently, more sophisticated Möbius chirality of expanded porphyrins [23] and bowl-shaped chirality of corannulenes [24], as well as supramolecular chirality of host-guest systems [25, 26], nanotubes [27], and nanorings [28], have been studied (Fig. 7.2). Small chiral molecules with structural rigidity are rather straightforward in preparation and subsequent property analyses, but they generally lack the conformational flexibility or dynamism as well as the handle or functionality for manipulating their chiroptical properties. In contrast, flexible chiral molecules are furnished with the dynamic chiroptical responses manipulatable through adequate structure tuning and environmental setting. In comparison to the conventional covalently bonded systems, chiral supramolecular systems are more facile and hence advantageous in controlling their chiroptical responses by various internal and external factors.

In this chapter, we will mainly describe the results of our recent experimental and theoretical studies on the propeller chirality of hexaarylbenzenes (HABs) and related molecules [7, 8, 29]. Some of these propeller-shaped molecules, when properly arranged, form supramolecular assemblies (e.g., dimer) to exhibit divergent chiroptical responses. The propeller chirality is fundamentally different in origin and dynamism from mechanical planar chirality in rotaxanes [30–33], although the correlation between propeller and thrust is conceptually similar to the wheel and axle relationship. In the former, the molecular structure is inherently chiral in conformation and the chiral sense is dynamically switchable, while the latter is supramolecularly induced by threading a wheel composed of three or more different segments with an unsymmetrical axle and the chiral sense is not manipulable or switched upon shuttling of the wheel. Indeed, the propeller chirality of HABs is not a result of simple or random combination of the six axially chiral aryls connected to the central benzene ring but is emerged by their synchronized unidirectional twists to the equilibrium angle to attain strong chiroptical responses. More crucially, the twist angle of synchronized propeller blades is dynamic in nature and hence susceptible to various environmental factors such as temperature, solvent polarity, viscosity, and pressure, eventually allowing us to manipulate the chiroptical properties. In the toroidal form, all the aromatic blades become orthogonal to the central benzene core and the chiroptical responses are practically diminished. In contrast to chiral supramolecular systems, the structures of propeller-chiral molecules presented in the following sections are precisely defined, but the extensive inter-blade interactions may dynamically alter the shape (and therefore the properties) by various environmental factors. As such, the combined experimental and theoretical studies to elucidate how and to what extent the chiroptical responses are affected by the intramolecular (inter-blade), intermolecular, and/or solvent interactions provide us with the substantial insights bridging the gap between molecular and supramolecular, as well as rigid and flexible, chirality.

## 7.2    Propeller Chirality in Hexaarylbenzenes

In designing propeller-shaped molecules, two- or three-bladed propeller may seem simpler, but firmly anchoring each propeller blade to a hub at a fixed twist angle is synthetically highly challenging. We chose a different strategy for contracting a flexible propeller molecule, in which multiple blades are anchored to a hub through a single bond to allow the blade to automatically adjust the twist angle by inter-blade interactions. In such an approach for studying propeller chirality, the HAB is one of the most ideal motifs [34]. In the HAB, all the six hydrogens of benzene are replaced by identical or divergent aromatic groups to endow the propeller structure, in which

**Fig. 7.3** Clockwise (*C*) and counterclockwise (*CC*) propeller chirality in hexaarylbenzene (HAB)

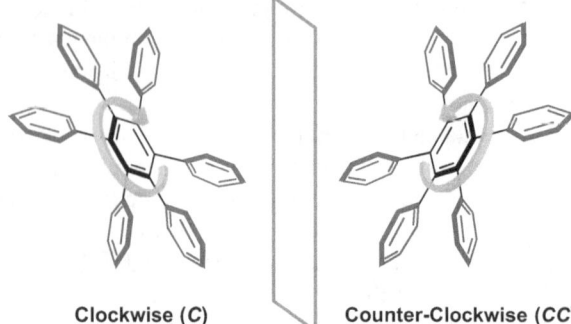

Clockwise (*C*)                    Counter-Clockwise (*CC*)

**Fig. 7.4** Schematic drawing of the toroidal interaction in HAB

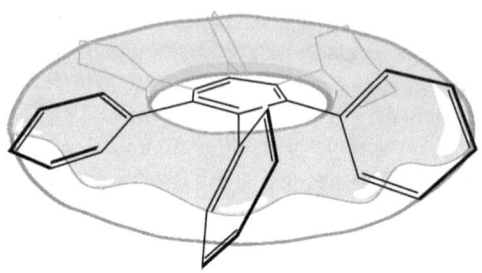

six biaryl units share the central benzene ring. Each two aromatic rings in the biaryl unit is in general not perpendicular to each other, due to the π-conjugation with the central benzene, which is counterbalanced by the steric conflict with the *ortho*-hydrogens (or substituents) of adjacent aromatic blades [13, 35–40]. This gives rise to the atropisomerism, the origin of axial chirality. The adjacent radial aromatic rings in HABs are separated by ca. 2.9 Å at the *ipso* positions, which is appreciably shorter than the van der Waals contact. Accordingly, the six aromatic blades synchronously tilt in one direction to give a propeller geometry, which is enantio-meric, being twisted in either clockwise (*C*) or counterclockwise (*CC*) man-ner (Fig. 7.3). The short inter-blade distance practically inhibits the independent rotation of a single blade but allows synchronized rotation of all the blades, which enables us to modulate the inter-blade interaction and thus the chiroptical responses.

The syntheses [41, 42] and unique properties of HABs and related compounds have attracted much attention for the practical applications in organic light-emitting diodes, photochemical switches, redox materials, molecular receptors, and liquid crystalline materials [34]. It is believed that the global electron and/or exciton delocalization over the entire radial aromatic rings occurs in the doughnut-shaped scaffold (Fig. 7.4), which is often referred to as "*toroidal interaction*" [43]. This unique interaction has been extensively studied in conjunction with the sequential donor-acceptor interaction, the electron/energy migration, and the π-delocalization by modulating the global interaction of radial aromatic rings [44–52]. Highlighting the toroidal interaction, all the radial aromatic rings are assumed to be orthogonally arranged in these studies, where the energetically more realistic propeller structure (with tilted blades) of HABs is essentially ignored. We have addressed this issue, at least in part, by showing that both the propeller and doughnut structures with tilted and orthogonal blades, respectively, are imperative in condensed phase, although the latter maximizes the toroidal interaction [7, 8]. The relative importance, however, can be controlled by changing the conditions employed.

## 7.3 Chiral Propeller Formation Driven by the Domino Effect

The intramolecular point-to-propeller chirality transfer has been investigated by introducing a point-chiral substituent to the *para*-position of 1–6 aromatic blades in hexaphenylbenzene and by subsequently examining the chiroptical consequences of the systematic peripheral modification [7, 8]. Thus, a series of HABs (**H1–H6**) incorporating either methoxy or chiral alkyloxy (R*O) group at the *para*-position of each aromatic blade (Fig. 7.5) are prepared by the transition metal-catalyzed trimerization of the corresponding acetylene derivatives or by the Diels-Alder

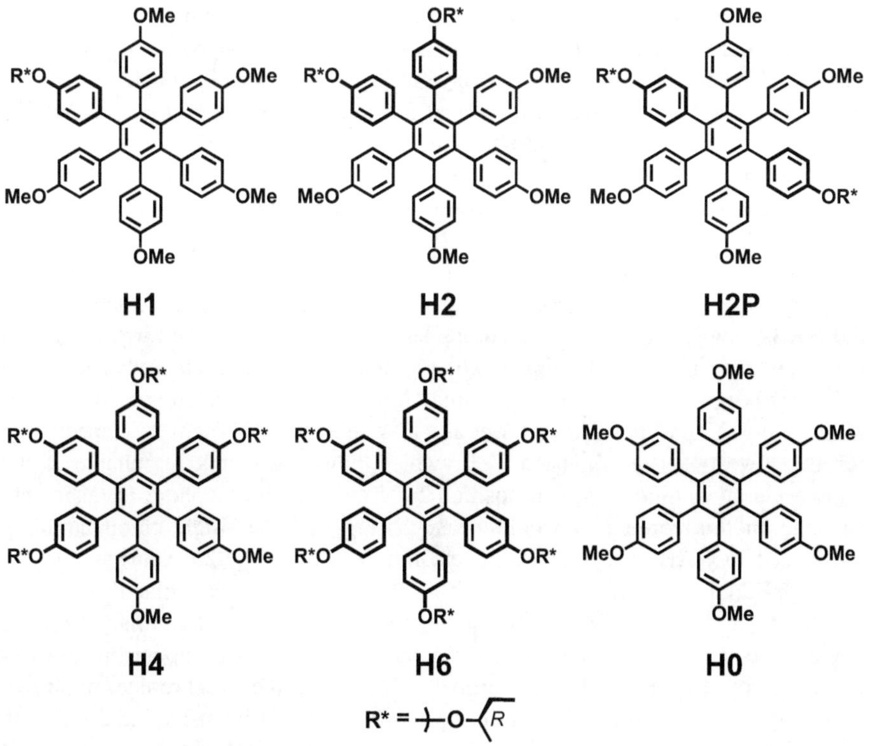

**Fig. 7.5** HABs with different number of chiral propeller blades

**Fig. 7.6** Comparison of CD
spectra of HABs **H1–H6** in
methylcyclohexane at 25 °C

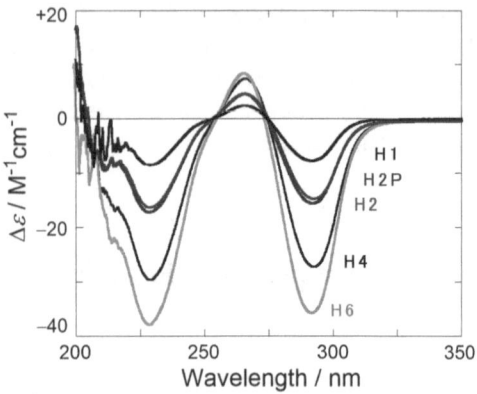

addition of an appropriate combination of tetraarylcyclopentadienone and
diarylacetylenes, followed by the aromatization via decarbonylation. Note that the
point-chiral alkyl group (R*) employed is the simplest, i.e., (R)-1-methylpropyl.
Although the chiroptical behavior of oligomeric solid materials containing HAB unit
has been reported [53], chiral HABs are still relatively rare and little is known about
the chiroptical properties of such molecules.

Circular dichroism (CD) spectra of chiral HABs with varying number of the
chiral auxiliary at the *para*-position of radial phenyls (**H1–H6**) are compared in
Fig. 7.6. The intense trisignate Cotton effects (with a negative-positive-negative
pattern for the R-configuration) observed are assignable to the $^1L_b$, $^1L_a$, and $^1B_b$
transitions, respectively, from low to high excitation energy. The CD spectrum of
**H6** shows no concentration dependence, indicating the absence of aggregation in
this system, which is in sharp contrast to the dimer formation of boron
dipyrromethenes (BODIPY) derivative (vide infra) [29]; probably the aggregation
or stacking is infeasible for such a highly substituted, less polarized compound as
**H6**. Crucially, the overall spectral shapes are practically indistinguishable for all
the HABs, suggesting that similar propeller geometries are evoked irrespective of
the number of introduced chiral auxiliary. It is to note that alkoxybenzene and
4,4′-dialkoxylbiphenyl carrying the same chiral auxiliary as reference compounds
afford molar CD values ($\Delta\varepsilon$) as low as −0.4 and −0.5 M$^{-1}$ cm$^{-1}$, respectively,
for the lowest-energy $^1L_b$ band [13], which however seems reasonable if the
remote chiral modification at a position far from the chromophore is taken into
account. In this context, it would be surprising that the single chiral auxiliary
introduced to HAB at a remote *para*-position enhances the $\Delta\varepsilon$ value of the $^1L_b$
band of **H1** up to −7.7 M$^{-1}$ cm$^{-1}$. This sudden increase of molar CD, though
difficult to explain as a direct influence of the peripheral chiral auxiliary alone, is
not unexpected but is rather taken as experimental evidence supporting the formation of the hypothetical chiral propeller driven by the synchronized unidirectional "*domino*" twisting of six radial phenyls. The CD intensity increases with
increasing number of the chiral auxiliary to afford nearly doubled $\Delta\varepsilon$ values for
**H2** and **H2P** and 4.6-fold larger −35.1 M$^{-1}$ cm$^{-1}$ for **H6**. This progressive CD

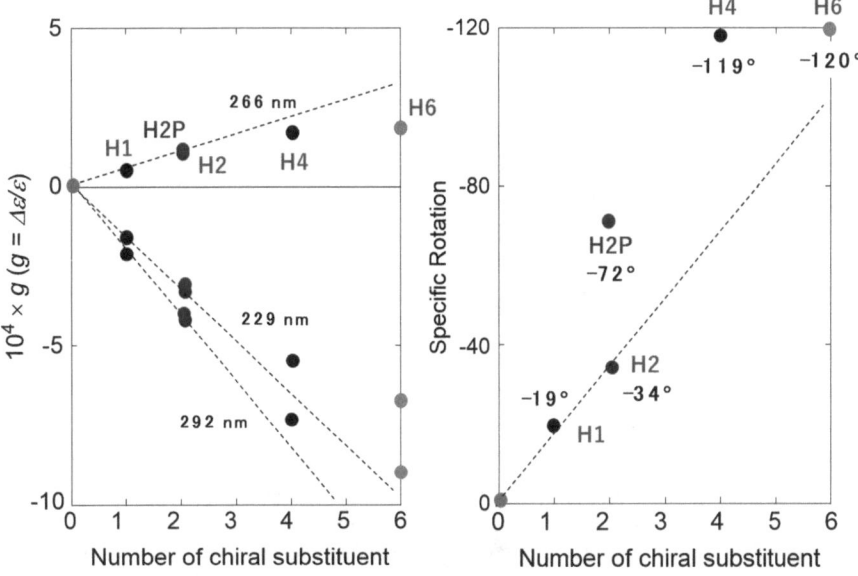

**Fig. 7.7** Plots of the dissymmetry ($g$) factor (left) and the specific rotation (right) against the number of chiral unit in HABs

augmentation, gradually saturating at higher substitutions, may indicate existence of the equilibrium between right- and left-handed (or $C$-$CC$) propellers in **H6**.

Because the UV-vis spectra of these HABs slightly differ from each other (presumably due to the different degrees of interaction between the blades), a quantitative comparison of the CD intensities of **H1**–**H6** (as a function of the number of chiral unit) is more accurately performed by using their dissymmetry factors ($g = \Delta\varepsilon/\varepsilon$). As can be seen from Fig. 7.7 (left), the $g$ factors for all the extrema at 229, 266, and 292 nm almost linearly increase with increasing number ($n$) of the chiral unit until $n$ reaches two (i.e., **H1**, **H2**, and **H2P**), but gradually saturate at $n = 4$–6 (**H4** and **H6**). It is interesting that the CD spectra of **H2** and **H2P** do not appreciably differ in shape and intensity (in methylcyclohexane), revealing that the relative position has almost no effect and only the number of chiral auxiliary is important in forming the chiral propeller. These results reveal that the impact of chiral modification at the periphery of propeller blade is not restricted to the adjacent blade but further propagates to more distant blades, being driven by the domino effect. In this system, the $C$ and $CC$ propellers are in equilibrium and the energy difference between them is incrementally enhanced with increasing number of the chiral modification to shift the equilibrium to the favored propeller with accompanying augmentation of CD intensity.

The specific rotations ($[\alpha]_D$) of **H1**–**H6** (in chloroform at 25 °C) progressively but somewhat irregularly increases from $-19°$ for **H1** to $-120°$ for **H6**, all of which are larger than $-13°$, that of ($R$)-1-methylpropanol, introduced as the chiral unit. As shown in Fig. 7.7 (right), the changing pattern of the $[\alpha]_D$ value substantially differs from that of the $g$ factor, exhibiting a nearly linear increase for **H1** and **H2**, but abrupt leaps for **H2P** and **H4** and a clear saturation for **H6**. Analogous, but much smaller, difference can be seen for the $g$ factors of **H2** and **H2P**, and the overall trends do not extremely differ between the changing patterns of $g$ factor and $[\alpha]_D$ value. It is to note that the optical rotation is a more complex function of molecular parameters, and the exact reasons for the rather exaggerated profile for observed $[\alpha]_D$ values remain to be elucidated.

## 7.4  Structure of Hexaarylbenzenes

Prior to discuss on the dynamics of propeller-shaped molecules, we analyze the relaxed structure of HABs with the aid of theoretical calculations. The geometries of HABs are optimized by the dispersion-corrected density functional theory at the DFT-D3(BJ)-TPSS/def2-TZVP level [54–56]. This level of theory, implementing the dispersion correction to simply augments the pairwise $-C_6/R^6$ potentials to describe the van der Waals interactions, has been successfully applied as a cost-efficient alternative to the more demanding electron correlation methods for a number of molecular systems, where the weak inter- and intramolecular interactions play significant roles [10, 13, 14, 16–20, 23, 57–59]. While typical biaryls are known to have the twist angles of ~40° [13, 60], the average tilt angle is consistently ~59° for all the HABs (**H0–H6**), due to the greater steric crowding in HABs compared with simple biaryls. Possessing the point-chiral ($R$)-auxiliary/ies, all the chiral HABs (**H1–H6**) prefer the $C$ rather than $CC$ propeller. Crucially, the core and radial benzene rings of **H0–H6** are essentially superimposable, despite the varying number of chiral auxiliaries at the periphery (Fig. 7.8). It is to note that the conformation of chiral methylpropyl group(s) as well as *syn/anti* orientation of alkoxy moiet(ies) may differ in chiral HABs, but the effect of such conformational variation is negligible on the calculated chiroptical properties [8].

X-ray crystallographic study provides a slightly different view on the structure of HABs. The crystal structure of hexakis($p$-hydroxyphenyl)benzene is not $C_6$-symmetrical, being expressively affected by the solvent molecules incorporated in the crystal. The twist angles for the DMF-solvated HAB are 91, 100, 77, 84, 76, and 98° with a fairly complicated hydrogen-bonding network, while those for the corresponding diethyl ether complex are 100, 68, and 97° (with centrosymmetry) [61]. In both the cases, the radial aromatic rings are not oriented in one direction (i.e., the $aR$ and $aS$ atropisomers coexist), suggesting that the peripheral aryls can rotate, in good agreement with the shallow potential for the inversion process (vide infra). It is also to note that each HAB molecule is sterically congested to form a channel structure without any appreciable interactions with the neighboring molecules. The crystal structure of fully chiral HAB **H6** reveals that the radial aromatic rings are twisted in one direction and their tilt angles fall in a narrow range ($\Delta = 9°$) with the individual values of 58, 67, 64, 60, 60, and 65° (Fig. 7.9). The conformations of the peripheral chiral alkyl groups are rather variable, primarily due to the packing

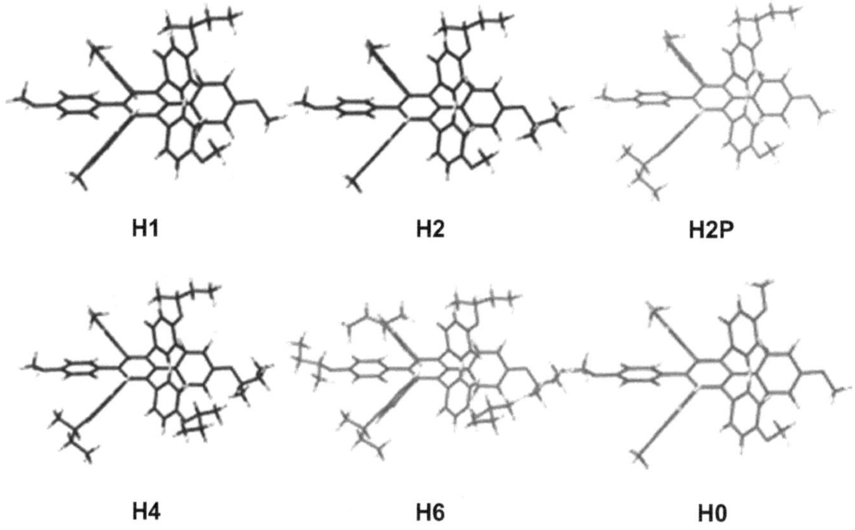

H1                          H2                          H2P

H4                          H6                          H0

**Fig. 7.8** Optimized structures of HABs calculated by the dispersion-corrected DFT method

requisite [8]. Nevertheless, the clockwise propeller chirality induced by the peripheral point-chiral (R)-alkoxy auxiliaries and the average twist angle of $62 \pm 2°$, as
well as the quasi-$C_6$ symmetric overall structure, well accord with the theoretical
predictions (vide supra). Possessing smaller numbers of chiral auxiliaries, **H2P** and
**H0** become less symmetrical (less uniform in tilt angle), as exemplified by a wider
range ($\Delta = 23°$) of the tilt angles, i.e., 66, 57, 77, 89, 74, and 66° (average: $72 \pm 5°$),
found in the crystal of **H0**. The crystal of **H2P** contains two independent but similar
structures. The range of the tilt angles becomes much broader ($\Delta = 41°$ and 32°) and
one of the radial aromatic rings is twisted counterclockwise. The actual tilt angles
widely vary, i.e., 111, 80, 70, 73, 71, and 77° and 99, 80, 67, 67, 69, and 79° (grand
average: $79 \pm 6°$), probably reflecting the high susceptibility to small perturbations
(e.g., packing forces) as a consequence of the shallow potentials for the inversion
process (vide infra) [7].

## 7.5   Solvent and Temperature Effects on Propeller Chirality

Figure 7.6 demonstrates very strong chiroptical responses of the propeller chirality in
HABs, where the point chirality is accumulated in the propeller geometry through
the cooperative or domino effect. However, such strong Cotton effects are observed
only in non-polar hydrocarbon solvents. In fact, the use of more polar solvents, such
as diethyl ether and acetonitrile, greatly reduces the CD intensity of **H6**, as shown in
Fig. 7.10. Obviously, a single solvent parameter, such as polarity, viscosity, or size
alone, cannot well rationalize such behaviors, for which the complex conformational

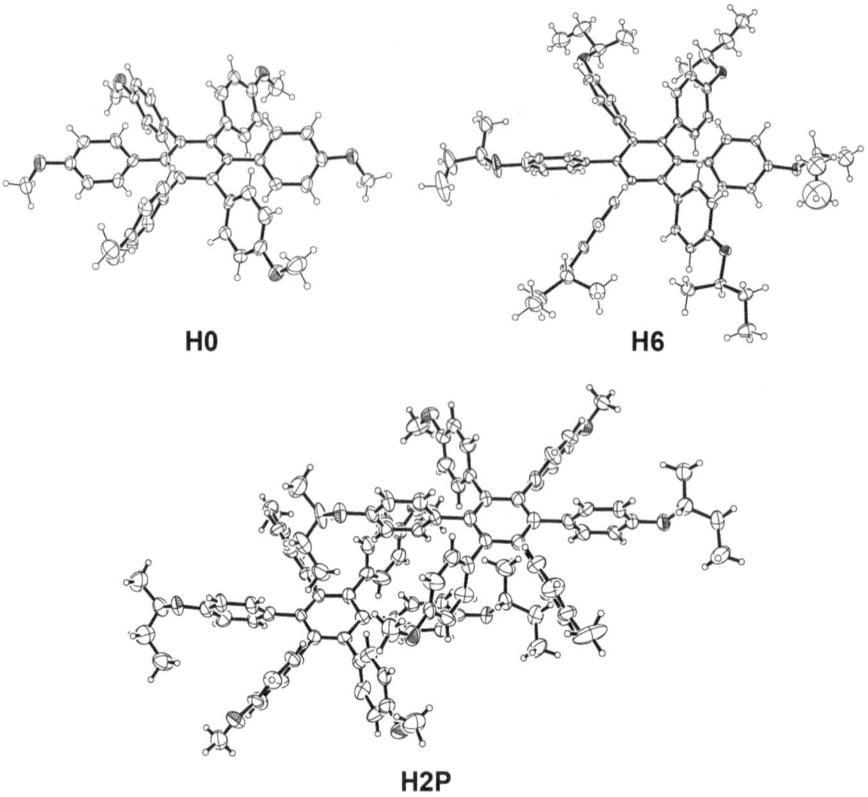

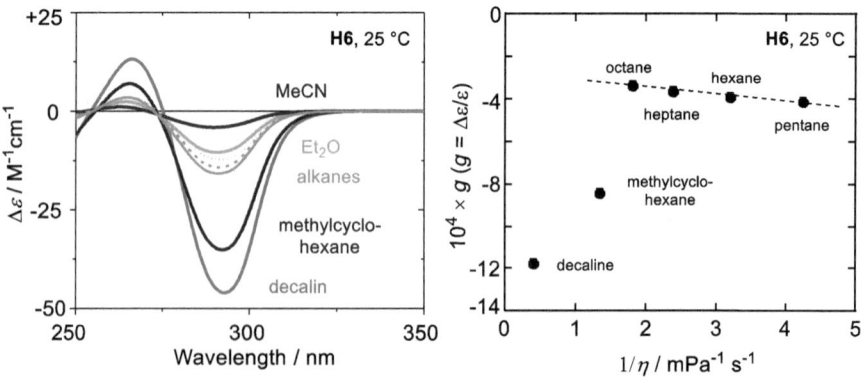

**Fig. 7.9** ORTEP drawings with 50% probability ellipsoid for crystal structures of some HABs. Note that two independent structures were found in the crystals of **H2P**

**Fig. 7.10** Solvent-dependent CD spectra of **H6**. The dotted, dashed, and solid lines marked "alkanes" are for octane, hexane, and pentane, respectively

**Fig. 7.11** Variable-temperature CD spectra of **H6**. The spectrum at $-150$ °C is obtained in isopentane, while those at other temperatures (+85 to $-120$ °C) are in methylcyclohexane

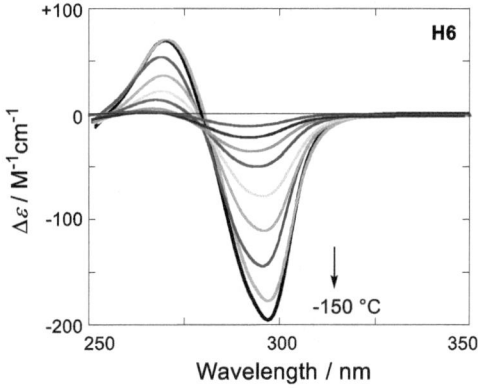

equilibrium operative in hexaarylbenzenes and/or the dynamic nature of propeller chirality should be responsible. The varied CD intensities are explained by the variation of effective tilt angles of radial aromatic blades against the central benzene ring, which is caused by several mechanisms. Under the condition without any perturbation by solvent, the angles are kept ideal at $\approx 60°$ (although either in clockwise or counterclockwise manner), where strong CE intensity is expected [8]. The reduced $g$ factors in smaller alkanes are explained by the (partial) penetration of solvent molecule(s) into the space between the radial aromatic blades, which increases the tilt angles near to perpendicular. The degree of penetration probably depends on the size of solvent molecule, which justifies the correlation among the linear alkanes. In larger cyclic alkanes (i.e., methylcyclohexane and decalin), $g$ factors are greatly enhanced as the penetration of solvent is not feasible. The reduced $g$ factors in polar solvents can be easily understood by conventional solvation at the periphery, which in turn increase the effective tilt angles of the radial aromatic blades. In addition, solvent viscosity also affects the dynamics between *C-CC* propeller equilibrium to reduce the contribution of *whizzing toroids*, circumventing the undesired reduction in CD intensity.

The CD intensity of flexible molecules generally increases at lower temperatures, due to the conformational freezing [12, 59]. The $^1L_b$ band is split to two transitions in HABs, but not in biaryls or teraryls [62], and shows bisignate Cotton effects in CD spectra, the magnitudes of which significantly increase with decreasing temperature (Fig. 7.11). Thus, the molar CD ($\Delta\varepsilon$) at the lowest-energy transition of **H6** is augmented to $-200$ $M^{-1}$ $cm^{-1}$ in isopentane at $-150$ °C, which is roughly six times larger than that observed at 25 °C in methylcyclohexane and equivalent to a 400 to 500-fold enhancement from the corresponding value for the parent chiral alkyl phenyl ether. Such an extraordinary CD augmentation by reducing temperature does not appear to arise from the conformational freezing alone. It is also to note that decreasing temperature does not cause peak sharpening but induces considerable red shifts of the CD extrema with appreciable band-broadening, suggesting existence of complex conformational equilibria independently affected by the temperature variation.

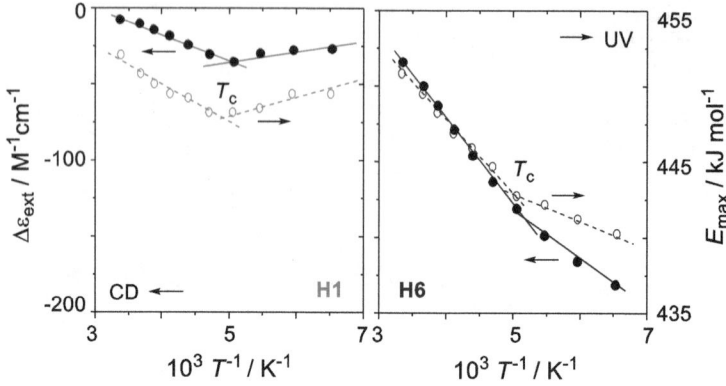

**Fig. 7.12** Temperature dependence of the excitation energy (absorption maximum) of the $^1L_b$ transition ($E_{max}$) and the CD intensity at the extremum ($\Delta\varepsilon_{ext}$) for **H1** (left) and **H6** (right) in methylcyclohexane. UV: open circle and dashed line. CD: solid circle and line

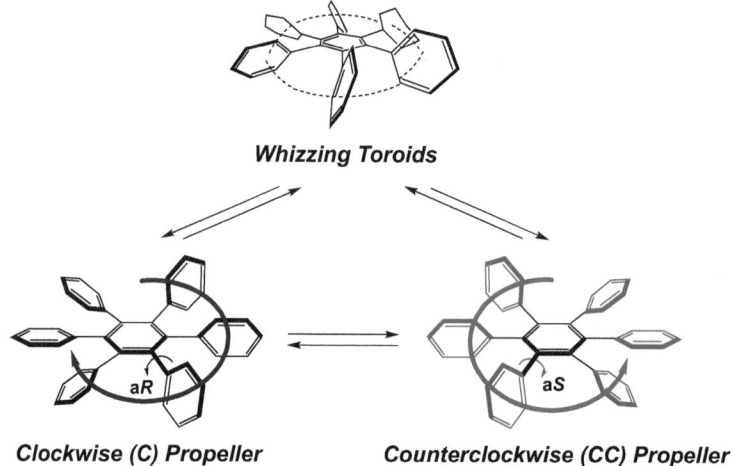

**Fig. 7.13** Conformer equilibrium in HABs

The Eyring-type analysis of temperature-dependent phenomenon often serves as an effective tool for elucidating the dynamic aspects of the system and indeed provides us with further insights into the dynamics of the propeller chirality of HABs. In Fig. 7.12, the CD intensity and the excitation energy (taken from the UV spectra) for the lowest-energy $^1L_b$ band of **H1** and **H6** are plotted against the reciprocal temperature ($T^{-1}$) to all afford a bent line or two kinds of correlations for each HAB, suggesting a switching of the major contributor at the critical temperature (that corresponds to the bending point observed): $T_c \approx -50\,°C$. We explain these observations by assuming the "*whizzing toroid*" conformation, in addition to the *C* and *CC* propeller geometries (Fig. 7.13).

In the temperature domain below $T_c$, the $C$ and $CC$ propellers are in equilibrium but the transient conformers of intermediate geometries (at higher vibrational levels) are less populated and hence less influential on the overall chiroptical responses. Accordingly, the intensity of CD is straightforwardly determined by the population of relaxed $C$ and $CC$ propellers at given temperature, and the experimental CD spectrum is well reproduced in intensity (and in pattern) by the theoretical calculations assuming the exclusive population to the $C$ conformation at the lowest temperature [8]. The relative preference for the more stable $C$ over $CC$ conformer explains the CD intensity of **H6** increased by lowering temperature (in this temperature domain). Along with the increase of Cotton effects, the excitation energy of **H6** is gradually blue-shifted with decreasing contribution of the $CC$ conformer at lower temperatures. In contrast, **H1** shows the opposite temperature-dependence behavior in this temperature domain. This is attributable, at least in part, to the different tilt angles between the relaxed $C$ and $CC$ conformers. Thus, more stable $C$ conformer is supposed to bear closer angles to the perpendicular tilt and accordingly leads to weaker CD intensity for this conformer of **H1**.

In the temperature domain above $T_c$, the CD intensity becomes more sensitive to temperature and the Cotton effects gradually fade out by increasing temperature for both **H1** and **H6**. The contribution of the *whizzing toroids*, which are maximized at the perpendicular blade angle, becomes more substantial with increasing population to the transient conformers in upper vibrational levels at higher temperatures. The excitonic coupling theory predicts that the chiroptical responses essentially vanish when two chromophores (more precisely, two transition moments) are perpendicular to each other [63–65], which is exactly the case with our observations. We consider that the *whizzing toroids* is not a discrete static conformation with six orthogonal radial aromatic rings but is better described as an ensemble of the intermediate geometries populated in between the $C$ and $CC$ conformers on the potential surface along the twist angle, contribution of which becomes larger at higher temperatures. Polar solvents drive the blade geometry analogous to the *whizzing toroids* by an effective solvation on the polar group(s) at the periphery. It has been already reported that the radial aromatic rings in HABs are perpendicular to the central ring at least in the NMR timescale in chloroform at an ambient temperature [66].

The toroidal interaction has been extensively investigated recently using a variety of HABs as model systems [34, 43]. The localization and delocalization of exciton, charge, and electron have been quantitatively discussed for this unique intramolecular π-interactions [44, 45, 48, 52]. In all the reported cases, the radial aromatic rings are assumed to align in perpendicular orientation and the rotation of rings is not considered. In our chiral HABs, the apparent blue-shifts observed for the UV and fluorescence excitation spectra of HABs in polar solvents and at higher temperatures are indicative of the H-type association of radial aromatic rings, where the propeller blades are more or less aligned face to face. In contrast, the partially π-overlapped J-type association becomes favored at lower temperatures in non-polar solvents, as evidenced by the bathochromic shift in UV and fluorescent excitation spectra [62, 67, 68]. The fluorescence lifetime examinations of **H6** also support the existence of the second conformer. All of the above observations are well comprehended by incorporating the *whizzing toroids* in the equilibrium of the *C* and *CC* propellers and indicate that their relative contribution as well as the interconversion dynamics play essential roles in determining the (chir)optical properties.

## 7.6 Propeller Inversion Dynamics

The energy profile for the propeller-chirality inversion of fully chiral HAB **H6** has been theoretically evaluated at the SCS-MP2/def2-TZVPP level incorporating the COSMO solvation model for methylcyclohexane. This level of theory has been proven to be quite accurate in describing the energy profiles of a variety of systems, especially where the weak interactions play critical roles [69–72]. A small energy difference of 0.8 kcal mol$^{-1}$ found for the *C* versus *CC* propeller geometries of **H6** corresponds to a *C*:*CC* ratio of 97:3, or 94% enantiomeric excess, at $-150$ °C. Careful kinetic studies on the blade inversion of substituted HABs have already demonstrated that the blades do not rotate simultaneously but rather one by one and the experimental free energy of activation for single inversion varies from 17 to 33 kcal mol$^{-1}$, depending on the substituent introduced [73]. In these studies, although the possible motion and distortion caused by the single blade rotation have been described, the cooperative effect has not been discussed any further.

Our theoretical investigation has revealed that the domino inversion is the most likely mechanism operative in propeller-inversion of HABs. Thus, the synchronous inversion, where all six radial aromatic blades rotate simultaneously, turned out to proceed with a fairly large rotation barrier of 8.4 kcal mol$^{-1}$, which is much larger than the barrier (4.2 kcal mol$^{-1}$) calculated for the single blade rotation assuming a fixed geometry for the remaining blades. Although this result clearly demonstrates the high preference for the single-blade rotation model (even without optimizing the geometries of the neighboring blades), it seems more probable that when one blade starts to rotate, the adjacent blades have to change the tilt angles to minimize the energy and such conformational adjustments propagate in a domino manner to the rest of the blades to eventually achieve the whole propeller inversion. This domino inversion model requires an activation energy as low as 2.1 kcal mol$^{-1}$ (by theoretical calculation) for the *C*-to-*CC* conversion in fully chiral HAB **H6**. Such a shallow potential surface well explains the diversity of the twist angles found in the crystal structures (vide supra) and the whizzing nature of aromatic blades in various HAB derivatives discussed in the earlier sections.

## 7.7 Propeller Chirality in Boron Heptaaryldipyrromethene

Recently, much attention has been paid to the preparation, analysis, and development of efficient circularly polarized luminescence (CPL) materials of small or simple organic molecules [74, 75]. Such molecules should be advantageous for

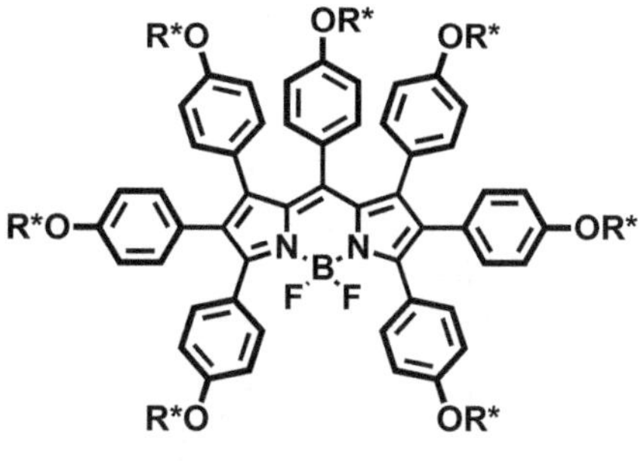

**Fig. 7.14** Propeller chiral
boron
heptaaryldipyrromethene
(heptaaryl-BODIPY)

**B7**

precisely controlling the emission wavelength as well as the luminescence efficiency through the rational design and sophisticated structural modification. So far, organic CPL molecules have shown relatively low luminescence dissymmetry factors ($g_{lum}$) in the order $10^{-2}$ to $10^{-5}$ ($g_{lum} = 2 \times (I_L - I_R) / (I_L + I_R)$, where $I_L$ and $I_R$ refer to the intensities of left- and right-handed circularly polarized luminescence observed upon excitation by unpolarized light).

Boron dipyrromethenes (BODIPYs), being highly fluorescent ($\Phi_{lum} = 0.93$ for parent BODIPY in methanol [76]), have been successfully employed as scaffolds for superior CPL materials [77]. The straightforward approach to introduce point-chiral auxiliary/ies to the parent structure does not appear to work due to the intrinsically small chiroptical responses of such BODIPY derivatives. Incorporating axial chirality in the BODIPY core to induce a helical twist to the chromophore has been proven more successful [77–79], but the fluorescence efficiency is significantly altered, mostly reduced, due to the deformation of the planar chromophore. In our approach to obtain less-distorted chiral BODIPYs, fully substituted BODIPY derivative **B7** was prepared by the condensation of chiral tetraarylcycopentadienes and the subsequent reaction with boron trifluoride in the presence of amine (Fig. 7.14). Indeed, the propeller chirality is attained in **B7**, but the magnitudes of chiroptical responses induced and their temperature-dependence behaviors are quite different from those of HAB **H6**.

The variable temperature CD spectra of boron heptaaryldipyrromethene **B7** in methylcyclohexane are shown in Fig. 7.15 (left). The bands at around 600 and 430 nm are assigned to the long- and short-axis transitions of the BODIPY chromophore, respectively, while that at 300 nm to a mixture of the $^1L_b$ and $^1L_a$ transitions of the radial aromatic rings. At 25 °C, practically no CD signals are seen for the BODIPY transitions at longer wavelengths, while two negative Cotton effects of similar strengths are observed at 230–340 nm. The Cotton effect pattern analogous to those reported for various chiral HABs suggests that the clockwise chirality is favored by this BODIPY-based propeller molecule incorporating the ($R$)-point chiral auxiliaries at the periphery of radial aromatic rings, as is the case with the chiral

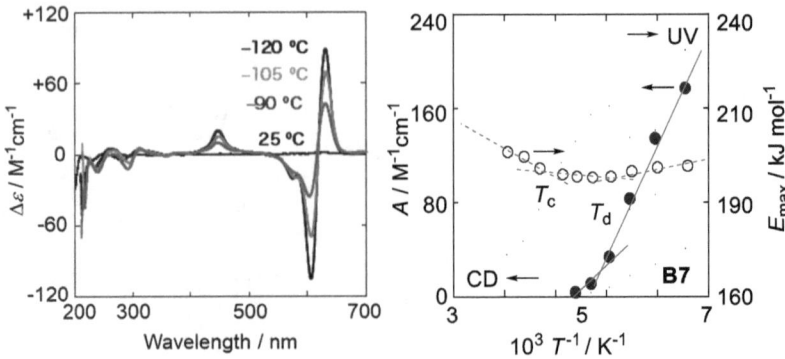

**Fig. 7.15** Left: variable temperature CD spectra of **B7** in methylcyclohexane. Right: plots of the CD couplet amplitude A (solid circle) and the excitation energy $E_{max}$ (open circle) evaluated from the absorption maximum of the long-axis transition of the BODIPY chromophore of **B7**

HAB propellers. This speculation is definitely supported by the theoretical calculations. The molar CD ($\Delta\varepsilon$) values for **B7** in methylcyclohexane at 25 °C are −5.2 and −4.3 $M^{-1}$ $cm^{-1}$ at 229 and 281 nm, respectively, both of which are roughly ten times smaller than those of **H6** under the comparable conditions, but still much larger than those of the reference compounds, i.e., chirally modified aryl ether and biphenyl (−0.4 and −0.5 $M^{-1}$ $cm^{-1}$) [13]. The following structural and electronic differences between **B7** and **H6** would be responsible for the tenfold smaller molar CD observed for the $^1L_b$ and $^1L_a$ transitions of **B7**. Although the radial aromatic rings interact with each other in **B7**, the domino effect is less efficient in **B7** than in **H6**, as a consequence of the larger separation between two adjacent radial aromatic rings, which measures 3.1 Å at the *ipso* carbon for **B7** but 2.9 Å for **H6**. The smaller twist angles ($\approx$55° on average) of radial aromatic rings should also diminish the CD intensity. More importantly, an interplane interaction is partially missing in **B7** at the boron atom, which should lead to a significant loss of CD intensity, as demonstrated in a study to elucidate how the toroidal interaction is affected by losing one or more blades from HAB [52].

The temperature dependence of the CD spectra of **B7** has been analyzed quantitatively. By decreasing the temperature, the Cotton effects at the $^1L_b$ and $^1L_a$ band region of radial aromatic rings (below 300 nm) gradually grows in intensity (until the transition temperature of $T_d = -70$ °C is reached) to give the extremal $\Delta\varepsilon$ of ca. −14 $M^{-1}$ $cm^{-1}$ at $T_d$ and then decreases. A closer examination reveals an additional inflection point at $T_c = +10$ °C in the positive-slope region, disclosing the existence of three different temperature domains for **B7** (instead of two for HABs). The variable temperature UV spectra of **B7** behave similarly to show three distinct temperature domains for the temperature-dependence profiles of both excitation wavelength and absorptivity (Fig. 7.15, right). The BODIPY's long-axis transition in the low-energy region exhibits more dramatic variable temperature CD spectral changes as shown in Fig. 7.15 (right). Crucially, this band is essentially CD silent at higher temperatures but starts to develop a positive excitonic couplet near

600 nm at $-75\,°\text{C}$, the intensity of which increases slowly with decreasing temperature down to $-120\,°\text{C}$ and leaps much more rapidly thereafter to achieve the largest amplitude $A$ of $193\ \text{M}^{-1}\ \text{cm}^{-1}$ and anisotropy factor $g_{\text{abs}}$ of $2.3 \times 10^{-3}$ at $-120\,°\text{C}$. The apparent excitonic coupling at lower temperatures indicates that two (or more) BODIPY units are positioned in close proximity under the low-temperature conditions employed (vide infra).

## 7.8 Supramolecular Behavior

In contrast to the two temperature domains (assignable to the propeller and toroidal domains) found for HABs, three discrete temperature domains are shown to exist for chiral BODIPY **B7**. The temperature-dependence behaviors of (chir)optical properties observed for **B7** in the higher temperature domains are quite analogous to those for **H6**. This allows us to assign the CD intensity of **B7** gradually increasing with decreasing temperature from $T_c = +10\,°\text{C}$ to $T_d = -70\,°\text{C}$ to the equilibrium shift of the *CC* to *C* propeller (Fig. 7.16, top). At temperatures above $T_c$, the rotation of propeller blades becomes more vigorous to render the contribution of *whizzing toroids* more significant. The much higher $T_c$ for **B7** ($+10\,°\text{C}$) than for **H6** ($-50\,°\text{C}$) is sensible in view of the larger inter-blade separation and the missing aromatic blade at the boron atom. As such, higher energy is required to effectively incorporate the quasi-toroidal interaction. The activation energy calculated for the domino inversion of **B7** propeller ($3.8\ \text{kcal mol}^{-1}$) is appreciably higher than that of **H6** ($2.1\ \text{kcal mol}^{-1}$), probably due to the irregular arrangement of propeller blades on the BODIPY core.

The temperature domain below $T_d$ ($-70\,°\text{C}$), where the CD couplet rapidly grows, is unique to heptaaryl-BODIPY **B7**, while HAB **H6** does not show any sign of such behavior at least down to $-150\,°\text{C}$. Concentration-dependence and diffusion-ordered NMR spectroscopic studies have revealed the supramolecular dimer formation of **B7** facilitated particularly in this temperature domain (Fig. 7.16, bottom). Theoretical investigation suggested the formation of a head-to-tail dimer in a plus-screw alignment ($\varphi \approx +15°$), where two **B7** molecules are stacked to each other in a face-to-face distance of $r = 5.1$ Å and a center-to-center displacement of $d = 2.4$ Å to avoid the sterical clashes yet mutually fill up the void space around the boron. The right-twisted arrangement of two **B7** should induce a positive excitonic coupling in the CD spectrum, which is in good agreement with the experimental observation (Fig. 7.15a). Covalently bonded chiral BODIPY dimers have been enthusiastically developed recently, either by tethering BODIPYs to a chiral scaffold or by directly connecting two BODIPY units to induce axial chirality [80–85]. The BODIPY derivative **B7** comprised from supramolecular dimer formation provides superior and well-defined chiroptical responses by the exciton coupled transition, which can be easily modulated by the environmental factors such as temperature. Moreover, the highly fluorescent nature of BODIPY core better serves as a switchable CPL materials that emits polarized light in visible and near-infrared region (vide infra).

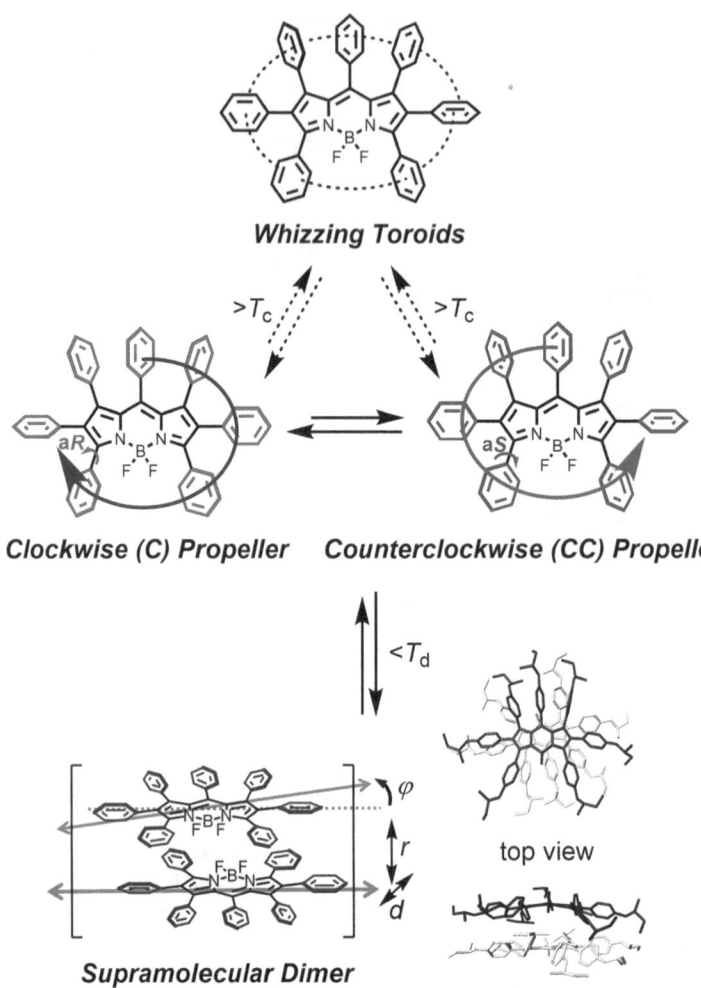

**Fig. 7.16** Conformer equilibria for boron heptaaryldipyrromethene **B7** and supramolecular dimer formation facilitated at lower temperatures; the geometry of supramolecular dimer (bottom) optimized at the DFT-D3(BJ)-TPSS/def2-SV(P) level

## 7.9 Circularly Polarized Luminescence in Propeller Chirality

The chiral HABs (**H1–H6**) are only weakly fluorescent. In contrast, chiral heptaaryl-BODIPY **B7** strongly fluoresces at around 640 nm with a relatively large Stokes shift of 850 cm$^{-1}$. The fluorescence quantum yield is 0.32 at 25 °C but is enhanced to 0.45 by reducing the temperature to −120 °C. It has been shown that J-type BODIPY dimers are moderately fluorescent, while H-type dimers are generally

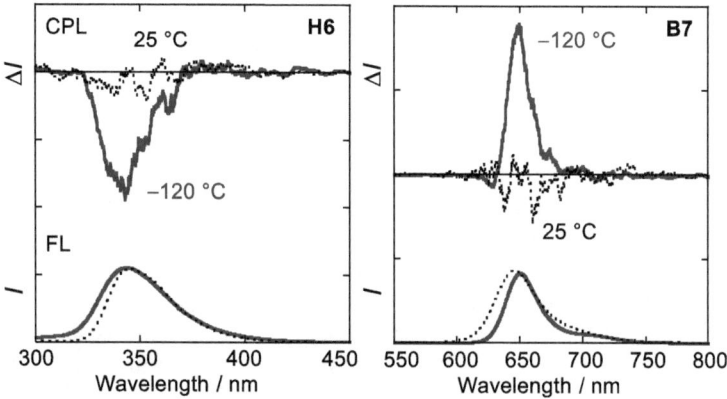

**Fig. 7.17** Circularly polarized luminescence and intensity-normalized total luminescence spectra of HAB **H6** and heptaaryl-BODIPY **B7** in methylcyclohexane at 25 °C (dotted lines) and −120 °C (solid lines). Excitation wavelengths: 260 and 440 nm for **H6** and **B7**, respectively

less fluorescent [80–85]. In this regard, the moderate fluorescence observed for **B7** at lower temperatures seems reasonable, as the H-type association is naturally avoided by the steric crowding of radial aromatic rings.

At 25 °C, where the contribution of *whizzing toroids* is substantial, nearly negligible CPL is observed for both **H6** and **B7** (Fig. 7.17), a behavior parallel to the CD spectra that only exhibit the weak Cotton effects (Figs. 7.11 and 7.15). This situation is remarkably improved in fully chiral HAB **H6** when the clockwise propeller chirality becomes dominant by reducing the temperature to −120 °C, affording positive CPL with a reasonably large $g_{lum}$ of $-6.0 \times 10^{-4}$ at 340 nm. On the other hand, having the same clockwise propeller chirality, fully chiral heptaaryl-BODIPY **B7** gives the oppositely signed, positive CPL with a much larger $g_{lum}$ of $2.0 \times 10^{-3}$ at 650 nm under the comparable conditions, which is rationalized by the formation of right-twisted supramolecular dimer (vide supra).

## 7.10  Perspective

Thanks to the sterical congestion of the adjacent aromatic rings, the conceptually new propeller chirality is induced in highly substituted benzenes and related aromatics. In this new type of chirality, an array of axially chiral aromatic blades attached to the hub behaves rather cooperatively to achieve very strong chiroptical responses, which are accordingly not static but highly dynamic and hence feasibly manipulatable in nature. We showed above such extraordinary features of the propeller chirality using the circular dichroism and circularly polarized luminescence of HABs and heptaaryl-BODIPY. By attaching a small chiral auxiliary at a periphery of radial aromatic blades, the whole propeller-shaped molecule of rich conformational diversity becomes highly susceptive to various environmental variants and easily driven to clockwise (*C*) or counterclockwise (*CC*) propeller. At low temperatures in less polar solvents, the minor toroidal conformer is almost negligible and the enantiomeric propeller, i.e., *C-CC* equilibrium becomes practically dominant in chiral HABs, while the propeller conformation is less significant in heptaaryl-BODIPY where the radial aromatic rings are mutually more separated in space. The enhanced chiroptical responses and the highly susceptive and manipulatable nature of propeller molecules will contribute to amplified chirality sensing as well as advanced chiroptical materials. The contribution of *whizzing toroids* becomes substantial at temperatures higher than critical temperature $T_c$ to affect the observed (chir)optical properties. Likewise, the propeller chirality is highly sensitive to the solvent employed. The smaller solvent may intervene between the radial blades in different degrees of penetration, depending on the size and shape of the solvent molecule. This will cause the increase of effective tilt angles of blades, which in turn reduces the observed chiroptical responses. A conventional solvation, feasible at the polar peripheries with polar solvent, also induces the increase of the blade angles to impact the observed CD. Recently, the propeller chirality and the corresponding CD responses have been demonstrated to be controlled also by applied pressure [86]. Such propeller inversion process and dynamics have to be taken into account in fine-tuning the structure and properties of propeller-shaped molecules and related materials. Remarkably, heptaaryl-BODIPY, lacking one blade in the propeller, was found to form supramolecular dimer at low temperatures in non-polar solvent. Thus, at temperatures lower than $T_d$, the bisignate Cotton effects become apparent, representing the strong excitonic coupling between the two BODIPY chromophores. Consequently, this supramolecular dimer affords intense CPL ($g_{lum} = 2.0 \times 10^{-3}$) with a good fluorescent quantum yield ($\Phi_{lum} = 0.45$), which is opposite in sign to those observed for the corresponding HAB propellers due to the different origins.

In conclusion, the propeller chirality of HABs and related molecules is not simply multiple combination of axial chirality. A domino-like corporation effect between the radial aromatic blades is crucial to harvest the highly enhanced chiroptical responses. More significantly, such chirality is dynamic in nature, which can be easily controlled by subtle environmental factors such as temperature, solvent, etc. We encourage further research on the propeller chirality in the relevant systems to boost the development of new design principle for advanced and expectantly superior chiroptical materials, especially those that can emit efficient circularly polarized light at desired wavelength and is easily switched on-off or modulated at will.

# References

1. von Delius M, Leigh DA (2011) Walking molecules. Chem Soc Rev 40:3656–3676. https://doi.org/10.1039/c1cs15005g
2. Michl J, Sykes ECH (2009) Molecular rotors and motors: recent advances and future challenges. ACS Nano 3:1042–1048. https://doi.org/10.1021/nn900411n

3. Browne WR, Feringa BL (2006) Making molecular machines work. Nat Nanotechnol 1:25–35. https://doi.org/10.1038/nnano.2006.45
4. Schliwa M, Woehlke G (2003) Molecular motors. Nature 422:759–765. https://doi.org/10.1038/nature01601
5. Kelly TR (2001) Progress toward a rationally designed molecular motor. Acc Chem Res 34:514–522. https://doi.org/10.1021/ar000167x
6. Feringa B, In L (2001) Control of motion: from molecular switches to molecular motors. Acc Chem Res 34:504–513. https://doi.org/10.1021/ar0001721
7. Kosaka T, Iwai S, Inoue Y, Moriuchi T, Mori T (2018) Solvent and temperature effects on dynamics and chiroptical properties of propeller chirality and toroidal interaction of hexaarylbenzenes. J Phys Chem A 122:7455–7463. https://doi.org/10.1021/acs.jpca.8b06535
8. Kosaka T, Inoue Y, Mori T (2016) Toroidal interaction and propeller chirality of hexaarylbenzenes. Dynamic domino inversion revealed by combined experimental and theoretical circular dichroism studies. J Phys Chem Lett 7:783–788. https://doi.org/10.1021/acs.jpclett.6b00179
9. Toda M, Matsumura C, Tsurukawa M, Okuno T, Nakano T, Inoue Y, Mori T (2012) Absolute configuration of atropisomeric polychlorinated biphenyl 183 enantiomerically enriched in human samples. J Phys Chem A 116:9340–9346. https://doi.org/10.1021/jp306363n
10. Nishizaka M, Mori T, Inoue Y (2011) Axial chirality of donor-donor, donor-acceptor, and tethered 1,1′-binaphthyls: a theoretical revisit with dynamics trajectories. J Phys Chem A 115:5488–5495. https://doi.org/10.1021/jp202776g
11. Nishizaka M, Mori T, Inoue Y (2010) Experimental and theoretical studies on the chiroptical properties of donor-acceptor binaphthyls. Effects of dynamic conformer population on circular dichroism. J Phys Chem Lett 1:1809–1812. https://doi.org/10.1021/jz100574e
12. Nishizaka M, Mori T, Inoue Y (2010) Conformation elucidation of tethered donor-acceptor binaphthyls from the anisotropy factor of a charge-transfer band. J Phys Chem Lett 1:2402–2405. https://doi.org/10.1021/jz100901n
13. Mori T, Inoue Y, Grimme S (2007) Experimental and theoretical study of the CD spectra and conformational properties of axially chiral 2,2′-, 3,3′-, and 4,4′-biphenol ethers. J Phys Chem A 111:4222–4234. https://doi.org/10.1021/jp071709w
14. Shimizu A, Inoue Y, Mori T (2017) Protonation-induced sign inversion of the cotton effects of pyridinophanes. A combined experimental and theoretical study. J Phys Chem A 121:977–985. https://doi.org/10.1021/acs.jpca.6b12287
15. Mori T, Inoue Y (2011) Recent theoretical and experimental advances in the electronic circular dichroisms of planar chiral cyclophanes. Top Curr Chem 298:99–128. https://doi.org/10.1007/128_2010_59
16. Wakai A, Fukasawa H, Yang C, Mori T, Inoue Y (2012) Theoretical and experimental investigations of circular dichroism and absolute configuration determination of chiral anthracene photodimers. J Am Chem Soc 134:4990–4997. https://doi.org/10.1021/ja300522y
17. Mori T, Inoue Y, Grimme S (2007) Quantum chemical study on the circular dichroism spectra and specific rotation of donor-acceptor cyclophanes. J Phys Chem A 111:7995–8006. https://doi.org/10.1021/jp073596m
18. Nakai Y, Mori T, Sato K, Inoue Y (2013) Theoretical and experimental studies of circular dichroism of mono- and diazonia[6]helicenes. J Phys Chem A 117:5082–5092. https://doi.org/10.1021/jp403426w
19. Nakai Y, Mori T, Inoue Y (2013) Circular dichroism of (di)methyl- and diaza[6]helicenes. A combined theoretical and experimental study. J Phys Chem A 117:83–93. https://doi.org/10.1021/jp3104084
20. Nakai Y, Mori T, Inoue Y (2012) Theoretical and experimental studies on circular dichroism of carbo[n]helicenes. J Phys Chem A 116:7372–7385. https://doi.org/10.1021/jp304576g
21. Tanaka H, Kato Y, Fujiki M, Inoue Y, Mori T (2018) Combined experimental and theoretical study on circular dichroism and circularly polarized luminescence of configurationally robust $D_3$-symmetric triple pentahelicene. J Phys Chem A 122:7378–7384. https://doi.org/10.1021/acs.jpca.8b05247

22. Tanaka H, Ikenosako M, Kato Y, Fujiki M, Inoue Y, Mori T (2018) Symmetry-based rational design for boosting chiroptical responses. Commun Chem 1:38. https://doi.org/10.1038/s42004-018-0035-x

23. Mori T, Tanaka T, Higashino T, Yoshida K, Osuka A (2016) Combined experimental and theoretical investigations on optical activities of Möbius aromatic and Möbius antiaromatic hexaphyrin phosphorus complexes. J Phys Chem A 120:4241–4248. https://doi.org/10.1021/acs.jpca.6b03978

24. Kang J, Miyajima D, Itoh Y, Mori T, Tanaka H, Yamauchi M, Inoue Y, Harada S, Aida T (2014) C5-Symmetric chiral corannulenes: desymmetrization of bowl inversion equilibrium via "intramolecular" hydrogen-bonding network. J Am Chem Soc 136:10640–10644. https://doi.org/10.1021/ja505941b

25. Mori T, Ko YH, Kim K, Inoue Y (2006) Circular dichroism of intra- and intermolecular charge-transfer complexes. enhancement of anisotropy factors by dimer formation and by confinement. J Org Chem 71:3232–3247. https://doi.org/10.1021/jo0602672

26. Mori T, Inoue Y (2005) Circular dichroism of a chiral tethered donor-acceptor system: enhanced anisotropy factors in charge-transfer transitions by dimer formation and by confinement. Angew Chem Int Ed 44:2582–2585. https://doi.org/10.1002/ange.200462071

27. Kang J, Miyajima D, Mori T, Inoue Y, Itoh Y, Aida T (2015) A rational strategy for the realization of chain-growth supramolecular polymerization. Science 347:646–651. https://doi.org/10.1126/science.aaa4249

28. Yamagishi H, Fukino T, Hashizume D, Mori T, Inoue Y, Hikima T, Takata M, Aida T (2015) metal-organic nanotube with helical and propeller-chiral motifs composed of a $C_{10}$-symmetric double-decker nanoring. J Am Chem Soc 137:7628–7631. https://doi.org/10.1021/jacs.5b04386

29. Toyoda M, Imai Y, Mori T (2017) Propeller chirality of boron heptaaryldipyrromethene: unprecedented supramolecular dimerization and chiroptical properties. J Phys Chem Lett 8:42–48. https://doi.org/10.1021/acs.jpclett.6b02492

30. Goldup SM (2016) Mechanical chirality: a chiral catalyst with a eing to it. Nat Chem 8:404–406. https://doi.org/10.1038/nchem.2509

31. Cakmak Y, Erbas-Cakmak S, Leigh DA (2016) Asymmetric catalysis with a mechanically point-chiral rotaxane. J Am Chem Soc 138:1749–1751. https://doi.org/10.1021/jacs.6b00303

32. Bordoli RJ, Goldup SM (2014) An efficient approach to mechanically planar chiral rotaxanes. J Am Chem Soc 136:4817–4820. https://doi.org/10.1021/ja412715m

33. Tachibana Y, Kihara N, Takata T (2004) Asymmetric benzoin condensation catalyzed by chiral rotaxanes tethering a thiazolium salt moiety via the cooperation of the component: can rotaxane be an effective reaction field? J Am Chem Soc 126:3438–3439. https://doi.org/10.1021/ja0394611

34. Vij V, Bhalla V, Kumar M (2016) Hexaarylbenzene: evolution of properties and applications of multitalented scaffold. Chem Rev 116:9565–9627. https://doi.org/10.1021/acs.chemrev.6b00144

35. Loxq P, Manoury E, Poli R, Deydier E, Labande A (2016) Synthesis of axially chiral biaryl compounds by asymmetric catalytic reactions with transition metals. Coord Chem Rev 308:131–190. https://doi.org/10.1016/j.ccr.2015.07.006

36. Kumarasamy E, Ayitou AJ-L, Vallavoju N, Raghunathan R, Iyer A, Clay A, Kandappa SK, Sivaguru J (2016) Tale of twisted molecules. Atroposelective photoreactions: taming light induced asymmetric transformations through non-biaryl atropisomers. Acc Chem Res 49:2713–2724. https://doi.org/10.1021/acs.accounts.6b00357

37. Cozzi PG, Emer E, Gualandi A (2011) Atroposelective organocatalysis. Angew Chem Int Ed 50:3847–3849. https://doi.org/10.1002/anie.201008031

38. Durairaj K (1994) Modern concepts and strategies in synthesis of biaryl compounds. Curr Sci 66:833–838

39. Bringmann G, Walter R, Weirich R (1990) Modern strategies for constructing biaryl compounds. Angew Chem 102:1006–1019

40. Crabbe P, Klyne W (1967) Optical rotatory dispersion and circular dichroism of aromatic compounds: a general survey. Tetrahedron 23:3449–3503. https://doi.org/10.1016/S0040-4020(01)92310-5
41. Lungerich D, Reger D, Hoelzel H, Riedel R, Martin MMJC, Hampel F, Jux N (2016) A strategy towards the multigram synthesis of uncommon hexaarylbenzenes. Angew Chem Int Ed 55:5602–5605. https://doi.org/10.1002/anie.201600841
42. Suzuki S, Segawa Y, Itami K, Yamaguchi J (2015) Synthesis and characterization of hexaarylbenzenes with five or six different substituents enabled by programmed synthesis. Nat Chem 7:227–233. https://doi.org/10.1038/nchem.2174
43. Lambert C (2005) Hexaarylbenzenes – prospects for toroidal delocalization of charge and energy. Angew Chem Int Ed 44:7337–7339. https://doi.org/10.1002/anie.200502105
44. Steeger M, Holzapfel M, Schmiedel A, Lambert C (2016) Energy redistribution dynamics in triarylamine-triarylborane containing hexaarylbenzenes. Phys Chem Chem Phys 18:13403–13412. https://doi.org/10.1039/C6CP01923D
45. Khan FA, Wang D, Pemberton B, Talipov MR, Rathore R (2016) Toroidal delocalization of a single electron through circularly-arrayed benzophenone chromophores in hexakis (4-benzoylphenyl)benzene. J Photochem Photobiol A Chem 331:153–159. https://doi.org/10.1016/j.jphotochem.2016.05.002
46. Steeger M, Griesbeck S, Schmiedel A, Holzapfel M, Krummenacher I, Braunschweig H, Lambert C (2015) On the relation of energy and electron transfer in multidimensional chromophores based on polychlorinated triphenylmethyl radicals and triarylamines. Phys Chem Chem Phys 17:11848–11867. https://doi.org/10.1039/C4CP05929H
47. Kim S, Lee S-H, Shin H, Kay K-Y, Park J (2014) New hole transporting materials based on hexaarylbenzene and aromatic amine moiety for organic light-emitting diodes. J Nanosci Nanotechnol 14:6382–6385. https://doi.org/10.1166/jnn.2014.8291
48. Steeger M, Lambert C (2012) Charge-transfer interactions in tris-donor-tris-acceptor hexaarylbenzene redox chromophores. Chem A Eur J 18:11937–11948. https://doi.org/10.1002/chem.201104020
49. Lambert C, Ehbets J, Rausch D, Steeger M (2012) Charge-transfer interactions in a multichromophoric hexaarylbenzene containing pyrene and triarylamines. J Org Chem 77:6147–6154. https://doi.org/10.1021/jo300924x
50. Rios C, Salcedo R (2014) Computational study of electron delocalization in hexaarylbenzenes. Molecules 19:3274–3296. https://doi.org/10.3390/molecules19033274
51. Tanaka Y, Koike T, Akita M (2010) 2-Dimensional molecular wiring based on toroidal delocalization of hexaarylbenzene. Chem Commun 46:4529–4531. https://doi.org/10.1039/c0cc00128g
52. Rosokha SV, Neretin IS, Sun D, Kochi JK (2006) Very fast electron migrations within p-doped aromatic cofacial arrays leading to three-dimensional (toroidal) π-delocalization. J Am Chem Soc 128:9394–9407. https://doi.org/10.1021/ja060393n
53. Xu H, Wolffs M, Tomovic Z, Meijer EW, Schenning APHJ, De Feyter S (2011) A multivalent hexapod having 24 stereogenic centers: chirality and conformational dynamics in homochiral and heterochiral systems. CrstEngComm 13:5584–5590. https://doi.org/10.1039/c1ce05433c
54. Reimers JR, Ford MJ, Goerigk L (2016) Problems, successes and challenges for the application of dispersion-corrected density-functional theory combined with dispersion-based implicit solvent models to large-scale hydrophobic self-assembly and polymorphism. Mol Simul 42:494–510. https://doi.org/10.1080/08927022.2015.1066504
55. Kim M, Kim WJ, Lee EK, Lebegue S, Kim H (2016) Recent development of atom-pairwise van der Waals corrections for density functional theory: from molecules to solids. Int J Quantum Chem 116:598–607. https://doi.org/10.1002/qua.25061
56. Cho Y, Cho WJ, Youn IS, Lee G, Singh NJ, Kim KS (2014) Density functional theory based study of molecular interactions, recognition, engineering, and quantum transport in π molecular systems. Acc Chem Res 47:3321–3330. https://doi.org/10.1021/ar400326q

57. Shimizu A, Inoue Y, Mori T (2017) A combined experimental and theoretical study on the circular dichroism of staggered and eclipsed forms of dimethoxy[2.2]-, [3.2]-, and [3.3] pyridinophanes and their protonated forms. J Phys Chem A 121:8389–8398. https://doi.org/10.1021/acs.jpca.7b08623

58. Shimizu A, Mori T, Inoue Y, Yamada S (2009) Combined experimental and quantum chemical investigation of chiroptical properties of nicotinamide derivatives with and without intramolecular cation-π interactions. J Phys Chem A 113:8754–8764. https://doi.org/10.1021/jp904243w

59. Mori T, Grimme S, Inoue Y (2007) A combined experimental and theoretical study on the conformation of multiarmed chiral aryl ethers. J Org Chem 72:6998–7010. https://doi.org/10.1021/jo071216n

60. Lovrecek B, Despic A, Bockris JOM (1959) Electrolytic junctions with rectifying properties. J Phys Chem 63:750–751. https://doi.org/10.1021/j150575a030

61. Kobayashi K, Shirasaka T, Sato A, Horn E, Furukawa N (1999) Self-assembly of a radially functionalized hexagonal molecule: hexakis(4-hydroxyphenyl)benzene. Angew Chem Int Ed 38:3483–3486. https://doi.org/10.1002/(SICI)1521-3773(19991203)38:23<3483::AID-ANIE3483>3.0.CO;2-A

62. Dale J (1957) Ultraviolet absorption spectra of ortho- and para-linked polyphenyls. Acta Chem Scand 11:650–659. https://doi.org/10.3891/acta.chem.scand.11-0650

63. Berova N, Bari LD, Pescitelli G (2007) Application of electronic circular dichroism in configurational and conformational analysis of organic compounds. Chem Soc Rev 36:914–931. https://doi.org/10.1039/B515476F

64. Sakai N, Talukdar P, Matile S (2006) Use of the exciton chirality method in the investigation of ligand-gated synthetic ion channels. Chirality 18:91–94. https://doi.org/10.1002/chir.20221

65. Berova N, Borhan B, Dong JG, Guo J, Huang X, Karnaukhova E, Kawamura A, Lou J, Matile S, Nakanishi K, Rickman B, Su J, Tan Q, Zanze I (1998) Solving challenging bioorganic problems by exciton coupled CD. Pure Appl Chem 70:377–383. https://doi.org/10.1351/pac199870020377

66. Gust D (1977) Restricted rotation in hexaarylbenzenes. J Am Chem Soc 99:6980–6982. https://doi.org/10.1021/ja00463a034

67. Kasha M, Rawls HR, El-Bayoumi MA (1965) Exciton model in molecular spectroscopy. Pure Appl Chem 11:371–392. https://doi.org/10.1351/pac196511030371

68. Jelley E, Spectral E (1936) Absorption and fluorescence of dyes in the molecular state. Nature 138:1009–1010. https://doi.org/10.1038/1381009a0

69. Hobza P (2012) Calculations on noncovalent interactions and databases of benchmark interaction energies. Acc Chem Res 45:663–672. https://doi.org/10.1021/ar200255p

70. Grimme S, Goerigk L, Fink RF (2012) Spin-component-scaled electron correlation methods. Wiley Interdiscip Rev Comput Mol Sci 2:886–906. https://doi.org/10.1002/wcms.1110

71. Schwabe T, Grimme S (2008) Theoretical thermodynamics for large molecules: walking the thin line between accuracy and computational cost. Acc Chem Res 41:569–579. https://doi.org/10.1021/ar700208h

72. Klamt A (2011) The COSMO and COSMO-RS solvation models. Wiley Interdiscip Rev Comput Mol Sci 1:699–709. https://doi.org/10.1002/wcms.56

73. Gust D, Patton A (1978) Dynamic stereochemistry of hexaarylbenzenes. J Am Chem Soc 100:8175–8181. https://doi.org/10.1021/ja00494a026

74. Sanchez-Carnerero EM, Agarrabeitia AR, Moreno F, Maroto BL, Muller G, Ortiz MJ, de la Moya S (2015) Circularly polarized luminescence from simple organic molecules. Chem A Eur J 21:13488–13500. https://doi.org/10.1002/chem.201501178

75. Tanaka H, Inoue Y, Mori T (2018) Circularly polarized luminescence and circular dichroisms in small organic molecules: correlation between excitation and emission dissymmetry factors. ChemPhotoChem 2:386–402. https://doi.org/10.1002/cptc.201800015

76. Arroyo IJ, Hu R, Merino G, Tang BZ, Peña-Cabrera E (2009) The smallest and one of the brightest. Efficient preparation and optical description of the parent borondipyrromethene system. J Org Chem 74:5719–5722. https://doi.org/10.1021/jo901014w

77. Jimenez J, Cerdan L, Moreno F, Maroto BL, Garcia-Moreno I, Lunkley JL, Muller G, de la Moya S (2017) Chiral organic dyes endowed with circularly polarized laser emission. J Phys Chem C 121:5287–5292. https://doi.org/10.1021/acs.jpcc.7b00654
78. Ray C, Sanchez-Carnerero EM, Moreno F, Maroto BL, Agarrabeitia AR, Ortiz MJ, Lopez-Arbeloa I, Banuelos J, Cohovi KD, Lunkley JL, Muller G, de la Moya S (2016) Bis (haloBODIPYs) with labile helicity: valuable simple organic molecules that enable circularly polarized luminescence. Chem A Eur J 22:8805–8808. https://doi.org/10.1002/chem. 201601463
79. Alnoman RB, Rihn S, O'Connor DC, Black FA, Costello B, Waddell PG, Clegg W, Peacock RD, Herrebout W, Knight JG, Hall MJ (2016) Circularly polarized luminescence from helically chiral N,N,O,O-boron-chelated dipyrromethenes. Chem A Eur J 22:93–96. https://doi.org/10. 1002/chem.201504484
80. Zinna F, Bruhn T, Guido CA, Ahrens J, Broering M, Di Bari L, Pescitelli G (2016) Circularly polarized luminescence from axially chiral BODIPY DYEmers: an experimental and computational study. Chem A Eur J 22:16089–16098. https://doi.org/10.1002/chem.201602684
81. Momeni MR, Brown A (2016) A local CC2 and TDA-DFT double hybrid study on BODIPY/aza-BODIPY dimers as heavy atom free triplet photosensitizers for photodynamic therapy applications. J Phys Chem A 120:2550–2560. https://doi.org/10.1021/acs.jpca.6b02883
82. Bruhn T, Pescitelli G, Witterauf F, Ahrens J, Funk M, Wolfram B, Schneider H, Radius U, Broering M (2016) Cryptochirality in 2,2′-coupled BODIPY DYEmers. Eur J Org Chem 2016:4236–4243. https://doi.org/10.1002/ejoc.201600585
83. Wang J, Wu Q, Wang S, Yu C, Li J, Hao E, Wei Y, Mu X, Jiao L (2015) Conformation-restricted partially and fully fused BODIPY dimers as highly stable near-infrared fluorescent dyes. Org Lett 17:5360–5363. https://doi.org/10.1021/acs.orglett.5b02717
84. Wang Y, Chen L, El-Shishtawy RM, Aziz SG, Muellen K (2014) Synthesis and optophysical properties of dimeric aza-BODIPY dyes with a push-pull benzodipyrrolidone core. Chem Commun 50:11540–11542. https://doi.org/10.1039/C4CC03759F
85. Misra R, Dhokale B, Jadhav T, Mobin SM (2014) meso-Aryloxy and meso-arylaza linked BODIPY dimers: synthesis, structures and properties. New J Chem 38:3579–3585. https://doi. org/10.1039/C4NJ00354C
86. Kosaka T, Iwai S, Fukuhara G, Imai Y, Mori T (2019) Hydrostatic pressure on toroidal interaction and propeller chirality of hexaarylbenzenes. Explicit solvent effects on differential volumes in methylcyclohexane and hexane. Chem A Eur J 25:2011–2018. https://doi.org/10. 1002/chem.201804688

# Chapter 8
# Photo-Switching of Circularly Polarized Luminescence

**Takuya Nakashima and Tsuyoshi Kawai**

**Abstract** Circularly polarized luminescence (CPL) emission from chiral molecular systems is readily switched by means of external stimuli such as temperature, solvent, chemicals, and light irradiation. Since CPL is one of the emission phenomena, it can be modulated in a similar manner to the emission switching. The ON-OFF switching of emission intensity of chiral molecular systems may simply lead to the modulation of CPL intensity. Apart from the modulation of emission intensity, the chiral structures including chiral arrangement of fluorophores or metal coordination geometries are switched by external stimuli, changing the dissymmetry factors, i.e., the quality of CPL. In this chapter, we review the design of chiroptical photo-switches based on photochromic molecules that modulate the CPL property in response to photo-irradiation in a dynamic manner.

## 8.1 Introduction

The CPL technique is expected to find potential applications in sensors, display, and advanced information technologies including anti-counterfeiting labeling. Recent research has made efforts in the manipulation and modulation of CPL activity to further develop these technologies including supramolecular approaches [1–11]. High-speed switching of CPL in a dynamic manner by external stimuli should be of particular interest for the application in cryptographic communication as a technology for information security [12–14]. Light is one of the external inputs for various practical applications, whereby one can control the physico-chemical properties of active materials with precise spatiotemporal control in a remote and noninvasive manner. In this context, chiroptical photo-switches are promising materials for the dynamic modulation of CPL property with imparting luminescent capability to photo-responsive molecules [15]. Photo-switching of luminescent

T. Nakashima (✉) · T. Kawai (✉)
Division of Materials Science, Graduate School of Science and Technology, Nara Institute of Science and Technology, Ikoma, Nara, Japan
e-mail: ntaku@ms.naist.jp; tkawai@ms.naist.jp

© Springer Nature Singapore Pte Ltd. 2020
T. Mori (ed.), *Circularly Polarized Luminescence of Isolated Small Organic Molecules*, https://doi.org/10.1007/978-981-15-2309-0_8

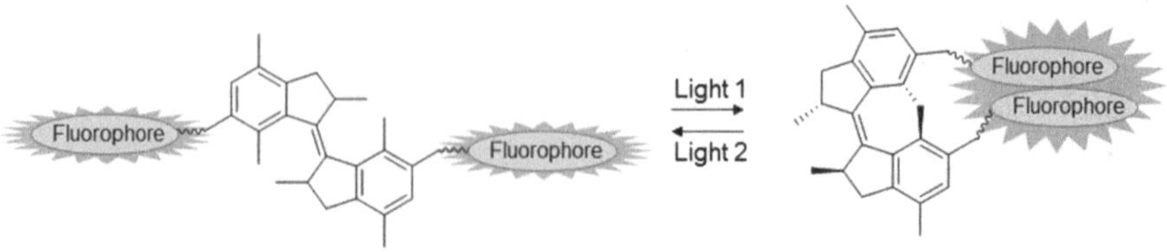

**Fig. 8.1** Change of the intramolecular interaction between fluorophores in response to the geometrical change of photochromic molecule

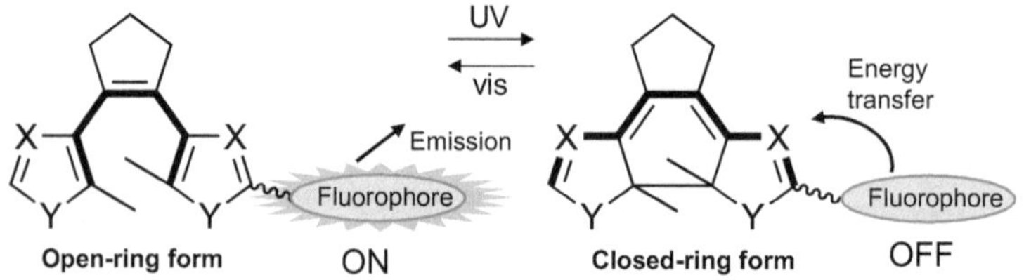

**Fig. 8.2** ON-OFF emission switching based on a change in the electronic structure of photochromic diarylethene

property has been demonstrated by exploiting changes in the electronic and geometric structures of photochromic molecules [16–18]. For example, *trans-cis* photo-isomerization of thioindigo [19] and overcrowded alkene [20] readily controlled the arrangement of two fluorophore units in an intramolecular manner, modulating their emission between the monomeric and dimeric (excimer) ones (Fig. 8.1). The photo-induced electrocylization reaction in diarylethenes [17] and fulgido [21] from the open-ring to the closed-ring form induces a dramatic red-shift of absorption band from UV to the visible region. This change in the electronic structure often quenches the emission of fluorophores attached to the photochromic unit based on the energy transfer with an appreciable spectral overlap between the emission band of fluorophores and the absorption band corresponding to the closed-ring form of photochromes (Fig. 8.2). Since most of photochromic reactions, especially the $6\pi$-based electro-cyclization reactions, proceed in a stereospecific manner with high quantum efficiency, the incorporation of photochromic units in a chiral fluorescent molecule is an effective way to control the chiroptical property including CPL emission of molecular system by means of light irradiation.

## 8.2 Quenching of CPL from Chiral π-Conjugated Polymers by Photochromic Reaction of Diarylethene

The first example of ON-OFF photo-switching of CPL was demonstrated by Akagi and coworkers [22]. Conjugated polymers based on poly(*p*-phenylene) and poly(bithenylene-phenylene) are linked with a photo-responsive diarylethene moiety

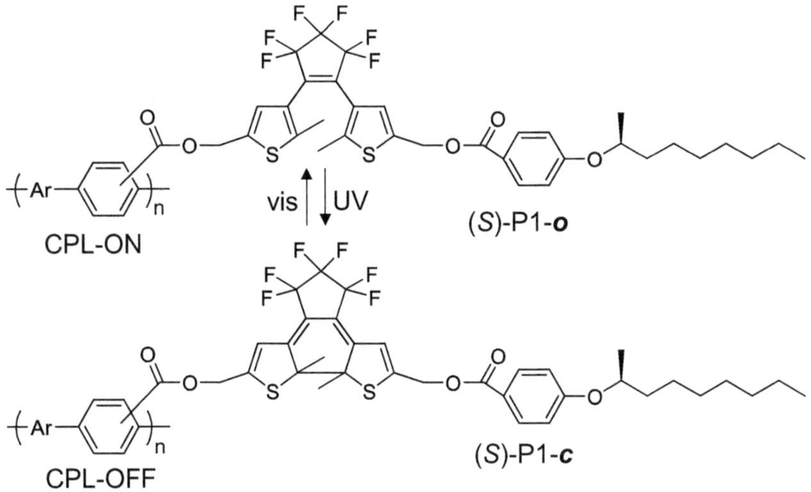

**Fig. 8.3** Structure and photoreaction of photo-responsive chiral conjugated polymer **P1**

tethering a chiral alkyl chain as a side-chain unit (**P1**, Fig. 8.3). The conjugated polymers exhibited an optical activity in the cast films with circular dichroism (CD) and CPL activities corresponding to the $\pi$-$\pi*$ transition in the $\pi$-conjugated main chain. The cast films afforded a bisignate split CD profile in the $\pi$-$\pi*$ transition region of the main chain, suggesting a one-handed helical inter-chain $\pi$-stacking structure based on the exciton coupling theory [23]. Although the chiral alkyl chain unit was introduced in the side chain through the photochromic moiety, the chirality information was transferred to the inter-polymer main-chain interactions, displaying the CPL activity in the film state in a similar manner to other conjugated polymers modified with chiral side-chain units [24–27]. The photochromic unit in the conjugated polymers **P1-*o*** maintained the photo-responsiveness in the cast films. The fluorescence and CPL of the $\pi$-conjugated main chain were readily quenched by means of UV irradiation through energy transfer from the main chain to the photo-generated closed-ring moiety of diarylethene in the side chain. The CPL activity was recovered by the cyclo-reversion reaction of the diarylethene moiety upon visible light irradiation and the ON-OFF photo-switching behavior was repeated over 10 cycles.

## 8.3 Photo-Switching of CPL Based on Pyrene-Bearing Photo-Responsive Foldamer

The inter-chromophoric interactions usually provide the most significant contributions to the chiroptical property in molecular systems as explained by the exciton coupling theory [23, 28]. As a rough framework of the theory, the chiral arrangement

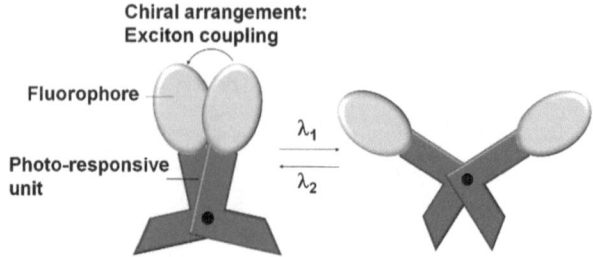

**Fig. 8.4** Schematic design of a CPL-photo-switch

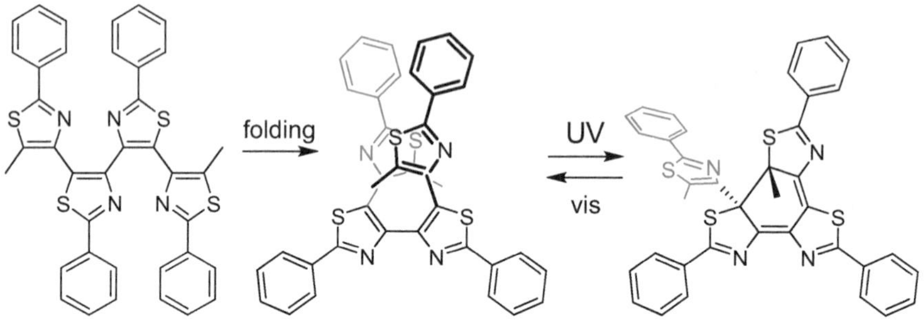

**Fig. 8.5** Conformational folding and photoreactions of tetra(2-phenylthiazole)

of multiple chromophores (or fluorophores) could be an important source of optical chirality including CD and CPL activities. On the basis of the theory, the chiroptical switch can be achieved if the external stimuli modulate the chiral chromophoric arrangement in molecular systems. Figure 8.4 depicts a schematic design of possible CPL-photo-switch. The photo-responsive unit with a chirally controlled geometry should carry multiple fluorophores, arranging them in a chiral manner so that the exciton coupling interaction operates on them. The photo-irradiation induces the structural change in the photo-responsive unit, leading to a change in the fluorophoric arrangement to modulate the inter-fluorophore exciton coupling.

Chiral photochromic scaffold with a helical secondary structure was employed as a platform for modulating CPL activity through a control of chiral fluorophore arrangement in an intramolecular manner. The central ethene part of diarylethene was replaced with a heteroaromatic ring to form a tetra-arylene scaffold with keeping the photochromic capability [29]. The use of oligo-heteroarylene framework enables the molecule to take advantage of inter-heteroaryl-unit interactions involving heteroatoms so that they can adopt the molecular conformation suitable for the electrocyclization reaction [30–33]. The extension of heteroaromatic units led to the formation of helical foldamers with photo-reactivity [34–36]. Tetra-arylenes composed of 2-phenylthiazole unit with a cis-cisoid-like connectivity form a one-turn helical secondary structure as the most stable conformation, which is also suitable for the stereo-specific electro-cyclization reaction (Fig. 8.5) [34–36]. The incorporation of chiral groups as a side-chain unit readily controlled the handedness of helicity. For the design of CPL photo-switch, a pyrene group was introduced at each

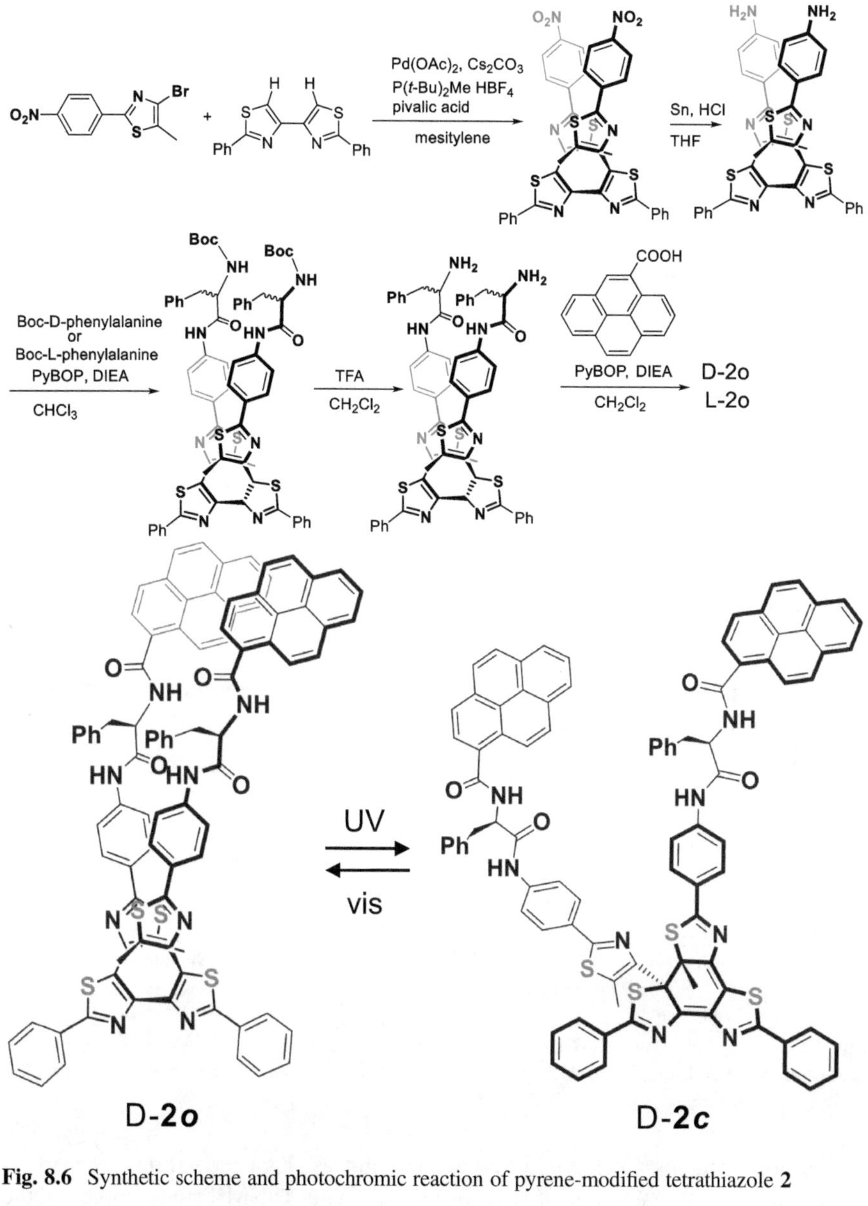

**Fig. 8.6** Synthetic scheme and photochromic reaction of pyrene-modified tetrathiazole **2**

end of tetra(2-phenylthiazole) through a chiral phenylalanine spacer (D- or L-**2o**, Fig. 8.6) [37]. The chiral spacer controls the handedness of a helix, arranging two pyrene units in a chiral manner. The photochromism in the tetrathiazole unit gives rise to a large structural change to form a ring-closed photoisomer **2c**, changing the position and orientation of two pyrene units.

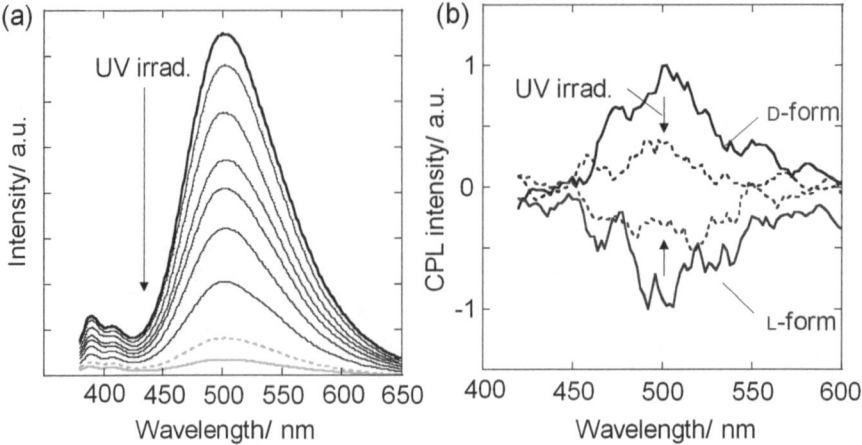

**Fig. 8.7** (**a**) Fluorescence and (**b**) CPL spectral change of **2o** upon UV (365 nm) irradiation in chloroform. Traces in (**a**) were measured with an irradiation interval of 5 s. Dashed line plots in (**b**) correspond to CPL spectra at the photostationary state (PSS) under UV irradiation

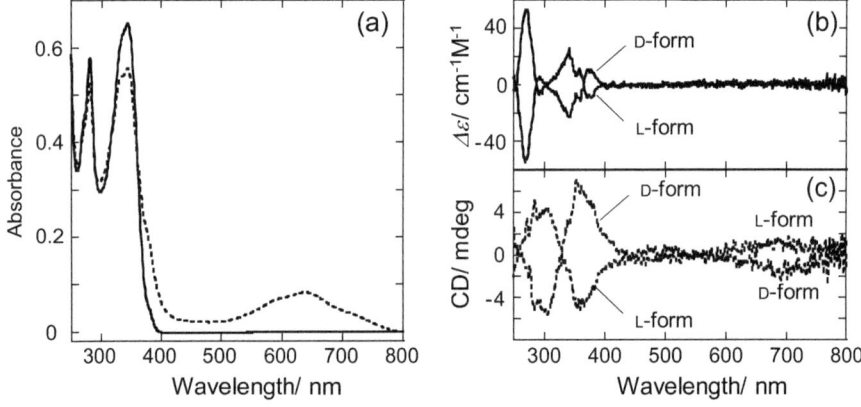

**Fig. 8.8** Absorption (**a**) and CD (**b**, **c**) spectral change of **2** in chloroform ($8.7 \times 10^{-6}$ M) upon UV irradiation. (**a**) Solid line, L-**2o**; dashed line, at photo-stationary state (PSS). (**b**, **c**) Solid lines, **2o**; dashed lines, at PSS

The conformational behavior of **2o** in solution was characterized by means of $^1$H NMR, CD, fluorescence, and CPL measurements. The $^1$H NMR study suggested the operation of intramolecular interactions including hydrogen bonding between the amide groups and π-π stacking between the pyrene units. Together with this result, the strong excimer-like emission observed at 500 nm and the suppressed monomer emission band at 390 nm (Fig. 8.7a) clearly support the preferential adoption of a compactly folded conformation as depicted in Fig. 8.6. The CD measurement confirms the chirality induction in the secondary structure of the main framework in **2o** (Fig. 8.8b). The positive- and negative-first Cotton effect signals for D- and

**Fig. 8.9** Plots of relative emission intensity in monomer (circle, at 390 nm) and excimer (square, at 500 nm) emission. $I_0$ corresponds to the intensity at a conversion of 0

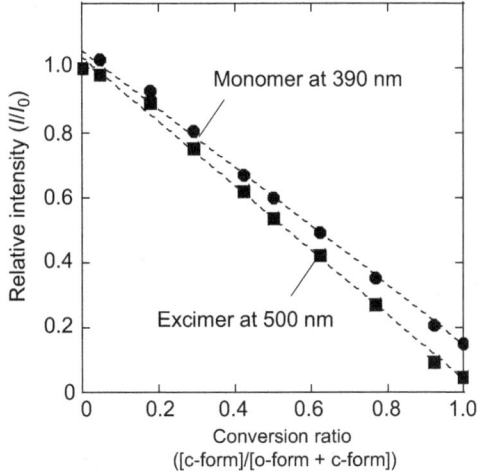

L-**2o**, respectively, suggest the preferential formation of right- and left-handed helices, respectively. In the helical molecular geometry of **2o**, the pyrene units were stacked in a chiral manner. Owing to this chiral arrangement, compounds **2o** are CPL active showing a positive and negative CPL signal in the region of excimer-like emission band for D- and L-isomers, respectively (Fig. 8.7b). The signs of CPL signal coincide with those of the first Cotton effect observed in the CD study. The anisotropy factor, which is given by the equation $g_{lum} = 2(I_L - I_R)/(I_L + I_R)$, where $I_L$ and $I_R$ are the intensities of the left- and right-handed circularly polarized emission, respectively, was estimated to be 0.01. This relatively high value for the small chiral organic molecule [38] is consistent with previously reported values for chiral pyrene excimers [39, 40].

UV irradiation to **2o** solution resulted in the emergence of an absorption band in the visible region extended to 800 nm due to the formation of **2c** (Fig. 8.8a). Meanwhile, a drastic change was recorded in the CD spectra, in which broad CD signals at the wavelength of the visible absorption band of photo-products emerged (Fig. 8.8c). The emergence of CD band in the visible region suggested the diastereo-selective photoreaction in the 6π-system. A combination of [1]H NMR and chiral HPLC studies for the photo-products indicated an absolute stereo-selective ring-cyclization reaction for **2**. The intensity of both monomer and excimer-like emission bands decreased with the progress of ring-cyclization reaction by the UV irradiation (Fig. 8.7a). Along with the decrease in the excimer-like emission intensity by the photo-isomerization, the CPL intensity decreased (Fig. 8.7b). Since the CPL measurement was performed at a relatively high concentration ($1.7 \times 10^{-4}$ M) in comparison to that for absorption spectra, the conversion ratio was not enough to achieve the completely off-state (Fig. 8.7b). In principle, the isolated **2c** gave a negligible emission quantum yield far less than 0.1%.

The emission from both the monomer and the excimer decreased in a liner fashion as a function of the conversion ratio to the colored form **2c** (Fig. 8.9). The quenching of monomer emission should be attributed to an energy transfer mechanism by the increased absorbance in the range of 380–410 nm in **2c**, which well overlaps with the pyrene monomer emission (Figs. 8.7a and 8.8a). The same mechanism partly explains the quenching of the excimer-like emission at 500 nm. Although the spectral overlap of excimer-like band at around 500 nm with absorption in **2c** is not as prominent as that of monomer emission band, the quenching of excimer-like emission was a little more significant than that of monomer emission (Fig. 8.9). This result might be attributed to the appreciable contribution of geometrical change in **2c** as depicted in Fig. 8.4. The structural change in the arrangement of pyrene units was supported by an upfield shift of amide-protons from **2o** to **2c** in the $^1$H NMR study, corresponding to the dissociation of intramolecular hydrogen bonding interactions.

## 8.4   Dynamic Switching of Hierarchical Chirality in Photo-Responsive Dinuclear Complexes

Unlike most chiral organic fluorophores affording CPL activity with a limited $|g_{lum}|$ value less than 0.05, large $|g_{lum}|$ values over 0.1 have been reported for the magnetic dipole transition in europium(III) complexes [41–45]. Eu(III) complexes have also been combined with 6$\pi$-based photochromic units to modulate the photoluminescence intensity [46–48]. Photochromic reaction is expected to modulate the emission property of Eu(III) ion through electronic and geometrical changes in the photochromic ligand. For the purpose of induction and modulation of chirality in the Eu(III) coordination cores, the pyrene unit in **2o** was replaced with a coordinating ligand [49]. Terpyridine (**terpy**) ligands were introduced at both ends of the helical tetrathiazole scaffold, forming **3o** with Eu(III) ions and $\beta$-diketonato ligands (**tta**) (Fig. 8.10). The local coordination site **Eu(terpy)(tta)₃** including asymmetric **tta** ligands is considered to have intrinsic chirality with eight possible chiral nine-coordination structures. The incorporation of **Eu(terpy)(tta)₃** sites in the chiral helical structure is expected to bring the complex sites close together in a chiral arrangement. This chiral arrangement should serve as a chiral perturbation to the coordination sites and one specific chiral coordination structure is preferentially formed among the eight possible ligand orientations, inducing optical activity in **3o**. Thus the point chirality in the amino acid spacer is expected to be transferred over hierarchy. The point chirality in the amino acid spacer was first introduced as a primary structure of a foldamer **3**, controlling the handedness in the helical conformation as a secondary structure. The handedness-controlled chiral helical conformation of **3** arranges the inherently chiral coordination cores in a chiral manner, inducing the biased formation of one-handed chiral ligand orientation in the coordination sites.

**3o** exhibited bright red emission from Eu(III)-centered f-f transitions with an apparent emission quantum yield ($\Phi_{lum}$) of 0.20 ($\lambda_{ex} = 360$ nm) and an emission lifetime of 0.55 ms in CDCl₃. The photoluminescence spectrum (Fig. 8.11a) gave

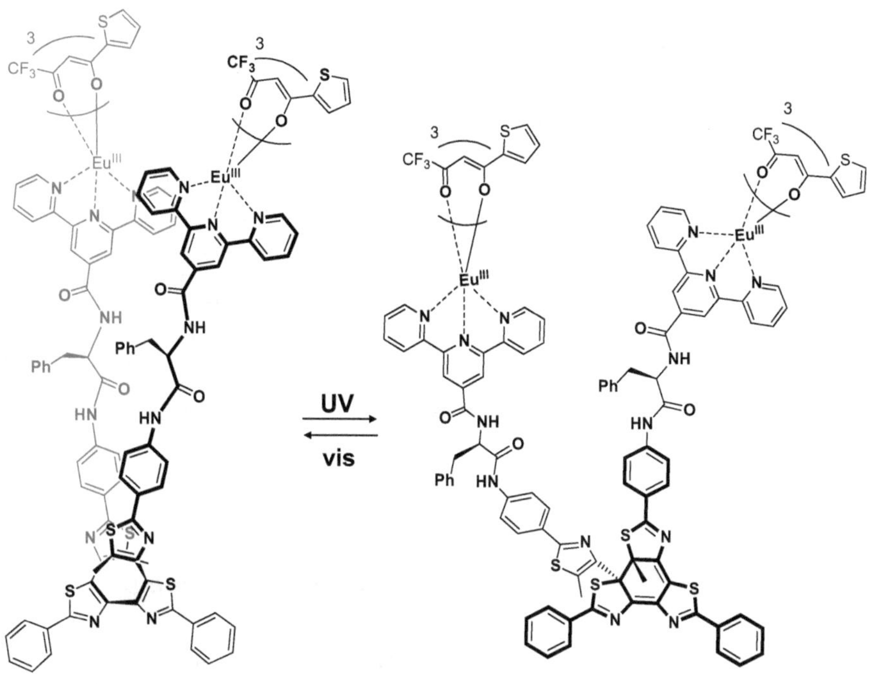

**Fig. 8.10** Photo-switching reaction of D-3

emission peaks at 580 ($^5D_0 \rightarrow {}^7F_0$), 595 ($^5D_0 \rightarrow {}^7F_1$), and 620 nm ($^5D_0 \rightarrow {}^7F_2$). Meanwhile, mirror-image CPL signals were obtained for the enantiomer pair of **3o** complexes at the bands of $^5D_0 \rightarrow {}^7F_1$ and $^5D_0 \rightarrow {}^7F_2$ transitions corresponding to magnetic and electronic dipole ones, respectively (Fig. 8.11b). The emergence of CPL signals clearly suggested the induction of optical activity in the dinuclear Eu (III) complexes. We obtained $|g_{lum}|$ values of 0.1 at the magnetic-dipole transition band, which are ten times larger than that of the pyrene-containing **2o**. The enantiomeric enrichment in the Eu(III)-complex sites was further confirmed with the induction of CD signal in the f-f transition absorption bands by replacing Eu(III) ions with Nd(III) ions (**3o(Nd)**). The mirror-image CD spectra over 500 nm corresponding to the f-f transition region in Nd(III) ion clearly support that the chirality is indeed induced as a coordination chirality in the local Ln(III) complex sites.

To further discuss the origin of optical activity, **tta** ligand in **3o** were replaced with a number of $\beta$-diketonato ligands both with symmetric and asymmetric structures. The Eu(III) complexes with asymmetric $\beta$-diketonato ligands having a trifluoromethyl ($CF_3$) and aromatic units showed much better photoluminescence and CPL efficiency compared with those with symmetric $\beta$-diketonato ligands. This result further confirms that the enriched population of an enantiomeric coordination

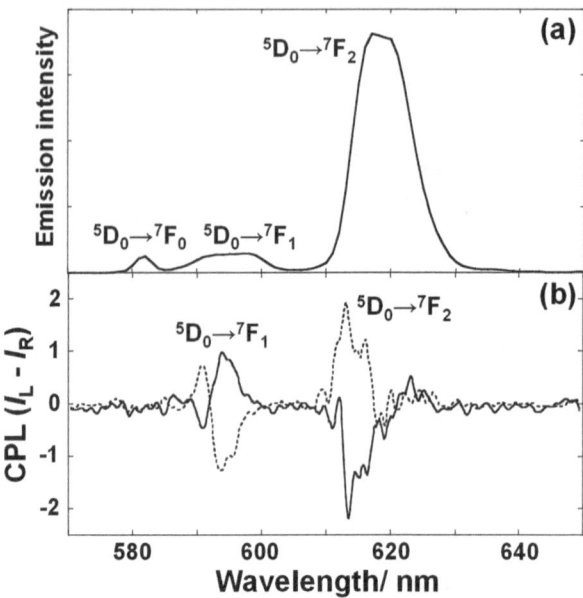

**Fig. 8.11** (a)
Photoluminescence and (b)
CPL spectra of D-(solid
lines) and L-**3o** (dashed
lines) in CHCl$_3$

structure in terms of chiral ligand orientation was the primary source as the origin of chirality induction in **3o**.

UV irradiation to a chloroform solution of **3o** resulted in the emergence of an absorption band in the visible region due to the formation of closed-ring isomer in the photochromic framework in a similar manner to **2**. The photo-reaction also induced a decrease in the emission intensity in the whole spectral region, and the $\Phi_{lum}$ value dropped to 0.03 at the PSS with a conversion ratio of 85% (Fig. 8.12a). The decrease in the emission intensity and lifetime were attributed to the Förster resonance energy transfer (FRET) mechanism from the excited state of Eu(III) to the photochromic center [46–48]. The photo-cyclization reaction of the photochromic backbone of **3** also promoted a decrease in CPL intensity (Fig. 8.12b). Given that the enantiomeric structure induced in the Eu(III) complex sites is maintained after the photoreaction, the $g_{lum}$ value should be unchanged even if the CPL intensity decreases. However, the $|g_{lum}|$ value decreased to <0.01, affording a large modulation amplitude in CPL dissymmetry, $|\Delta g_{lum}| > 0.09$. The decrease in the dissymmetry factor should be attributed to the cancellation in the enrichment of one-hand chiral coordination structure in the Eu(III) complex sites with the aid of labile coordination character [50].

Meanwhile, UV irradiation to **3o(Nd)** gave rise to an emergence of a broad absorption band in the visible band which is superimposed with the $4I_{9/2} \rightarrow 4G_{5/2}$ transition at 580 nm in Nd(III) ion (Fig. 8.12c). This spectral change accompanies the gradual progression of broad negative and positive CD signals over 500 nm for D- and L-**3(Nd)**, respectively, supporting the diastereo-selective formation of the closed-ring isomer (*c*-forms). The positive and negative CD signals corresponding to $4I_{9/2} \rightarrow 4G_{5/2}$ transition band at 580 nm observed for D- and L-**3o(Nd)**, respectively,

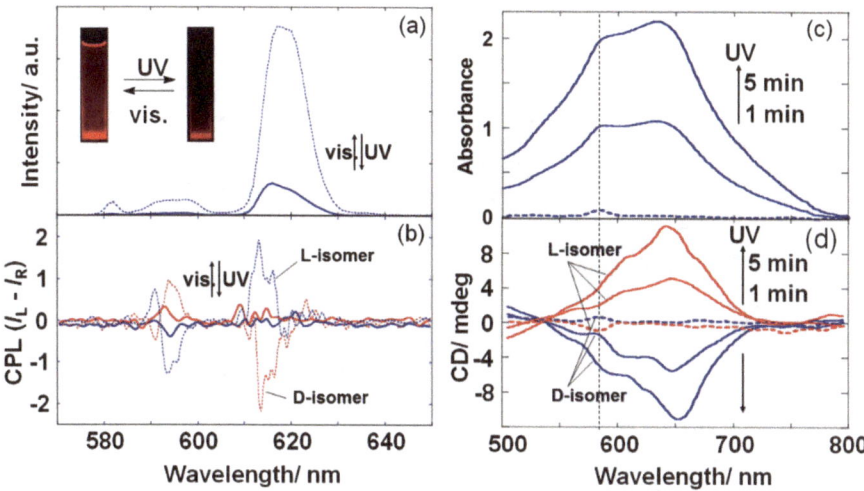

**Fig. 8.12** (**a**) Photoluminescence and (**b**) CPL spectral change of **3** in CHCl₃ before (broken lines) and after photo-irradiation to achieve PSS under irradiation at 365 nm (solid lines). (**c**) Absorption and (**d**) CD spectral change of **3(Nd)** before (broken lines) and after UV irradiation (solid lines) for 1 and 5 min in CHCl₃

remained in the CD spectra after 1 min of irradiation (Fig. 8.12d). The apparent features of these positive and negative signals at 580 nm diminished as the photo-reaction proceeded after 5 min of irradiation while the sharp absorption bump was still apparent in the absorption profile (Fig. 8.12c). This result suggests that the preferential formation of enantiomeric coordination structure in the Ln(III)-complex sites was effective only in the helical structure, wherein the close contact between the complex sites is operative. The photo-induced helical to non-helical transformation of the main framework switches off the close arrangement of the coordination units. The labile coordination character of lanthanide ions should negate the enrichment of enantiomeric coordination structure, diminishing the optical activity (Fig. 8.13).

## 8.5   Photo-Switching of CPL in Photo-Responsive Molecular Self-Assemblies

Chiroptical properties have often been demonstrated to enhance in chiral supramolecular assemblies as supramolecular chirality [51, 52]. The CPL property was highly dependent on the supramolecular morphologies, wherein the ordering of building unit varies [4, 5]. Photo-responsive azobenzene units were incorporated in a building block of a chiral supramolecular assembly (**4**; Fig. 8.14) [53]. (*S*)-(*E*, *E*)-**4** formed entangled helical fibers with a right-handed twist as a self-assembling structure, exhibiting an intense positive CD signal at 464 nm. (*R*)-**4** gave a similar morphology in the self-assembling state but with opposite twist and CD signal to

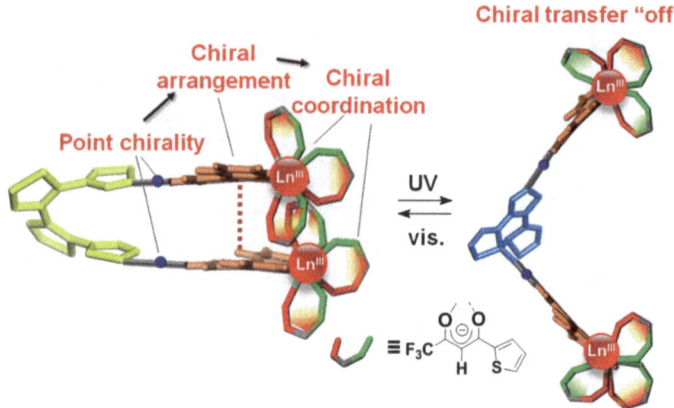

**Fig. 8.13** Schematic illustration of hierarchical chirality transfer and its reversible modulation in the photo-responsive chiral dinuclear complex

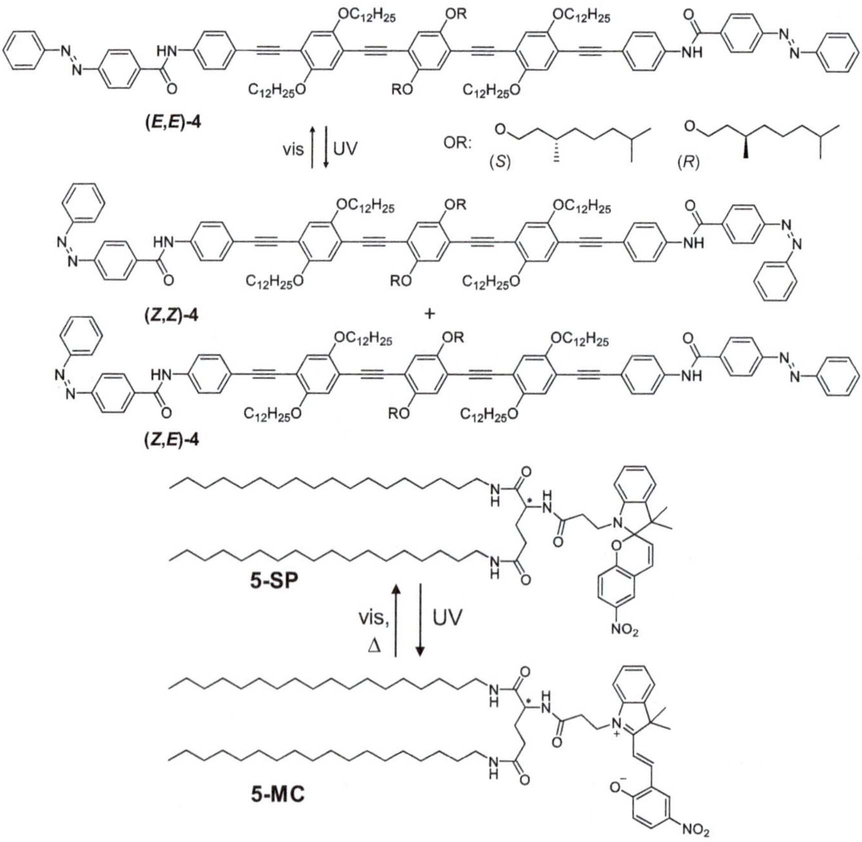

**Fig. 8.14** Structures and photoreactions of photo-responsive chiral self-assembling molecules

(*S*)-**4**. Thus the chirality in the alkyl side chains was transferred as a helical molecular arrangement in the self-assemblies, affording an emergence of supramolecular chirality. Owing to the supramolecular chirality, the self-assembly of (*S*)-(*E*, *E*)-**4** gave a positive CPL signal with the $g_{lum}$ value of 0.008. Although the photo-reaction of azobenzene moiety could not be induced in the self-assembling state, the dissociation of self-assembly with heating could enable the *E*-to-*Z* photo-isomerization. The photo-irradiated solution was cooled below 300 K and the inversion of CD signal sign was induced, thus reversing the supramolecular chirality. Along with the supramolecular chirality inversion, the CPL signal became negative with the $g_{lum}$ value of −0.002. After the photo-irradiation, the supramolecular helicity in the fibrous assembly was also reversed. Photo-reactive spiropyran moiety was also introduced in a self-assembling chiral glutamate organogelator **5** (Fig. 8.14) [54]. **5-SP** formed helical nanofibers to gelatinize organic solvents, while fluorescent property was absent. Upon UV irradiation, the colorless organogel of **5-SP** turned blue with the formation of a merocyanine form (**5-MC**) with changing the micro-structures. **5-MC** exhibited red-color fluorescence, switching on the CPL activity. Alternate UV and visible light irradiations reversibly switched the CPL activity of organogel of **5** between ON and OFF states, respectively. A similar chiral glutamate compound having a photo-reactive cinnamic acid moiety was also demonstrated to reverse CPL sign in response to photo-dimerization reaction [55].

The photo-isomerization of azobenzene unit involves the rotational motion around the N–N bond in the excited state, which may prohibit the *trans*-to-*cis* isomerization in the self-assembling state of (*E*,*E*)-**4**. Therefore, dynamic in situ photo-switching of CPL activity in a supramolecular system had still remained a challenge. In comparison to the photo-reaction of azobenzene, the photo-cyclization reaction of 6π-system requires a smaller structural change, affording very efficient solid-state reactivity [17]. The CPL photo-switching molecules D- and L-**2o** were employed as a building unit of self-assembly [56]. The emission turn-off behavior is expected to be amplified in such aggregation systems because the energy transfer based emission quenching operates not only in intramolecular but also in intermolecular manners [57, 58]. Compound **2o** was molecularly dispersed in chloroform, while it gave dispersion of nanoparticle assemblies in a mixture of chloroform/methylcyclohexane(MCH) (1:9 or 1:99) solvents. The transmission electron microscopy (TEM) observation revealed the formation of nanoparticle aggregates with an average diameter of ca. 250 nm in the chloroform/MCH (1:9) mixture at a concentration of $1.0 \times 10^{-4}$ M (Fig. 8.15a). CD spectra exhibited mirror-image profiles in response to the stereochemistry in the amino-acid spacer with the same sign of first Cotton effect regardless of the solvent composition (Fig. 8.15b). The peak splitting patterns of MCH-rich solutions differ from those in molecularly dispersed chloroform solutions, suggesting the difference in the exciton coupling mode. The CD spectra for the MCH-rich solution should include intermolecular exciton coupling effect in the aggregates. The molar CD ($\Delta\varepsilon$), which is the parameter of CD amplitude, exhibited a pronounced increase in the MCH-rich solution compared to that in molecularly dispersed state.

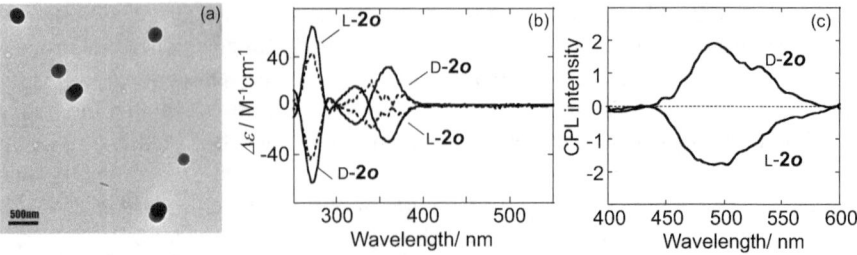

**Fig. 8.15** (**a**) TEM image of aggregates formed by D-**2o** in chloroform/MCH (1:9) mixture solvent. (**b**) CD spectra of 2o in chloroform (broken lines) and chloroform/MCH (1:9) mixture solvent (solid lines). (**c**) CPL spectra of **2o** in chloroform/MCH (1:9) mixture solvent (concentration: $1.0 \times 10^{-4}$ M)

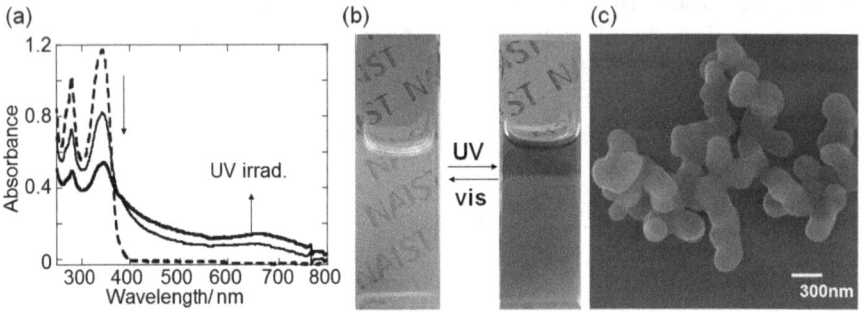

**Fig. 8.16** (**a**) Absorption spectral change and (**b**) picture of photochromic reaction of D-**2o** upon UV irradiation in chloroform/MCH (1:9) ($1.0 \times 10^{-4}$ M). (**c**) SEM image of aggregates formed at the PSS of D-**2** in chloroform/MCH (1:9) ($1.0 \times 10^{-4}$ M)

The nanoparticle aggregates afforded appreciable CPL signals at 490 nm, which showed a shift to shorter wavelength in comparison to that in chloroform due to the less stabilization of excimer-like state in the less polar MCH-rich solvent (Fig. 8.15c). The dissymmetry factor $|g_{lum}|$ was a little enhanced to be 0.017 compared to that in molecularly dispersed state of 0.01, showing a good agreement with the enhanced $\Delta\varepsilon$ value observed in the CD measurement.

UV irradiation at 365 nm to the MCH-rich solution of **2o** induced the photo-isomerization with the emergence of an absorption band at 650 nm corresponding to the formation of colored isomer **2c** (Fig. 8.16a). Interestingly, upon UV irradiation, the MCH-rich solution turned turbid at room temperature, suggesting the photo-induced secondary aggregation of nanoparticles (Fig. 8.16b). The clouding behavior was apparent in the absorption spectral change as an increase of background scattering (Fig. 8.16a). The colored isomer **2c** has a more rigid structure with an additional covalent bond compared to **2o** with a flexible conformation, leading to a decrease in the solubility of compound in MCH-rich solutions. In comparison to the increase in the absorption band in the visible region, the decrease of the absorption in the UV region was more prominent. This significant decrease should be attributed to

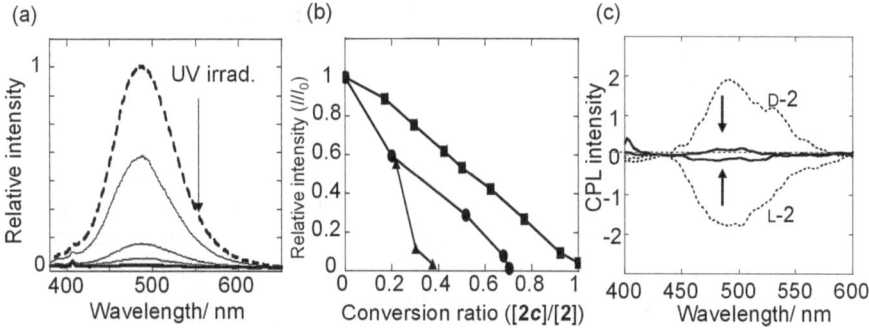

**Fig. 8.17** (a) Emission spectral change of D-**2o** upon UV irradiation in chloroform/MCH (1:99) solution ($1.0 \times 10^{-4}$ M). Before UV irradiation, broken line; at PSS achieved by the excitation at 365 nm, thick black line. (b) Plots of relative emission intensity as a function of a conversion ratio between D-**2o** and **2c**. Square, in chloroform; circle, in chloroform/MCH (1:9); triangle, in chloroform/MCH (1:99) (concentration: $1.0 \times 10^{-5}$ M). (c) CPL spectral change of D- and L-**2** before (broken lines) and after the UV irradiation at the PSS (solid lines) in chloroform/MCH (1:9) (concentration: $1.0 \times 10^{-5}$ M)

the decrease of the solution concentration upon the secondary aggregation accompanied by the photo-cyclization reaction of **2**. The visible light irradiation above 440 nm bleach the solution color and the visible absorption band disappeared reversibly to give an absorption spectrum identical to that of **2o** with transparency in the visible range, indicating the dissolution of secondary aggregates.

SEM observation after the UV light irradiation clearly demonstrated the formation of secondary aggregates together with an increase in the size of nanoparticles (Fig. 8.16c). The photo-reaction seems to take place for the molecules on the surface of nanoparticles in the early stage of photo-conversion. The less solubility of the colored isomer **2c** in the MCH-rich solvent could reduce the colloidal stability of nanoparticles, leading to the further aggregation and fusion of nanoparticles. Intermolecular hydrogen bonding interactions are also possible in the colored form **2c**, which may also drive the inter-nanoparticle aggregation.

The photoreaction also induced a decrease in the emission intensity in the whole spectral range (Fig. 8.17a). In the chloroform solution, the emission intensity decreased linearly as a function of conversion ratio (Fig. 8.17b). The more dramatic decrease in the emission intensity was observed in the MCH-rich solutions in comparison to that in the molecularly dispersed chloroform solution. The emission was almost completely quenched in the small conversion ratio values of 70 and 39% in chloroform/MCH (1:9) and (1:99) solvents, respectively (Fig. 8.17b). In a similar manner to the molecularly dispersed state in chloroform, the emission quenching is due to the FRET process from the excited state of pyrene moiety to the photochrome ring-closed part in **2c**. The energy transfer process also take place in an intermolecular manner in the aggregates, enhancing the emission quenching efficiency in MCH-rich solvents. The FRET process from the pyrene moiety in the open-ring state **2o** to the ring-closed photochromic part in **2c** can apparently contribute to the emission quenching [57, 58]. Furthermore, the secondary aggregation might also increase the probability of intermolecular energy transfer.

The reversible modulation of CPL intensity upon photo-irradiations was performed (Fig. 8.17c). Along with the emission quenching, the CPL signal also diminished. As demonstrated in the emission quenching study, the CPL signal intensity basically responds to the conversion ratio between $2o$ and $2c$ in a linear fashion in chloroform. However, the appreciable emission quenching did not need a high conversion ratio in the aggregate state (Fig. 8.17b), proposing the more efficient and high-contrast CPL photo-switching system. The CPL intensity ($I_L - I_R$) was reversibly switched between ON and OFF states for 8 cycles with consecutive UV-visible light irradiations.

## 8.6  Summary and Outlook

In this chapter, we introduced a design concept and several examples of CPL photo-switches which modulate CPL activity in response to light irradiation. Photo-responsive units are combined with fluorophore units in a chiral molecular system. The CPL intensity can be controlled by means of energy transfer quenching by the formation of a quencher unit upon photochromic reaction. The CPL activity is also controlled in terms of the exciton coupling modulation in response to the chiral geometrical change induced by the stereo-specific photo-reaction. The photo-switching effect could be enhanced in the chiral supramolecular systems.

Highly pure CPL could be generated by the use of selective reflection in a cholesteric liquid crystalline phase [59, 60]. Helical pitch, handedness, and arrangement of mesogens in such chiral liquid crystalline system can be modulated by the photo-reaction of chiral photochromic dopants [61–64]. High speed and dynamic modulation of highly pure CPL could find applications for 3D displays, energy-saving organic light-emitting diode (OLED) devices, switchable lasers, and optical memory devices [65].

## References

1. Maeda H, Bando Y, Shimomura K, Yamada I, Naito M, Nobusawa K, Tsumatori H, Kawai T (2011) Chemical-stimuli-controllable circularly polarized luminescence from anion-responsive π-conjugated molecules. J Am Chem Soc 133:9266–9269
2. Saleh N, Moore B, Srebro M, Vanthuyne N, Toupet L, Williams JA, Roussel C, Deol KK, Muller G, Autschbach J, Crassous J (2015) Acid/base-triggered switching of circularly polarized luminescence and electronic circular dichroism in organic and organometallic helicenes. Chem Eur J 21:1673–1681
3. Amako T, Nakabayashi K, Mori T, Inoue Y, Fujiki M, Imai Y (2014) Sign inversion of circularly polarized luminescence by geometry manipulation of four naphthalene units introduced into a tartaric acid scaffold. Chem Commun 50:12836–12839

4. Kumar J, Nakashima T, Tsumatori H, Kawai T (2014) Circularly polarized luminescence in chiral aggregates: dependence of morphology on luminescence dissymmetry. J Phys Chem Lett 5:316–321
5. Kumar J, Tsumatori H, Yuasa J, Kawai T, Nakashima T (2015) Self-discriminating termination of chiral supramolecular polymerization: tuning the length of nanofibers. Angew Chem Int Ed 54:5943–5947
6. Sethy R, Kumar J, Métivier R, Louis M, Nakatani K, Mecheri NMT, Subhakumari A, Thomas KG, Kawai T, Nakashima T (2017) Enantioselective light harvesting with perylenediimide guests on self-assembled chiral naphthalenediimide nanofibers. Angew Chem Int Ed 56:15053–15057
7. Kumar J, Kawai T, Nakashima T (2017) Circularly polarized luminescence in chiral silver nanoclusters. Chem Commun 53:1269–1272
8. Morisue M, Yumura T, Sawada R, Naito M, Kuroda Y, Chujo Y (2016) Oligoamylose-entwined porphyrin: excited-state induced-fit for chirality induction. Chem Commun 52:2481–2484
9. Yuasa J, Ueno H, Kawai T (2014) Sign reversal of a large circularly polarized luminescence signal by the twisting motion of a bidentate ligand. Chem Eur J 20:8621–8627
10. Imai Y, Nakano Y, Kawai T, Yuasa J (2018) A smart sensing method for object identification using circularly polarized luminescence from coordination-driven self-assembly. Angew Chem Int Ed 57:8973–8978
11. Nagata Y, Nishikawa T, Suginome M (2014) Chirality-switchable circularly polarized luminescence in solution based on the solvent-dependent helix inversion of poly(quinoxaline-2,3-diyl)s. Chem Commun 50:9951–9953
12. Zhang YJ, Oka T, Suzuki R, Ye JT, Iwasa Y (2014) Electrically switchable chiral light-emitting transistor. Science 344:725–728
13. Nishizawa N, Nishibayashi K, Munekata H (2017) Pure circular polarization electroluminescence at room temperature with spin-polarized light-emitting diodes. Proc Natl Acad Sci U S A 114:1783–1788
14. Sherson JF, Krauter H, Olsson RK, Julsgaard B, Hammerer K, Cirac I, Polzik ES (2006) Quantum teleportation between light and matter. Nature 443:557–560
15. Feringa BL, van Delden RA, Koumura N, Geertsema EM (2000) Chiroptical molecular switches. Chem Rev 100:1789–1816
16. Yildiz I, Deniz E, Raymo FM (2009) Fluorescence modulation with photochromic switches in nanostructured constructs. Chem Soc Rev 38:1859–1867
17. Irie M, Fukaminato T, Matsuda K, Kobatake S (2014) Photochromism of diarylethene molecules and crystals: memories, switches, and actuators. Chem Rev 114:12174–12277
18. Fukaminato T (2011) Single-molecule fluorescence photoswitching: design and synthesis of photoswitchable fluorescent molecules. J Photochem Photobiol C 12:177–208
19. Saika T, Iyoda T, Honda K, Shimidzu T (1992) Emission control of a pyrene-thioindigo compound. J Chem Soc Chem Commun 1992:591–592
20. Wang J, Kulago A, Browne WR, Feringa BL (2010) Photoswitchable intramolecular H-stacking of perylenebisimide. J Am Chem Soc 132:4191–4192
21. Heller HG, Oliver S (1981) Photochromic heterocyclic fulgides. 1. Rearrangement reactions of (E)-alpha-3-furylethylidene(isopropylidene)succinic anhydride. J Chem Soc Perkin Trans 1:197–202
22. Hayasaka H, Miyashita T, Tamura K, Akagi K (2010) Helically π-stacked conjugated polymers bearing photoresponsive and chiral moieties in side chains: reversible photoisomerization-enforced switching between emission and quenching of circularly polarized fluorescence. Adv Funct Mater 20:1243–1250
23. Berova N, Nakanishi K, Woody RW (2000) Circular dichroism: principal and applications, 2nd edn. Wiley-VCH, New York

24. Peeters M, Christiaans MPT, Janssen RAJ, Schoo HFM, Dekkers HPJM, Meijer EW (1997) Circularly polarized electroluminescence from a polymer light-emitting diode. J Am Chem Soc 119:9909–9910

25. Geng Y, Trajkovska A, Katsis D, Ou JJ, Culligan SW, Chen SH (2002) Synthesis, characterization, and optical properties of monodisperse chiral oligofluorenes. J Am Chem Soc 124:8337–8347

26. Wilson JN, Steffen W, McKenzie TG, Lieser G, Oda M, Neher D, Bunz UHF (2002) Chiroptical properties of poly(p-phenyleneethynylene) copolymers in thin films: large g-values. J Am Chem Soc 124:6830–6831

27. Satrijo A, Meskers SCJ, Swager TM (2006) Probing a conjugated polymer's transfer on organization-dependent properties from solutions to films. J Am Chem Soc 128:9030–9031

28. Berova N, Di Bari L, Pescitelli G (2007) Application of electronic circular dichroism in configuration and conformational analysis of organic compounds. Chem Soc Rev 36:914–931

29. Nakashima T, Atsumi K, Kawai S, Nakagawa T, Hasegawa Y, Kawai T (2007) Photochromism of thiazole-containing triangle terarylenes. Eur J Org Chem 2007:3212–3218

30. Fukumoto S, Nakashima T, Kawai T (2011) Photon-quantitative reaction of a dithiazolylarylene in solution. Angew Chem Int Ed 50:1565–1568

31. Fukumoto S, Nakashima T, Kawai T (2011) Intramolecular hydrogen bonding in a triangular dithiazolyl-azaindole for efficient photoreactivity in polar and nonpolar solvents. Eur J Org Chem 2011:5047–5053

32. Li R, Nakashima T, Galangau O, Iijima S, Kanazawa R, Kawai T (2015) Photon-quantitative 6p-electrocyclization of a diarylbenzo[b]thiophene in polar medium. Chem Asian J 10:1725–1730

33. Nakagawa T, Miyasaka Y, Yokoyama Y (2018) Photochromism of a spiro-functionalized diarylethene derivative: multi-colour fluorescence modulation with a photon-quantitative photocyclization reactivity. Chem Commun 54:3207–3210

34. Gavrel G, Yu P, Léaustic A, Gillot R, Métivier R, Nakatani K (2012) 4,4′-Bithiazole-based tetraarylenes: new photochromes with unique photoreactive patterns. Chem Commun 48:10111–10113

35. Nakashima T, Yamamoto K, Kimura Y, Kawai T (2013) Chiral photoresponsive tetrathiazoles that provide snapshots of folding states. Chem Eur J 19:16972–16980

36. Nakashima T, Imamura K, Yamamoto K, Kimura Y, Katao S, Hashimoto Y, Kawai T (2014) Synthesis, structure, and properties of α,β-linked oligothiazoles with controlled sequence. Chem Eur J 20:13722–13729

37. Hashimoto Y, Nakashima T, Shimizu D, Kawai T (2016) Photoswitching of an intramolecular chiral stack in a helical tetrathiazole. Chem Commun 52:5171–5174

38. Sanchez-Cárnerero EM, Agarrabeitia AR, Moreno F, Maroto BL, Muller G, Ortiz MJ, de la Moya S (2015) Circularly polarized luminescence from simple organic molecules. Chem Eur J 21:13488–13500

39. Brittain H, Ambrozich DL, Saburi M, Fendler JH (1980) Enhanced optical activity associated with chiral 1-(1-hydroxyhexyl)pyrene excimer formation. J Am Chem Soc 102:6372–6374

40. Kano K, Matsumoto H, Hashimoto S, Sisido M, Imanishi Y (1985) Chiral pyrene excimer in the γ-cyclodextrin cavity. J Am Chem Soc 107:6117–6118

41. Richardson FS, Riehl JP (1977) Circularly polarized luminescence spectroscopy. Chem Rev 77:773–792

42. Lunkley JL, Shirotani D, Yamanari K, Kaizaki S, Muller G (2008) Extraordinary circularly polarized luminescence activity exhibited by cesium tetrakis(3-heptafluoro-butylryl-(+)-camphorato) Eu(III) complexes in EtOH and CHCl₃ solutions. J Am Chem Soc 130:13814–13815

43. Carr R, Evans NH, Parker D (2012) Lanthanide complexes as chiral probes exploiting circularly polarized luminescence. Chem Soc Rev 41:7673–7686

44. Aspinall HC (2002) Chiral lanthanide complexes: coordination chemistry and applications. Chem Rev 102:1807–1850

45. Bing TY, Kawai T, Yuasa J (2018) Ligand-to-ligand interactions that direct formation of $D_2$-symmetrical alternating circular helicate. J Am Chem Soc 140:3683–3689
46. Hasegawa Y, Nakagawa T, Kawai T (2010) Recent progress of luminescent metal complexes with photochromic units. Coord Chem Rev 254:2643–2651
47. Cheng HB, Hu GF, Zhang ZH, Gao L, Gao X, Wu HC (2016) Photocontrolled reversible luminescent lanthanide molecular switch based on a diarylethene-europium dyad. Inorg Chem 55:7962–7968
48. He X, Norel L, Hervault YM, Métivier R, D'Aleo A, Maury O, Rigaut S (2016) Modulation of Eu(III) and Yb(III) luminescence using a DTE photochromic ligand. Inorg Chem 55:12635–12643
49. Hashimoto Y, Nakashima T, Yamada M, Yuasa J, Rapenne G, Kawai T (2018) Hierarchical emergence and dynamic control of chirality in a photoresponsive dinuclear complex. J Phys Chem Lett 9:2151–2157
50. Metcalf DH, Snyder SW, Demas JN, Richardson FS (1990) Chiral dynamics in the excited state of a stereochemically labile metal complex. J Phys Chem 94:7143–7153
51. Liu M, Zhang L, Wang T (2015) Supramolecular chirality in self-assembled systems. Chem Rev 115:7304–7397
52. Kumar J, Nakashima T, Kawai T (2015) Circularly polarized luminescence in chiral molecules and supramolecular assemblies. J Phys Chem Lett 6:3445–3452
53. Gopal A, Hifsudheen M, Furumi S, Takeuchi M, Ajayaghosh A (2012) Thermally assisted photonic inversion of supramolecular handedness. Angew Chem Int Ed 51:10505–10509
54. Miao W, Wang S, Liu M (2017) Reversible quadruple switching with optical, chiroptical, helicity, and macropattern in self-assembled spiropyran gels. Adv Funct Mater 27:1701368
55. Jiang H, Jiang Y, Han J, Zhang L, Liu M (2018) Helical nanostructures: chirality transfer and a photodriven transformation from superhelix to nanokebab. Angew Chem Int Ed 58:785–790
56. Hashimoto Y, Nakashima T, Kuno J, Yamada M, Kawai T (2018) Dynamic modulation of circularly polarized luminescence in photoresponsive assemblies. ChemNanoMat 4:815–820
57. Bu J, Watanabe K, Hayasaka H, Akagi K (2014) Photochemically colour-tuneable white fluorescence illuminants consisting of conjugated polymer nanospheres. Nat Commun 5:3799-1–3799-8
58. Su J, Fukaminato T, Placial JP, Onodera T, Suzuki R, Oikawa H, Brosseau A, Brisset F, Pansu R, Nakatani K, Métivier R (2015) Giant amplification of photoswitching by a few photons in fluorescent photochromic organic nanoparticles. Angew Chem Int Ed 55:3662–3666
59. Chen SH, Katsis D, Schmid AW, Mastrangelo JC, Tsutsui T, Blanton TN (1999) Circularly polarized light generated by photoexcitation of luminophores in glassy liquid-crystal films. Nature 397:506–508
60. San Jose BA, Yan J, Akagi K (2014) Dynamic switching of the circularly polarized luminescence of disubstituted polyacetylene by selective transmission through a thermotropic chiral nematic liquid crystal. Angew Chem Int Ed 53:10641–10644
61. Vicario J, Katsonis N, Ramon BS, Bastiaansen CWM, Broer DJ, Feringa BL (1999) Nanomotor rotates microscale objects. Nature 440:163
62. Hayasaka H, Miyashita T, Nakayama M, Kuwada K, Akagi K (2012) Dynamic photoswitching of helical inversion in liquid crystals containing photoresponsive axially chiral dopants. J Am Chem Soc 134:3758–3765
63. Li Y, Urbas A, Li Q (2012) Reversible light-directed red, green, and blue reflection with thermal stability enabled by a self-organized helical superstructure. J Am Chem Soc 134:9573–9576
64. Zheng ZG, Li Y, Bisoyi HK, Wang L, Bunning TJ, Li Q (2016) Three-dimensional control of the helical axis of a chiral nematic liquid crystal by light. Nature 531:352–356
65. Brandt JR, Salerno F, Fuchter MJ (2017) The added value of small-molecule chirality in technological applications. Nat Rev Chem 1:0045-1–0045-12

# Chapter 9
# Circularly Polarized Luminescence of Chirally Arranged Achiral Organic Luminophores by Covalent and Supramolecular Methods

**Toshiaki Ikeda and Takeharu Haino**

**Abstract** Circularly polarized luminescence (CPL) produced by achiral organic luminophores is described. Achiral organic luminophores can exhibit CPL by the chiral arrangement of the achiral luminophores. Chiral arrangement of achiral luminophores can be constructed through a covalently linked chiral spacer like a binaphthyl moiety. A helical supramolecular assembly also provides chiral environment on an achiral luminophore. The helically stacked assemblies of achiral luminophores are excellent for realizing CPL of the achiral luminophore since the highly assembled structure in the helical assembly provides good CPL activity. The stimuli-responsivity of supramolecular systems provides stimuli-responsive CPL.

## 9.1 Introduction

In the field of circularly polarized luminescence (CPL), chiral lanthanide complexes have been dominant due to their large CPL dissymmetry factor, $g_{lum}$ [1]. In recent years, CPL-active organic compounds have attracted growing interest toward their potential applications in optoelectronic devices [2, 3]. The early research on CPL-active organic molecules started with chiral ketones in the 1960s [4]. After that, various CPL-active organic compounds were developed, and some of them exhibit large $g_{lum}$ values comparable to those of chiral lanthanide complexes.

A key issue for constructing CPL-active materials is how to prepare an asymmetric environment on luminophores. A straightforward strategy for realizing CPL-active materials is the direct introduction of chirality on the luminophore. Actually, many lanthanide complexes possessing chiral ligands have been reported

T. Ikeda
Department of Chemistry, School of Science, Tokai University, Hiratsuka, Kanagawa, Japan

T. Haino (✉)
Department of Chemistry, Graduate School of Science, Hiroshima University, Hiroshima, Japan
e-mail: haino@hiroshima-u.ac.jp

© Springer Nature Singapore Pte Ltd. 2020
T. Mori (ed.), *Circularly Polarized Luminescence of Isolated Small Organic Molecules*, https://doi.org/10.1007/978-981-15-2309-0_9

to exhibit CPL with the luminescence of the metal center [1]. In the same manner, chiral organic dyes such as helicenes also exhibit CPL [3]. In the field of polymers, chiral polymers are reported as CPL-active materials [5–7]. The alternative way to achieve CPL-active organic compounds is the chiral arrangement of achiral luminophores. If two or more luminophores take a chiral orientation, the luminophores chirally interact each other. Then, the electronic state of the luminophore is chirally perturbed. Thus, the emission of the luminophore gets circularly polarized despite the luminophore being achiral. Such chiral orientation can be constructed through a covalently linked chiral spacer. Another way to achieve chiral orientation of achiral luminophores is the use of a chiral supramolecular assembly. In this chapter, CPL properties of small organic compounds and their supramolecular assemblies are described (Fig. 9.1).

## 9.2 CPL of Chiral Organic Luminophores

The early CPL of organic compounds was produced by fluorescence based on the $\pi* \to n$ transition of the carbonyl group of chiral ketones (Fig. 9.2) [4, 8–11]. However, the $\pi* \to n$ transition of the carbonyl group is weak and limited at short

(a)                                    (b)                                    (c)

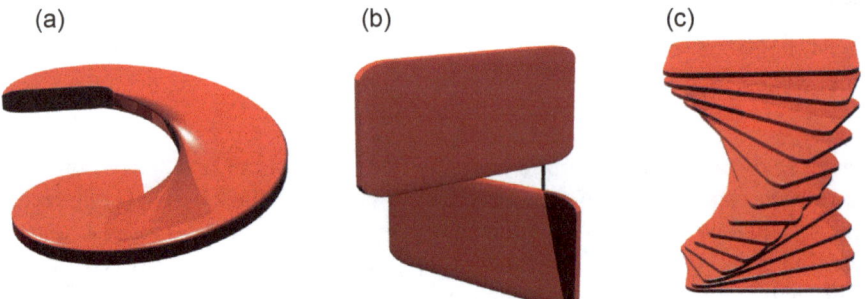

**Fig. 9.1** CPL-active organic compounds: (**a**) chiral luminophore, (**b**) chirally arranged achiral luminophore, and (**c**) supramolecular helical assembly

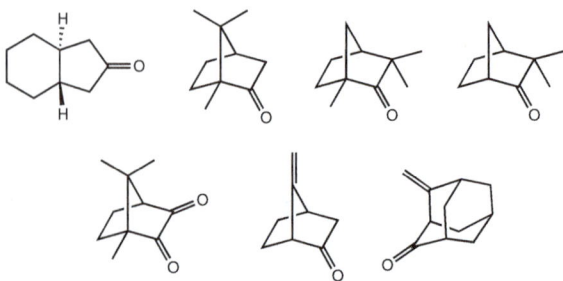

**Fig. 9.2** CPL-active chiral ketones

wavelengths (~360 nm). The use of chiral π-conjugated luminophores that exhibit fluorescence by the π* → π transition as CPL-active organic compounds is desirable due to their emission in the visible region with a high emission quantum yield. The most popular chiral π-conjugated molecule is a helicene. A helicene is an ortho-ring-fused polycyclic aromatic compound, in which aromatic rings are angularly annulated to give a helical structure. Helicenes have been widely investigated from the 1950s, and the CPL of a helicene was reported by Katz and coworkers in 2001 [12]. They synthesized [7]helicene-like compound 1 (Fig. 9.3). The molecule exhibited polarized fluorescence at 440 nm in diluted dodecane solution ($2 \times 10^{-6}$ mol L$^{-1}$), but CPL was not reported under these conditions. However, 1 displayed CPL in the condensed dodecane solution ($>1 \times 10^{-3}$ mol L$^{-1}$). The authors considered that the formation of aggregated species triggers the CPL.

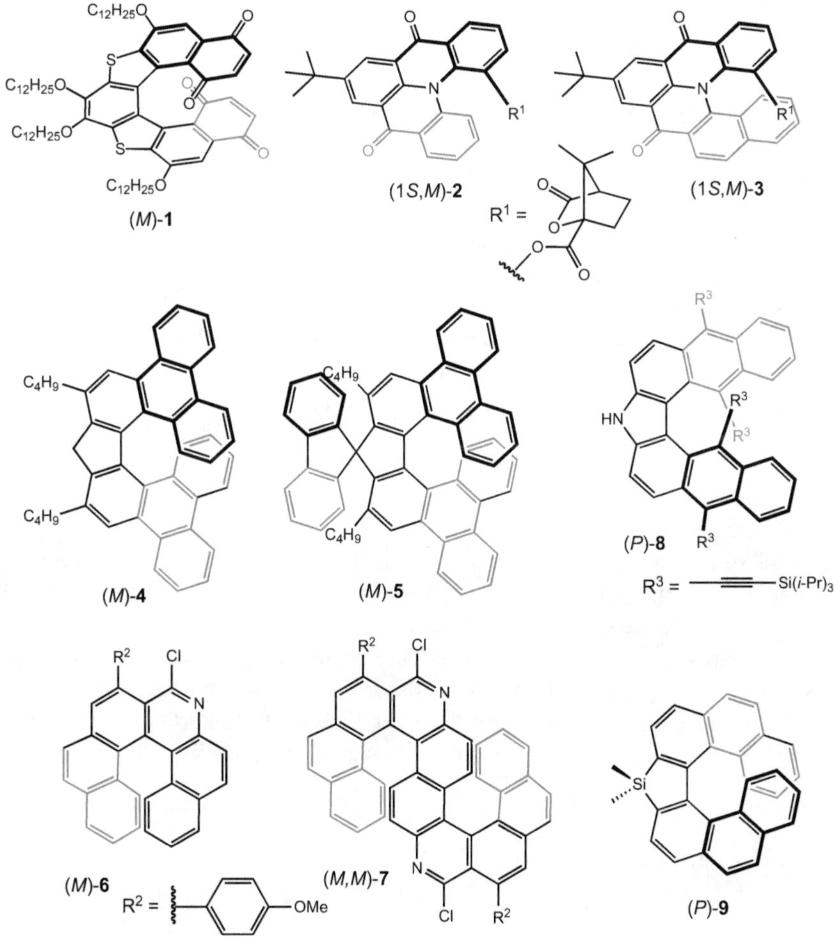

Fig. 9.3 CPL-active helicenes

The first observation of CPL of a nonaggregated helicene was reported by Venkataraman and coworkers in 2003. They synthesized a racemic pair of a helicene, and then, the helicene was reacted with (1$S$)-camphanate to give a corresponding diastereomeric pair of the helicene ((1$S$,$M$)- and (1$S$,$P$)-**2**, Fig. 9.3) [13]. Both of the diastereomers exhibited CPL on the monomeric fluorescence with a $g_{lum}$ of ca. 1 × 10$^{-3}$. Interestingly, the CPL spectrum of (1$S$,$M$)-**2** was a mirror-image of that of (1$S$,$P$)-**2** in spite of a diastereomer, indicating that the (1$S$)-camphanate attached on the helicene core does not perturb the CPL of the helicene. They also synthesized the more extended [5]helicene-like compounds (1$S$,$M$)- and (1$S$,$P$)-**3** that display CPL too.

After the pioneering work by Venkataraman and coworkers, various chiral helicenes exhibiting CPL have been investigated. Tanaka and coworkers have developed several [7]helicene-like compounds displaying CPL since 2012 (Fig. 9.3) [14, 15]. For example, ($M$)-**4** and ($M$)-**5** exhibited CPL on its monomeric fluorescence in CHCl$_3$ with a $g_{lum}$ of ca. 3 × 10$^{-2}$, which is a high value for an organic compound [15]. Tanaka and coworkers have also developed the CPL of aza[6]helicene-like compounds [14]. They synthesized aza[6]helicene ($M$)-**6** and S-shaped double aza[6]helicene-like compound ($M$,$M$)-**7**. Interestingly, the CPL of ($M$,$M$)-**7** ($g_{lum} = 1.1 \times 10^{-2}$) was enhanced compared to that of ($M$)-**6** ($g_{lum} < 1 \times 10^{-3}$). Shinokubo and coworkers reported the CPL of aza[7]helicenes ($M$)- and ($P$)-**8** in 2012 (Fig. 9.3) [16]. They exhibited CPL with a $g_{lum}$ of 3 × 10$^{-3}$ in CH$_2$Cl$_2$. Nozaki and coworkers have developed CPL-active sila[7]helicenes ($M$)- and ($P$)-**9** that exhibited CPL with a $g_{lum}$ of 3.5 × 10$^{-3}$ in CH$_2$Cl$_2$ (Fig. 9.3) [17].

Takeuchi and coworkers have reported an interesting CPL behavior of a phthalhydrazide-functionalized [7]helicene-like compound **10** (Fig. 9.4) [18]. **10** formed a trimeric disk in chloroform via the hydrogen-bonding interaction of a phthalhydrazide moiety. Further aggregation of the trimeric disk resulted in the formation of a one-dimensional fiber-like assembly. The CPL property of **10** was found in chloroform, but not in methanol, which suggests that the trimeric disk was dissociated by breaking the hydrogen bonds. The $g_{lum}$ values of ($M$)-**10** were estimated to be −3.5 × 10$^{-2}$ in chloroform and −2.1 × 10$^{-2}$ in methanol (4 × 10$^{-4}$ mol L$^{-1}$), respectively. The formation of the trimeric disk and one-dimensional fiber-like assembly may increase the $g_{lum}$ value of ($M$)-**10** in chloroform compared to that in methanol.

Thus, various CPL-active helicene-like compounds have been reported. However, the fluorescence quantum yields of CPL-active helicenes are moderate (32% for **4**, 36% for **8**, and 23% for **9**, respectively). The distortion of the $\pi$-plane may decrease the effective fluorescence of planar $\pi$-conjugated luminophores.

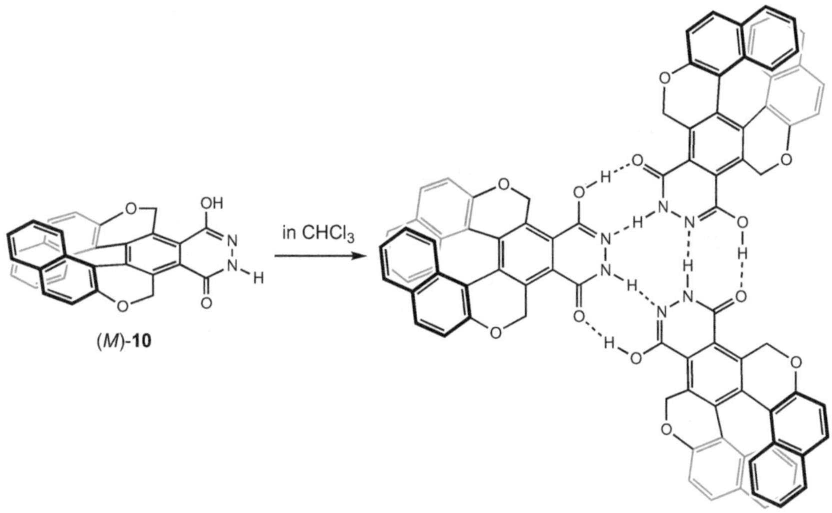

**Fig. 9.4** CPL-active trimeric disk of the phthalhydrazide-functionalized [7]helicene-like compound (*M*)-**10**

## 9.3 CPL of Chirally Arranged Achiral Luminophores Through Chiral Spacers

An achiral luminophore has an advantage with regard to the fluorescence efficiency compared to a chirally distorted luminophore such as a helicene. A strategy to provide a chiral environment on an achiral luminophore is the chiral arrangement of two or more luminophores. When two luminophores are linked through a chiral spacer, the luminophores take a chiral orientation. Then, the luminophores interact each other, and the electronic state of a luminophore is chirally perturbed by the other luminophore. Various skeletons providing chiral orientation have been developed in the field of asymmetric catalysts. The most famous one is a chiral binaphthyl moiety. A binaphthyl has axial chirality since the free rotation about the bond linking the naphthyl rings is restricted due to the steric effect of hydrogen atoms at the 8 and 8′ positions. Thus, each naphthalene takes a chiral orientation. Fujiki, Imai, and coworkers have developed CPL of simple and extended binaphthyls **11–18** (Fig. 9.5) [19–27]. The CPL properties of the binaphthyl derivatives depend on the dihedral angle of binaphthyl, the topology of the neighboring groups, the position of binaphthyl linkage, and the π-extension of the binaphthyl unit. The binaphthyl derivatives other than **14** (nonfluorescent) and **17** (CPL-inactive) exhibited CPL with $g_{lum}$ values within the range of $(1.5–0.8) \times 10^{-3}$ and with fluorescent quantum yields within the range of 15–25%.

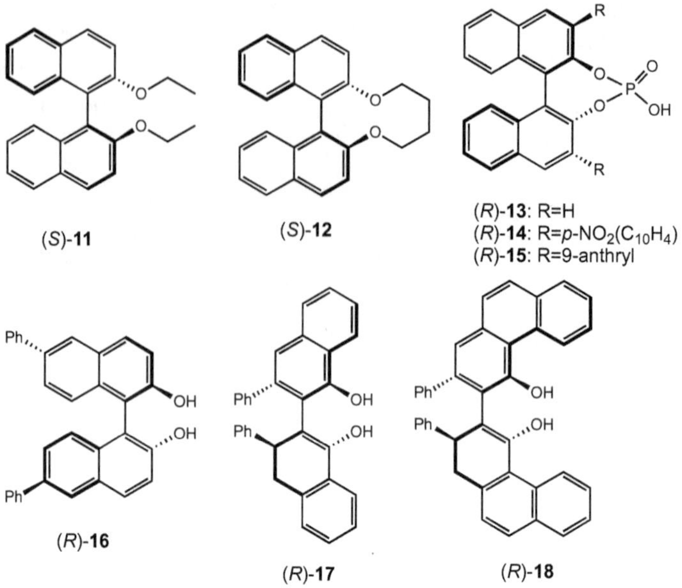

(S)-11          (S)-12

(R)-13: R=H
(R)-14: R=p-NO$_2$(C$_{10}$H$_4$)
(R)-15: R=9-anthryl

(R)-16

(R)-17          (R)-18

**Fig. 9.5** CPL-active chiral binaphthyls

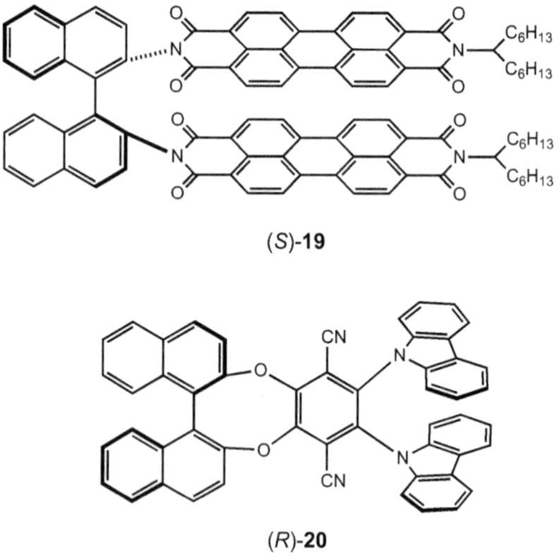

(S)-19

(R)-20

**Fig. 9.6** CPL-active achiral luminophores attached onto binaphthyls

The introduction of two luminophores onto binaphthyl is effective for inducing the CPL of the achiral luminophore. The luminophores substituted on each naphthalene take a chiral orientation reflecting the axial chirality of binaphthyl. Kawai and coworkers reported the CPL of perylene bisimide (PBI) linked by a binaphthyl unit (Fig. 9.6) [28]. PBI is one of the most known organic luminophores that exhibits its fluorescence at approximately 550 nm with a high emission quantum yield. They synthesized (*R*)- and (*S*)-**19**, with two PBI moieties attached to the 2 and 2′ positions of the binaphthyl. (*R*)- and (*S*)-**19** exhibited characteristic π–π∗ absorption and emission for the pair of PBI units. Chiroptical dissymmetry was observed in the circular dichroism (CD) spectrum, indicating that the two PBI moieties take a chiral orientation. (*S*)-**19** displayed CPL with a $g_{lum}$ of $2 \times 10^{-3}$ in their diluted solution ($1 \times 10^{-3}$ mol L$^{-1}$). The $g_{lum}$ value increased as the concentration of (*S*)-**19** increased and reached $6 \times 10^{-3}$ at the concentration of $1 \times 10^{-3}$ mol L$^{-1}$, where (*S*)-**19** formed an opaque colloidal solution. The authors suggest that the formation of the aggregate in the condensed solution results in an increase in the $g_{lum}$ value.

Pieters and coworkers developed a CPL-active delayed fluorescence system [29]. In thermally activated delayed fluorescence (TADF) emitters, the energy gap between their singlet and triplet states is so small that reverse intersystem crossing processes easily occur. Thus, both singlet and triplet excitons can be harvested for their fluorescence from the singlet excited state. This property has an advantage in developing motivated organic light emitting diodes (OLEDs) because of the possibility to overcome the theoretical maximum efficiency. They synthesized TADF emitter **20** possessing two achiral carbazoles as luminophores and a binaphthyl (Fig. 9.6). The TADF character of **20** was demonstrated using time-resolved fluorescence analysis. **20** exhibited CPL with a $g_{lum}$ of $1.3 \times 10^{-3}$. The combination of CPL and TADF may provide a remarkable CPL-OLED material.

The use of a chiral cyclophane is also effective for creating CPL-active organic materials. A cyclophane is a cyclic compound that includes aromatic moieties as an integral part of its structure. [2.2]Paracyclophane has two benzene rings linked by two ethylenes at the 1,1′ and 4,4′ positions. The orientations of two benzene rings of a [2.2]paracyclophane are fixed since the benzene rings cannot invert. Thus, chirally substituted [2.2]paracyclophanes provide a chiral arrangement of luminophores that exhibit chiroptical properties. Recently, CPL-active [2.2] paracyclophanes have been developed by Morisaki, Chujo, and coworkers (Fig. 9.7) [30]. They synthesized propeller-shaped [2,2]paracyclophane derivatives (*R*)- and (*S*)-**21** and their precursors (*R*)- and (*S*)-**22**. (*S*)-**21** exhibited CPL with a $g_{lum}$ of $1.1 \times 10^{-2}$ and fluorescence quantum yield of 45%. These values are quite good compared to ordinary chiral organic luminophores. Interestingly, the $g_{lum}$ value of (*S*)-**21** is approximately ten-fold higher than that of (*S*)-**22** ($g_{lum} = 1.1 \times 10^{-3}$). The good CPL activity of (*S*)-**21** comes from its highly extended and crisscrossed delocalized structure. The distortion of the π-conjugated plane in (*S*)-**21** may hinder its fluorescence efficiency, but it is clear that the distortion causes the high $g_{lum}$ value of (*S*)-**21** compared to that of (*S*)-**22**. Morisaki, Chujo, and coworkers also synthesized planar chiral tetra-substituted [2.2]paracyclophanes (*R*)- and (*S*)-**23** and **24** [31]. **23** and **24** have a chirally oriented two *para*-phenylene-ethynylene luminophore that is not distorted. **24** has a more extended π-conjugated structure compared to **23**. **23** and **24** displayed optical dissymmetry in their emissions with good fluorescence quantum yields (65% for **23** and 87% for **24**) in diluted chloroform solution ($1 \times 10^{-6}$ mol L$^{-1}$). The $g_{lum}$ values of (*R*)-**23** and (*R*)-**24** in the diluted solution were $-1.7 \times 10^{-3}$ and $-1.2 \times 10^{-3}$, respectively. Interestingly,

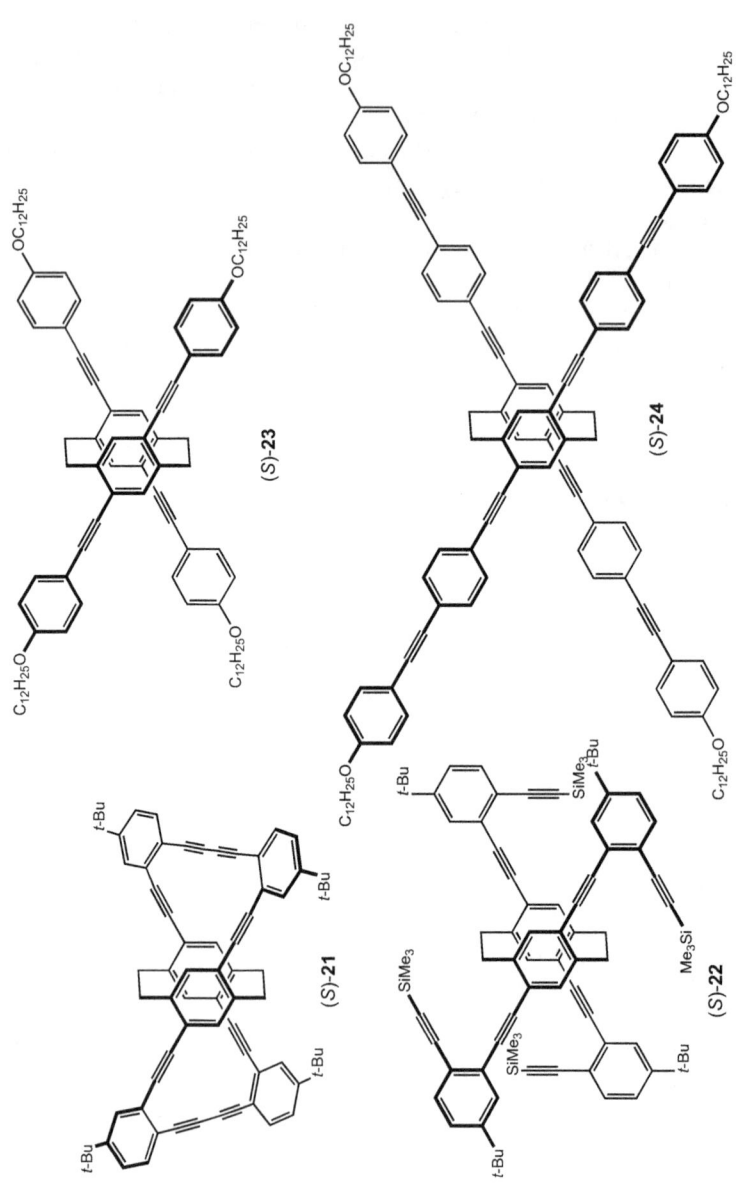

**Fig. 9.7** CPL-active chiral cyclophanes

**Table 9.1**  CPL properties of the films of **23** and **24**

| Film state | Spin-coated film | | Drop-casted thin film | | Drop-casted thick film | |
|---|---|---|---|---|---|---|
| | Before annealing | After annealing | Before annealing | After annealing | Before annealing | After annealing |
| (*R*)-**23** | −0.0061 | −0.0087 | −0.012 | −0.026 | −0.075 | −0.0043 |
| (*S*)-**23** | +0.0056 | +0.010 | +0.0096 | +0.034 | +0.056 | +0.0015 |
| (*R*)-**24** | +0.021 | −0.12 | −0.0086 | −0.17 | −0.030 | −0.25 |
| (*S*)-**24** | −0.014 | +0.13 | +0.016 | +0.13 | +0.011 | +0.27 |

the CPL activities of **23** and **24** drastically changed when the molecules formed films. They made spin-coated thin films, drop-casted thin films, and drop-casted thick films of **23** and **24**. The films were annealed at 65 °C (for **23**) and 90 °C (for **24**) for 3 h. The CPL properties of the films are summarized in Table 9.1. The $g_{lum}$ values of **24** were increased after annealing, whereas those of **23** were not increased as much. For example, the $g_{lum}$ value of the drop-casted thick film of (*R*)-**24** was $-3.0 \times 10^{-2}$ before annealing and $-2.5 \times 10^{-1}$ after annealing. The value after annealing is quite large for organic compounds. The authors proposed that the self-assembly of (*R*)-**24** in the annealed film results in the increase of the $g_{lum}$.

Thus far, the CPL properties of chiral organic luminophores and chirally arranged achiral luminophores through covalently linked spacers have been summarized. Three important findings are obtained from the results: (1) the distortion of the π-plane is effective for CPL activity, but it hinders the fluorescence efficiency of the planar π-conjugated luminophore; (2) a chiral arrangement of an achiral luminophore is effective for CPL activity; and (3) the aggregation of CPL-active organic compounds often increases their CPL activity. These findings encourage the use of the supramolecular assembly of achiral luminophores as CPL-active materials.

## 9.4   CPL Produced by Helical Supramolecular Assemblies

A supramolecular assembly is a well-defined molecular assembly held by noncovalent bonds such as hydrogen-bonding, π–π stacking, dipole–dipole, and hydrophilic/hydrophobic interactions. A finely designed supramolecular system provides a highly ordered structure of the assembly like the double helix of DNA. Recently, various highly ordered supramolecular assemblies have been developed [32–34]. Among them, supramolecular assemblies equipped with chirality have attracted attention due to their potential applications in the fields of asymmetric catalysts, chiral sensors, and chiroptical materials. Here, it is important to remember that the CPL properties of some chiral organic compounds increase in the condensed conditions compared to in the diluted solutions as mentioned above [12, 28, 35]. The increase would result from the formation of aggregates in the condensed conditions. Then, the idea to use supramolecular methods for constructing CPL-active organic materials is attractive since the luminophores form highly assembled structures in highly ordered supramolecular assemblies [2]. The helix is one of the most frequently appearing motifs in chiral supramolecular systems. The right-handed helix is a mirror-image of the left-handed one, and thus, the helix is chiral. Various types of helical assemblies have been developed: chiral and achiral small molecules, oligomers, and polymers form helical assemblies [36]. Considering the CPL properties, helically stacked assemblies of planar π-conjugated luminophores are optimal to balance the highly assembled structure with the chiral orientation of a luminophore.

To form an assembly of a small π-conjugated molecule, the appropriate molecular design is indispensable. If a π-conjugated molecule forms a helical stacked assembly, a racemic mixture of *P*-helix and *M*-helix would be obtained without any chiral source. However, the introduction of a chiral source makes the *P*-helix and *M*-helix diastereomers. The diastereomers have differences in their Gibbs free-energy, and thus one (*P* or *M*) is more stable than the other (*M* or *P*) (Fig. 9.8). Therefore, the chiral source, a chiral alkyl side-chain in many cases, must be introduced into the molecule.

The assembled structure of a small π-conjugated molecule is stabilized by π–π stacking interactions, but other intermolecular interactions are needed to form a stable stacked assembly. The hydrogen-bonding interactions of amide moieties are often employed to form stacked assemblies. The hydrogen bonds are formed between hydrogen atoms attached to nitrogen and carbonyl oxygen. The consecutive hydrogen bonds of amide moieties form a one-dimensional network of hydrogen bonds (Fig. 9.9). When the amide moiety is attached to the aromatic rings of a π-conjugated luminophore, the amide moiety takes a twisted conformation to the π-conjugated plane of the aromatic ring. Thus, one-dimensional consecutive

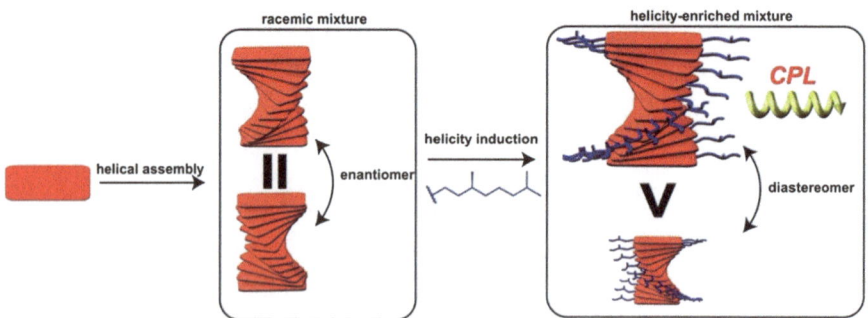

**Fig. 9.8** CPL produced by a helical supramolecular assembly of an achiral luminophore

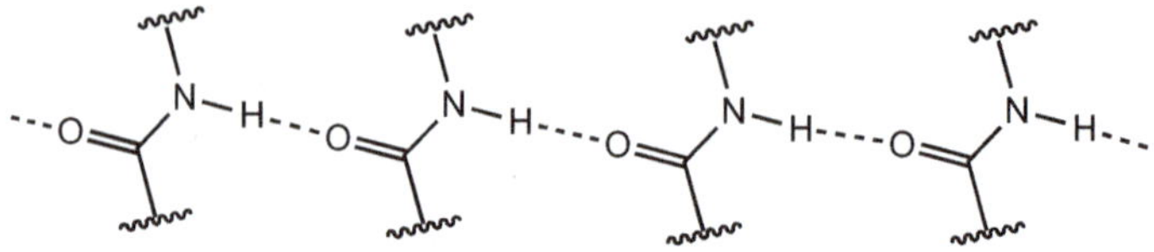

**Fig. 9.9** One-dimensional hydrogen bond network of amide moieties

hydrogen bonds are formed to stabilize the stacked assembly of the π-conjugated molecules.

Various helical stacked assemblies via hydrogen-bonding interactions have been reported, and some of them exhibited CPL activity. Ajayaghosh and coworkers reported the self-assembly behavior as well as optical and chiroptical properties of **25**, an oligo-p-phenylene-ethynylene derivative possessing an amide moiety and chiral side-chains (Fig. 9.10) [37]. The molecule mostly existed as a monomer in tetrahydrofuran (THF) solution ($1 \times 10^{-5}$ mol $L^{-1}$), whereas **25** formed stacked assemblies in methylcyclohexane (MCH) solution ($1 \times 10^{-5}$ mol $L^{-1}$). **25** is equipped with an azobenzene moiety that exhibits photo-induced isomerization. In fact, photoisomerization from the (E,E)-isomer to (Z,E)- and (Z,Z)-isomers was observed with the irradiation of ultraviolet (UV) light. The formation of a helical assembly was confirmed by using CD spectroscopy. The assembly of the (E,E)-isomer of (S)-**25** exhibited a positive CD signal at 464 nm with two negative CD signals at 407 nm and 322 nm, corresponding to the π → π* transition of the oligo-p-phenylene-ethynylene moiety. Interestingly, the CD signals inverted in the CD spectrum of the assembly of the (Z,Z)-isomer of (S)-**25**. This suggests that the helical sense of the assembly inverted during the photoisomerization of **25**. The helical assembly of the (E,E)-isomer of (S)-**25** exhibited CPL on its emission at 503 nm with a $g_{lum}$ of $8 \times 10^{-3}$. The photoisomerization of (S)-**25** inverted the CPL as in the case of the CD. The helical assembly of the (Z,Z)-isomer displayed a negative CPL signal with a $g_{lum}$ of $-2 \times 10^{-3}$. (R)-**25** gives a mirror-image CD and CPL of (S)-**25**.

Recently, Takeda, Akutagawa, and coworkers reported a CPL-active assembly of a pyrene derivative **26** (Fig. 9.11) [38]. They synthesized a pyrene possessing four amide moieties, and a chiral side-chain was introduced onto the amides. The simple molecule formed a helical assembly in chloroform, THF, and MCH through the one-dimensional hydrogen-bonding interactions of the amide moieties, which was confirmed by UV-vis and CD spectroscopy techniques. The helical assembly exhibited CPL on its excimer emission band at 500 nm in chloroform, THF, and MCH ($1.0 \times 10^{-4}$ mol $L^{-1}$). The $g_{lum}$ value of the assembly in MCH ($3.0 \times 10^{-2}$) was one order of magnitude higher than those in chloroform and THF (ca. $2 \times 10^{-3}$). Furthermore, the CPL signal of the helical assembly in chloroform was inverted

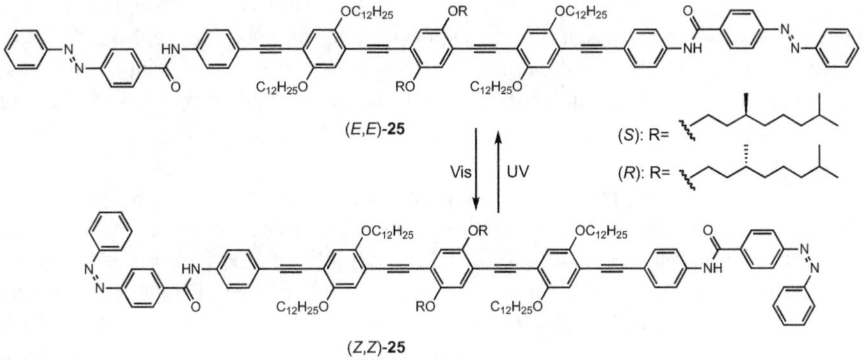

**Fig. 9.10** CPL-active helical assembly of a photoresponsive oligo-p-phenylene-ethynylene derivative

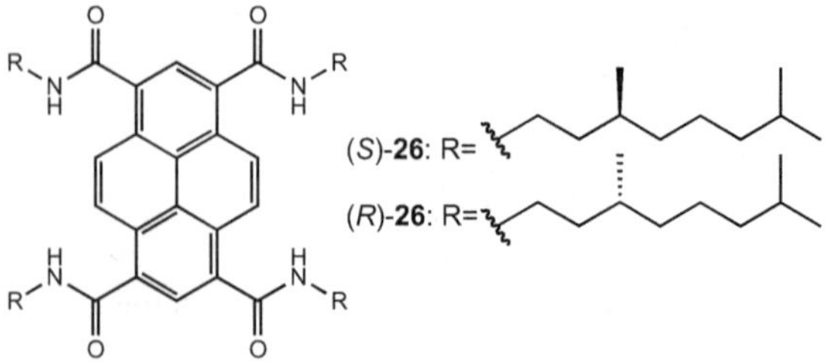

**Fig. 9.11** CPL-active
helical assembly of pyrene
derivatives

(S)-**26**: R=

(R)-**26**: R=

compared to those in THF and MCH. The assemblies in chloroform, THF, and MCH exhibited good fluorescence quantum yields (42%, 31%, and 29%, respectively).

A helical assembly can be formed via intermolecular interactions other than hydrogen-bonding interactions. Dipole–dipole interactions are fruitful to form a helical assembly since $C_3$ symmetrically arranged dipole moments tend to form a helical stacked assembly [39]. Haino and coworkers have reported the helical assembly behavior of $C_3$ symmetric 1,3,5-tris(4-alkoxyphenylisoxazolyl)benzene **27**, which possesses three isoxazoles that provide the local dipole moment (Fig. 9.12) [40, 41]. The directional circular arrangement of the isoxazole rings is directed by the head-to-tail dipole–dipole interaction of the local dipole of isoxazoles. Molecular modeling studies for the hexameric assemblies of 1,3,5-tris (phenylisoxazolyl)benzene revealed that the hexameric assembly has two major geometries, helical and eclipsed. In the former geometry, the local dipoles of isoxazole align in a head-to-tail fashion, whereas they take an antiparallel conformation in the latter geometry. The optical and chiroptical properties of **27** in MCH were considered by using UV-vis and CD spectroscopy techniques. Monomeric **27** in the diluted MCH solution exhibited a monomeric absorption band at 278 nm, whereas the assembly of **27** in the concentrated MCH solution displayed the absorption maximum at 310 nm. The redshift of the absorption suggests the formation of the *J*-type aggregate. The monomeric **27** displayed no CD signals, indicating that the chiral side-chain of **27** does not perturb the $\pi \rightarrow \pi*$ transition. On the other hand, the assembly of (S)- and (R)-**27** formed in the concentrated MCH solution exhibited CD spectra that have a mirror-image relationship. This suggests that not an antiparallel but helical assembly is formed, and the helicity was determined by the chirality of the side-chain. By using exciton coupling theory, the helical sense of the assembly of (S)-**27** was determined to be right-handedness. From these results, the dipole–dipole interaction of the isoxazole rings drives the helical assembly of small molecules. Unfortunately, the CPL properties of **27** were not reported, but the authors have reported the CPL properties of some luminophores possessing phenylisoxazoles.

Haino and coworkers have reported the helical assembly of PBI possessing phenylisoxazoles **28** and their optical and chiroptical properties (Fig. 9.13) [42]. The formation of the supramolecular assembly of PBI perturbs its $\pi \rightarrow \pi*$ absorption and $\pi* \rightarrow \pi$ emission. The introduction of tris(phenylisoxazolyl)benzene onto a nitrogen atom of PBI resulted in the formation of a helical assembly, in which the $\pi \rightarrow \pi*$ and $\pi* \rightarrow \pi$ transitions are chirally perturbed to give chiroptical

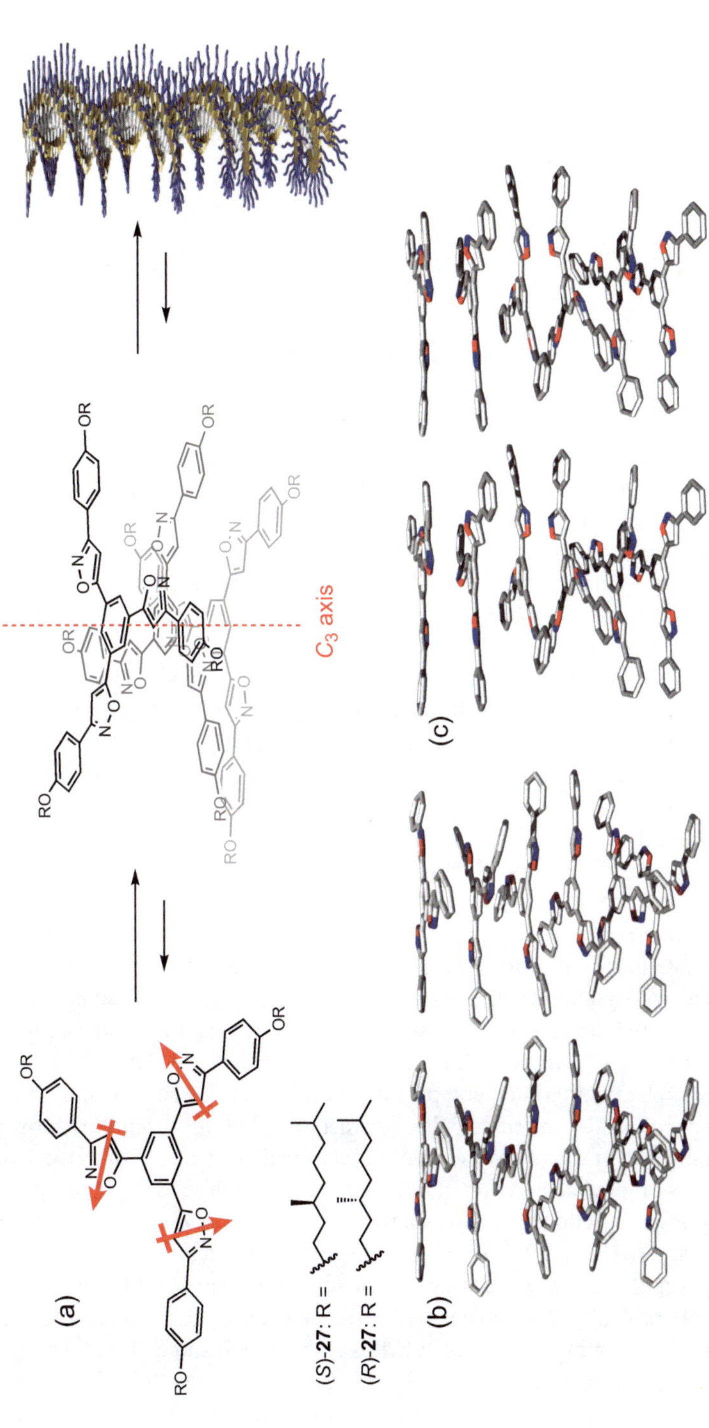

**Fig. 9.12** (a) Helical assembly of tris(phenylisoxazolyl)benzene derivatives **27**. (**b**, **c**) Stereoplots of two local minimum geometries for the hexamers obtained from a conformation search by the MacroModel program: (**a**) helical arrangement of the local dipoles and (**b**) antiparallel arrangement of them. Adapted with permission from Sato et al. [35]. Copyright 2011 American Chemical Society

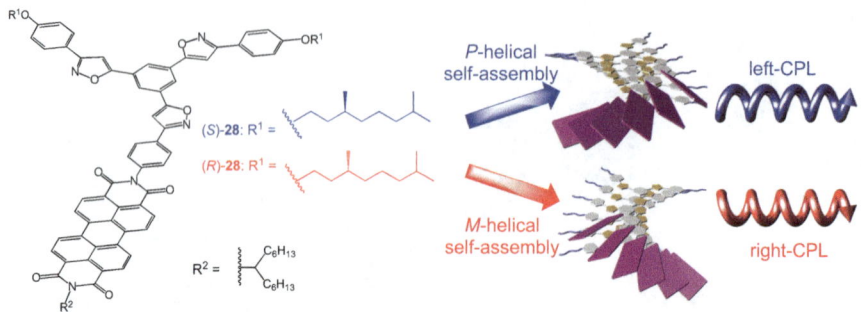

**Fig. 9.13** CPL-active helical assembly of PBI derivatives possessing tris(phenylisoxazolyl)benzene **28**. Adapted with permission from Satrijo et al. [5]. Copyright 2012 The Royal Society of Chemistry

properties. The self-assembly behavior of **28** was considered by using $^1$H NMR, UV-vis, and CD spectroscopy techniques. **28** formed a $J$-type assembly that was confirmed by the redshift of the absorption band of the assembly. The assembly of **28** in decalin solution displayed strong CD with a $g_{abs}$ of $1.4 \times 10^{-3}$. The right-handedness of the helix was assigned by the plus-to-minus patterns observed in the ascending energy in the CD spectrum of **27**. In the emission spectrum of the assembly of **28** in decalin, a broad and redshifted band of the assembly was observed in addition to the sharp emission band of the monomeric species. CPL was observed on the emission of the assembly with a $g_{lum}$ of $7 \times 10^{-3}$, but no CPL was observed on the monomeric emission. Furthermore, CPL was not observed on the emission of **28** in chloroform, in which most **28** does not form an assembly. ($S$)- and ($R$)-**28** displayed mirror-image CPL spectra. These results clearly suggest that the formation of a helical assembly leads to the CPL activity of **28**.

A square planar Pt(II) complex tends to form a stacked assembly creating a one-dimensional metal array through metallophilic (Pt–Pt) interactions [43]. The Pt(II) phenylbipyridine complex is known as a luminophore exhibiting phosphorescence that comes from a triplet metal-to-ligand charge transfer ($^3$MLCT) transition. The phosphorescence property of the Pt(II) phenylbipyridine complex is perturbed by the formation of a stacked assembly via Pt–Pt interactions to exhibit a triplet metal-metal-to-ligand charge transfer ($^3$MMLCT) transition [44]. The Pt(II) phenylbipyridine complex does not form a helical assembly without any assistance of other intermolecular interactions, but the complex formed a helical assembly when the phenylisoxazole moiety was introduced onto the ligand. Haino and coworkers have reported the optical and chiroptical properties of a Pt(II) phenylbipyridine complex possessing a 3,5-bis(phenylisoxazolyl)phenylethynyl ligand (($S$)- and ($R$)-**29**, Fig. 9.14) [45]. **29** effectively formed a stacked assembly via Pt–Pt, dipole–dipole, and π–π stacking interactions. Interestingly, the self-assembly behavior of **29** drastically changed depending on the solvent effect. In chloroform, **29** formed a stacked assembly exhibiting $^3$MMLCT absorption and emission bands, but the assembly displayed no CD and CPL. It turns out that the assembly of **29** formed in chloroform is not helical, most likely due to the strong solvation that prevents the

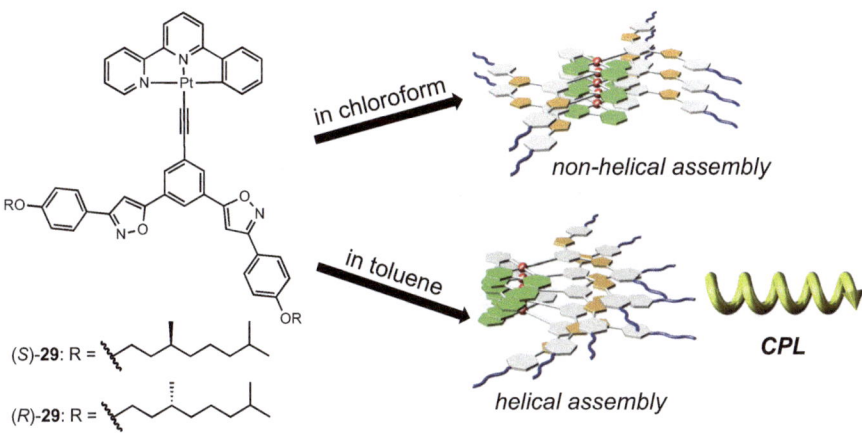

**Fig. 9.14** CPL-active helical assembly of **29**, a Pt(II)phenylbipyridine complex possessing phenylisoxazoles. Adapted with permission from Schippers et al. [10]. Copyright 2015 The Royal Society of Chemistry

dipole–dipole interaction between the isoxazole rings. On the other hand, the assembly of **29** formed in toluene displayed completely different characteristics compared to that in chloroform. The assembly formed in toluene was CD active, and mirror-image CD signals were observed upon the assembly of (S)- and (R)-**29**. Thus, helical assemblies are formed, and the helical senses were directed by the chiral side-chains. The assembly in toluene exhibited aggregation-induced emission enhancement (AIEE); the emission is strongly enhanced in the assembled form compared to that of the monomeric species. The molecular motion, presumably the rotation of the triple bond, is prevented through the formation of the assembly to reduce the nonradiative decay of the excited state, which results in the increase in the emission. The monomeric **29** did not display any CD and CPL even in toluene, but the assembly formed in toluene exhibited CPL with a $g_{lum}$ of $1.0 \times 10^{-2}$.

Spano, Meskers, and coworkers have developed the self-assembly and chiroptical properties of oligo-$p$-phenylene-vinylene derivative **30** (Fig. 9.15) [46]. **30** dimerized through the complementary quadruple hydrogen bond of the triazine group, and then, the dimer assembled to form a helical stacked assembly via π–π stacking. The helical assembly exhibited CPL on its fluorescence with a $g_{lum}$ on the order of $10^{-3}$. Wong, Tang, and coworkers have developed the CPL of tetraphenylsilole derivatives. Tetraphenylsiloles are known as good aggregation-induced emission (AIE) active luminophores; the luminophore is not emissive in its monomeric form, but the aggregate of the luminophore is highly emissive. For example, **31**, a tetraphenylsilole derivative possessing chiral sugar pendants, formed a helical assembly displaying CPL in its suspension and solid state (Fig. 9.15) [47]. The $g_{lum}$ values of **31** depend on the state of the assembly. In a heterogeneous suspension, a neat static cast film, and 10 wt.% Poly(methyl methacrylate) (PMMA) matrix, the $g_{lum}$ values were ca. $-0.12$, $-0.08$, and $-0.17$, respectively. Interestingly, the fabricated pattern obtained by the evaporation of a dichloromethane/toluene solution of **31** in Teflon-based microfluidic channels

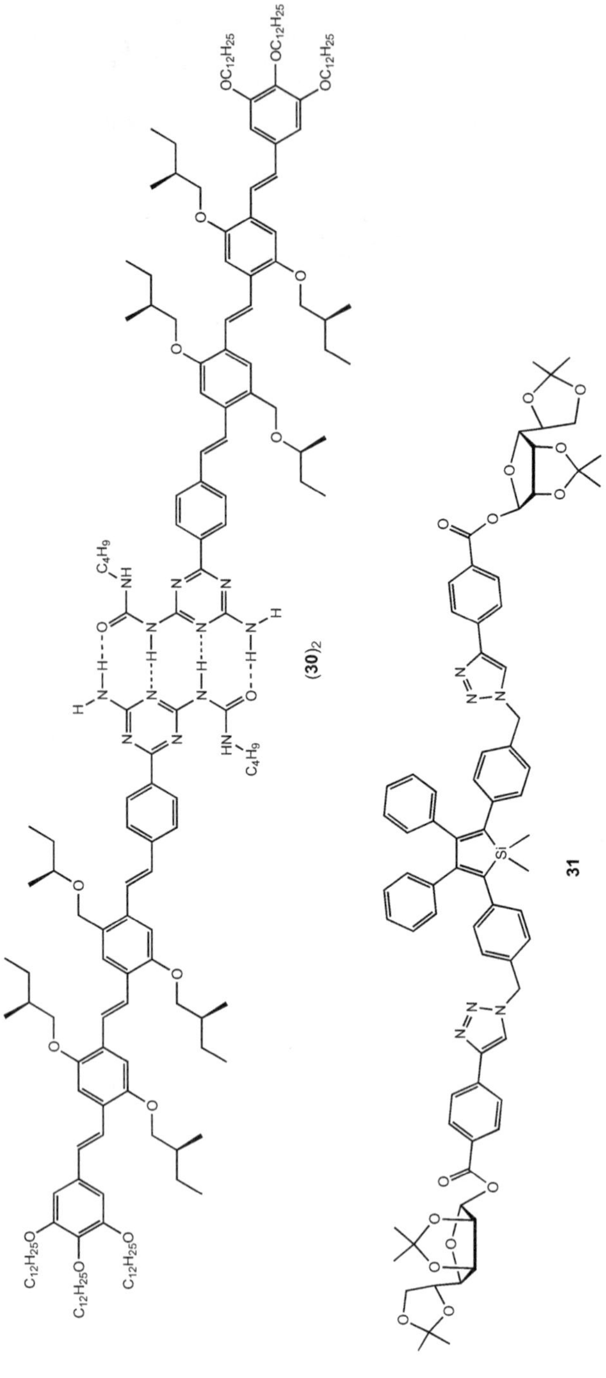


$(\mathbf{30})_2$

**31**

**Fig. 9.15** CPL-active helical assemblies of an achiral luminophore exhibited CPL with a $g_{lum}$ value of $-0.32$, which is quite higher than those of other solid samples.

As described herein, supramolecular assemblies that have chirality, especially the helical assemblies, exhibited excellent CPL properties. Highly assembled structures of the luminophore bring about high $g_{lum}$ values [2].

## 9.5 Stimuli-Responsive CPL Using a Supramolecular Assembly

A notable feature of CPL-active organic compounds is the stimuli-responsiveness of CPL. Organic compounds that exhibit stimuli-responsivity have been widely developed since organic compounds have flexible conformations. For example, **32** and **33**, photoresponsive CPL-active polymers possessing the photochromic dithienylethene moiety, have been developed by Akagi and coworkers (Fig. 9.16) [48]. A supramolecular assembly is constructed by reversible intermolecular interactions, and thus, one can control the assembly and disassembly of the supramolecular system by external stimuli. Therefore,

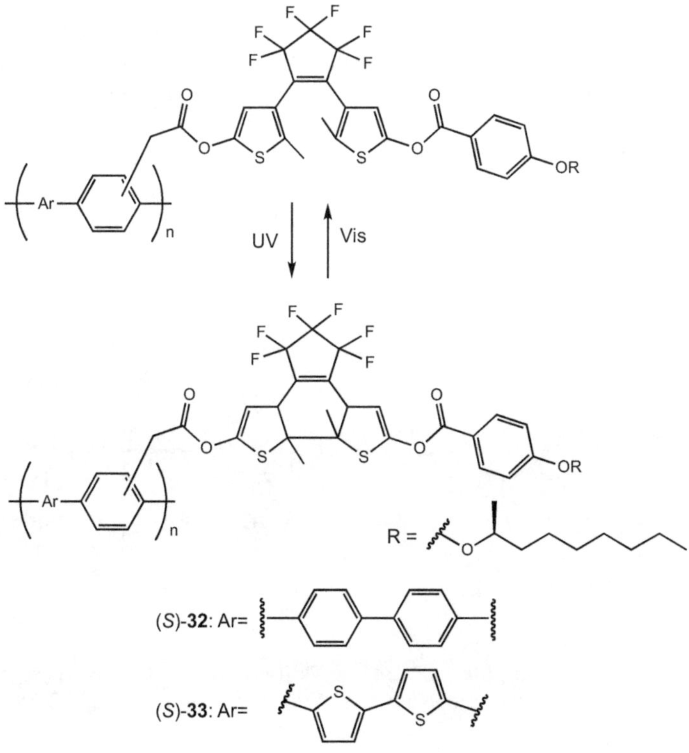

**Fig. 9.16** Photoresponsive CPL-active polymer

stimuli-responsive supramolecular assemblies have been widely investigated so far [49]. Many of the CPL-active supramolecular assemblies mentioned above exhibited concentration- and temperature-dependent CPL; the monomeric form of the luminophore that exists in the low concentration or high temperature shows no CPL, whereas the helical assembly formed in the high concentration or low temperature exhibits CPL [38, 42, 45–47]. The CPL signal of compound **25** is photoresponsive, reflecting the photoisomerization of the azobenzene unit [37]. Here, other stimuli-responsive CPL of organic luminophores is described.

Haino and coworkers reported the gelation-induced CPL of Pt(II) phenylbipyridine complex **34** (Fig. 9.17) [50]. The complex has a similar structure to complex **29** mentioned above, but **34** possesses multiple long alkyl side-chains. **34** formed a stacked assembly as **29** did, but the assembly formed in MCH did not display any CD and CPL. The long alkyl side-chains provide the interassembly interaction to afford a three-dimensional network of the one-dimensional stacked assembly. The - three-dimensional network holds the solvent molecules in its vacancies to form organogel. **34** formed luminescent organogel in 1-decanol at the concentration of $9.2 \times 10^{-3}$ mol $L^{-1}$ with decreasing temperature from 50 to 10 °C. At 50 °C, no CPL signal was observed, but the CPL signals appeared at 500 nm with the gelation. The $g_{lum}$ value at 10 °C was $1.1 \times 10^{-2}$. It is noteworthy that the gelation of **34** also triggered the AIEE; the emission intensity increased ca. 50-fold from 50 to 10 °C.

Maeda and coworkers have developed pyrrole-based anion receptors, $BF_2$ complexes of 1,3-dipyrrolyl-1,3-propanedione [51]. The two pyrrole NH are located at the side of the carbonyl oxygen without any anion, whereas the molecule forms an anion complex through the $N-H \cdots X^-$ interaction of the flipped pyrroles and the $C-H \cdots X^-$ interaction. The structural change of the luminophore before and after anion-binding can be applied for the anion-responsive CPL system (Fig. 9.18). Anion-responsive CPL was observed for chiral anion-receptor **35** upon complexation with anions such as $Cl^-$ and tetrabutylammonium (TBA) salts [52]. **35** possesses a chiral 1,1′-bi-2-naphthol (BINOL) moiety as the ligand of boron. **35** displayed CPL with a

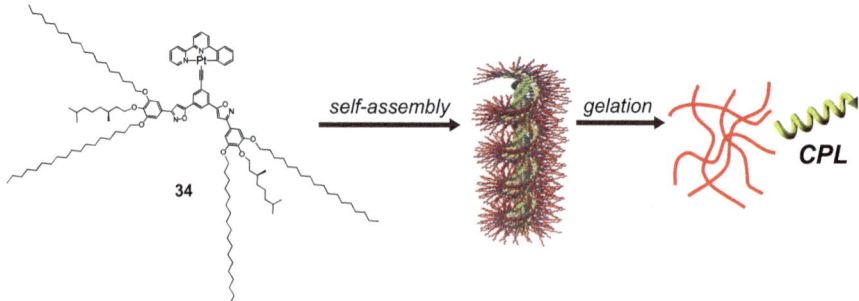

**Fig. 9.17** Gelation-induced CPL of **34**, a Pt(II)phenylbipyridine complex possessing phenylisoxazoles. Adapted with permission from Wilson et al. [6]. Copyright 2018 The Royal Society of Chemistry

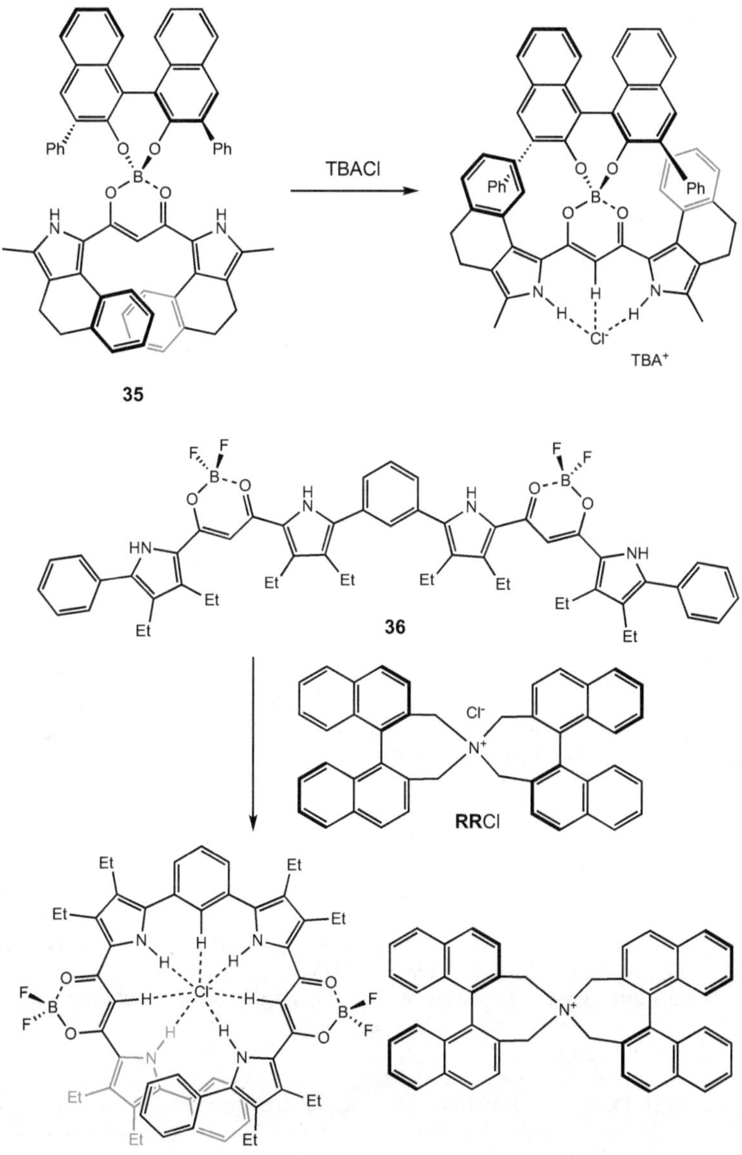

**Fig. 9.18** Anion-responsive CPL of pyrrole-based anion receptors

$g_{lum}$ of $1.0 \times 10^{-2}$ in the presence of $Cl^-$, whereas almost no CPL was observed without $Cl^-$. *meta*-Phenylene-bridged dimeric anion-receptor **35** also displayed anion-responsive CPL [53]. **36** formed a helical complex with $Cl^-$. Unlike the chiral anion-receptor **35**, **36** is an achiral molecule, and thus, the helical complex of **36·$Cl^-$** is a racemic mixture that exhibits no CPL without any chiral source. However, once $Cl^-$ was added as a salt of chiral binaphthylammonium (**RR$^+$**), the helicity of the complex was induced since the helical anion complex formed an ion pair with the chiral cation in the dichloromethane solution. The ion pair of **36**·Cl⁻·**RR**⁺ exhibited CPL with a $g_{lum}$ of $1.8 \times 10^{-2}$ in dichloromethane/octane solution (1:1, $1 \times 10^{-3}$ mol L⁻¹) at –70 °C.

## 9.6 Summary

An achiral π-conjugated luminophore can exhibit CPL by the chiral arrangement of the achiral luminophore. The helical supramolecular assembly of an achiral luminophore is excellent for realizing CPL of the achiral luminophore since the highly assembled structure in the helical assembly provides good CPL activity. Furthermore, the supramolecular assembly has an advantage in its stimuli-responsiveness. The development of CPL-active supramolecular assemblies and their applications are in progress all over the world.

## References

1. Bünzli J-CG, Piguet C (2005) Taking advantage of luminescent lanthanide ions. Chem Soc Rev 34:1048–1077
2. Kumar J et al (2015) Circularly polarized luminescence in chiral molecules and supramolecular assemblies. J Phys Chem Lett 6:3445–3452
3. Sánchez-Carnerero EM et al (2015) Circularly polarized luminescence from simple organic molecules. Chem Eur J 21:13488–13500
4. Emeis CA, Oosterhoff LJ (1967) Emission of circularly-polarised radiation by optically-active compounds. Chem Phys Lett 1:129–132
5. Satrijo A et al (2006) Probing a conjugated polymer's transfer of organization-dependent properties from solutions to films. J Am Chem Soc 128:9030–9031
6. Wilson JN et al (2002) Chiroptical properties of poly(p-phenyleneethynylene) copolymers in thin films: large g-values. J Am Chem Soc 124:6830–6831
7. Zhao Y et al (2016) Supramolecular chirality in achiral polyfluorene: chiral gelation, memory of chirality, and chiral sensing property. Macromolecules 49:3214–3221
8. Dekkers HPJM, Closs LE (1976) The optical activity of low-symmetry ketones in absorption and emission. J Am Chem Soc 98:2210–2219
9. Schippers PH, Dekkers HPJM (1983) Circular polarization of luminescence as a probe for intramolecular 1nπ* energy transfer in meso-diketones. J Am Chem Soc 105:145–146
10. Schippers PH et al (1983) Circular polarization in the fluorescence of β γ-enones: distortion in the 1nπ* state. J Am Chem Soc 105:84–89
11. Steinberg N et al (1981) Measurement of the optical activity of triplet-singlet transitions. The circular polarization of phosphorescence of camphorquinone and benzophenone. J Am Chem Soc 103:1636–1640
12. Phillips KES et al (2001) Synthesis and properties of an aggregating heterocyclic helicene. J Am Chem Soc 123:11899–11907
13. Field JE et al (2003) Circularly polarized luminescence from bridged triarylamine helicenes. J Am Chem Soc 125:11808–11809
14. Nakamura K et al (2014) Enantioselective synthesis and enhanced circularly polarized luminescence of S-shaped double azahelicenes. J Am Chem Soc 136:5555–5558

15. Sawada Y et al (2012) Rhodium-catalyzed enantioselective synthesis, crystal structures, and photophysical properties of helically chiral 1,1′-bitriphenylenes. J Am Chem Soc 134:4080–4083
16. Goto K et al (2012) Intermolecular oxidative annulation of 2-aminoanthracenes to diazaacenes and aza[7]helicenes. Angew Chem Int Ed 51:10333–10336
17. Oyama H et al (2013) Facile synthetic route to highly luminescent sila[7]helicene. Org Lett 15:2104–2107
18. Kaseyama T et al (2011) Hierarchical assembly of a phthalhydrazide-functionalized helicene. Angew Chem Int Ed 50:3684–3687
19. Amako T et al (2013) Solid-state circularly polarised luminescence and circular dichroism of viscous binaphthyl compounds. RSC Adv 3:23508–23513
20. Amako T et al (2013) A comparison of circularly polarized luminescence (CPL) and circular dichroism (CD) characteristics of four axially chiral binaphthyl-2,2′-diyl hydrogen phosphate derivatives. Tetrahedron 69:2753–2757
21. Amako T et al (2013) Dependence of circularly polarized luminescence due to the neighboring effects of binaphthyl units with the same axial chirality. RSC Adv 3:6939–6944
22. Kimoto T et al (2012) Control of circularly polarized luminescence by using open- and closed-type binaphthyl derivatives with the same axial chirality. Chemistry 7:2836–2841
23. Kinuta T et al (2011) Solid-state chiral optical properties of axially chiral binaphthyl acid derivatives. J Photochem Photobiol A Chem 220:134–138
24. Kinuta T et al (2012) Control of circularly polarized photoluminescent property via dihedral angle of binaphthyl derivatives. Tetrahedron 68:4791–4796
25. Kitayama Y et al (2014) Enhancing circularly polarised luminescence by extending the π-conjugation of axially chiral compounds. Org Biomol Chem 12:4342–4346
26. Kitayama Y et al (2015) Circularly polarized luminescence of biaryl atropisomers: subtle but significant structural dependency. RSC Adv 5:410–415
27. Nakabayashi K et al (2014) Nonclassical dual control of circularly polarized luminescence modes of binaphthyl–pyrene organic fluorophores in fluidic and glassy media. Chem Commun 50:13228–13230
28. Tsumatori H et al (2010) Observation of chiral aggregate growth of perylene derivative in opaque solution by circularly polarized luminescence. Org Lett 12:2362–2365
29. Feuillastre S et al (2016) Design and synthesis of new circularly polarized thermally activated delayed fluorescence emitters. J Am Chem Soc 138:3990–3993
30. Morisaki Y et al (2014) Planar chiral tetrasubstituted [2.2]paracyclophane: optical resolution and functionalization. J Am Chem Soc 136:3350–3353
31. Gon M et al (2017) Enhancement and controlling the signal of circularly polarized luminescence based on a planar chiral tetrasubstituted [2.2]paracyclophane framework in aggregation system. Macromolecules 50:1790–1802
32. De Greef TFA et al (2009) Supramolecular polymerization. Chem Rev 109:5687–5754
33. Haino T (2015) Supramolecular polymerization engineered with molecular recognition. Chem Rec 15:837–853
34. Hoeben FJM et al (2005) About supramolecular assemblies of pi-conjugated systems. Chem Rev 105:1491–1546
35. Sato S et al (2017) Chiral intertwined spirals and magnetic transition dipole moments dictated by cylinder helicity. Proc Natl Acad Sci U S A 114:13097–13101
36. Yashima E et al (2016) Supramolecular helical systems: helical assemblies of small molecules, foldamers, and polymers with chiral amplification and their functions. Chem Rev 116:13752–13990
37. Gopal A et al (2012) Thermally assisted photonic inversion of supramolecular handedness. Angew Chem Int Ed 51:10505–10509
38. Anetai H et al (2018) Circular polarized luminescence of hydrogen-bonded molecular assemblies of chiral pyrene derivatives. J Phys Chem C 122:6323–6331
39. Ikeda T, Haino T (2017) Supramolecular polymeric assemblies of pi-conjugated molecules possessing phenylisoxazoles. Polymer 128:243–256

40. Haino T, Saito H (2009) A new organogelator based on 1,3,5-tris(phenylisoxazolyl)benzene. Synth Met 159:821–826

41. Tanaka M et al (2011) Self-assembly and gelation behavior of tris(phenylisoxazolyl)benzenes. J Org Chem 76:5082–5091

42. Ikeda T et al (2012) Circular dichroism and circularly polarized luminescence triggered by self-assembly of tris(phenylisoxazolyl) benzenes possessing a perylenebisimide moiety. Chem Commun 48:6025–6027

43. Eryazici I et al (2008) Square-planar Pd(II), Pt(II), and Au(III) terpyridine complexes: their syntheses, physical properties, supramolecular constructs, and biomedical activities. Chem Rev 108:1834–1895

44. Wong KM-C, Yam VW-W (2011) Self-assembly of luminescent alkynylplatinum(II) terpyridyl complexes: modulation of photophysical properties through aggregation behavior. Acc Chem Res 44:424–434

45. Ikeda T et al (2015) Novel helical assembly of a Pt(II) phenylbipyridine complex directed by metal-metal interaction and aggregation-induced circularly polarized emission. Dalton Trans 44:13156–13162

46. Spano FC et al (2007) Probing excitation delocalization in supramolecular chiral stacks by means of circularly polarized light: experiment and modeling. J Am Chem Soc 129:7044–7054

47. Liu J et al (2012) What makes efficient circularly polarised luminescence in the condensed phase: aggregation-induced circular dichroism and light emission. Chem Sci 3:2737–2747

48. Hayasaka H et al (2010) Helically π-stacked conjugated polymers bearing photoresponsive and chiral moieties in side chains: reversible photoisomerization-enforced switching between emission and quenching of circularly polarized fluorescence. Adv Funct Mater 20:1243–1250

49. Maeda H, Bando Y (2013) Recent progress in research on stimuli-responsive circularly polarized luminescence based on pi-conjugated molecules. Pure Appl Chem 85:1967–1978

50. Ikeda T et al (2018) A circularly polarized luminescent organogel based on a Pt(II) complex possessing phenylisoxazoles. Mater Chem Front 2:468–474

51. Haketa Y, Maeda H (2017) Dimension-controlled ion-pairing assemblies based on π-electronic charged species. Chem Commun 53:2894–2909

52. Maeda H et al (2011) Chemical-stimuli-controllable circularly polarized luminescence from anion-responsive pi-conjugated molecules. J Am Chem Soc 133:9266–9269

53. Haketa Y et al (2012) Asymmetric induction in the preparation of helical receptor-anion complexes: ion-pair formation with chiral cations. Angew Chem Int Ed 51:7967–7971

# Chapter 10
# Structural and Electronic Information Drawn from the Circularly Polarized Luminescence Spectra: Many Questions and Some Answers for Simple Organic Molecules, Polymers, and Molecular Aggregates

Giovanna Longhi and Sergio Abbate

**Abstract** In the last ten years, circularly polarized luminescence (CPL) has greatly advanced: lots of data have been collected and many compounds have been synthesized with the aim of enhancing this chiroptical response. We review here a few aspects with illustrative examples. After examination of the relation of circular dichroism (CD) and CPL signals, we investigate how CPL can be used to probe environment: aggregation phenomena or the presence of metal ions or pH variations. We also study the CPL of inherently dissymmetric chromophores and of metal complexes. We finally touch upon whether CPL originates from molecules or from aggregated inhomogeneous systems.

## 10.1  Introduction

Circularly polarized luminescence (CPL) has a long history, which, schematically, may be divided, with some degree of arbitrariness, into three stages: we may define the seminal work by Emeis and Osterhoof [1] as the starting point of CPL history for organic compounds. They concentrated on the systematic investigation of ketones, which exhibit large circular dichroism (CD), associated with $n \rightarrow \pi*$ transitions, the latter being electric dipole moment-forbidden/magnetic dipole moment-allowed, as of the definition of the rotational strength:

G. Longhi (✉) · S. Abbate
Dipartimento di Medicina Molecolare e Traslazionale, Università di Brescia, Brescia, Italy
e-mail: giovanna.longhi@unibs.it; sergio.abbate@unibs.it

© Springer Nature Singapore Pte Ltd. 2020
T. Mori (ed.), *Circularly Polarized Luminescence of Isolated Small Organic Molecules*, https://doi.org/10.1007/978-981-15-2309-0_10

$$R = Im\langle 1|\vec{\mu}|0\rangle \cdot \langle 0|\vec{m}|1\rangle$$

where |0> and |1> are the ground and excited states under consideration, *Im* stands for the imaginary part of the subsequent function, and μ and m are the electric and magnetic dipole moment operators, respectively. In the second part of the CPL history, a few labs undertook the development of the technique: the same Dutch school, mainly Dekkers and Osterhoof [2, 3] and in Israel Gafni and Steinberg [4] improved the experimental schemes; Richardson, Riehl and Dekkers [5, 6] developed the theory of CPL and treated a few interesting prototypical cases, advocating the contribution from fine theoretical chemists [7]; recently, physical chemists like Spano and Yamagata [8] focused on the treatment of aggregates. We would also like to mention the contributions of Parker et al. [9, 10] and Brittain [11]. However, it is in the last part of the CPL history that lot of interest was aroused among spectroscopists and physical chemists. This was originated by to several factors: (a) the availability of novel optical elements and new design or update of the Gafni–Steinberg scheme [12–14], with the appearance of commercial apparatuses [15, 16]; (b) the possibility to run quantum mechanical calculations [17–19]; and (c) vivid interest from material scientists, among the others [20, 21]. The recent advancement and expansion of the technique, which approximately started around 2010, is now contemplating each year about 200–300 papers reporting CPL spectra; the literature to be cited is interesting and quite sizeable. We have no intention to be exhaustive here, but rather we wish to cite, at the beginning, some recent review articles, written in this last period covering several aspects and attempting at defining general themes in the set of problems defined through the CPL technique.

A separate treatment, while considering CPL, can be deserved to lanthanide complexes; we may refer to reviews on the subject, among the others [9, 10, 22, 23]; we will not describe their work, since it is outside the focus of the present chapter and we are mostly interested in small organic molecules and polymers or aggregates thereof. For this reason we will refer to the review of Sánchez-Carnerero et al. [24], who presented the first rather exhaustive collection of CPL data on purely organic molecules and concluded that the maximum values for the observed *g* factor, namely the ratio of the difference in circularly polarized emitted intensities to the total emitted intensities ($g_{lum} = \Delta I/I$, $\Delta I = (I_L - I_R)$ and $I = (I_L + I_R)/2$), be of the order of $10^{-2}$. An updated systematic classification of the CPL data of organic molecules was provided recently by Mori et al. [25]. In that report the molecules were classified in five groups, namely ketones, paracyclophane-based molecules endowed with planar chirality, axially chiral biaryls, helicenes/helicenoids, and chiral boron-dipyrromethene (BODIPY) derivatives. For each one of the above groups of molecules, the authors looked not only at the magnitude of the $g_{lum}$ ratio, but also at the ratio $g_{lum}/g_{abs}$ ($g_{abs} = \Delta\varepsilon/\varepsilon$, $\varepsilon$ = molecular absorption coefficient); that review article may be defined as the first effort to discover a common motif in the scattered behaviors of CPL data, namely that the closer the latter ratio is to 1, the more undistorted the excited state is from the ground state. Another recent review-type work has been presented by Fujiki et al. [26]. It is also worth mentioning here the recent perspective article by Kawai et al. [27], which, in its brevity, has though the virtue of a general review paper, tackling the issue of molecular aggregates, of polymers and of solvent dependence of CPL spectra. In this introductory part, we mention also that Kawai group is carrying on, with great determination, the issue of setting up the design of a CPL-based instrument for spatially resolved imaging of emitting bodies [14]. Finally, we mention our own review [28], where we discuss experimental and theoretical/computational aspects of the CPL technique and data. At this stage, our feeling is that, notwithstanding the 50-year history we have roughly outlined above, the technique is still rather "young" and poses questions more than providing answers. In other words, we find that puzzling behaviors are encountered by analyzing CPL data, even in the sets of molecules just discussed. For this reason, in the discussion section we will present a few examples (from Sects. 10.2.1 to 10.2.5) of some issues from CPL spectroscopy, namely in Sect. 10.2.1, whether CPL simply allows one to measure the geometrical distortion in the structure of the excited state or to check also finer electronic features thereof. In Sect. 10.2.2, we will discuss: the use of fluorescence probes, even achiral ones like thioflavin, to monitor dissymmetric fibril-bundling and self-aggregation of biomolecules; the switching on/off fluorescence and CPL into *ortho*-oligo-phenylene-ethynylene (*o*-OPE) molecules interacting with metals; and the dependence of CPL on pH. In these cases, we will discuss the efficacy of CPL with respect to the concomitant CD phenomenon; particularly in the case of *o*-OPEs, we discuss if a ratiometric probe can be built in the ad hoc designed molecules bearing two distinct CPL transitions. In Sect. 10.2.3 we will then consider CPL data for molecules useful for material science, encompassing inherently dissymmetric thiophene oligomers, substituted helicenes, and chiral peropyrenes, the latter ones mimicking the behavior of short twisted graphenes. In Sect. 10.2.4, as anticipated above and as amply treated by Crassous et al. [29], we find it interesting to deal with the association of metals to organic ligands, or chiral organometallic compounds, related to well-known and amply studied transition metal (Pt and Ir) complexes. In Sect. 10.2.5, finally, we will touch on how to handle CPL data of mesomorphic chiral materials.

## 10.2   Discussion

### 10.2.1   CPL and CD: A Sometimes Difficult Relationship

In the large majority of cases, the observed CPL spectra consist in a monosignate band. This stems from Kasha's rule, which states that fluorescence originates from the first excited electronic state in its minimum geometry. Consequently, fluorescence and CPL depend on the same electronic transition moments as absorption and CD, with due account of the different starting geometry. This should allow one to conclude that the observed sign for the CPL band is the same as for the CD band observed at the highest wavelength. As recalled above, Mori et al. [25] moved a step further, by adding information on the absolute values of $g_{lum}$, namely $|g_{lum}|/|g_{abs}| = 0.14$ for chiral ketones, $|g_{lum}|/|g_{abs}| = 0.94$ for chiral cyclophanes, $|g_{lum}|/|g_{abs}| = 0.93$ for axially chiral biaryls, $|g_{lum}|/|g_{abs}| = 0.83$ for helicenes/helicenoids, and $|g_{lum}|/|g_{abs}| = 1.02$ for BODIPY-type compounds. These values for $|g_{lum}|/|g_{abs}|$ show a good correlation coefficient for chiral cyclophanes, while for the other classes of molecules the values of $g_{lum}$ and $g_{abs}$ span such a large variety within each class of molecules that the $r^2$ value is sometimes fairly small. The authors of the paper then relate such ratio to the conformational relaxation in the emissive excited state.

In line with the discussion in this subsection, which is about whether it is possible to define relations between the sign of CPL band and one of the lowest energy CD band, beyond what is obvious and just discussed above, we wish to present some exceptions and caveats regarding the rule of monosignate CPL maintaining the sign of the corresponding CD band.

Care must be taken when comparing the sign of CPL to the sign of the CD band at lowest energy (red edge), which may present low rotational strengths: in some cases the first CD band is due to a nearly forbidden transition gaining intensity from vibronic contributions (see, for example, simple helicenes [30, 31]); in other cases, some examples can be found in molecules with axial chirality, rotational strength is strongly dependent on the mutual orientation of the two moieties that can be tuned by different substituents [32]. Usually the sign is conserved considering $S_0 \rightarrow S_1$ and $S_1 \rightarrow S_0$, meaning that $S_0$ and $S_1$ geometries are not so different as to cause a sign change of rotational strength.

Well-known counterexamples can be found in some ketones: not only camphor [33] (Fig. 10.1) but also (1S,3R)-4-methyleneadamantan-2-one [34] and other similar examples. In particular the adamantanone derivative, reported by Dekkers and Closs [35], shows inversion of sign; on the contrary camphor exhibits two oppositely signed CPL bands. The sign of the strongest CPL band for camphor is opposite to the sign of CD band at lowest energy, with $|g_{abs}| \approx 0.03$ and $|g_{lum}| \approx 2.8 \times 10^{-3}$ so that $|g_{lum}|/|g_{abs}| \approx 0.09$, being the same order of magnitude as the average behavior commented by Mori et al. for chiral ketones. These examples, belonging to ketone class, can be explained considering that $S_1$ (giving rise to CPL) presents two possible geometries with slightly different energy, while the ground state geometry can be assumed as the one corresponding to the interconversion barrier between the two. In some cases (camphor) both excited state structures are observed, in other cases only one dominates (adamantanone derivative). We notice, by the way, that for camphor the less stable geometry, with CPL at lower wavelength, has the same octant-configuration (as defined by Moscowitz [36] and by Lightner and Gurst [37]) as the ground state geometry and thus its CPL has the same sign as CD. Instead, the lowest energy one, at higher wavelength, has opposite octant-configuration and thus its CPL has opposite sign to CD (see Ref. [33] for a discussion).

In general, one of the following circumstances is expected: (1) One dominant conformer in its ground state $S_0$ is excited and readjusts to the energy minimum for $S_1$, which usually has a slightly different geometry, such that relative orientations of electric and magnetic dipole transition moments are not so different as to change rotational strength sign. (2) One dominant conformer in its ground state $S_0$ reaches the first excited state; however, if the ground state geometry is a saddle point for $S_1$, the geometry of the emissive transition has two possible structures eventually generating rotational strengths of different signs. That is the case of camphor and adamantanone derivative. (3) Many conformers are present, each one with its own $S_0$

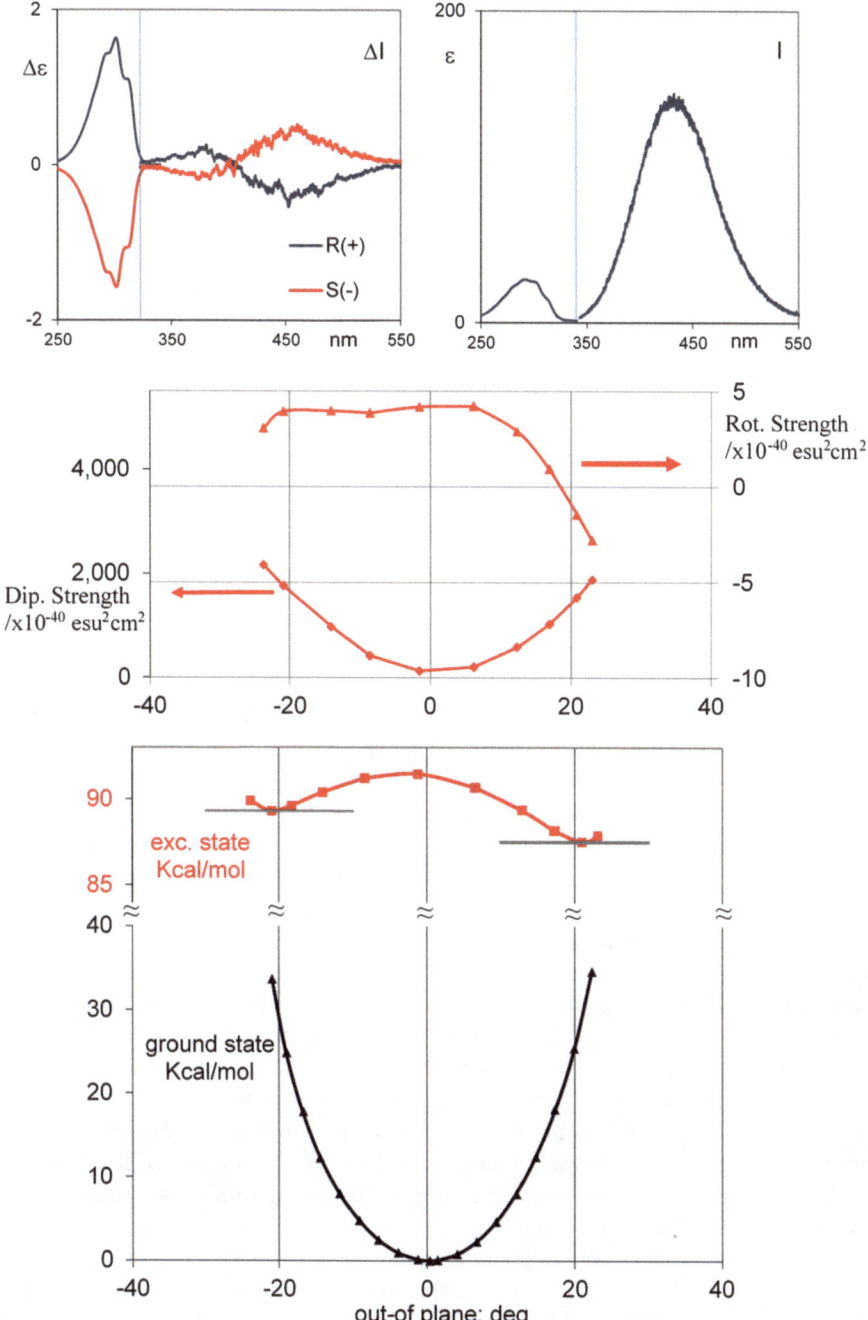

**Fig. 10.1** (Top): Superimposed CD and CPL spectra of (*R*)- and (*S*)-camphor in chloroform solution [33] (left); UV-absorption and fluorescence spectra of (*S*)-camphor in chloroform solution (right). (Middle): Dipole and rotational strengths calculated at different out of plane CO angle for the excited state $S_1$. (Bottom): $S_0$ and $S_1$ energies at different out of plane CO angles

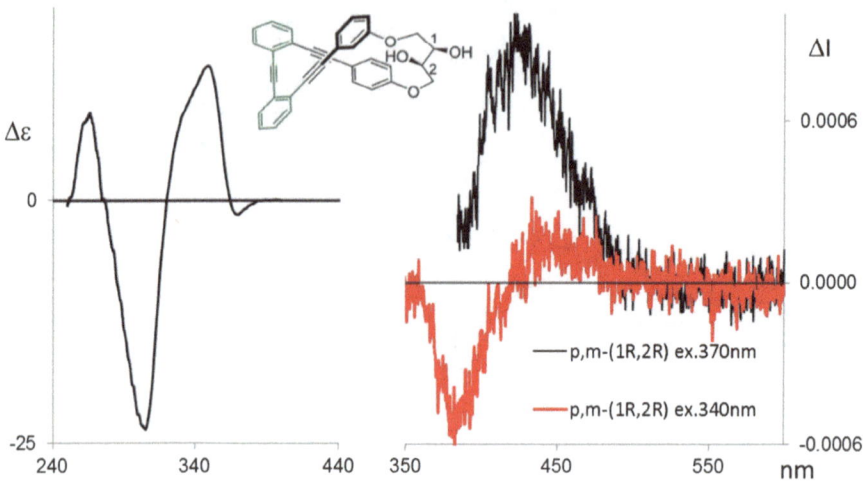

**Fig. 10.2** $p,m$-($1R,2R$)-OPE diol: example of a molecule with several conformers presenting opposite CD signals; this has a counterpart in $S_1$ states of different structures giving emission CPL bands of different sign also depending on excitation wavelength (see Ref. [38])

geometry giving rise to different CD bands and possibly arriving to different $S_1$ structures. Also, in this case, one may have different components contributing to emission; these components could be eventually discriminated by time resolved fluorescence spectroscopy. An example of the last circumstance is represented by $p$, $m$-($1R,2R$)-OPE diol (Fig. 10.2). From ground state calculations [38] it is known that different conformers with opposite CD contribute to the observed CD spectrum: in this case, it has been possible to observe that different excitation wavelengths are capable to evidence a minus–plus CPL couplet or a monosignated positive band. Other interesting examples are present in the literature signaling a different geometry between ground and excited states (see, for example, cryptochirality for the CD technique, but manifesting in CPL [39, 40]). (4) A fourth case that is worthwhile mentioning, when one considers relations between ground and emissive states, is represented by systems bearing two different fluorophoric electronically independent groups, examples of which may consist of OPE backbone with pyrene pendant groups, studied in Ref. [41] (a better description will be given later).

A similar but not identical aspect is worthwhile to mention: to understand the sign of the observed signal, it is important to consider differences not only in structures, but also in electronic properties, for example, due to a peripheral substituent with electron-withdrawing or electron-donating character. An instructive example is an azabora[5]helicene dye with different substituents [42] (see Fig. 10.3).

Depending on the electron-donor substituent one is dealing with, one observes for the $M$-isomer: a positive band (lowest energy electronic CD [ECD] transition and dominant CPL feature) for the first compound, and a negative one (first ECD and CPL transition) for the other two compounds. Calculations suggest nearly superimposable structures for the three molecules; the origin of the sign change, accounted for by the relative orientation of magnetic and electrical dipole transition moments,

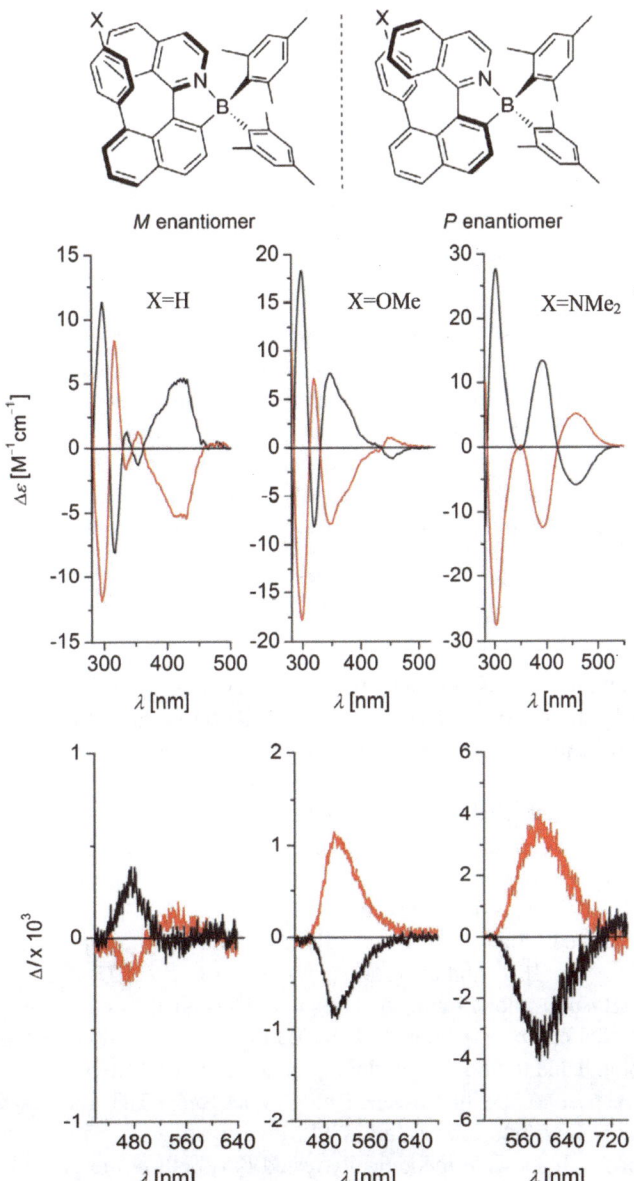

**Fig. 10.3** (Top): Azabora[5]helicenes with different substituents. (Middle): ECD spectra of compounds with $X = H$ (left), $X = OMe$ (middle), and $X = NMe_2$ (right). (Bottom): corresponding CPL spectra. The red spectra correspond to the $P$-isomer and the black spectra to the $M$-isomer (see Ref. [42])

can be attributed to the different degrees of charge transfer character in the three compounds that differ for electron-donor character of the substituent and not for geometry. This is a nice example of how peripheral substituents can modulate a number of CPL relevant parameters: transition wavelength, Stokes shift, rotational strength sign, and magnitude.

## 10.2.2   Fluorescence and CPL Probes

In this section, we investigate how a fluorescent probe is able to report on molecular or supramolecular phenomena that are either determined by chirality or related to it. Sensing in these circumstances is caused by a noncovalent interaction of the fluorescent molecule with the environment: aggregates, metal ions, pH; such interaction either turns on or turns off chiroptical responses, like CD or CPL. We present three examples of the phenomenon and will compare the CPL response with other chiroptical responses, mainly electronic CD (ECD), activated at the same time.

### 10.2.2.1   Sensing Fibrils

The first example regards sensing of fibril aggregation by a fluorescent nonchiral probe, namely thioflavin T (ThT). Fibril aggregation is one of the numerous phenomena of macromolecular assembling in which two types of ordering, $M$- and $P$-insulin fibrils in this case, coexist, and can be separated through a bifurcation process [43, 44]. In the present case the two species were indeed separated depending on temperature and pH condition while vortexing [43]. In Fig. 10.4, we report the ECD and UV-absorption spectra and the CPL and fluorescence spectra of ThT staining insulin fibrils of either $P$ or $M$ helicity.

One may see that, for the 1:10 (ThT:Insulin) molar ratio, $g_{abs} = (\Delta Abs/Abs) = 1 \times 10^{-2}$, and also for $g_{lum} = (\Delta I/I) = 1 \times 10^{-2}$, thus $g_{lum}/g_{abs} = 1$. For 1:2 (ThT:Insulin) molar ratio, comparison of CD and absorption spectra shows an evident shift that can be explained by considering that absorption contains contributions from bound and nonbound thioflavin; on the contrary there is no shift of the CPL band with respect to the fluorescence band since only ThT molecules bound to amyloid exhibit both fluorescence and CPL. Investigating the origin of the phenomenon, two possibilities were theoretically tested [45] both in the ground and excited state, namely: a chiral twist of ThT is stabilized by the fibril or particular orientations of (even flat) thioflavin with respect to nearby aromatic amino acids; the two situations provided theoretical chiral responses of the same order of magnitude.

### 10.2.2.2   Sensing Metal Cations

The second example regards the possibility to reveal the presence of metal ions in a solution of special organic molecules or oligomers based on *ortho*-oligo-phenylene-

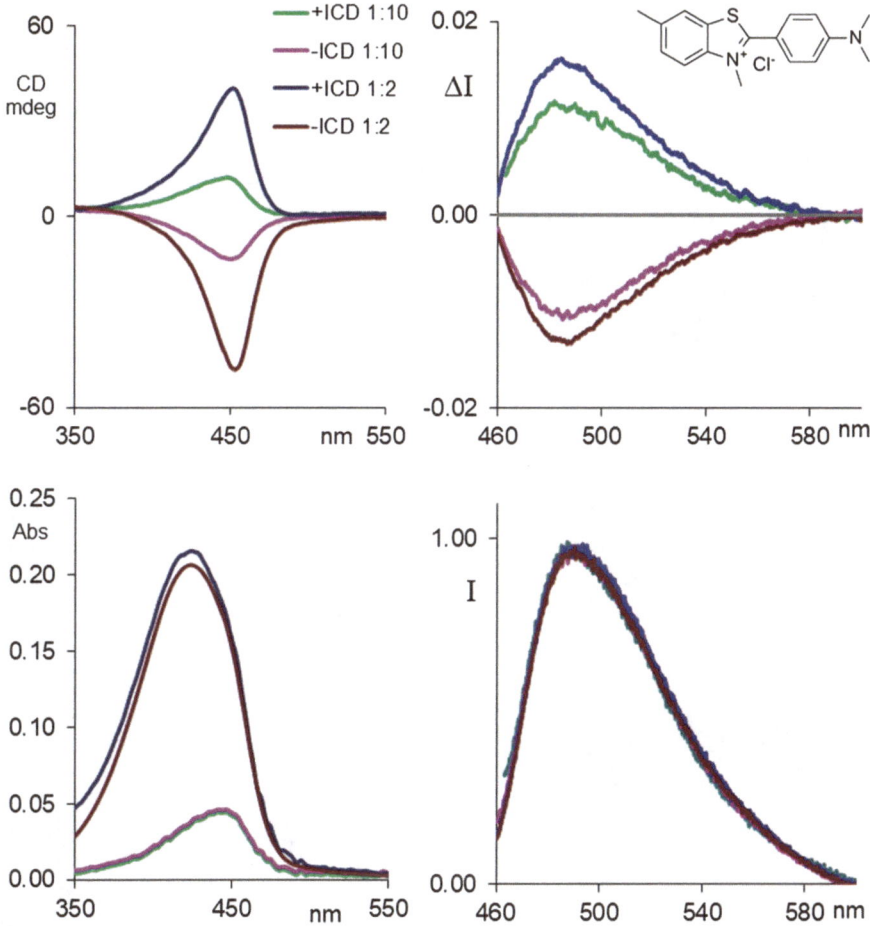

**Fig. 10.4** (Left): ECD (top) and UV/absorption (bottom) spectra. (Right): CPL (top) and fluorescence (bottom) spectra for aqueous suspensions of +ICD and −ICD insulin fibrils stained with ThT, in aqueous suspensions. The traces correspond to molar ratio ThT:insulin = 1:2 and 1:10. Excitation wavelength 440 nm (see Ref. [45])

ethynylene (*o*-OPE) moiety. Such moiety, besides being an active UV-chromophore, is quite prone to chiral folding in the presence of chemical substituents imparting chiral twists. Besides, the special carbophilic interaction of some metals with the same moiety makes it belong to the class of metallo-foldamers.

In Fig. 10.5, we report the CPL spectra of three examples of *o*-OPE compounds, recently studied in collaboration with Cuerva et al. The first compound (A) is stapled with a 2,3-butanediol fragment in para position to the *o*-OPE system and the two stereogenic centers impart a *M*-helicity to the *o*-OPEs moiety [38]; the second one (B) is terminated by two (*S*)-phenyl-sulfoxides still imparting *M*-helicity to the *o*-OPE system. For the second system also longer *o*-OPE terms, containing up to eight *o*-OPE units were investigated [46]. The affinity to Ag in the two cases is 12,000 $M^{-1}$

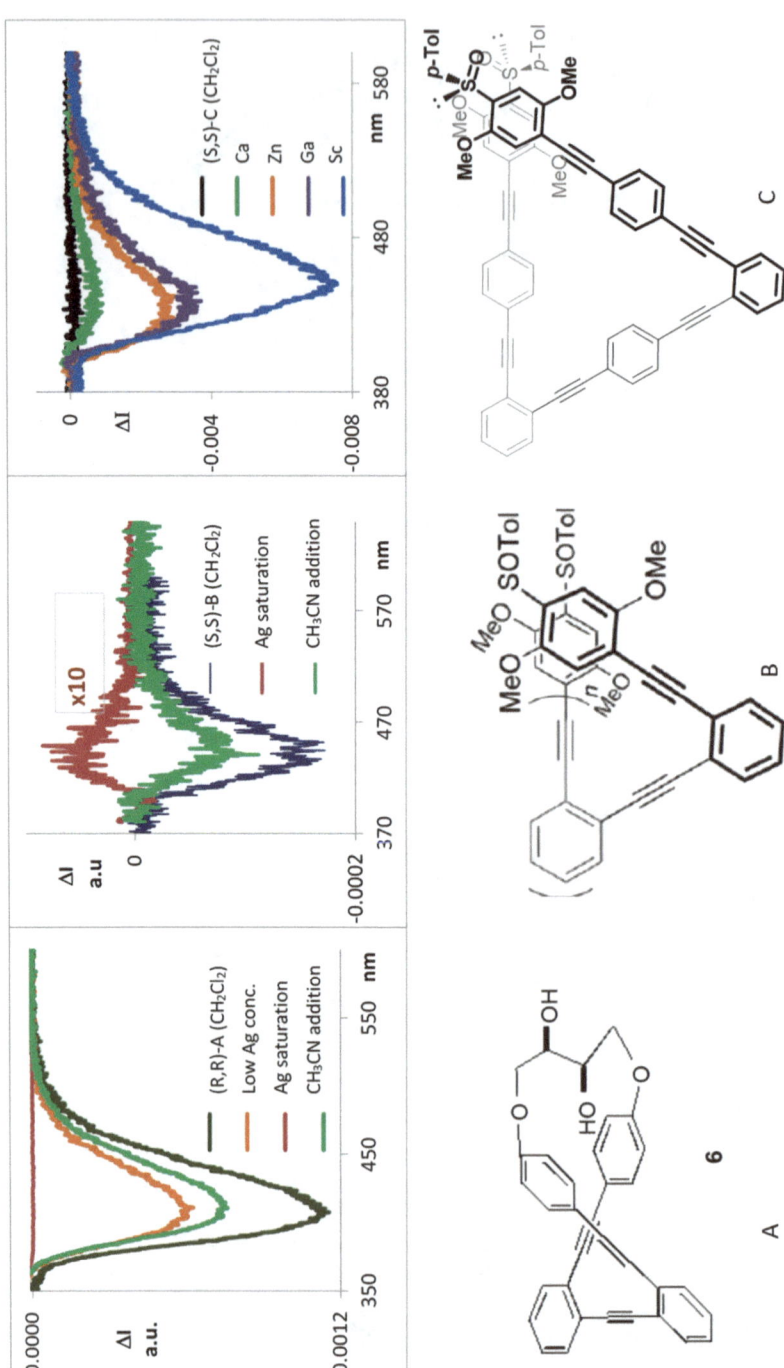

**Fig. 10.5** (Left): CPL spectra of $p,p$-(1R,2R)-diol-$o$-OPE (A) in $CH_2Cl_2$, with addition of $AgBF_4$ salt and after acetonitrile has been added to the mixed Ag:$p,p$-diol-$o$-OPE system [38]. (Middle): CPL spectra of (S,S)-$o$-OPE-phenyl-sulfoxide (B) in $CH_2Cl_2$, with addition of $AgBF_4$ salt and after acetonitrile has been added to the mixed Ag:(S,S)-$o$-OPE-phenyl-sulfoxide system [46]. (Right): CPL spectra of compound: (S,S)-$o$,$p$-OPE-phenyl-sulfoxide (C) in $CH_2Cl_2$, and in saturated conditions with X-triflate salt, X = Ca, Zn, Ga, Sc [47]

and 5000 M$^{-1}$, respectively. The effect of Ag is to silence the CPL reversibly, the system without Ag exhibiting a remarkable $g_{lum}/g_{abs}$ ratio, namely 1.15 in the first case and 1 in the second. Besides, in the first case also $g_{lum}$ is particularly large (0.012). Such large CPL values in principle provide a wide range for sensing quite different quantities of Ag and thus make the two systems as optimal probes for sensing silver. Compound (B), and the analogous compound (C) with longer lateral arms presented in Fig. 10.5 (right panel), can interact also with other metals via oxophilic mechanism through sulfoxide groups, reversible forming complexes, and with no quenching of fluorescence. Many metals have been tested [47] giving different CPL responses, some of them quite intense: compound (C) is particularly interesting since it presents quite low CPL before complexation (CH$_2$Cl$_2$ solution) and CPL with different dissymmetry ratio depending on the associated metal.

An interesting sensing system, which we discuss below, regards the possibility of probing Ag cation with a CPL feature, while recording in the same CPL spectrum a second CPL band not sensitive to Ag, which thus can be used as internal reference [41]. In Fig. 10.6 we report the data for the molecule that was designed and synthesized to the scope by taking advantage of the property of high sensitivity to silver of compound $p,p$-(1$S$,2$S$)-diol-$o$-OPE (the enantiomer of compound A of Fig. 10.5), and substituting the H hydroxyl atoms with two other fluorescent groups, namely two (CH$_2$)-pyrene units. The CPL spectrum is composed of two bands, a negative one centered at ca. 520 nm, the intensity of which is constant with addition of Ag(I), and a positive one centered at ca. 410 nm, the intensity of which instead is linearly dependent on the concentration of Ag(I); Ag has 6230 M$^{-1}$ binding constant to this compound. This means that the ratio of the two CPL bands is linearly dependent on Ag concentration, making the CPL band at 520 nm an internal constant reference and the idea of measuring the concentration of Ag in absolute terms really working. The lower part of Fig. 10.6 provides a possible theoretical justification to the different roles of the pyrene 520 nm transition and of the $o$-OPE 410 nm transition showing an indicative representation of molecular orbitals involved in the transition localized on the pyrene moiety (calculations were run on the ground state geometry).

### 10.2.2.3 Sensing pH

The final example we propose in this paragraph is the sensing of basicity conditions: it was found that (1$R$,2$R$)- and (1$S$,2$S$)-$trans$-cyclohexane diesters and diamides exhibit rather strong chiroptical ECD and vibrational CD (VCD) spectra, with characteristic bisignate features, which were proven to be associated with excitons. Such excitons, electronic and vibrational in the two cases, exhibit bathochromic and hyperchromic behaviors with addition of NaOH, leaving the shape of the ECD unaltered. Correspondingly, while in the neutral condition fluorescence is so low that a CPL spectrum is unobservable, at high pH a CPL band is recorded. As observed on other $C_2$ symmetric compounds presenting CD excitonic couplets (see also the next paragraph), concordant with Kasha's rule the sign of the CPL

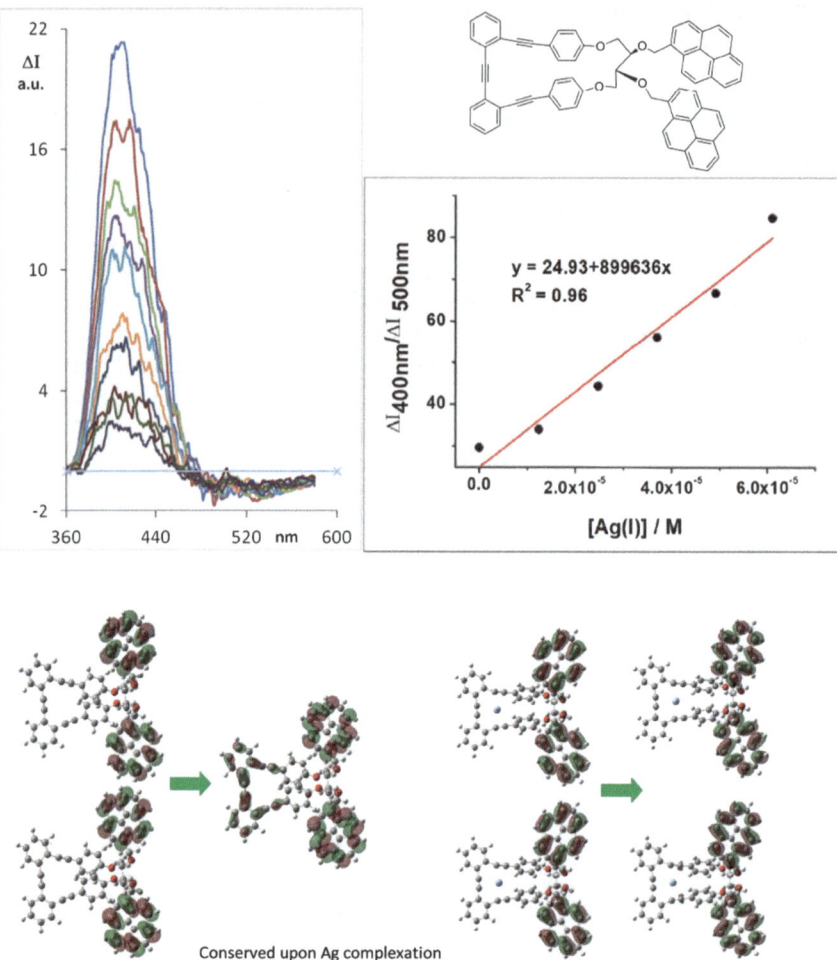

**Fig. 10.6** (Top): CPL spectra of the pyrene-*o*-OPE ether compound (*p,p*)-(1*S*,2*S*) in CH$_2$Cl$_2$ solutions with different amounts of Ag(I): on the right the linearity plot of the ratio $\Delta I_{410nm}/\Delta I_{520nm}$ of the *o*-OPE CPL emission to the pyrene CPL emission is reported. (Bottom): Molecular orbital representation of the principal orbitals involved in bands assigned to pyrene moieties and conserved upon Ag complexation, in absence (left) or presence (right) of Ag cation (see Ref. [41])

band is the same as the sign of the longer wavelength ECD component of the exciton feature described above (see Fig. 10.7) [48]. Furthermore, there is an important solvent dependence of the phenomena: (1) the absorption and ECD spectra undergo a bathochromic shift in going from methanol to dimethyl sulfoxide (DMSO) both in the neutral and in the basic solution, while fluorescence and CPL spectra of the basic solutions shift correspondingly by ca. 25 nm. (2) For diesters we have $g_{lum} < 10^{-3}$ in MeOH solution and $g_{lum} \approx 1.5 \times 10^{-3}$ in DMSO solution, while correspondingly $g_{abs} \approx 0.8 \times 10^{-3}$ in MeOH and in DMSO. Consequently, for cyclohexane diesters

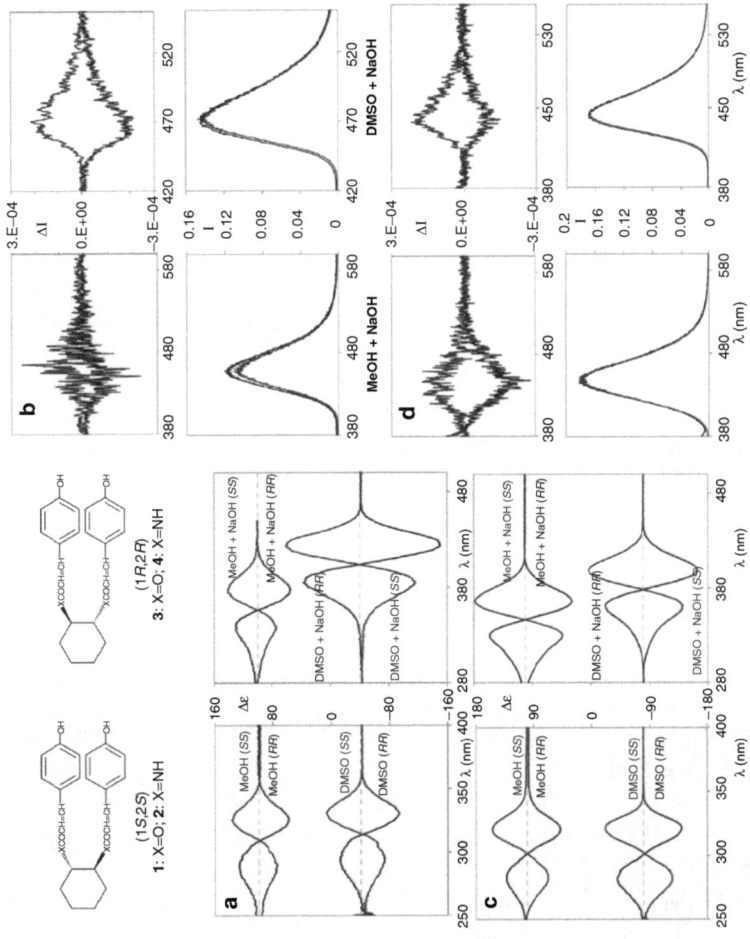

**Fig. 10.7** ECD, fluorescence, and CPL spectra of the diesters and diamides (structures in the top-left panel). (**a**): ECD spectra of (*S,S*)- and (*R,R*)-cyclohexane diesters (**1** and **3**) in methanol and dimethyl sulfoxide solution with and without addition of NaOH. (**b**): CPL and fluorescence spectra of (*S,S*)- and (*R,R*)-cyclohexane diesters (**1** and **3**) in methanol and dimethyl sulfoxide solution with addition of NaOH. (**c**): ECD spectra of (*S,S*)- and (*R,R*)-cyclohexane diamide (**2** and **4**) in methanol and dimethyl sulfoxide solution with and without addition of NaOH. (**d**): CPL and fluorescence spectra of (*S,S*)- and (*R,R*)-cyclohexane diamide (**2** and **4**) in methanol and dimethyl sulfoxide solution with addition of NaOH (see Ref. [48])

we have in MeOH solution $g_{lum}/g_{abs} \approx 1$ and in DMSO solution $g_{lum}/g_{abs} \approx 1.7$. Similar, though nonidentical, behaviors are observed for cyclohexane diamides. As stated in Ref. [48], since only from the charged state does one observe the fluorescence and CPL phenomena, the latter may become a yes-or-no probe of the charged state. Such effectiveness of the CPL technique may be related to the large Stokes shift.

In conclusion, common motifs of the effectiveness of the CPL signal to sense different sorts of phenomena, as in the three examples presented above, namely aggregation, presence of ions, pH changes, appear large Stokes shifts and values of $g_{lum}/g_{abs}$ greater or equal to 1.

### 10.2.3 Do Inherently Dissymmetric Chromophores Exhibit Large CPL?

Inherently dissymmetric chromophores may be defined as molecular moieties bearing groups exhibiting large transition amplitudes in the UV-vis and having a chiral shape; generally "these compounds have in common a dissymmetrically twisted π-system which essentially constitutes the entire chromophore" [49]. Such molecules have large specific optical rotations (ORs), of the order of several thousands; for the ECD spectra, transitions with such large amplitude are both electrically and magnetically allowed. The situation of the CPL spectra is more variegated. We examine here three classes of molecules in which this variety is manifest.

#### 10.2.3.1 C2-Symmetric Bi-chromophoric Systems

The first example is the class of $C_2$-chrompohores composed of two equal moieties tied with a single bond, the two equal parts containing in the present examples bithianaphthene units, with various pendants. In Fig. 10.8 we report the ECD and CPL spectra of three such molecules, namely (S)-2,2'-bis(2,2'-bithiophene-5-yl)-3,3'-bithianaphthene (BT2-T4), 4H-cyclopenta[2,1-b3:4b']dithiophene-3,3'--bithianaphthene (BT2-CPDT2), and dithieno[3,3-b:2',3'-d]pyrrole-3,3'--bithianaphthene (BT2-DTP2).

The values of $g_{lum}$ for the three inherently dissymmetric molecules BT2-T4, BT2-CPDT2, and BT2-DTP2 are slightly smaller but close to $10^{-2}$, and the ratios $g_{lum}/g_{abs}$ are 4.7, 7.1, and 4.2, respectively (we considered the longest wavelength transition in the ECD spectrum, in accordance to Kasha's rule). In all cases such ratios are pretty big, among the biggest in Mori et al. [25] classification, and indeed, as pointed out by those Authors, these large values are associated with pretty large values of the Stokes shifts (ca. 130 nm) indicating great excited state relaxation.

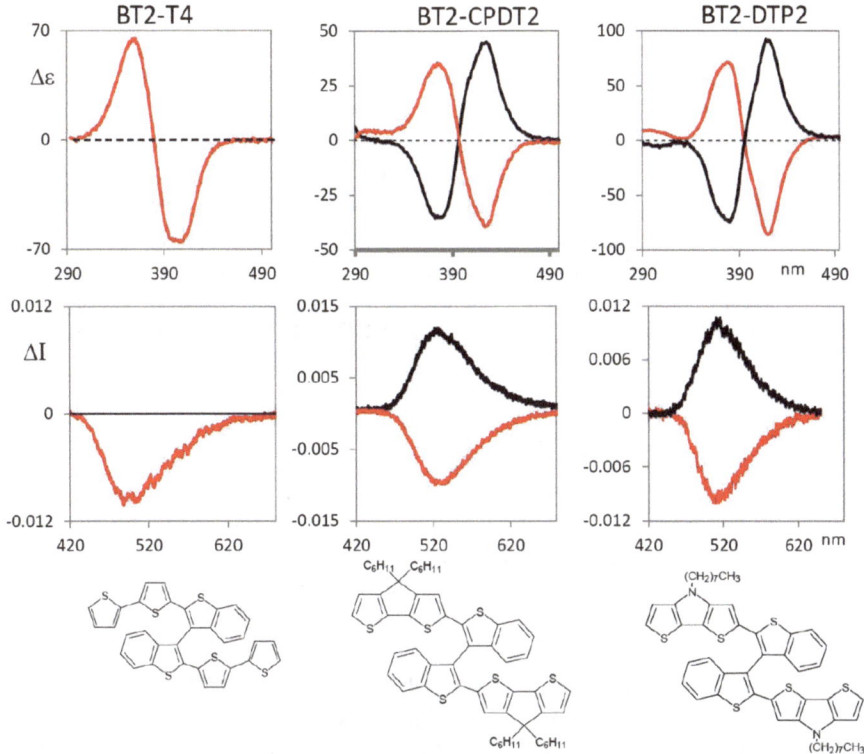

**Fig. 10.8** (Left): ECD and CPL spectra of (*S*)-2,2'-bis(2,2'-bithiophene-5-yl)-3,3'--bithianaphthene (BT2-T4). (Center): ECD and CPL spectra of 4H-cyclopenta[2,1-b3:4b'] dithiophene-3,3'-bithianaphthene (BT2-CPDT2). (Right): ECD and CPL spectra of dithieno [3,3-b:2',3'-d]pyrrole-3,3'-bithianaphthene (BT2-DTP2). Corresponding fluorescence spectra have been normalized: (*S,S*)-(red) and (*R,R*)-(black) (see Refs. [50, 51])

Additionally, in Refs. [50, 51], excited state calculations were carried out and the CPL spectra for the three molecules were computed: it was found that in the excited state a large variation occurs in the inter-unit dihedral angle with respect to the ground state, by an amount varying in the three molecules from ca. 30° to ca. 50°. Still the sense of rotation of one BT unit to the other BT unit is univocally related to chirality *M* or *P* for the molecule, which does not change from the ground to the excited state. The latter characteristic is related to the fact that the sign of the CD observed at the longest wavelength is the same as the sign of the unique CPL band. Another difference from the ground to the excited state is the flattening of the different conjugated parts in the two moieties. Similar $g_{lum}$ and $g_{abs}$ values were observed for the oligomers of BT2-T4 and of BT2-CPDT2 [52]. It is worthwhile to recall that, in presence of exciton couplets, the observed $g_{abs}$ may be influenced by partial mutual cancellation of the two components of the couplet depending on bandwidth and exciton wavelength separation; on the contrary, $g_{lum}$ is determined by just one component of the couplet.

## 10.2.3.2  Helicenes

The second example consists of hexahelicenes and of helicene-based molecules: through the years, helicenes have been thought of as the prototypes of inherently dissymmetric chromophores due to the chemical structure of such molecules, which are at the same time strongly conjugated and dissymmetrically distorted, being constituted by ortho-fused benzene rings. Experimentally, it has been known for a long time that helicene molecules possess large specific rotations (OR) at 589 nm, of the order of thousands, and this fact generated the expectation that all their chiroptical properties would be large and easy to measure. Indeed for 2-Br-hexahelicene, besides OR, also ECD, VCD, ROA, and CPL spectra were measured. The last spectra, though, were particularly weak also due to heavy atom effect quenching fluorescence. In Fig. 10.9 we report three cases of hexahelicene-based molecules, namely carbo-hexahelicene (HEX), 5-aza-hexahelicene (5N-HEX) [30], and thia-bridged triarylamine-hetero [6]-helicene (THIA-HEX) [53]. The spectra are for both enantiomers of each molecule.

The ECD spectra of HEX and 5N-HEX are overall pretty similar; they differ in the two tiny bands at ca. 410 nm, which are the lowest frequency ones and are ca. 30 times as intense in the second case with respect to the first case. Due to Kasha's rule, the lowest energy band is the one corresponding to the CPL band, which indeed has the same sign and is more intense for 5N-HEX. $g_{abs}$ and $g_{lum}$ approximately triple from HEX to 5-N-HEX, however maintaining the ratio $g_{lum}/g_{abs} \approx 1.1$ in both cases (see Table 10.1). We can conclude that CPL is quite weak a phenomenon for simple hexahelicene systems, since the CPL, as well as the ECD lowest energy band, originates from an $L_b$ transition that presents not only weak rotational strength, but also weak dipole strength, being dipole forbidden, and gains intensity from vibronic contributions [31]. In case of nearly forbidden transitions (hexahelicene and methylhexahelicene), absorption, emission, CD, and CPL are weak and the sign of the bands is sensitive to substituents as observed on many CD spectra of this kind of molecules in Ref. [54]. As a further characteristic of these systems, dipole and rotational strengths are sensitive to heteroatoms: the presence of nitrogen in 5N-HEX (the geometries of the ground state and excited state of HEX and 5N-HEX are quite similar) not only increases electric dipole transition moment but also changes the relative orientation of magnetic and electric dipole moments giving a higher rotational strength and dissymmetry ratio both in absorption and in emission. The role of heteroatom is even more important in the case of TRIA-HEX system (Fig. 10.9): a further increase in both $g_{lum}$ and $g_{abs}$ can be noticed, while the ratio $g_{lum}/g_{abs}$ is maintained at ~1.1. Comparing the Stokes shift, it increases from about 15 nm, in HEX and 5N-HEX, to ca. 100 nm, in TRIA-HEX; in this last case, the CPL spectrum loses evident vibrational features.

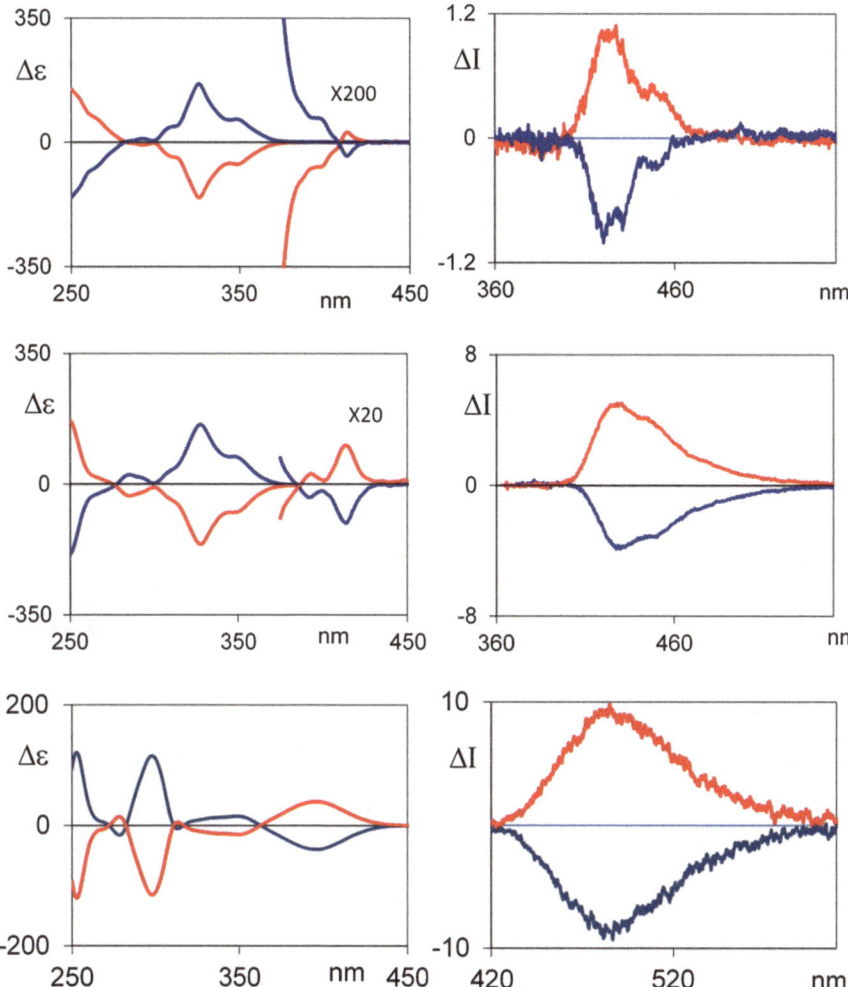

**Fig. 10.9** (Top): ECD and CPL spectra of hexahelicene (HEX). (Center): ECD and CPL spectra of 5-aza-hexahelicene (5N-HEX). (Bottom): ECD and CPL spectra of thia-bridged triarylamine-hetero [6]-helicene (TRIA-HEX). See Refs. [30, 53]

### 10.2.3.3   Distorted Graphenes

The third and final example we wish to discuss here is a peropyrene system that is chirally distorted by opportune substitutions (Fig. 10.10).

The chemical structure of the investigated molecule is quite interesting: oppositely oriented distortions in the two opposite ends of the molecule take place due to

**Table 10.1** Comparison of experimental $g_{abs}$, $g_{lum}$, and $g_{lum}/g_{abs}$ values and of calculated electric and magnetic dipole transition moments (atomic units) and angle formed by them (see Refs. [30, 53]) for HEX, 5N-HEX, and TRIA-HEX. Calculations are at CAM-B3LYP/TZVP level

|     | nm | $g_{abs}$, $g_{lum}$ | $g_{abs}/g_{lum}$ | $|\mu|$ | $|m|$ | Angle |
|-----|-----|------------------------|---------------------|----------|--------|-------|
| _HEX_ |
| abs | 414 | $\sim 1.3 \times 10^{-3}$ | | 0.16 | 0.19 | 96 |
| lum | 422 | $1.4 \times 10^{-3}$ | $\sim 1.1$ | 0.10 | 0.07 | 130 |
| _5N-HEX_ |
| abs | 415 | $5 \times 10^{-3}$ | | 0.32 | 0.23 | 114 |
| lum | 430 | $5.5 \times 10^{-3}$ | $\sim 1.1$ | 0.40 | 0.20 | 161 |
| _TRIA-HEX_ |
| abs | 340 | $8.3 \times 10^{-3}$ | | 1.07 | 1.03 | 39 |
| lum | 442 | $9.1 \times 10^{-3}$ | $\sim 1.1$ | 0.69 | 1.50 | 67 |

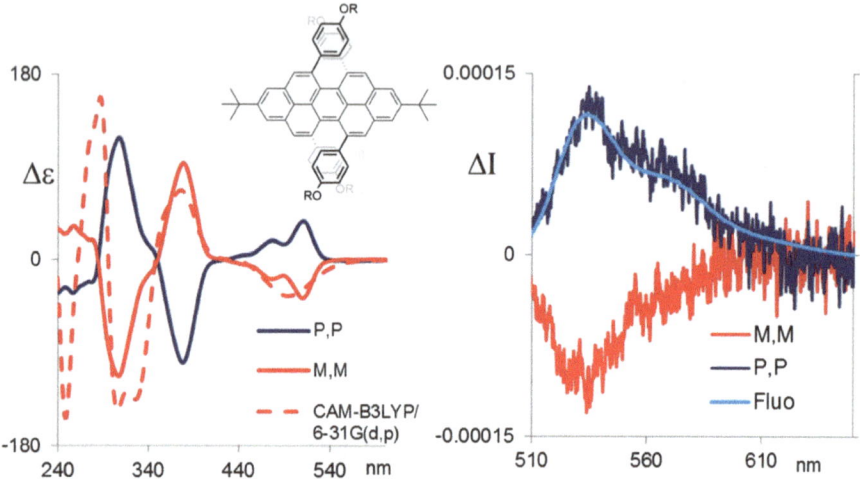

**Fig. 10.10** (Left): ECD spectra of the chiral peropyrene, OR = O-hexyl groups (O-methyl for the calculated spectrum). (Right): CPL spectra of the same compounds. Corresponding fluorescence has been normalized (see Ref. [55])

steric hindrance of four OR-phenyl substituents facing each other. The induced twist is what makes the molecule chiral giving rise to a chiral graphene-type system [55]. Furthermore, a central $C_2$-symmetry axis is preserved, with the system being comparable to two connected helicenes. The experimental ECD spectrum presents intense bands of alternating signs at high energies, while at low energy there is a low intensity CD feature, which has been demonstrated to be shaped by vibronic effects analyzed theoretically in Ref. [55]. The fluorescence spectrum is quite intense, while the CPL spectrum is weak, showing vibronic structuring as well, as often typical of weak CPL spectra. Indeed, in the present case we have $g_{abs} = 1.8 \times 10^{-3}$, while $g_{lum} \leq 10^{-3}$: thus $g_{lum}/g_{abs} \leq 0.5$, which is in accord to Mori's characterization; also, the Stokes shift is quite small. Similar distorted nanographenes [56] present low $g_{lum}$ and small Stokes shift.

## 10.2.4 CPL of Transition Metal Complexes

Below we investigate how CPL reports information about the coordination of organic ligands around two different transition metal atoms, namely Iridium and Platinum, which sit close by in the periodic table.

The first cases are two types of complexes of organic ligands with Ir(III) (Fig. 10.11). For the first one, which is commercially available, though not in separated enantiomers, none of the organic ligands is chiral [57], while for the second type one of the three ligands was chiral and we were able to obtain CPL data for all diastereoisomers [58]. In the two cases, the three ligands are organized in a chiral octahedral way around the Iridium ion, which is thus either $\Lambda$ or $\Delta$: the sign of the CPL band is univocally determined by arrangement around the Ir(III) ion and does not depend on the structure of the ligands. Even when the latter are chiral, their configuration has no influence on the sign of the CPL band. There is a dependence on the precise chemical nature of the substituents for the center wavelength of the fluorescence and CPL band and for the band-shape, i.e., the vibronic features. The observed CPL spectra are weak, being the $g_{lum}$ ratio of the order of $10^{-3}$, and the $g_{lum}/g_{abs}$ ratio of the order of 1; the presence of vibronic features can be also noticed. In these systems contributions from triplet states and thus phosphorescence are present, which makes these systems challenging from a computational point of view [59].

The second example consists of Pt organometallic complexes. These complexes are square-planar chiral systems with the Pt atom serving as stereo center, either $(S)$ or $(R)$. In the first case [60] the spectra were measured in water solution, in the second one [61] in $CH_2Cl_2$. The sign rule that the CPL sign should be the same as the one for the longest wavelength CD feature is obeyed. CPL spectra are weak in both cases: in the first case they are one order of magnitude lower than the Ir complexes commented above, while in the second case they are of the same order of magnitude (Fig. 10.12). Coming to the quantitative evaluation of the spectra, we had for the [(ppy)Pt((R)-Campy)] complex $g_{abs} \approx 10^{-3}$, while $g_{lum} \approx 10^{-4}$. Thus $g_{lum}/g_{abs} \approx 10^{-1}$, and the Stokes shift is considerable (ca. 150 nm). For the dichelated [PtL$^{CN}_2$] *trans-2* complex we have $g_{abs} \approx 5 \times 10^{-3}$, while $g_{lum} \approx 10^{-3}$. Thus $g_{lum}/g_{abs}$ is slightly bigger, and once again the Stokes shift is large, ca. 100 nm. The weakness of CPL is due to the square-planar arrangement of the complex. The special arrangement of the square-planar coordination in the second type of compounds, whereby two different planar moieties coordinate at right angle to the central ion, helps in boosting all chiroptical properties, from ECD to CPL. For either complex, the luminescence is associated to a $T_1 \rightarrow S_0$ transition. (In the Supplementary information of Ref. [61], a calculation of the ECD spectra considering singlet states is provided, which successfully compares to the experimental one.)

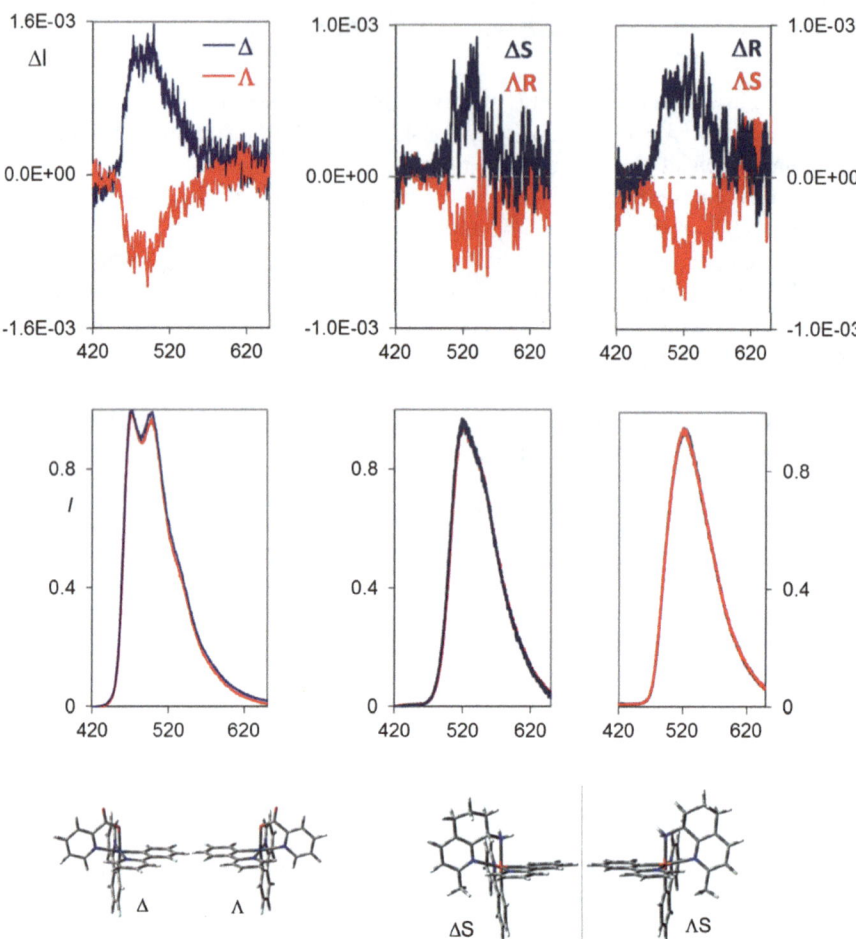

**Fig. 10.11** CPL (top) and fluorescence spectra (middle) of three enantiomeric pairs of Ir(III) complexes. On the left, spectra of (FIrpic): *fac-trans*-N–N,bis-[2-(4,6-difluorophenyl)pyridinato-C2,N](picolinato)iridium(III); on the right, spectra of Δ,Λ-(*R*,*S*)-[(ppy)2Ir-(Me-Campy)]Cl, where ppy = 2-phenylpyridine and Me-Campy = -methyl-5,6,7,8-tetrahydroquinolin-8-amine. (Bottom): DFT calculated structures of Λ-FIrpic and Δ-FIrpic and of Δ-(*S*)-[(ppy)2Ir-(Me-Campy)]Cl and Λ-(*S*)-[(ppy)2Ir-(Me-Campy)]Cl (see Refs. [57, 58])

## 10.2.5   *CPL of Molecular Origin and Other Effects*

In all chiroptical spectroscopies, care should be taken to understand exactly what one is observing and, to this instance, the analysis of all optical phenomena involved is particularly important when considering CPL. Since the early days of the technique,

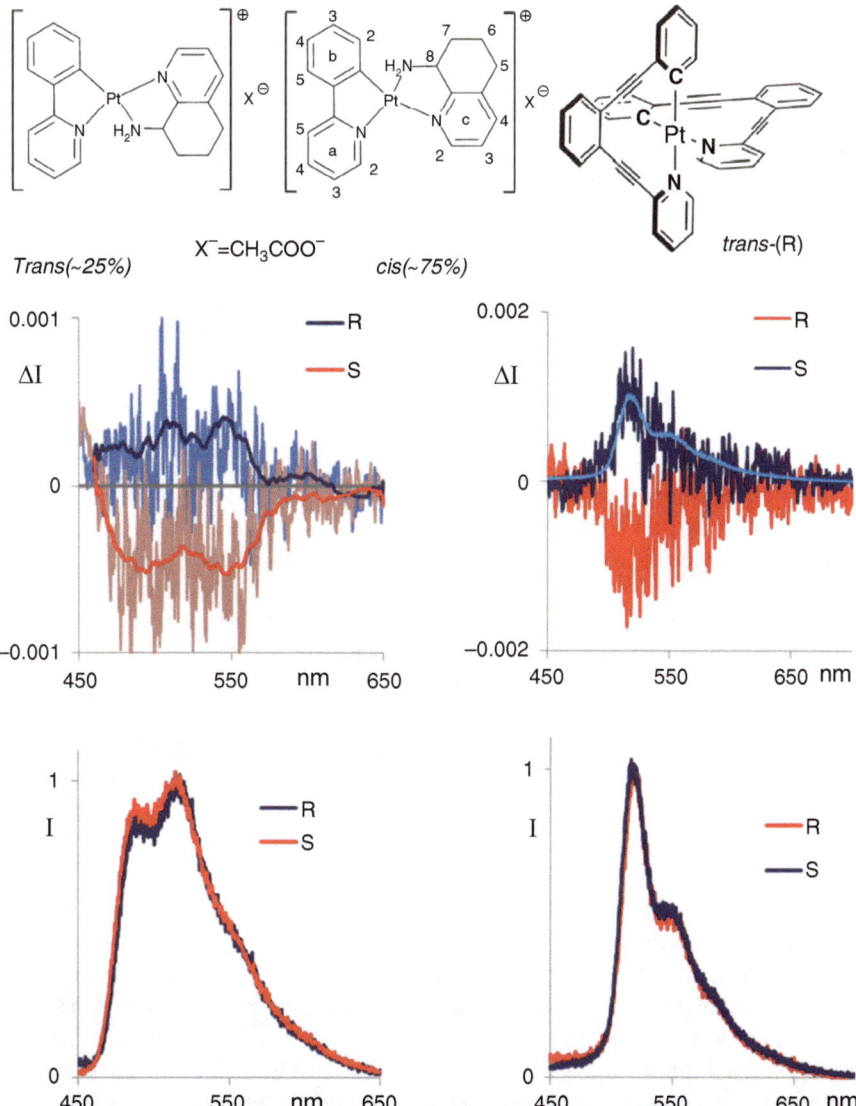

**Fig. 10.12** (Left): Chemical structure of [(ppy)Pt((R)-Campy)]$^+$CH$_3$COO$^-$ (ppy = phenyl pyridine; Campy = 5,6,7,8-tetrahydroquinolin-8-amine), CPL and fluorescence spectra in water solution, and a mixture of *trans* and *cis* isomers are presented. (Right): Chemical structure of the dichelated *trans*-[PtL$^{CN}$$_2$], and CPL and fluorescence spectra are presented

it has been well known that attention must be paid to photoselection, linear dichroism, and birefringence effects: important caveats with relative examples have been illustrated, among the others, in Refs. [62-64]. As stated by Steinberg and Ehrenberg in 1974 [62] "if Brownian motion is not fast enough during the lifetime of the excited state, the polarization properties of the light emitted from the excited molecules will depend on polarization characteristics of the excitation beam." The use of a highly fluorescent achiral molecule like fluorescein in solvents of different viscosity shows the presence of CPL signals which are the leaking of linearly polarized components induced by photoselection and observed in CPL measurements due to nonideal optics of the system [63].

In case of different concomitant optical phenomena, detailed analysis of each contribution can be conducted through the analysis of Mueller matrices [65–68]; recently such analysis has been undertaken also for the development of innovative experimental settings working also at the microscopic level [69, 70]. A review of Mueller analysis is beyond the scope of this chapter, so we just draw the attention here to common (sometimes overlooked) effects.

An often disregarded problem is the possibility of auto-absorption: in such a situation, it may be possible to observe a circularly polarized signal that is difficult to attribute to absorption (CD) or emission (CPL). The results can be particularly tricky when several species are present with differences in absorption wavelength and in emissive properties. An observed dependence of the shape of the signal on concentration or on path-length and inversion of the expected CPL signal (compared to CD) difficult to explain are important symptoms to analyze.

This problem is often encountered when aggregation phenomena occur in solution, for example, induced by appropriate solvent mixture or temperature variations, a situation often studied because it may originate enhancement of chiroptical properties [71, 72]. In several cases, genuine CPL is obtained. One of the first example is given by a chiral polyalkylthiophene by Meijer et al. [73]; in other cases absorption and emission effects are co-present with either possible cancellation or reinforcement: the emitted light (with circularly polarized components: CPL) can be reabsorbed differently in its right and left circularly polarized components (CD) so that it can be difficult to discriminate which phenomenon one is observing. We met with a clear example considering another chiral polyalkylthiophene in Ref. [74]: upon increasing the aggregated component, by changing solvent mixture, the system presents absorption at longer wavelength that interferes with the emission from the nonaggregated specie, giving rise to a sort of CPL-detected CD with sign opposite to that of the lowest energy recorded CD band. A simple analysis can be conducted using both transmittance and CD data, together with CPL and fluorescence measurements. In this way, fluorescence and CPL data can be corrected with the following expression:

$$\text{Fluo}_{\text{corr}} = \frac{\text{Fluo}_{\text{obs}}}{T}$$

$$\text{CPL}_{\text{corr}} = \frac{1}{T} \text{Fluo}_{\text{obs}} \left( \alpha \cdot \ln (10) \cdot \overline{\Delta A} \right) + \frac{1}{T} \text{CPL}_{\text{obs}}$$

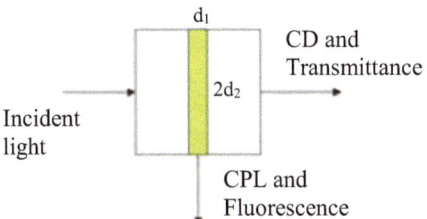

where $Fluo_{obs}$ and $CPL_{obs}$ are the actually observed fluorescence and CPL signals, $T$ is transmittance, $\Delta A$ is CD in absorbance units, $\alpha$ accounts for the average path-length of the emitted light inside the cuvette: it is the ratio between the average path of the emitted light within the cuvette (namely $d_2$) and the path of the CD/absorption experiment (namely $d_1$).

Finally, we recall that interest is growing in considering cooperative effects present in aggregate states [27, 75], particularly the solid state, since supramolecular chiral structures greatly enhance chiroptical response [76]. In these cases, spectroscopic results should be considered more a property of the whole material rather than a molecular property: one may speak of local versus long-range ordering effects, or intrinsic versus extrinsic effects. The second case presents a $g$ factor depending on sample thickness. Many examples are present in the literature where the enhancement of chiroptical response is favored by cholesteric phases doped with fluorophores or fluorescent achiral polymers assembled in chiral supramolecular structures induced by a chiral dopant [76–79].

For this purpose, we studied an interesting material showing high chiroptical response in various spectroscopic ranges: a chiroptically active syndiotactic polystyrene (s-PS) crystallized after sorption–desorption of chiral agents like limonene or carvone. This material appears to behave as a metamaterial that influences light polarization properties to a huge extent. Important manifestations have been observed by VCD spectroscopy; in Ref. [80] a particularly meaningful observation was made, that is to say, once the polymer had received the imprinting from a molecule of chosen chirality, the sign of the observed VCD bands is determined once and for all. In fact, once the polymer had been prepared by sorbing–desorbing (R)-carvone, it exhibited the same VCD apparent spectrum when hosting either (R)-carvone or (S)-carvone. This evidences that one is not observing an enhancement of molecular host VCD signal but a modulation of the molecular host absorption due to the guest material; in other words s-PS is a medium acting differently on the transmitted left and right circularly polarized components of light. Concerning the UV-vis region, the difference in left and right circularly polarized transmissions of these films is quite large over a wide spectral range (Fig. 10.13a). The existence of a CD signal does not necessarily imply circular dichroism at the molecular level, rather one can state that the material shows different transmission for left and right circularly polarized lights. We noticed that no absorption band is present in this region (280–600 nm) and we checked that the data are independent from rotation of

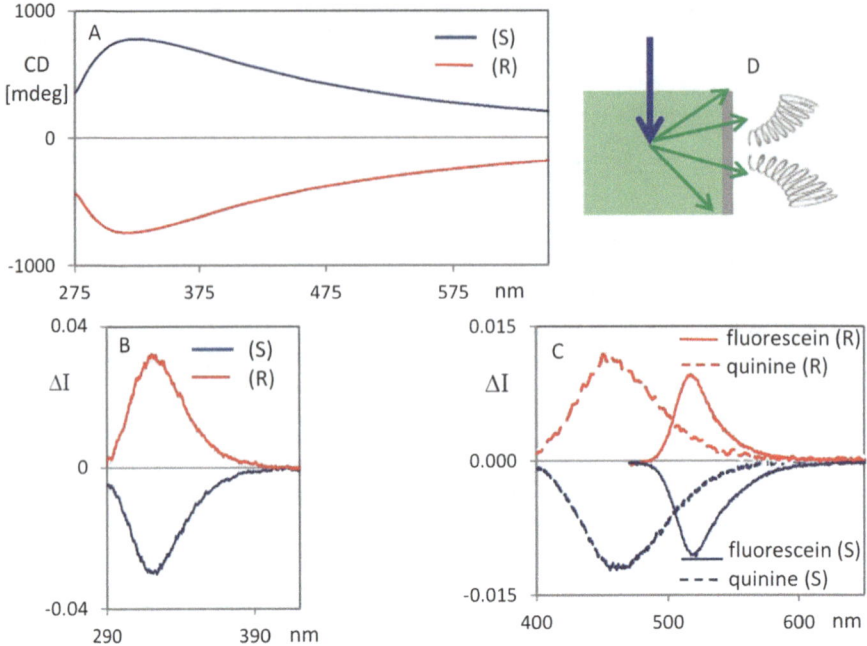

**Fig. 10.13** CD spectra (**a**) and CPL spectra (**b**) (obtained upon 265 nm excitation) of extruded s-PS films, after sorption/desorption of (*R*)-carvone (red curve) and (*S*)-carvone (blue curve); emission of fluorescein solution (exc. 450 nm) and quinine solution (exc. 365 nm): the emitted light is "filtered" through s-PS film sorbed/desorbed with (*R*)- and (*S*)-carvone (**c**); geometry illustrated in (**d**). See Ref. [81]

the sample and from changing back and front side of the film [81]. Regarding fluorescence emitted by s-PS upon irradiation [82], it is difficult to analyze the origin of the chiral components of the emitted light, since from CD experiments it is shown that transmitted light is highly circularly polarized. We observe that the sign of the CPL band is opposite with respect to CD absorption. This is compatible with the interpretation that fluorescence emission at 320 nm from s-PS exhibits circularly polarized components undergoing also different transmissions through the sample (note that excitation light has quite lower wavelength to avoid superposition). The observation of chiral components of emitted light (Fig. 10.13b), paralleling the observation of chiral transmission (Fig. 10.13a), suggests a possible use of these films to obtain chiral light from emission by (achiral) fluorophores, thus avoiding the need of inducing chirality on the fluorophore.

With a very simple experiment, we showed how a fluorescent solution, placed before the optically active s-PS film, can be used to obtain emitted light with highly chiral components: left and right circularly polarized components of fluorescence are differently "filtered" by the dichroic s-PS films. In particular, we measured the circularly polarized components of light emitted by a fluorescein solution or a quinine solution, "filtered" by a s-PS film sorbed–desorbed with (R)- and (S)-carvone (Fig. 10.13c). The dissymmetry factor is somehow related to the apparent CD signal of the s-PS film in the emissive spectroscopic region of the fluorescent solution and it is thus of extrinsic origin.

In conclusion, s-PS films with induced chirality, due to sorption–desorption of chiral molecules, exhibit chiroptical properties that resemble those of liquid crystals or polyfluorene films studied in the literature (see, for example, Refs. [78, 83, 84]). The behavior is reminiscent of metamaterials constituted of intrinsically chiral holes, since CD signals do not significantly change with rotation of the film around a perpendicular axis [85]. Besides that, also considering change of front and back sides of the film, little variations occur, so cross terms between linear birefringence and linear dichroism [63, 86] do not seem important enough to explain the observed behavior. The observation of a broad CD band, in absence of a real absorption band, is strongly suggestive of possible different reflection/scattering of right and left circularly polarized light, as observed/hypothesized, for example, in Ref. [87], and proved on polyfluorene films by Meskers [84] with an ad hoc instrumental setup. This experience hints at possible applications of chiral polymer films coupled with light emitting devices—the advantage for s-PS treated with volatile chiral species being that one has no need to add chiral units by chemical synthesis. This experience suggests similar strategies to obtain and testing materials generating CPL [88].

## 10.3 Concluding Remarks

CPL is a thrilling phenomenon that has still a high degree of novelty and deserves attention and concerted efforts from theoreticians and experimentalists. The first ones will help in finding adequate protocols to improve the description of luminescence from singlet excited states and to find new ways of dealing with phosphorescence phenomena from triplet excited states of chiral molecules; finally, an effort toward a better description of metal ions in chiral complexes is desirable. The experimentalists are expected to synthesize new molecular systems as well as to discover complex systems, like polymers, fibrils, or mesoscopic systems, where CPL is co-present with other chiroptical phenomena to be disentangled from.

## References

1. Emeis CA, Oosterhoff LJ (1967) Emission of circularly polarized radiation by optically active compounds. Chem Phys Lett 1:129–132
2. Dekkers HPJM, Emeis CA, Oosterhoff LJ (1969) Measurement of optical activity in racemic mixtures. J Am Chem Soc 91:4590–4590
3. Schippers PH, van den Beukel A, Dekkers HPJM (1982) An accurate digital instrument for the measurement of circular polarisation of luminescence. J Phys E Sci Instrum 15:945–950

4. Steinberg IZ, Gafni A (1972) Sensitive instrument for the study of circular polarization of luminescence. Rev Sci Instrum 43:409–413
5. Richardson FS, Riehl JP (1977) Circularly polarized luminescence spectroscopy. Chem Rev 77:773–792
6. Dekkers HPJM (2000) Circularly polarized luminescence: a probe for chirality in the excited state. In: Berova N, Nakanishi K, Woody RW (eds) Circular dichroism: principles and applications, 2nd edn. Wiley, New York, pp 185–215
7. Moffit W, Moscowitz A (1959) Optical activity in absorbing media. J Chem Phys 30:648–660
8. Spano FC, Yamagata H (2011) Vibronic coupling in J-aggregates and beyond: a direct means of determining the exciton coherence length from the photoluminescence spectrum. J Phys Chem B 115:5133–5143
9. Carr R, Evans NH, Parker D (2012) Lanthanide complexes as chiral probes exploiting circularly polarized luminescence. Chem Soc Rev 41:7673–7686
10. Carr R, Puckrin R, McMahon BK, Pal R, Parker D, Pålsson L-O (2014) Induced circularly polarized luminescence arising from anion or protein binding to racemic emissive lanthanide complexes. Methods Appl Fluoresc 2:024007. (7 pp)
11. Brittain HG (1985) Excited-state optical activity. In: Molecular luminescence spectroscopy methods and applications. Part I. Wiley Interscience, New York; Chapter 6
12. Castiglioni E, Abbate S, Longhi G (2010) Revisiting with updated hardware and old spectroscopic technique: circularly polarized luminescence. Appl Spectrosc 64:1416–1419
13. Castiglioni E, Abbate S, Lebon F, Longhi G (2014) Chiroptical spectroscopic techniques based on fluorescence. Methods Appl Fluoresc 2:024006. (7 pages)
14. Imai Y, Nakano Y, Kawai T, Yuasa J (2018) A smart sensing method for object identification using circularly polarized luminescence from coordination-driven self-assembly. Angew Chem Int Ed 57:8973–8978
15. Jasco International Company CPL-300 product (2018) https://jascoinc.com/products/spectros copy/circularly-polarized-luminescence-cpl-300/download-cpl/
16. Circularly Polarized Luminescence (CPL) with OLIS DSM 172 (2018) http://olisweb.com/portfolio-item/dsm-172/
17. Pecul M, Ruud K (2011) The optical activity of beta, gamma-enones in ground and excited states using circular dichroism and circularly polarized luminescence. Phys Chem Chem Phys 13:643–650
18. Pritchard B, Autschbach J (2010) Calculation of vibrationally resolved circularly polarized luminescence of d-camphorquinone and (S,S)-trans-β-hydrindanone. ChemPhysChem 11:2409–2415
19. McAlexander HR, Crawford TD (2015) Simulation of circularly polarized luminescence spectra using coupled cluster theory. J Chem Phys 142:154101
20. Saleh N, Moore B, Srebro M, Vanthuyne N, Toupet L, Williams JAG, Roussel C, Deol KK, Muller G, Autschbach J, Crassous J (2015) Acid/base-triggered switching of circularly polarized luminescence and electronic circular dichroism in organic and organometallic helicenes. Chemistry 21:1673–1681
21. Fujiki M, Kawagoe Y, Nakano Y, Nakao A (2013) Mirror-symmetry-breaking in poly[(9,9-di-n-octylfluorenyl- 2,7-diyl)-alt-biphenyl] (PF8P2) is susceptible to terpene chirality, achiral solvents, and mechanical stirring. Molecules 18:7035–7057
22. Zinna F, Di Bari L (2015) Lanthanide circularly polarized luminescence: bases and applications. Chirality 27:1–13
23. Wu T, Kapitàn J, Bouř P (2015) Detection of circularly polarized luminescence of a Cs-EuIII complex in Raman optical activity experiments. Angew Chem Int Ed 54:14933–14936
24. Sánchez-Carnerero E, Agarrabeitia AR, Moreno F, Maroto BL, Muller G, Ortiz MJ, de la Moya S (2015) Circularly polarized luminescence from simple organic molecules. Chemistry 21:13488–13500

25. Tanaka H, Inoue Y, Mori T (2018) Circularly polarized luminescence and circular dichroisms in small organic molecules: correlation between excitation and emission dissymmetry factors. ChemPhotoChem 2:386–402

26. Kitayama Y, Amako T, Suzuki N, Fujiki M, Imai Y (2014) Enhancing circularly polarised luminescence by extending the π-conjugation of axially chiral compounds. Org Biomol Chem 12:4342–4346

27. Kumar J, Nakashima T, Kawai T (2015) Circularly polarized luminescence in chiral molecules and supramolecular assemblies. J Phys Chem Lett 6:3445–3452

28. Longhi G, Castiglioni E, Koshoubu J, Mazzeo G, Abbate S (2016) Circularly polarized luminescence: a review of experimental and theoretical aspects. Chirality 28:696–707

29. Saleh N, Srebro M, Reynaldo T, Vanthuyne N, Toupet L, Chang VI, Muller G, Williams G, Roussel C, Autschbach J, Crassous J (2015) Enantio-enriched CPL-active helicene–bipyridine–rhenium complexes. Chem Commun 51:3754–3757

30. Abbate S, Longhi G, Lebon F, Castiglioni E, Superchi S, Pisani L, Fontana F, Torricelli F, Caronna T, Villani C, Sabia R, Tommasini M, Lucotti A, Mendola D, Mele A, Lightner DA (2014) Helical sense-responsive and substituent-sensitive features in vibrational and electronic circular dichroism, in circularly polarized luminescence and in Raman spectra of some simple optically active hexahelicenes. J Phys Chem C 118:1682–1695

31. Liu Y, Cerezo J, Mazzeo G, Lin N, Zhao X, Longhi G, Abbate S, Santoro F (2016) Vibronic coupling explains the different aspect of electronic circular dichroism and of circularly polarized luminescence spectra of hexahelicenes. J Chem Theory Comput 12:2799–2819

32. Sato T, Tajima N, Ueno H, Harada T, Fujiki M, Imai Y (2017) Binaphthyl luminophores with triphenylsilyl groups: sign inversion of circularly polarized luminescence and circular dichroism. Tetrahedron 72:7032–7038

33. Longhi G, Castiglioni E, Abbate S, Lebon F, Lightner DA (2013) Experimental and calculated CPL spectra and related spectroscopic data of camphor and other simple chiral bicyclic ketones. Chirality 25:589–599

34. Schippers PH, van der Ploeg JPM, Dekkers HPJM (1983) Circular polarization in the Fluorescence of β,γ-enones: distortion in the $^1n\pi*$ states. J Am Chem Soc 105:84–89

35. Dekkers HPJM, Closs LEJ (1976) The optical activity of low symmetry ketones in absorption and emission. J Am Chem Soc 98:2210–2219

36. Moscowitz A (1960) Theory and analysis of rotatory dispersion curves. In: Djerassi C (ed) Optical rotatory dispersion. McGraw-Hill, New York, pp 150–177

37. Lightner DA, Gurst JE (2000) Organic conformational analysis and stereochemistry from circular dichroism spectroscopy. Wiley, New York, Chapter 4

38. Morcillo SP, Miguel D, Alvarez de Cienfuegos L, Justicia J, Abbate S, Castiglioni E, Bour C, Ribagorda M, Cardenas DJ, Paredes JM, Crovetto J, Choquesillo-Lazarte D, Mota AJ, Carreño MC, Longhi G, Cuerva JM (2016) Stapled helical o-OPE foldamers as new circularly polarized luminescence emitters based on carbophilic interactions with Ag(I)-sensitivity. Chem Sci 7:5663–5670

39. Amako T, Nakabayashi K, Suzuki N, Guo S, Rahim NAA, Harada T, Fujiki M, Imai Y (2015) Pyrene magic: chiroptical enciphering and deciphering 1,3-dioxolane bearing two wirepullings to drive two remote pyrenes. Chem Commun 51:8237–8240

40. Hara N, Yanai M, Kaji D, Shizuma M, Tajima N, Fujiki M, Imai Y (2000) A pivotal biaryl rotamer bearing two floppy pyrenes that exhibits cryptochiral characteristics in the ground state. ChemistrySelect 3:9970

41. Reiné P, Justicia J, Morcillo SP, Abbate S, Vaz B, Ribagorda M, Orte Á, Álvarez de Cienfuegos L, Longhi G, Campaña AG, Miguel D, Cuerva JM (2018) Pyrene-containing ortho-oligo(phenylene)ethynylene foldamer as a ratiometric probe based on circularly polarized luminescence. J Org Chem 83:4455–4463

42. Domínguez Z, López-Rodríguez R, Álvarez E, Abbate S, Longhi G, Pischel U, Ros A (2018) Azabora[5]helicene charge-transfer dyes show efficient and spectrally variable circularly polarized luminescence. Chem Eur J 24:12660–12668

43. Loksztejn A, Dzwolak W (2008) Chiral bifurcation in aggregating insulin: an induced circular dichroism study. J Mol Biol 379:9–16
44. Dzwolak W (2010) Vortex-induced chiral bifurcation in aggregating insulin. Chirality 2:E154–E160
45. Rybicka A, Longhi G, Castiglioni E, Abbate S, Dzwolak W, Babenko V, Pecul M, Thioflavin T (2016) Thioflavin T: Electronic circular dichroism and circularly polarized luminescence induced by amyloid fibrils. ChemPhysChem 17: 2931-2937
46. Cuerva JM, Resa S, Miguel D, Guisán-Ceinos S, Mazzeo G, Choquesillo-Lazarte D, Abbate S, Crovetto L, Cárdenas DJ, Carreño MC, Ribagorda M, Longhi G, Mota AJ, Álvarez de Cenfuegos L (2018) Sulfoxide-induced homochiral folding of o-OPEs by Ag(I) templating: structure and chiroptical properties. Chem Eur J 24:2653–2662
47. Reiné P, Ortuño AM, Resa S, Álvarez de Cienfuegos L, Blanco V, Ruedas-Rama MJ, Mazzeo G, Abbate S, Lucotti A, Tommasini M, Guisán-Ceinos S, Ribagorda M, Campaña AG, Mota A, Longhi G, Miguel D, Cuerva JM (2018) OFF/ON switching of circularly polarized luminescence by oxophilic interaction of homochiral sulfoxide-containing o-OPEs with metal cations. Chem Commun 54:13985–13988
48. Mazzeo G, Abbate S, Longhi G, Castiglioni E, Boiadjiev SE, Lightner DA (2016) pH dependent chiroptical properties of (1R,2R)- and (1S,2S)-trans-cyclohexane diesters and diamides from VCD, ECD, and CPL spectroscopy. J Phys Chem B 120:2380–2387
49. Mislow K (1965) Introduction to stereochemistry. WA Benjamin, New York, pp 65–66
50. Longhi G, Abbate S, Mazzeo G, Castiglioni E, Mussini PR, Benincori T, Martinazzo R, Sannicolò F (2014) Structural and optical properties of inherently chiral polythiophenes: a combined CD-electrochemistry, circularly polarized luminescence and TDDFT investigation. J Phys Chem C 118:16019–16027
51. Benincori T, Appoloni G, Mussini PR, Arnaboldi S, Cirilli R, Quartapelle Procopio E, Panigati M, Abbate S, Mazzeo G, Longhi G (2018) Searching for models exhibiting high circularly polarized luminescence: the electroactive inherently chiral oligothiophenes. Chemistry 24:11082–11093
52. Sannicolò F, Mussini PR, Benincori T, Cirilli R, Abbate S, Arnaboldi S, Casolo S, Castiglioni E, Longhi G, Martinazzo R, Panigati M, Pappini M, Quartapelle Procopio E, Rizzo S (2014) Inherently chiral macrocyclic oligothiophenes: easily accessible electrosensitive cavities with outstanding enantioselection performances. Chemistry 20:15298–15302
53. Longhi G, Castiglioni E, Villani C, Sabia R, Menichetti S, Viglianisi C, Devlin F, Abbate S (2016) Chiroptical properties of the ground and excited states of two thia-bridged triarylamine heterohelicenes. J Photochem Photobiol A Chem 331:138–145
54. Nakai Y, Mori T, Inoue Y (2012) Theoretical and experimental studies on circular dichroism of carbo[n]helicenes. J Phys Chem A 116:7372–7385
55. Yang W, Longhi G, Abbate S, Lucotti A, Tommasini M, Villani C, Catalano VJ, Lykhin AO, Varganov SA, Chalifoux WA (2017) Chiral peropyrene: synthesis, structure, and properties. J Am Chem Soc 139:13102–13109
56. Cruz CM, Márquez IR, Mariz IFA, Blanco V, Sánchez-Sánchez C, Sobrado JM, Martín-Gago JA, Cuerva JM, Maçôas E, Campaña A (2018) Enantiopure distorted ribbon-shaped nanographene combining two-photon absorption based upconversion and circularly polarized luminescence. Chem Sci 9:3917–3924
57. Citti C, Battisti UM, Ciccarella G, Maiorano V, Gigli G, Abbate S, Mazzeo G, Castiglioni E, Longhi G, Cannazza G (2016) Analytical and preparative enantioseparation and main chiroptical properties of Iridium(III) bis(4,6-difluorophenylpyridinato)picolinato. J Chromatogr A 1467:335–346
58. Mazzeo G, Fusè M, Longhi G, Rimoldi I, Cesarotti E, Crispini A, Abbate S (2016) Vibrational circular dichroism and chiroptical properties of chiral Ir(III) luminescent complexes. Dalton Trans 45:992–999

59. Kamiński M, Cukras J, Pecul M, Rizzo A, Coriani S (2015) A computational protocol for the study of circularly polarized phosphorescence and circular dichroism in spin-forbidden absorption. Phys Chem Chem Phys 17:19079–19086

60. Ionescu A, Godbert N, Ricciardi L, La Deda M, Aiello I, Ghedini M, Rimoldi I, Cesarotti E, Facchetti G, Mazzeo G, Longhi G, Abbate S, Fusè M (2017) Luminescent water-soluble cycloplatinated complexes: structural, photophysical, electrochemical and chiroptical properties. Inorg Chim Acta 461:267–274

61. Schulte TR, Holstein JH, Krause L, Michel R, Stalke D, Sakuda E, Umakoshi K, Longhi G, Abbate S, Clever GH (2017) Chiral-at-metal phosphorescent square-planar Pt(II)-complexes from an achiral organometallic ligand. J Am Chem Soc 139:6863–6866

62. Steinberg IZ, Ehrenberg B (1974) A theoretical evaluation of the effect of photoselection on the measurement of the circular polarization of luminescence. J Chem Phys 61:3382–3386

63. Shindo Y, Nakagawa M (1985) On the artifacts in circularly polarized emission spectroscopy. Appl Spectrosc 39:32–38

64. Dekkers HPJM, Moraal PF, Timper JM, Riehl JP (1985) Optical artifacts in circularly polarized luminescence spectroscopy. Appl Spectrosc 39:818–821

65. Shindo Y, Oda Y (1992) Mueller matrix approach to fluorescence spectroscopy. Part I: mueller matrix expressions for fluorescent samples and their application to problems of circularly polarized emission spectroscopy. Appl Spectrosc 46:1251–1259

66. Kuroda R, Harada T, Shindo Y (2001) A solid-state dedicated circular dichroism spectrophotometer: development and application. Rev Sci Instrum 72:3802–3810

67. Thomas A, Chervy T, Azzini S, Li M, George J, Genet C, Ebbesen TW (2018) Mueller polarimetry of chiral supramolecular assembly. J Phys Chem C 122:14205–14212

68. Harada T, Hayakawa H, Watanabe M, Takamoto M (2016) A solid-state dedicated circularly polarized luminescence spectrophotometer: development and application. Rev Sci Instrum 87:075102

69. Katayama K, Hirata S, Vacha M (2014) Circularly polarized luminescence from individual microstructures of conjugated polymer aggregates with solvent-induced chirality. Phys Chem Chem Phys 7:17983–17987

70. Tsumatori H, Harada T, Yuasa J, Hasegawa Y, Kawai T (2011) Circularly polarized light from chiral lanthanide(III) complexes in single crystals. Appl Phys Express 4:011601

71. Zhao B, Pan K, Deng J (2018) Intense circularly polarized luminescence contributed by helical chirality of monosubstituted polyacetylenes. Macromolecules 51:7104–7111

72. Fujiki M, Donguri Y, Zhao Y, Nakao A, Suzuki N, Yoshida K, Zhang W (2016) Photon magic: chiroptical polarisation, depolarisation, inversion, retention and switching of non-photochromic light-emitting polymers in optofluidic medium. Polym Chem 6:1627–1638

73. Langeveld-Voss BMW, Janssen RAJ, Christiaans MPT, Meskers SCJ, Dekkers HPJM, Meijer EW (1996) Circular dichroism and circular polarization of photoluminescence of highly ordered poly{3,4-di[(S)-2-methylbutoxy]thiophene}. J Am Chem Soc 118:4908–4909

74. Castiglioni E, Abbate S, Lebon F, Longhi G (2012) UV, CD, fluorescence and CPL spectra of regioregular poly-[3-((S)-2-methylbutyl)-thiophene] in solution. Chirality 24:725–730

75. Roose J, Tang BZ, Wong KS (2016) Circularly-polarized luminescence (CPL) from chiral AIE molecules and macrostructures. Small 12:6495–6512

76. Chen SH, Katsis D, Schmid AW, Mastrangelo JC, Tsutsui T, Blanton TN (1999) Circularly polarized light generated by photoexcitation of luminophores in glassy liquid-crystal films. Nature 397:506–508

77. Kulkarni C, Meskers SCJ, Palmans ARA, Meijer EW (2018) Amplifying chiroptical properties of conjugated polymer thin-film using an achiral additive. Macromolecules 51:5883–5890

78. Watanabe K, Osaka I, Yorozuya S, Akagi K (2012) Helically π-stacked thiophene-based copolymers with circularly polarized fluorescence: high dissymmetry factors enhanced by self-ordering in chiral nematic liquid crystal phase. Chem Mater 24:1011–1024

79. Craig MR, Jonkheijm P, Meskers SCJ, Schenning APHJ, Meijer EW (2003) The chiroptical properties of a thermally annealed film of chiral substituted polyfluorene depend on film thickness. Adv Mater 15:1435–1438
80. Rizzo P, Lepera E, Guerra G (2014) Enantiomeric guests with the same signs of chiral optical responses. Chem Commun 50:8185–8188
81. Rizzo P, Abbate S, Longhi G, Guerra G (2017) Circularly polarized luminescence of syndiotactic polystyrene. Opt Mater 73:595–601
82. Sano T, Uchiyama A, Sago T, Itagaki H (2017) Fluorescence behavior of syndiotactic polystyrene and its derivative: formation of a ground-state dimer in the solid state. Eur Polym J 90:114–121
83. Shindo Y, Ohmi Y (1985) Problems of CD spectrometers. 3. Critical comments on liquid crystal induced circular dichroism. J Am Chem Soc 107:91–97
84. Lakhwani G, Meskers SCJ, Janssen RAJ (2007) Circular differential scattering of light in films of chiral polyfluorene. J Phys Chem B 111:5124–5131
85. Maoz BM, BenMoshe A, Vestler D, Bar-Elli O, Markovich G (2012) Chiroptical effects in planar achiral plasmonic oriented nanohole arrays. Nano Lett 12:2357–2361
86. Harada T, Moriyama H (2013) Solid-state circular dichroism spectroscopy. In: Encyclopedia of polymer science and technology. Wiley, Hoboken, NJ, pp 1–29
87. Tachibana T, Mori T, Hori K (1979) New type of twisted mesophase in jellies and solid films of chiral 12-hydroxyoctadecanoic acid. Nature 278:578–579
88. Zhao B, Pan K, Deng J (2019) Combining chiral helical polymer with achiral luminophores for generating full-color, on−off, and switchable circularly polarized luminescence. Macromolecules 52:376–384

# Chapter 11
# Circularly Polarized Luminescence from Gelator Molecules: From Isolated Molecules to Assemblies

Tonghan Zhao, Pengfei Duan, and Minghua Liu

**Abstract** Currently, molecular gels have become one kind of the fascinating candidates for fabricating chiroptical materials with circularly polarized luminescence (CPL) properties, due to the tunable and modifiable structure, simple yet facile synthesis, controlled and reversible assembly, and so on. Since self-assembly approach has been regarded as an efficient way for amplifying the chirality, supramolecular gelation provides a remarkable method for fabricating CPL materials with high dissymmetry factor ($g_{lum}$). Various gel systems, including chiral-, achiral-, organic-inorganic hybrid systems can be endowed with CPL activities through supramolecular gelatinization, possessing excellent luminescence circular polarization. This chapter summarizes and reviews the present status and progress of supramolecular gel systems with CPL activity.

T. Zhao · P. Duan
CAS Center for Excellence in Nanoscience, CAS Key Laboratory of Nanosystem and Hierarchical Fabrication, National Center for Nanoscience and Technology (NCNST), Beijing, P. R. China

University of Chinese Academy of Sciences, Beijing, P. R. China

M. Liu (✉)
CAS Center for Excellence in Nanoscience, CAS Key Laboratory of Nanosystem and Hierarchical Fabrication, National Center for Nanoscience and Technology (NCNST), Beijing, P. R. China

University of Chinese Academy of Sciences, Beijing, P. R. China

Beijing National Laboratory for Molecular Science, CAS Key Laboratory of Colloid, Interface and Chemical Thermodynamics, Institute of Chemistry, Chinese Academy of Sciences, Beijing, P. R. China
e-mail: liumh@iccas.ac.cn

© Springer Nature Singapore Pte Ltd. 2020                                      249
T. Mori (ed.), *Circularly Polarized Luminescence of Isolated Small Organic Molecules*, https://doi.org/10.1007/978-981-15-2309-0_11

## 11.1 Introduction

Chiroptical materials with circularly polarized luminescence (CPL) activity have been drawing extensive attentions owing to their potential applications in the fields of photoelectric devices [1–3], 3D displays [4], security systems [5], chiral sensing [6–8], asymmetric catalysis [9–13], and so on. In order to get the CPL, it is generally necessary that both the chirality and luminescent chromophore are integrated in one molecule. Thus, various molecules possessing such elements are developed and some typical molecules exhibiting CPL are shown in Fig. 11.1.

It is seen that organic molecules with different kinds of molecular chirality, i.e., point chirality, axial chirality, and planar and helical chirality, can be designed as CPL-active molecules [14–17]. So far, a large number of the organic molecules have been reported to show CPL. In those molecules, the dissymmetry factor defined as $g_{lum} = 2 \times (I_L - I_R)/(I_L + I_R)$, where $I_L$ and $I_R$ are the intensity of left-handed and right-handed circularly polarized light, respectively, is in the range of $10^{-5}$ to $10^{-3}$. Since most of these molecules have been discussed in other chapters, we will not discuss them in detail here.

Among various organic molecules, low-molecular-weight gelator (LMWG) or simply gelator is a unique kind of molecules. The gelator molecules can self-assemble into nanostructures via noncovalent bonds and immobilize certain organic solvents (organogels) or water (hydrogels), as shown in Fig. 11.2. Interestingly, many of the gelator molecules are chiral and some of them show CPL. These gelator molecules have two states, i.e., an isolated molecular state in dilute solution and an assembly state in gels. Interestingly, many of the assembly states showed more intense CPL than that of their isolated molecular state. One of the important directions of CPL research is for materials. Since many materials are in solid or gel state, the research on the CPL of aggregation molecules cannot be avoided. Herein, recent progress in the CPL of the gelator molecules, from isolated molecules to their assembly state, will be discussed.

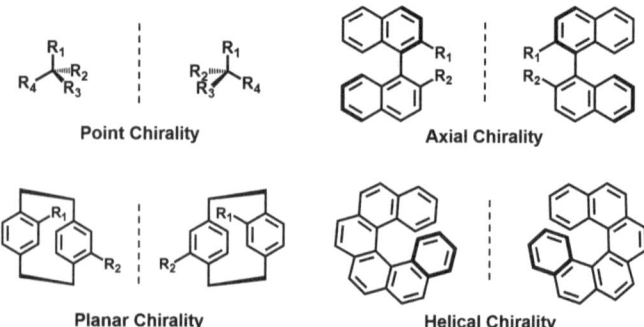

**Fig. 11.1** Some typical CPL emitters integrated the chirality and luminescence chromophores. In the case of point chirality, one of the substituent groups is fluorophore

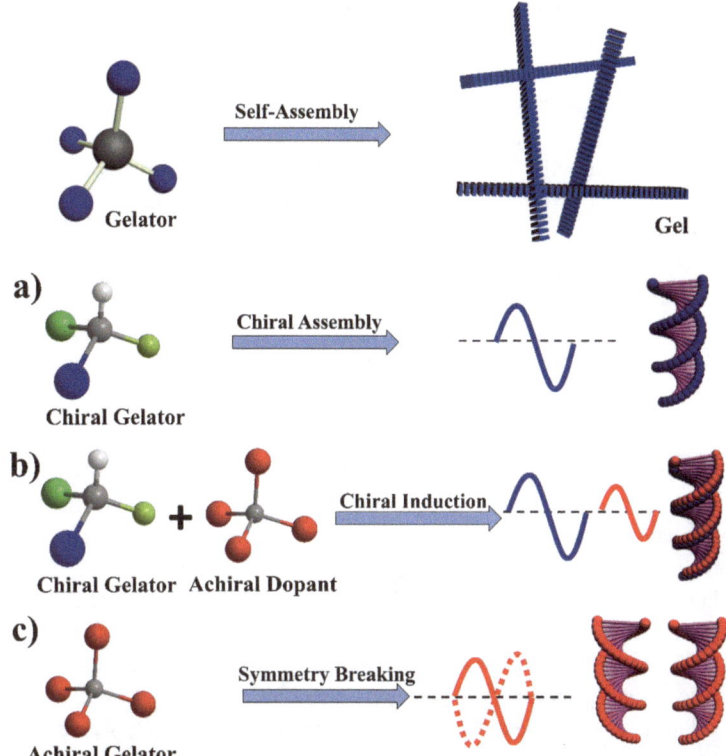

**Fig. 11.2** Schematics of gelation induced by self-assembly of gelator (top) and three types of gelation-induced supramolecular chirality (bottom). Chiral gels from (**a**) exclusive chiral gelators, (**b**) the co-assembly of chiral gelator and achiral dopants, and (**c**) exclusive achiral gelators [18]. Copyright 2014, The Royal Society of Chemistry

For the CPL of gelator molecules in a dilute molecular state, it is similar to most of organic molecules. However, in the design of CPL molecules with point chirality, the distance between the chiral center and the chromophore is very important. If the spacer between the chiral center and the chromophore is too long, you cannot get CPL. In the case of CPL from the gelators in assemblies, another phenomenon called aggregation caused quench (ACQ) and aggregation-induced emission (AIE) will frequently appear, which will greatly alter the CPL of the system. These are very important issues which should be considered in the design of the CPL-active gelators.

As illustrated in Fig. 11.2, there are three cases for the design of gelator molecules in order to get the CPL, similar to the formation of the supramolecular chiral gels [18].

The first case is that the isolated gelator shows CPL and the gel showed enhanced or diminished CPL due to the self-assembly. The second case is that the fluorophore does not have any chiral unit. However, when co-assembled with the chiral gelator, the assemblies showed CPL. In such case, the chiral element is from the nonemissive gelator, while the fluorescence (comes or originates) from the nonchiral fluorophore. The third case is the achiral chromophore without any chiral unit. In molecular state, it does not show any CPL activity. However, upon gelation, symmetry breaking occurs and the CPL could also be obtained.

In this chapter we will start from the design of chiral emissive gelator molecules and then provide the examples of CPL from several typical gelator molecules and their assemblies.

## 11.2   Design of the Chiral Gelator

The discovery of the gelator was fortuitous in the early stage of the development on supramolecular gels. However, it becomes now possible to design the gelator molecules on purpose. Hinted from the notion of supramolecular synthon in crystal engineering and synthon in organic synthesis, we proposed a concept of gelaton to the design of the gelator [19]. A gelator molecule can be illustrated as in Fig. 11.3.

A gelaton is defined as a special molecule or a structural unit that can be used to formulate the supramolecular gelator and/or gels. Through the covalent bond of the gelaton with certain fluorophore via a linker, it is easier to design the chiral gelator for CPL. Fortunately, many of the gelatons are chiral. Figure 11.3 illustrates several typical gelatons for the design of the gelator molecules, which will be useful for

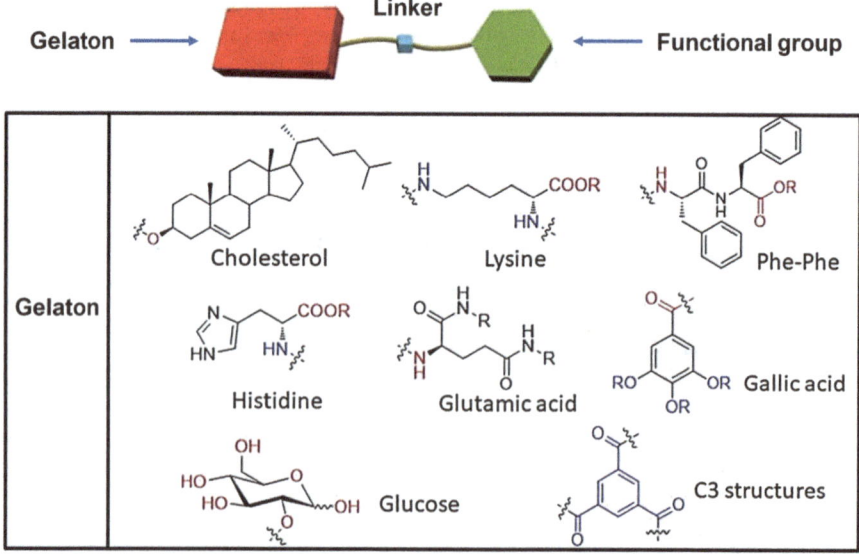

**Fig. 11.3**  Three main parts of gelator and some typical gelatons [19]. Copyright 2018, The Royal Society of Chemistry

developing various kinds of molecules for CPL research. Among these gelatons, sugar, cholesterol and amino acid derivatives are frequently used. However, only one kind of enantiomers of sugar and cholesterol could be easily obtained, while it is relatively easier to obtain both the *L*- and *D*-enantiomers for amino acids. Thus, CPL-active materials derived from amino acids are extensively investigated.

## 11.3 CPL from Gelator Molecules and Their Gels

### 11.3.1 CPL-Active Simple Molecules with Enhanced Circular Polarization via Self-Assembly

For fabricating chiral luminescent molecules, introduction of a luminescent moiety to nonluminous chiral moiety through covalent bond is a general strategy [14]. In this case, lots of chiral moieties bonding with aromatic π-conjugated chromophore exhibited CPL activity have been widely investigated. For example, Kawai and Nakashima et al. reported the CPL from an axial chiral binaphthyl derivatives, in which two perylene bisimides were bonded. The monomeric state in chloroform (molecule **1**, Fig. 11.4a) showed the $g_{lum}$ value as low as $3 \times 10^{-3}$ [15]. However, the assembly of **1** resulted in almost one order of magnitude higher $g_{lum}$ value. In addition, the morphology of the assemblies played a crucial role in deciding the luminescence dissymmetry. These studies elegantly demonstrated that chiral assemblies could act as an efficient approach to amplify the luminescent dissymmetry factor of simple organic molecules. $C_2$-symmetrical chiral 1,2-diaminocyclohexane was also used as a chiral source for designing the CPL-active compounds. Strong CPL activity was reported from molecules **2** and **3** (Fig. 11.4b) [20]. For the properties of CPL we could get, the force of the noncovalent interactions, the competition of the chiral and achiral interactions, and the length of fluorophore to the chiral center all have effect. Chiral simple molecules with CPL activity have attracted great interests in recent years and extensive

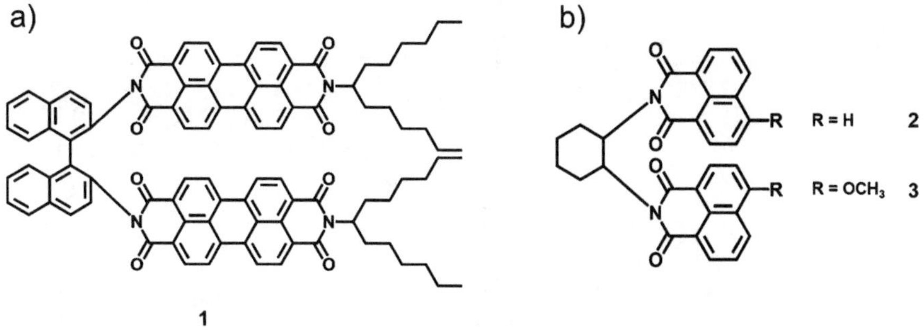

**Fig. 11.4** (**a**) Molecular structure of chiral binaphthyl derivatives **1**. (**b**) Molecular structures of chiral diaminocyclohexane derivatives **2** and **3**

accounts and reviews are available; thus, this review will not highlight such a case. We will focus on the recent developments in the field of chiral gelators and chiral supramolecular gels [21].

## 11.3.2 CPL from Chiral Gelators

### 11.3.2.1 CPL from Amino Acid Derivative Gelators

In supramolecular gel system, amino acids are excellent gelaton candidates because of its natural chirality, easy accessibility, and wide diversity [19]. For the CPL-active amino acid derivatives, the gelator molecule usually does not show CPL because the spacer between the chromophore and the chiral center is too long. However, upon gelation, the chirality can be transferred to the assemblies which show intense CPL activity. For example, Ihara et al. designed a glutamic acid gelaton-linked anthracene gelator **4** (Fig. 11.5) [22]. Interestingly, no CPL signal was detected for the monomeric state in tetrahydrofuran (THF). However, when **4** formed supramolecular gel through self-assembly in n-hexane/THF (50:1), the excimer of anthracene formed and emitted CPL at 25 °C. It was found that the formation of assembly state of **4** could be controlled by temperature. At low temperature,

**Fig. 11.5** (**a**) Schematic illustration of the excimer formed by molecule **4** in an n-hexane/THF (50:1) mixed solution at 25 °C. (**b**) CD (blue line) and CPL (red line) spectra of the **4** gel in an n-hexane/THF (50:1) mixed solution (solid line) and in THF (dash line) at 25 °C. Reproduced with permission [22]. Copyright 2015, The Royal Society of Chemistry

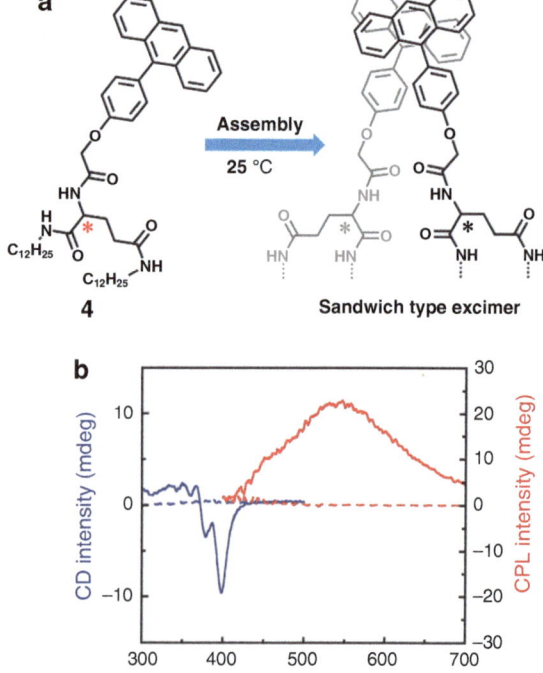

intermolecular H-bond was stronger than π–π stacking interaction. Thus, the anthracene chromophores formed twisted stacks with partially overlapped excimer. At 25 °C, the main driving force was π–π stacking for self-assembly. Thus, a Sandwich-type excimer was formed by the anthracene fluorophores in the excited state (Fig. 11.5a). Meanwhile, the obtained gel from gelator **4** with the Sandwich-type excimer can emit intense CPL, with a dissymmetry factor of $3.2 \times 10^{-3}$ (Fig. 11.5b). In addition, intermolecular interaction became weak at 45 °C; therefore, most anthracene fluorophores could not form an excimer and no CPL could be observed. Therefore, temperature-controlled fluorescence and CPL of organogel was accomplished.

Photon upconversion, in which the higher-energy excited state could be populated by lower-energy light, provides a novel view for achieving higher-energy emission. Duan and Liu et al. proposed an idea to modulate triplet–triplet annihilation-based photon upconversion (TTA-UC) in chiral assembly system, in which they designed a self-assembly system based on anthracene-derived chiral gelator (*L-/D-***5**) as acceptor and achiral $Pd^{II}$ octaethylporphyrin derivative (**6**) as sensitizer [23]. Interestingly, the co-assembly of **6** and *L-/D-***5** via co-gelation could not only generate chirality transfer from *L-/D-***5** to **6**, but also realize triplet–triplet energy transfer (TTET) from **6** to *L-/D-***5**. Thus, dual upconverted (460 nm) and downconverted CPL (550–750 nm) emission was detected in the co-gels under excitation of 532 nm laser (Fig. 11.6a). As shown in Fig. 11.6b, *L-***5/6** and *D-***5/6** co-gels in deaerated toluene showed the mirror-imaged upconverted circularly polarized luminescence (UC-CPL) signal at 460 nm. More interestingly, under high excitation laser power, strong CPL signal could be observed in the wide emission range of 550–750 nm (Fig. 11.6c). This strongly suggested that a downconverted CPL could be emitted from **6** in the co-gels at deaerated condition. Thus, a dual circularly polarized light emission involving upconverted CPL at 460 nm and downconverted CPL was realized in co-gels. Two channels of chirality and energy-transfer process were successfully integrated, and the interplay of energy and chirality transfer to produce a dual CPL emission was revealed simultaneously.

Except for glutamic acid-based gelator, an *L*-histidine-derived gelator was designed and the CPL property has been investigated in Fig. 11.7 [24]. As shown in Fig. 11.7a, it was found that a supramolecular gel with CPL properties could be formed by gelator **7** upon sonication. The calculated $g_{lum}$ was $\pm 5.0 \times 10^{-4}$ at 370 nm. In addition, combined with achiral benzoic acids, **7** could form two-component co-gels. Although the benzoic acid is achiral, the co-gels exhibited unexpected enhanced CPL and one order of magnitude amplification of the largest $g_{lum}$ value is detected ($\pm 3.0 \times 10^{-3}$). The possible mechanism for this CPL enhancement was illustrated in Fig. 11.7b. A bilayer structure could be constructed by gelator **7** due to the H-bond between the imidazole moieties and urea and π–π stacking between the naphthalene. For the co-assembly, a $C_3$-like secondary unit was firstly formed by **7** and benzoic acid, then this unit further self-assembly into hexagonal structures. The tight π–π interaction in this hexagonal stacking resulted in better chirality transfer from gelator **7** to the supramolecular assembles, which

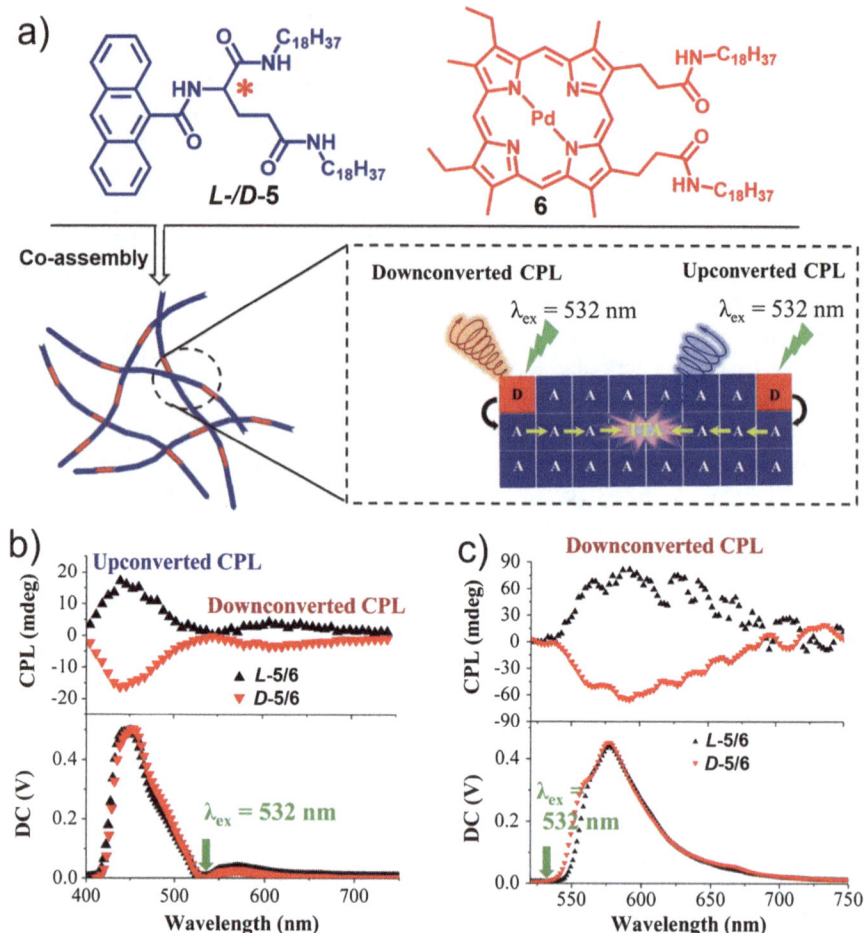

**Fig. 11.6** (**a**) Schematics of energy acceptor *L*-**5** or *D*-**5** and energy donor **6** in TTA-UC process. In the co-assembly system, the chirality can be transferred from gel **5** to **6**. (**b**) CPL spectra of *L*-**5**/**6** and *D*-**5**/**6** in deaerated toluene, $\lambda_{ex}$ = 532 nm laser. (**c**) CPL spectra of *L*-**5**/**6** and *D*-**5**/**6** in deaerated toluene, $\lambda_{ex}$ = 532 nm laser. Reproduced with permission [23]. Copyright 2018, WILEY-VCH

might further profit the $(I_L - I_R)$ values. In addition, the total fluorescence intensity $(I_L + I_R)$ decreased after adding benzoic acid. Based on the $g_{lum} = 2 \times (I_L - I_R)/(I_L + I_R)$, an amplification of the CPL resulted from lower $(I_L + I_R)$ values could be observed.

### 11.3.2.2 Aggregation-Induced Emission (AIE) Triggered CPL in Gels

As stated above, the regulated assembly of chiral emitters would be one of the general and simple ways adopted for enhancing $g_{lum}$. However, due to the aggregation-caused quenching (ACQ) effect, which leads to a much lower or

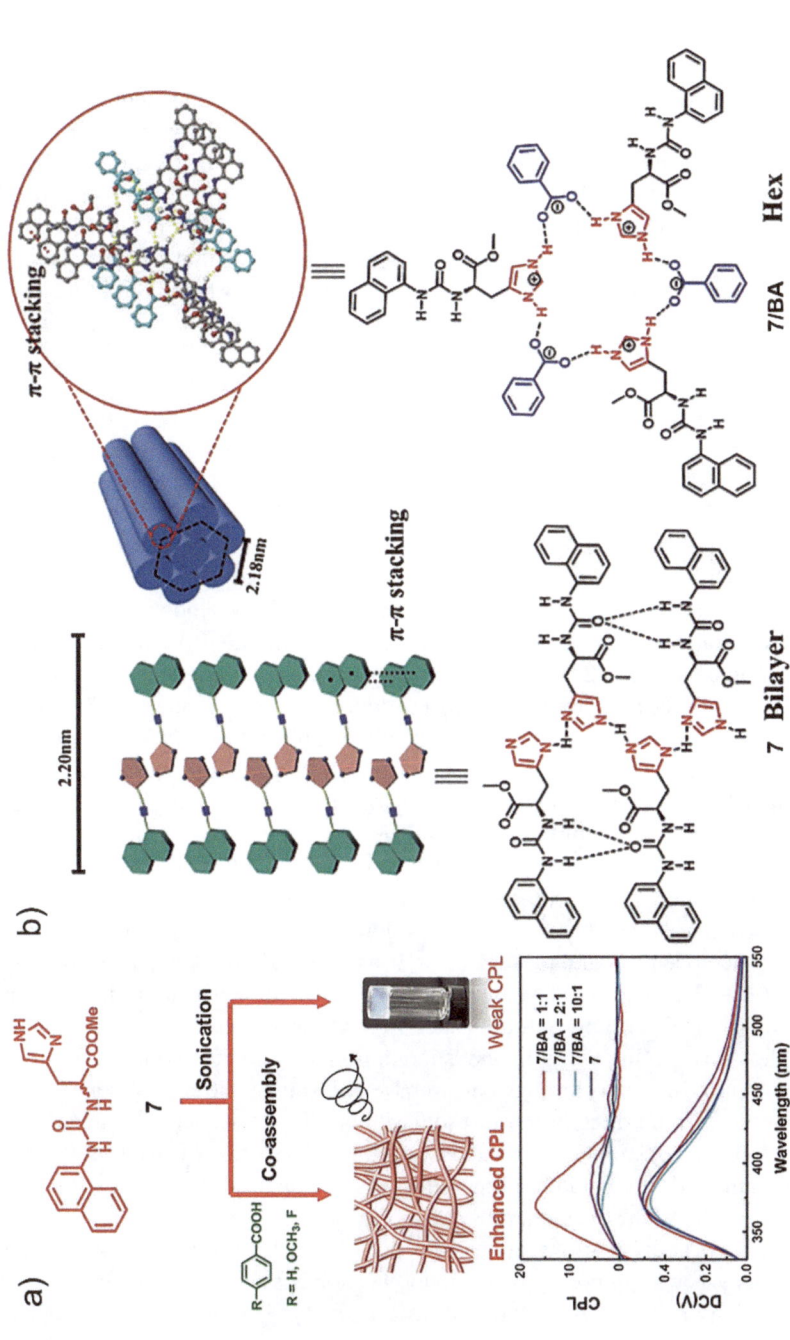

**Fig. 11.7** (a) Schematics of the achiral molecule-boosted CPL in histidine-derived naphthalene organogels (up) and CPL spectra of **7** (gel) and co-assemblies of **7** with BA in different ratios (bottom). (b) Schematic illustration of the molecule packing modes of **7** self-assemblies and the **7/BA** co-assemblies. Reproduced with permission [24]. Copyright 2018, The Royal Society of Chemistry

even no emission performance of luminophores in the aggregated state, the CPL performance of most luminophores becomes even worse from isolate molecule to aggregated state. Fortunately, Tang's group advanced an excellent idea to solve this problem by combined aggregation-induced emission (AIE) effect with chiral assembly [25]. AIE luminophore incorporated with the chiral component to fabricate novel CPL-active materials has become an efficient way to achieve both large $g_{lum}$ values and high emission efficiency in condensed phase [26]. Cyano-substituted stilbene (CNSB) is a well-known compound with AIE property. When CNSB conjugated with a glutamic-derived gelaton, an AIE gelator was obtained (*L-/D*-**8**) [27]. As shown in Fig. 11.8, the chiral donor (gelator **8**) and achiral acceptor (BPEA) could form a composite nanohelix through co-assembly, in which both energy and chirality transfer were observed simultaneously. Amazingly, not only the chirality transfer happened in the complex, but also the dissymmetry of CPL was significantly amplified during the energy transfer. Thanks to the $\pi$–$\pi$ stacking of CNSB and the H-bond between the amide groups, an ordered nanohelix structure could be obtained from gelator **8** through self-assembly. During the self-assembly process, not only the emission intensity of **8** remarkably increased, but also the molecular level chirality transferred to the supramolecular level, resulting in the excellent CPL properties. When **8** co-assembled with achiral BPEA, the acceptor BPEA was inserted into the nanohelix through the weak $\pi$–$\pi$ stacking to form the co-gel (Fig. 11.8a). Interestingly, the achiral acceptors could be endowed with CPL caused by chirality transfer from the nanohelix. As shown in Fig. 11.8b, the detected CPL $g_{lum}$ value was $\pm 1.2 \times 10^{-3}$ by directly exciting the acceptor (at 400 nm), while the $g_{lum}$ for the acceptor by exciting the donor (at 320 nm) showed a significant enhancement (up to $\pm 3 \times 10^{-3}$). This result exhibited that $g_{lum}$ value of CPL was amplified more than 2.5 times through the energy-transfer process. This might be resulted from the enhancement of acceptor emission via the energy transfer, which seems to further amplify the $g_{lum}$ values.

Encouraged by the result of energy transfer amplified CPL, Liu et al. further investigated a cooperative chirality and sequential energy transfer in a supramolecular co-assembly system to explore its mechanism. In this work, a cyanostilbene-appended glutamate gelator **9** and two kinds of achiral acceptors, thioflavin T (ThT) and acridine orange (AO), were employed to from a co-gel [28]. Similar to gelator **8**, the chiral gelator **9** could form supramolecular nanotubes with CPL activity. In addition, the supramolecular chirality could transfer to these two achiral acceptors through co-assembly. Meantime, the excited-state energy of **9** nanotubes could directly transfer to ThT but only be sequentially transferred to AO. More interestingly, compared with CPL from directly exciting AO, a stepwise amplified CPL could be observed when exciting the donor **9** or intermediate donor ThT in the **9**/ThT/AO ternary system (Fig. 11.9).

It should be noted that energy transfer boosted CPL was considerable, and a possible analysis was proposed. When donor **9** nanotubes are excited by unpolarized light, a CPL is obtained due to its intrinsic chirality. However, when acceptors were added, the excited-state energy with chiral information will transfer to acceptors, resulting in a new CPL from the acceptor. Based on a theoretical

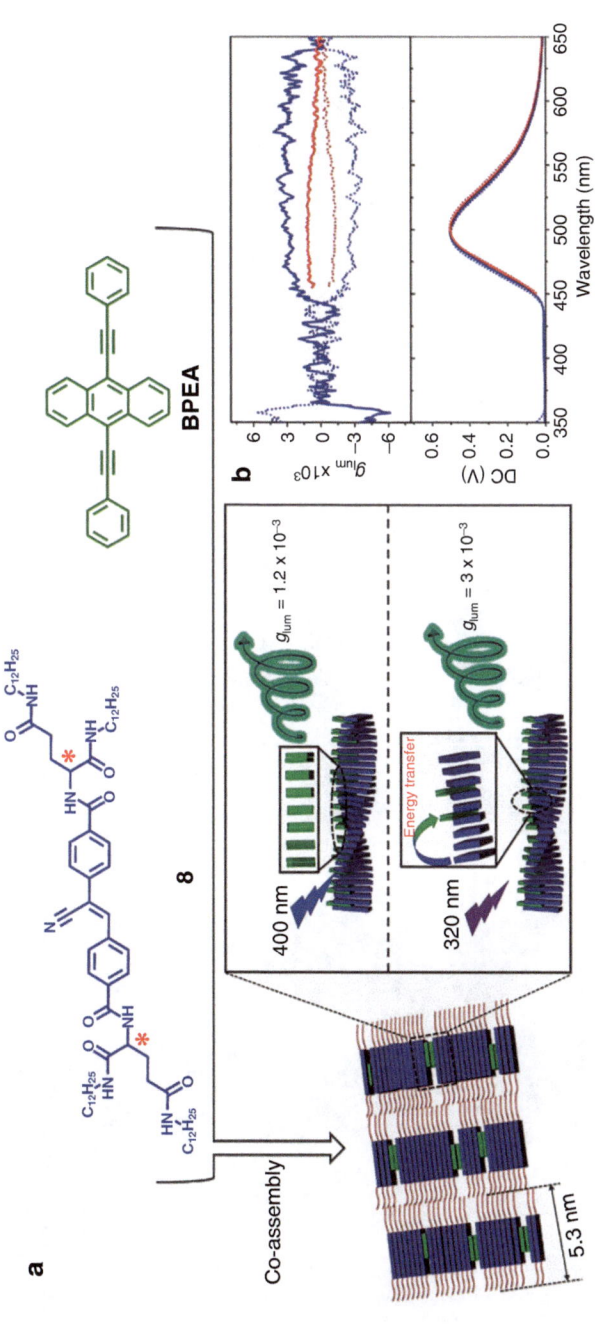

**Fig. 11.8** (a) Chemical structures of **8** and BPEA, and schematic illustration of energy transfer amplified CPL in co-assembly nanohelix. Under excitation at 400 nm, the composite nanohelix showed green CPL with $g_{lum} = 1.2 \times 10^{-3}$ (top pattern), while energy transfer boosted CPL with a relatively large value $g_{lum} = 3.0 \times 10^{-3}$ under excitation at 320 nm (bottom pattern). (b) CPL dissymmetry factor $g_{lum}$ versus wavelength, $\lambda_{ex} = 320$ nm (blue line) and $\lambda_{ex} = 400$ nm (red line). Reproduced with permission [27]. Copyright 2017, Nature Publishing Group

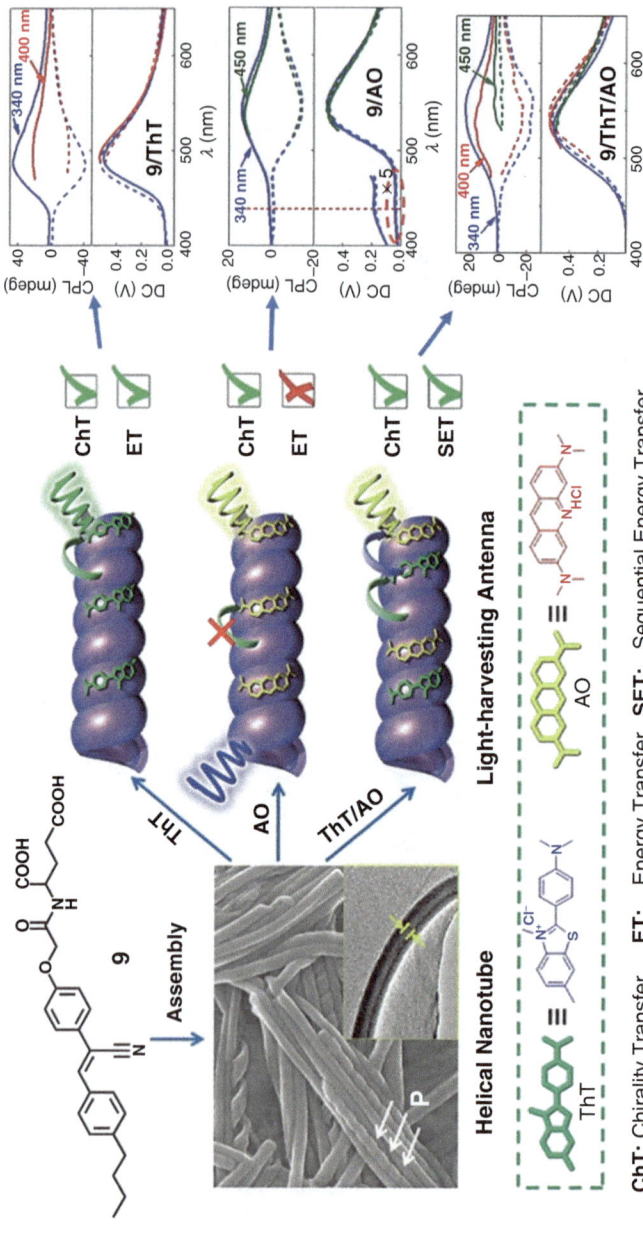

**Fig. 11.9** Left: Schematics of the self-assembly **9** nanotubes, and different chirality and energy-transfer modes of the co-assembly of **9**/ThT, **9**/AO, and **9**/ThT/ AO. Insets are scanning electron microscopy (SEM) and tunneling electron microscopy (TEM) images of the **9** nanotubes. Right: CPL spectra of **9**/ThT, **9**/AO, and **9**/ThT/AO under different excitation wavelength ($\lambda_{ex} = 340$ nm for **9**, $\lambda_{ex} = 400$ nm for ThT, and $\lambda_{ex} = 450$ nm for AO). Reproduced with permission [28]. Copyright 2019, Wiley-VCH

calculation by Bene and co-workers, in a helicity and energy-transfer process (hFRET), the helicity in fluorescence from a rotating donor dipole could be preserved [29]. In the instance of **9**/ThT/AO system, the helicity of luminescent donor **9** even ThT may be sequentially transferred to acceptor AO and amplified the $g_{lum}$ of AO. For experimental aspect, the *L*-**9**/ThT system exhibited an amplified $g_{lum}$ value ($1.89 \times 10^{-2}$) when excited by circularly polarized light contrasted to the $g_{lum}$ value excited by unpolarized light ($g_{lum} = 3.3 \times 10^{-3}$). Thus, the helicity in the donor excited state could be transferred to the acceptor and subsequently amplified the $g_{lum}$ was indirectly illustrated.

#### 11.3.2.3   CPL from Chiral π-Conjugated Gelators

In supramolecular gel system, π-conjugated gelators also play an important role because of their unique electronic and optical properties. Ajayaghosh's group has reported the design of oligo(*p*-phenylenevinylene) (OPV)-derived gelators and their hierarchical self-assembly properties [30]. Furthermore, the CPL property of a photosensitive π-conjugated gelator azobenzene-linked phenyleneethynylene (**10**) was also been investigated in Fig. 11.10 [31]. As shown in Fig. 11.10a, the assembly of **10** showed a positive CPL before photoisomerization. Interestingly, the signal of CPL was inverted after irradiation. A possible mechanism for the observed photoinduced CPL inversion resulted from helicity inversion was proposed in Fig. 11.10b. The assembly of **10** groups in methylcyclohexane (MCH) shows *P*-helical. After ultraviolet (UV) photoirradiation, the remaining *E,E* isomers were low in the photostationary state and the excess homonuclear *M*-helical YY aggregates will nucleate and control the self-assembly. Therefore, the photoinduced CPL signal was reversed.

### 11.3.3   CPL from Achiral Luminophores

#### 11.3.3.1   Symmetry Breaking Triggered CPL

In the above description, luminophores conjugated with chiral gelaton moieties are required for the fabrication of CPL-active gels. Nevertheless, not only chiral gelators but chiral supramolecular gels can be formed by completely achiral gelators through assembly, thus provided the possibility for the exclusive achiral fluorophores to fabricate CPL materials [32]. If an asymmetric environment is provided during self-assembly, some achiral gelators can result spontaneous symmetry breaking to form chiral supramolecular assemblies in this situation. As showed in Fig. 11.11, CPL activities could be observed if an achiral ionic polymer **11** and Rhodamine B were under mechanical stirring [33]. Moreover, the stir direction could determine the CPL signal: a positive sign was resulted from counterclockwise (CCW) stirring, while clockwise (CW) stirring induced a negative sign (Fig. 11.11a). In order to reveal the origin of the CPL, CPL spectra from four different angles of the cuvette

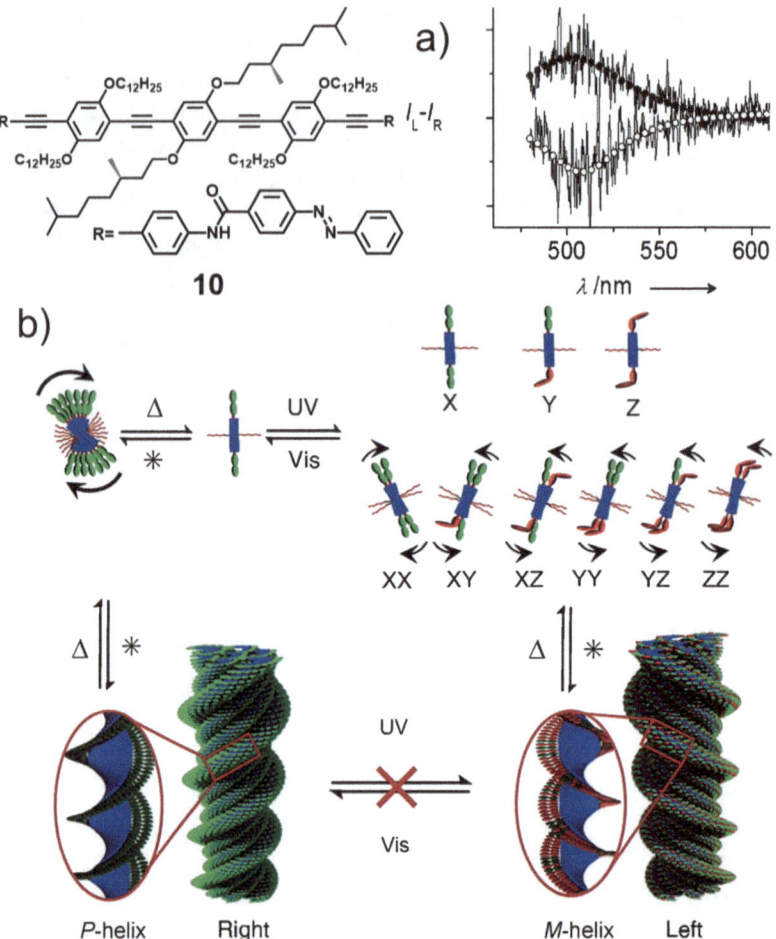

**Fig. 11.10** Chemical structure of **10** and (**a**) CPL spectra for **10** before (filled circle) and after (open circle) UV irradiation. (**b**) Schematics of supramolecular helical assembly of **10**'s reversible helicity inversion via a disassembly/reassembly process accompanying with *E/Z* photoisomerization of azobenzene moieties. $\Delta$ = heating; $*$ = cooling. Reproduced with permission [31]. Copyright 2012, Wiley-VCH

were detected (Fig. 11.11b). The CPL spectra detected from the four different angles showed a same sign with similar shape in a series of evaluation, indicating that the chirality of Rhodamine B was induced by stir on a macroscopic level. Nevertheless, it should be noted that such CPL only can be observed during the stirring process. Once the stirring stopped, the sample became CPL silent.

In the supramolecular gel systems, an achiral $C_3$-symmetric gelator was also found to display fascinating CPL performance (Fig. 11.12) [34]. It was found that when the achiral gelators formed the organogels, it showed strong emission as well as the CPL. The handedness of the CPL appeared randomly, suggesting

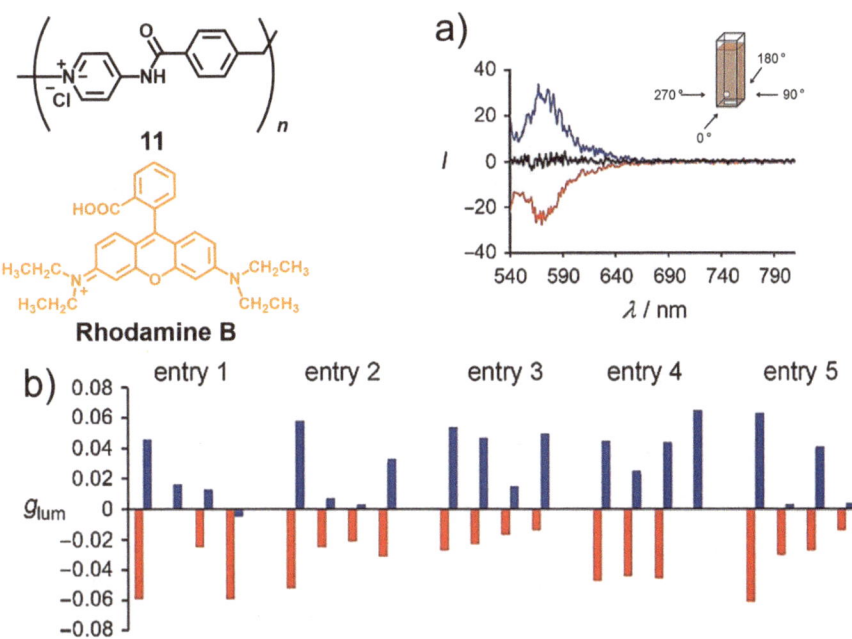

**Fig. 11.11** Chemical structures of **11**, Rhodamine B and (**a**) CPL spectra recorded with counterclockwise (CCW, blue line) stirring, clockwise (CW, red line) stirring, and no stirring (black line). The inset shows the four faces of the cuvette detected for the CPL measurements. (**b**) Statistical distributions of $g_{lum}$ values in five different samples under CCW (blue) and CW (red) stirring preparation in four faces of the sample cuvette. Reproduced with permission [33]. Copyright 2011, Wiley-VCH

a spontaneous symmetry breaking and the $g_{lum}$ value was $\pm 0.8 \times 10^{-2}$, as shown in Fig. 11.12a. Interestingly, in such a system mechanical stirring could enhance the $g_{lum}$ values during the supramolecular gelation process (Fig. 11.12b). In addition, the obtained gel dispersion was quite stable and the CPL remained even after stopping the stirring. However, the direction of the CPL signals could not be controlled by the stirring direction. The direction of CPL signals can be readily regulated by adding some simple chiral dopants (Fig. 11.12c). More interestingly, the $g_{lum}$ value was also amplified by chiral dopants ($\pm 2.3 \times 10^{-2}$).

In some cases, the CPL from the nanoassemblies based on achiral molecules showed a morphology dependence. As illustrated in Fig. 11.13, twisted ribbons, nanobelts and trumpet-like nanostructures can be formed from an achiral $C_3$-symmetric molecule via the assembly in a mixed DMF/water solvent [35]. At a unity mixing ratio, nanobelts were observed. Upon increasing the amount of dimethylformamide (DMF), nanotwists and nanotrumpets were formed by such nanobelts through twisting and rolling, respectively. Intriguingly, the nanotwists showed supramolecular chirality with relatively strong CPL performance ($g_{lum} = \pm 2.1 \times 10^{-2}$) although the component compound is achiral, while the other nanostructure could not.

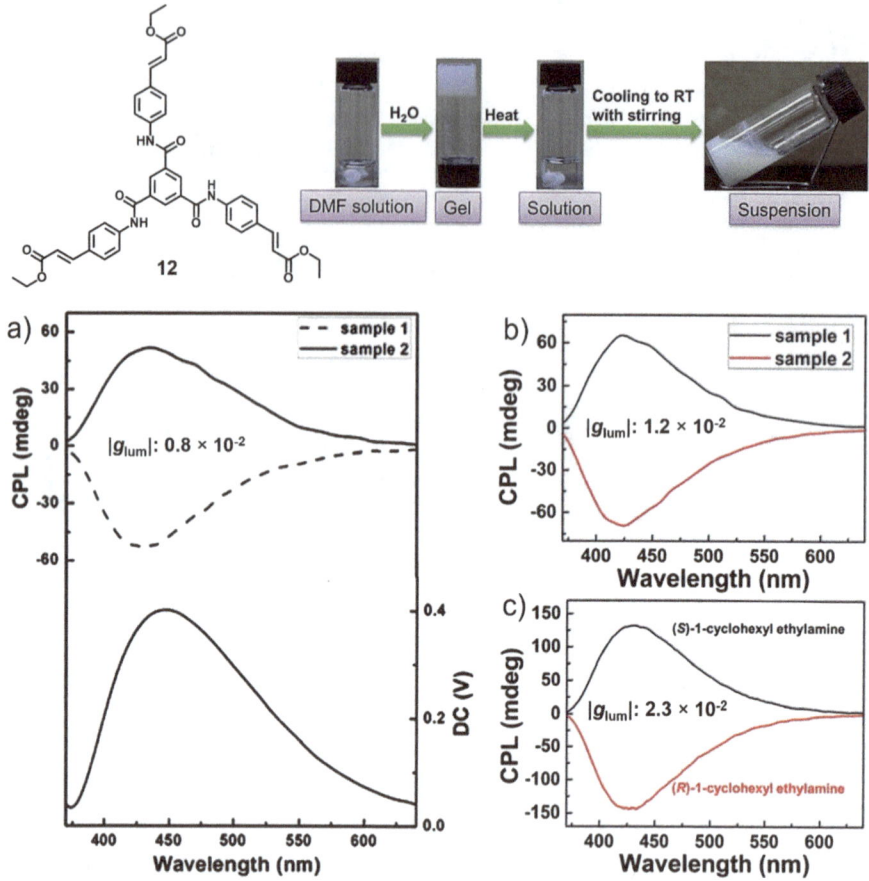

**Fig. 11.12** Chemical structure of **12** and photographs showing **12** assemblies in *N,N*-dimethylformamide/H$_2$O (DMF/H$_2$O) upon stirring during preparation of the gels (top). (**a**) CPL spectra (left axis) and fluorescence spectra (right axis) of **12** gels in DMF/H$_2$O. CPL spectra of **12** assemblies after 900 rpm clockwise stirring during gelation (**b**) and containing 900 mol% chiral 1-cyclohexyl ethylamine (**c**) in DMF/H$_2$O. Reproduced with permission [34]. Copyright 2015, The Royal Society of Chemistry

### 11.3.3.2　CPL from Chiral Gelator and Achiral Luminophores

As stated above, exclusive achiral luminescent gelators could exhibit CPL through self-assembly in asymmetric environment. In addition, another important approach for the fabrication of CPL-active materials from achiral luminophores is the co-assembly. The chirality transfer from chiral components to achiral luminophores causing induced chirality is of utmost importance. It is not only having potential to endow almost all luminophores with CPL, but also simplify the strategy to produce various CPL-active materials instead of the tedious organic synthesis.

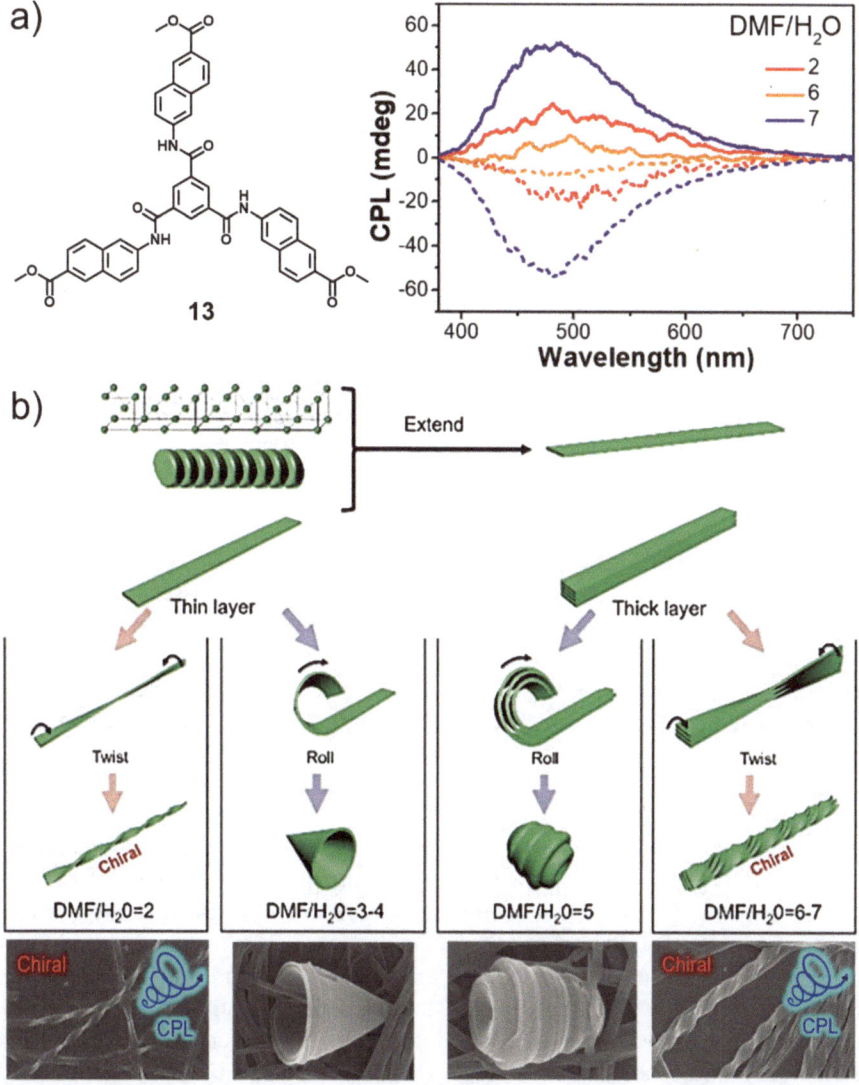

**Fig. 11.13** (a) Molecular structure for the gelator **13** and CPL spectra of **13** at a volume ratio of DMF/$H_2O$ = 2, 6, and 7. (b) Schematics of the formation of nanotwists and nanotrumpets from nanobelts. Bottom was the SEM of the nanotwists and nanotrumpets. The nanotwists showed CPL activities, while the nanotrumpets did not. Reproduced with permission [35]. Copyright 2018, The Royal Society of Chemistry

For the chirality induction of the achiral luminophores, the interaction between the chiral molecules and the achiral luminophores plays a very important role. Noncovalent interactions, such as electrostatic interactions, hydrophobic interactions, hydrogen bonds, and host–guest interactions, could be employed to regulate the chirality induction.

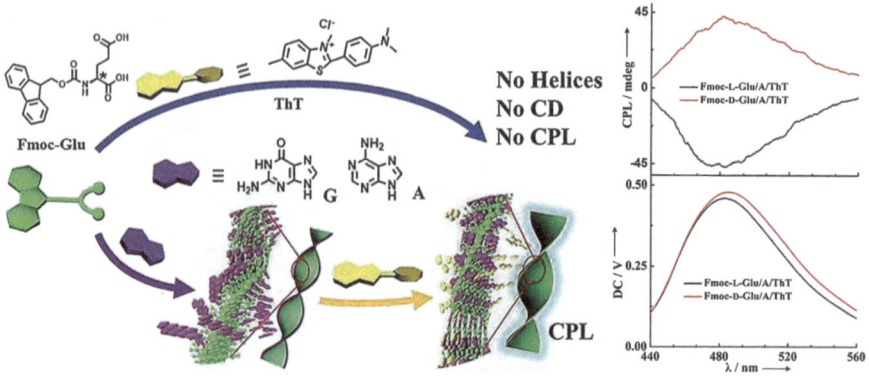

**Fig. 11.14** Left: Schematic illustration of the achiral nucleobase-assisted helical self-assembly based on the Fmoc-Glu and supramolecular chirality transfer from Fmoc-Glu to achiral ThT. Right: CPL spectra of Fmoc-(*L/D*)-Glu/A/ThT. Reproduced with permission [36]. Copyright 2016, Wiley-VCH

An example of a three-component co-assembly for induced CPL is shown in Fig. 11.14 [36]. Interestingly, the helix of Fmoc-Glu assembly could be triggered by the achiral nucleobase (guanine (G) or adenine (A)). Furthermore, this supramolecular chirality transferred to the achiral cationic dye ThT. It should be noted that only the second achiral molecule is introduced; the co-assembly of Fmoc-Glu and ThT exhibits helicity at the nanoscale level, finally leading to a distinct CPL.

On the other hand, chiral confined spaces or environments can also endow achiral components with chirality. Figure 11.15 shows a general approach to fabricate CPL-active assembly nanotubes through loading achiral AIE luminophores (AIEgens) [37]. As illustrated in Fig. 11.15a, the hexagonal nanotube structures can be constructed by $C_3$ symmetric chiral gelators *L-/D*-**14**, and the intrinsic chirality of the substituted glutamate moieties can transfer to the supramolecular nanotubes during the self-assembly process. Furthermore, the achiral AIE luminophores could be embedded into the confined nanotubes via co-assembly, and achiral AIE dyes aggregated during the gelation process, which showed enhanced fluorescence intensity (Fig. 11.15b) and distinct circularly polarized luminescence by direct excitation. Meantime, the polarization of the CPL is regulated by the supramolecular chirality of the nanotubes. As shown in Fig. 11.15c, through simply altering the doped dyes, mirror-imaged CPL signals from 425 to 595 nm, covering the full-color from blue via green and yellow to orange-red color, can be tuned.

A proton-triggered CPL switch in a co-assembled gel system consisted of a chiral gelator and an achiral fluorophore was showed in Fig. 11.16 [38]. A co-gel could be formed by chiral gelator *L-/D*-**15** and achiral perylene bisimide (PBI) in ethanol and gelation-induced chirality of PBI was transferred from *L-/D*-**15**. Due to low luminescence efficiency of PBI, no CPL signal could be observed. However, after exposing the co-gel to an acid atmosphere, significantly increased emission intensity

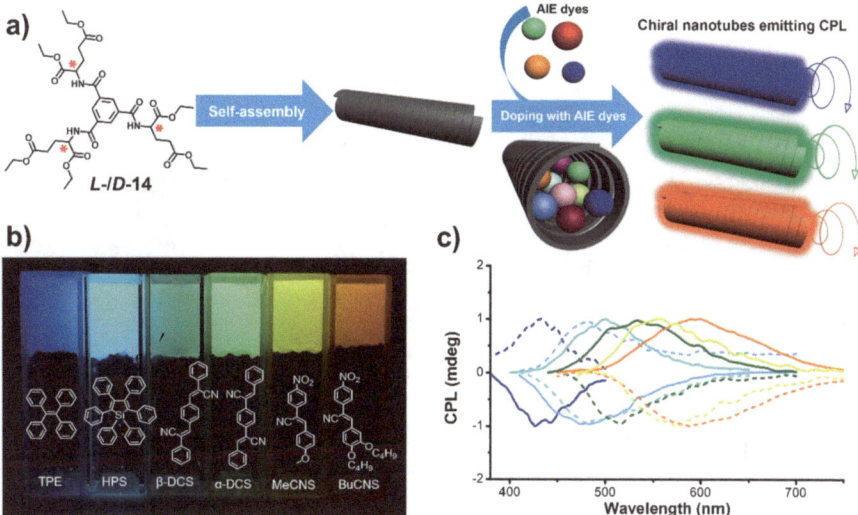

**Fig. 11.15** (**a**) Schematic representation of chiral nanotubes formed by *L-/D-*14 encapsulated different AIE luminophores. (**b**) The photograph of AIEgens-loaded co-gels under UV light irradiation. (**c**) Mirror image CPL spectra of TPE ($\lambda_{ex}$ = 300 nm), HPS ($\lambda_{ex}$ = 365 nm), $\beta$-DCS ($\lambda_{ex}$ = 354 nm), $\alpha$-DCS ($\lambda_{ex}$ = 363 nm), MeCNS ($\lambda_{ex}$ = 376 nm), BuCNS ($\lambda_{ex}$ = 376 nm) in *L-*14 (solid lines) or *D-*14 (dash lines) host gels; all the co-gel samples were made in DMSO/H$_2$O (1/1, v/v) mixing solvents. Reproduced with permission [37]. Copyright 2017, Wiley-VCH

as well as a CPL signal could be detected in the co-gel. Additionally, if the co-gel was exposed to an ammonia atmosphere, both fluorescence and CPL could be switched off. Therefore, this co-gel system achieved a dual switch of fluorescence and CPL (Fig. 11.16a). Because of protonation by acetic acid, the co-gel showed strong fluorescence when under excitation at 447 nm. Meanwhile, CPL was also detected, as verified by the mirror image CPL spectra in Fig. 11.16b. As shown in Fig. 11.16c, when the *L*-15/PBI film was exposed to HCl atmosphere, appearing the fluorescence and CPL curve. However, if exposing the film to an ammonia atmosphere, the fluorescence and CPL will be wiped out. The $g_{lum}$ value showed stable repeatability after several cycle treatments. Therefore, a chiroptical switch based on acid–base exposure regulated could be fabricated.

In the co-assembly system, gelator with photochromic properties could also construct the CPL switch. As illustrated in Fig. 11.17, a photo-switched nanostructure, formed by photosensitive cinnamic acid derivatives (*L*- or *D*-16) through self-assembly, can successfully change from superhelix to nanokebab by light irradiation, resulting from the dimerization of cinnamic acid. Meanwhile, the supramolecular chirality could be reversed [39]. Furthermore, fluorescent achiral dyes (CBS) could embed into these helical nanostructures and induced chirality was observed as well as CPL properties. These nanohelixes could turn into nanokebabs by irradiated with UV light (365 nm), and both the inversion of helicity and CPL were observed. Interestingly, the handedness of CPL showed that following the supramolecular chirality of the nanoassemblies rather than the intrinsic molecular chirality. Based

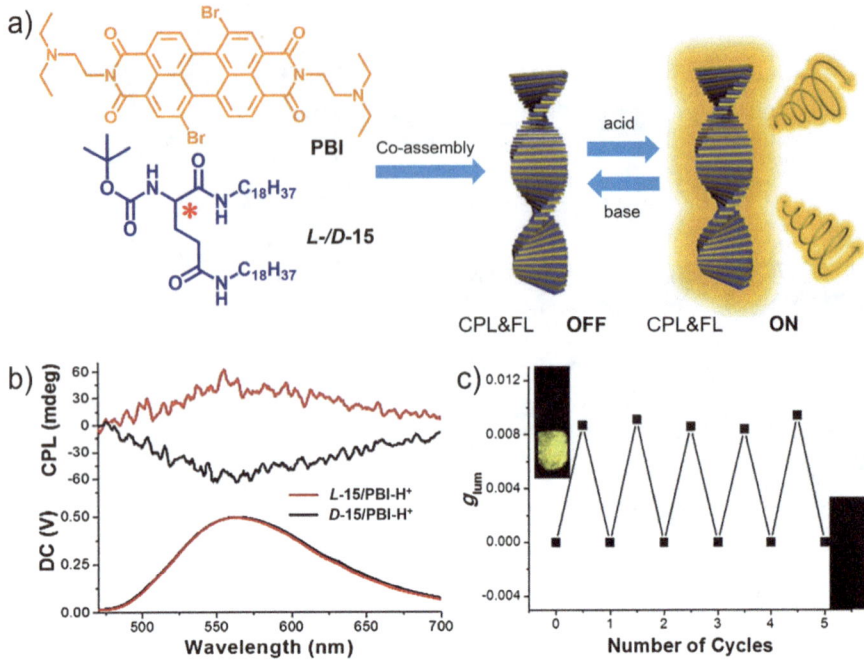

**Fig. 11.16** (a) Molecular structures of *L-/D-*15 and PBI and schematic illustration of the fabrication of co-gel composed of *L-/D-*15 and achiral PBI. Circularly polarized light emission could be switched by changing acid–base exposure. (b) CPL spectra of protonated co-gels (*L-*15/PBI and *D-*15/PBI) excited at 447 nm. (c) Intensity of the CPL $g_{lum}$ value of the co-gel *L-*15/PBI against the repeated acid–base exposure cycles. The inset pictures are the gel on quartz under a UV lamp. Reproduced with permission [38]. Copyright 2018, the Royal Society of Chemistry

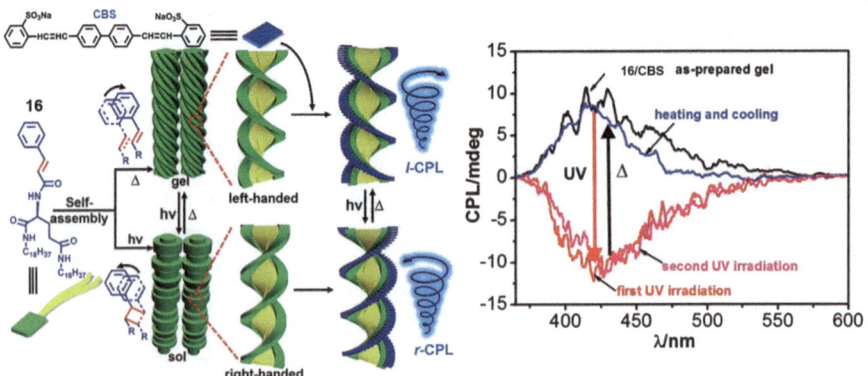

**Fig. 11.17** Left: Schematic illustration of the self-assembly nanohelices formed by **16** and co-assembly of **16** with a chiral fluorescent molecule (CBS). The chirality, morphology of formed nanostructures, and induced CPL inversion are triggered by alternating UV irradiation and heating/cooling. Right: CPL spectra of **16**/CBS methanol co-gel under repeated treatment of UV irradiation and heating/cooling. Reproduced with permission [39]. Copyright 2019, Wiley-VCH

on these results, alternative photoirradiation and heating/cooling procedures were utilized to treat the supramolecular nanoassemblies; thus, the inverted helicity and CPL performances could be reversibly repeated during the treatment processes. Therefore, the switched inversion of supramolecular chirality and CPL are successfully realized in one kind of chiral nanostructures by using the co-assembly strategy.

## 11.4 Conclusion

CPL-active materials have potential applications in various research fields. To achieve this goal, molecular gelator could be regarded as one of the better candidates because of the precisely adjustable intermolecular interactions and enhanced performances in relating to CPL [40–41]. Now, the introduction of new fluorescent materials and the concept of self-assembly have led to the rapid development of CPL-active materials [42–44]. For the research of materials with CPL activities, supramolecular gels provide a broad platform. Besides the self-assembly from chiral luminophores, chiral supramolecular gels could be fabricated from many of the building blocks either chiral or achiral. On the basis of various noncovalent interactions such as H-bond, $\pi$–$\pi$ stacking, host–guest interaction and so on, non-CPL-active isolated chiral gelators could assembly into supramolecular gels exhibiting CPL performance. Analogously, chiral gel matrixes can load or encapsulate achiral luminophores; the achiral luminophores thereby are endowed with CPL activity. This is beneficial not only for organic system but also for inorganic system. Except for the construction of CPL-active materials through gelation, supramolecular gels also can further develop their functions such as energy transfer, CPL switch, and so on. The energy-transfer amplified dissymmetry in gel systems will serve as an excellent platform for expanding the fabrication of highly efficient CPL-active materials. It is worth further research in the future. Although CPL-active gels have shown their application in some cases, a future effort for the CPL application is still in its infancy.

## References

1. Josse P, Favereau L, Shen C, Dabos-Seignon S, Blanchard P, Cabanetos C, Crassous J (2017) Enantiopure versus racemic naphthalimide end-capped helicenic non-fullerene electron acceptors: impact on organic photovoltaics performance. Chem Eur J 23:6277–6281
2. Li M, Li SH, Zhang DD, Cai MH, Duan L, Fung MK, Chen CF (2018) Stable enantiomers displaying thermally activated delayed fluorescence: efficient OLEDs with circularly polarized electroluminescence. Angew Chem Int Ed 57:2889–2893
3. Hellou N, Srebro-Hooper M, Favereau L, Zinna F, Caytan E, Toupet L, Dorcet V, Jean M, Vanthuyne N, Williams JAG, Di Bari L, Autschbach J, Crassous J (2017) Enantiopure cycloiridiated complexes bearing a pentahelicenic N-heterocyclic carbene and displaying long-lived circularly polarized phosphorescence. Angew Chem Int Ed 56:8236–8239

4. Schadt M (1997) Liquid crystal materials and liquid crystal displays. Annu Rev Mater Sci 27:305–379
5. Zheng H, Li W, Li W, Wang X, Tang Z, Zhang SX-A, Xu Y (2018) Uncovering the circular polarization potential of chiral photonic cellulose films for photonic applications. Adv Mater 30:1705948
6. Song F, Wei G, Jiang X, Li F, Zhu C, Cheng Y (2013) Chiral sensing for induced circularly polarized luminescence using an Eu(III)-containing polymer and D- or L-proline. Chem Commun 49:5772–5774
7. Yang Y, da Costa RC, Fuchter MJ, Campbell AJ (2013) Circularly polarized light detection by a chiral organic semiconductor transistor. Nat Photonics 7:634
8. Shuvaev S, Suturina EA, Mason K, Parker D (2018) Chiral probes for α1-AGP reporting by species-specific induced circularly polarised luminescence. Chem Sci 9:2996–3003
9. Wang L, Yin L, Zhang W, Zhu X, Fujiki M (2017) Circularly polarized light with sense and wavelengths to regulate azobenzene supramolecular chirality in optofluidic medium. J Am Chem Soc 139:13218–13226
10. Kawasaki T, Sato M, Ishiguro S, Saito T, Morishita Y, Sato I, Nishino H, Inoue Y, Soai K (2005) Enantioselective synthesis of near enantiopure compound by asymmetric autocatalysis triggered by asymmetric photolysis with circularly polarized light. J Am Chem Soc 127:3274–3275
11. Richardson RD, Baud MGJ, Weston CE, Rzepa HS, Kuimova MK, Fuchter MJ (2015) Dual wavelength asymmetric photochemical synthesis with circularly polarized light. Chem Sci 6:3853–3862
12. Yeom J, Yeom B, Chan H, Smith KW, Dominguez-Medina S, Bahng Joong H, Zhao G, Chang W-S, Chang S-J, Chuvilin A, Melnikau D, Rogach AL, Zhang P, Link S, Král P, Kotov NA (2014) Chiral templating of self-assembling nanostructures by circularly polarized light. Nat Mater 14:66
13. He C, Yang G, Kuai Y, Shan S, Yang L, Hu J, Zhang D, Zhang Q, Zou G (2018) Dissymmetry enhancement in enantioselective synthesis of helical polydiacetylene by application of superchiral light. Nat Commun 9:5117
14. Sanchez-Carnerero EM, Agarrabeitia AR, Moreno F, Maroto BL, Muller G, Ortiz MJ, de la Moya S (2015) Circularly polarized luminescence from simple organic molecules. Chem Eur J 21:13488–13500
15. Kumar J, Nakashima T, Tsumatori H, Kawai T (2014) Circularly polarized luminescence in chiral aggregates: dependence of morphology on luminescence dissymmetry. J Phys Chem Lett 5:316–321
16. Dhbaibi K, Favereau L, Srebro-Hooper M, Jean M, Vanthuyne N, Zinna F, Jamoussi B, Di Bari L, Autschbach J, Crassous J (2018) Exciton coupling in diketopyrrolopyrrole-helicene derivatives leads to red and near-infrared circularly polarized luminescence. Chem Sci 9:735–742
17. Gon M, Morisaki Y, Chujo Y (2017) Optically active phenylethene dimers based on planar chiral tetrasubstituted [2.2]paracyclophane. Chem Eur J 23:6323–6329
18. Duan P, Cao H, Zhang L, Liu M (2014) Gelation induced supramolecular chirality: chirality transfer, amplification and application. Soft Matter 10:5428–5448
19. Liu M, Ouyang G, Niu D, Sang Y (2018) Supramolecular gelatons: towards the design of molecular gels. Org Chem Front 5:2885–2900
20. Sheng Y, Ma J, Liu S, Wang Y, Zhu C, Cheng Y (2016) Strong and reversible circularly polarized luminescence emission of a chiral 1,8-naphthalimide fluorophore induced by excimer emission and orderly aggregation. Chem Eur J 22:9519–9522
21. Kumar J, Nakashima T, Kawai T (2015) Circularly polarized luminescence in chiral molecules and supramolecular assemblies. J Phys Chem Lett 6:3445–3452
22. Jintoku H, Kao M-T, Del Guerzo A, Yoshigashima Y, Masunaga T, Takafuji M, Ihara H (2015) Tunable stokes shift and circularly polarized luminescence by supramolecular gel. J Mater Chem C 3:5970–5975

23. Yang D, Duan P, Liu M (2018) Dual upconverted and downconverted circularly polarized luminescence in donor–acceptor assemblies. Angew Chem Int Ed 130:9501–9505
24. Niu D, Ji L, Ouyang G, Liu M (2018) Achiral non-fluorescent molecule assisted enhancement of circularly polarized luminescence in naphthalene substituted histidine organogels. Chem Commun 54:1137–1140
25. Liu J, Su H, Meng L, Zhao Y, Deng C, Ng JCY, Lu P, Faisal M, Lam JWY, Huang X, Wu H, Wong KS, Tang BZ (2012) What makes efficient circularly polarised luminescence in the condensed phase: aggregation-induced circular dichroism and light emission. Chem Sci 3:2737–2747
26. Li HK, Xue S, Su HM, Shen B, Cheng ZH, Lam JWY, Wong KS, Wu HK, Li BS, Tang BZ (2016) Click synthesis, aggregation-induced emission and chirality, circularly polarized luminescence, and helical self-assembly of a Leucine-containing silole. Small 12:6593–6601
27. Yang D, Duan P, Zhang L, Liu M (2017) Chirality and energy transfer amplified circularly polarized luminescence in composite nanohelix. Nat Commun 8:15727
28. Ji L, Sang Y, Ouyang G, Yang D, Duan P, Jiang Y, Liu M (2019) Cooperative chirality and sequential energy transfer in a supramolecular light-harvesting nanotube. Angew Chem Int Ed 58:844–848
29. Bene L, Bagdány M, Damjanovich L (2018) Checkpoint for helicity conservation in fluorescence at the nanoscale: energy and helicity transfer (hFRET) from a rotating donor dipole. Biophys Chem 239:38–53
30. Ajayaghosh A, Praveen VK, Srinivasan S, Varghese R (2007) Quadrupolar π-gels: sol–gel tunable red–green–blue emission in donor–acceptor-type oligo(p-phenylenevinylene)s. Adv Mater 19:411–415
31. Gopal A, Hifsudheen M, Furumi S, Takeuchi M, Ajayaghosh A (2012) Thermally assisted photonic inversion of supramolecular handedness. Angew Chem Int Ed 51:10505–10509
32. Sang Y, Yang D, Duan P, Liu M (2019) Towards homochiral supramolecular entities from achiral molecules by vortex mixing-accompanied self-assembly. Chem Sci 10:2718–2724
33. Okano K, Taguchi M, Fujiki M, Yamashita T (2011) Circularly polarized luminescence of rhodamine B in a supramolecular chiral medium formed by a vortex flow. Angew Chem Int Ed 50:12474–12477
34. Shen Z, Wang T, Shi L, Tang Z, Liu M (2015) Strong circularly polarized luminescence from the supramolecular gels of an achiral gelator: tunable intensity and handedness. Chem Sci 6:4267–4272
35. Sang YT, Duan PF, Liu MH (2018) Nanotrumpets and circularly polarized luminescent nanotwists hierarchically self-assembled from an achiral C-3-symmetric ester. Chem Commun 54:4025–4028
36. Deng M, Zhang L, Jiang Y, Liu M (2016) Role of achiral nucleobases in multicomponent chiral self-assembly: purine-triggered helix and chirality transfer. Angew Chem Int Ed 55:15062–15066
37. Han J, You J, Li X, Duan P, Liu M (2017) Full-color tunable circularly polarized luminescent nanoassemblies of achiral AIEgens in confined chiral nanotubes. Adv Mater 29:1606503
38. Han DX, Han JL, Huo SW, Qu ZM, Jiao TF, Liu MH, Duan PF (2018) Proton triggered circularly polarized luminescence in orthogonal- and co-assemblies of chiral gelators with achiral perylene bisimide. Chem Commun 54:5630–5633
39. Jiang H, Jiang Y, Han J, Zhang L, Liu M (2019) Helical nanostructures: chirality transfer and a photodriven transformation from superhelix to nanokebab. Angew Chem Int Ed 58:785–790
40. Yang D, Zhao Y, Lv K, Wang X, Zhang W, Zhang L, Liu M (2016) A strategy for tuning achiral main-chain polymers into helical assemblies and chiral memory systems. Soft Matter 12:1170–1175
41. Zhao Y, Abdul Rahim NA, Xia Y, Fujiki M, Song B, Zhang Z, Zhang W, Zhu X (2016) Supramolecular chirality in achiral polyfluorene: chiral gelation, memory of chirality, and chiral sensing property. Macromolecules 49:3214–3221

42. Huo SW, Duan PF, Jiao TF, Peng QM, Liu MH (2017) Self-assembled luminescent quantum dots to generate full-color and white circularly polarized light. Angew Chem Int Ed 56:12174–12178
43. Goto T, Okazaki Y, Ueki M, Kuwahara Y, Takafuji M, Oda R, Ihara H (2017) Induction of strong and tunable circularly polarized luminescence of nonchiral, nonmetal, low-molecular-weight fluorophores using chiral nanotemplates. Angew Chem Int Ed 56:2989–2993
44. Shi Y, Duan P, Huo S, Li Y, Liu M (2018) Endowing perovskite nanocrystals with circularly polarized luminescence. Adv Mater 30:1705011

# Chapter 12
# Circularly Polarized Luminescence from Intramolecular Excimers

Francesco Zinna, Elodie Brun, Alexandre Homberg, and Jérôme Lacour

**Abstract** In this chapter, examples of circularly polarized luminescence (CPL) stemming from intramolecularly formed excimers will be reviewed. Emission from excimers has peculiar photophysical properties with respect to fluorescence of regular monomers. In addition, if the fluorophoric couple forming the excimer in the excited state is mounted on a chiral scaffold, a strong CPL can be usually observed. Examples of chiral scaffolds include oligopeptides, macrocycles, binaphthyl, and diaminocyclohexane derivatives. CPL from excimers has mainly been observed from pyrenes but other molecules are also able to give rise to such phenomenon, e.g., perylenes and 1,8-naphthalene monoimide. Excimer CPL can provide important information about the conformation of a molecule in the excited state and how it evolves depending on the environment (e.g., solvent and temperature) or external stimuli (e.g., light irradiation and cation addition). Moreover, thanks to the peculiar photophysical nature of excimers, the degree of circular polarization associated with excimer emission is usually much larger than the one associated with the absorption (electronic circular dichroism, ECD) for the same molecule. This allows to study chiroptical emission properties of molecules which are ECD-silent (ground state cryptochirality). As a whole, excimer CPL is an interesting and useful strategy to develop organic molecular systems endowed with bright and highly polarized luminescence.

F. Zinna (✉)
Department of Organic Chemistry, University of Geneva, Geneva, Switzerland

Dipartimento di Chimica e Chimica Industriale, Università di Pisa, Pisa, Italy
e-mail: francesco.zinna@unipi.it

E. Brun · A. Homberg · J. Lacour (✉)
Department of Organic Chemistry, University of Geneva, Geneva, Switzerland
e-mail: jerome.lacour@unige.ch

© Springer Nature Singapore Pte Ltd. 2020                                    273
T. Mori (ed.), *Circularly Polarized Luminescence of Isolated Small Organic Molecules*, https://doi.org/10.1007/978-981-15-2309-0_12

## 12.1 Introduction

According to IUPAC Gold Book an *excimer* is *"an electronically excited dimer,
'non-bonding' in the ground state. For example, a complex formed by the interaction of an excited molecular entity with a ground state partner of the same structure*
[1]." Typically, in the realm of organic molecules, an excimer is formed between a
pair of aromatic moieties ($M$) loosely or non-interacting in the ground state. Upon
excitation, a strong $\pi$–$\pi$ interaction takes place between the excited fluorophore ($M*$)
and the one in ground state, generating thus an excited molecular complex, denoted
as (MM)$*$ (Eq. 12.1) [2]. Emission from such species is called *excimer emission*.
Such bands are red-shifted with respect to monomer emission and are broad and
structureless, since the ground state is dissociative (Fig. 12.1) [3]. As for other chiral
luminescent systems [4], excimer emission can be circularly polarized when using
enantioenriched molecules or in non-racemic environments.

$$M + M \rightarrow M * + M \rightarrow (\mathrm{MM})* \qquad (12.1)$$

When dealing with purely organic fluorescent compounds, circularly polarized
luminescence (CPL) activity is usually observed for intrinsically chiral and chirally
perturbed fluorophores. In these cases, most of the time, a single CPL band is
observed, if Kasha's rule applies. This band corresponds to the fluorescence stemming from the lowest singlet excited state. If the geometry of the emitting excited
state is not significantly different from that of the ground state, such a band has the
same sign of the most red-shifted Cotton effect (which corresponds to the ECD of the
same transition observed in absorption). Thus the $g_{\mathrm{lum}}$ factor is similar to the
absorption dissymmetry factor ($g_{\mathrm{abs}}$) of the corresponding ECD transitions
(in terms of order of magnitude and sign) [5]. On the other hand and by definition,
an excimer is a state existing only in the excited state. As a consequence, CPL

**Fig. 12.1** General scheme
of electronic states involved
in exciter emission.
Reproduced from reference
[3] with permission from
The Royal Society of
Chemistry

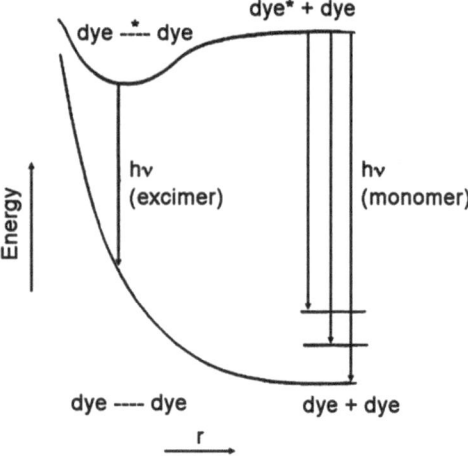

properties of such transitions cannot thus be inferred from the ECD bands. For example, the sign observed for the first ECD and CPL bands are generally not related, indicating that the excimer geometry can be totally different to that of the ground state. In this way CPL gives new information which is complementary to that extracted from absorption spectroscopies.

To the best of our knowledge, in all the cases reported, the $g_{lum}$ factor measured for excimer transitions is higher than $g_{abs}$ by one or two orders of magnitude and typically falls in the range of $10^{-2}$. A possible explanation for such high $g_{lum}$ can be the fact that an excimer is a single, extended intrinsically chiral fluorophore. As a comparison, we note that most of chirally perturbed chromophores/fluorophores or exciton systems display $g_{lum}$ in the $10^{-4}$–$10^{-3}$ range. The n–π∗ transition from carbonyl groups in chiral molecules has typically a $g_{lum}$ factor around $10^{-2}$, as it is magnetically allowed but electrically forbidden, but for the same reason very low quantum yields are observed in these cases [4]. In order to fully compare the overall polarization efficiency, it is convenient to take into account other relevant photophysical parameters beside dissymmetry factor. To this purpose, *circular polarization brightness* ($B_{CP}$) was proposed (Eq. 12.2) in which $\varepsilon_\lambda$ is the extinction coefficient at the excitation wavelength, $\Phi$ is the emission quantum yield, and $B$ is the resulting brightness [6].

$$B_{CP} = \varepsilon_\lambda \cdot \Phi \cdot \frac{|g_{lum}|}{2} = B \cdot \frac{|g_{lum}|}{2} \tag{12.2}$$

Considering, as a prototypical example, the circularly polarized excimer emission from pyrene, it is possible to elaborate the following numbers: $\varepsilon_\lambda$ ~4 × $10^4$ $M^{-1}cm^{-1}$, quantum yield ~0.3, $g_{lum}$ ~$10^{-2}$. These numbers yield a $B_{CP}$ around 60 $M^{-1}cm^{-1}$, which is around one order of magnitude higher than the values typical for most non-aggregated organic systems, and it approaches the figures of some lanthanide chiral complexes [7, 8].

Typical intermolecular excimer formation is a diffusion-controlled process. For this reason, excimer allied CPL is often observed in solid state or in aggregates [9] of non-racemic fluorophoric small molecules or oligomers and polymers. An effective strategy to alleviate this dependence is to allow intramolecular excimer formation by linking two or more fluorophores to a chiral molecular scaffold through suitable chains. In such cases, if the scaffold or the chains connecting the fluorophore units are scalemic, CPL can be observed with the typical $g_{lum}$ factors discussed above.

In this chapter, we will review examples of CPL from intramolecularly formed excimers only, in different molecular systems and different contexts. A focus will be given on systems based on pyrene, perylene, and perylene bisimide and 1,8-naphthalene monoimide (NMI) moieties.

## 12.2 Pyrene-Based Systems

A privileged candidate for excimer formation is pyrene. The little overlap between the main $S_0 \rightarrow S_2$ absorption transition with the emission spectrum minimizes the chance of fluorescence resonance energy transfer (FRET), thus ensuring a low excimer dissociation constant [10]. This often results in an intense excimer emission with the maximum around 480–500 nm. Tailored systems with defined geometries bringing pyrene moieties in close proximity to each other thus allow very intense excimer emission. If the scaffold on which the fluorophores are mounted is scalemic, or more generally the environment surrounding the targeted system, then excimer CPL can be observed depending on the geometry described by the mutual arrangement of the pyrene moieties in the excimer state.

The first observation of excimer CPL on pyrene was carried out by Kano and Sisido's groups in 1985 from a self-assembly of a pyrene dimer in the chiral cavity of a γ-cyclodextrin (γ-CDx) [11]. In this study, the authors were able to measure a strong CPL signal centered around 490 nm with a $g_{lum}$ of $|1.2 \times 10^{-2}|$, thanks to the asymmetrically twisted configuration of the two pyrene molecules, acquired in the excimer state inside the γ-CDx chiral cavity. Interestingly, the dissymmetry factor measured in absorption for the first Cotton effect in the ECD spectrum was $|6 \times 10^{-5}|$, indicating very weak asymmetry of the arrangement of the two pyrenes in the ground state.

Recently, Inouye's group has taken advantage of the same concept [12]. Two substituted pyrenes bearing PEGylated chains to ensure an overall water solubility were included into a γ-CDx. Then, through Sonogashira couplings, 3,5-diaryl substituted phenyls were linked to the pyrenes at their extremities obtaining compound **pyr-1** (Fig. 12.2). By proceeding in this manner, the two pyrenes were locked physically inside the γ-CDx. A clear CPL associated to excimer emission was recorded (Fig. 12.2), with a $g_{lum}$ of $|1.5 \times 10^{-2}|$, similar to that measured in the first experiment by Kano and Sisido (see above) [11].

### 12.2.1 Poly- and Oligopeptides Bearing Pyrene Units

In another context, that of synthetic polypeptides, pyrene CPL excimers were also studied to elucidate the changes of secondary/tertiary structures upon varying conditions, such as temperature and solvent. The chosen strategy was to functionalize different polypeptides with pyrene moieties and to exploit intramolecular formation of excimers to obtain information on the folding in different conditions. Thanks to the chirality of the polypeptidic backbone, excimer CPL could be induced and measured. Importantly, as chiroptical signals are extremely sensitive to the surrounding environment, even minor changes in the secondary structure caused relevant changes in the CPL response.

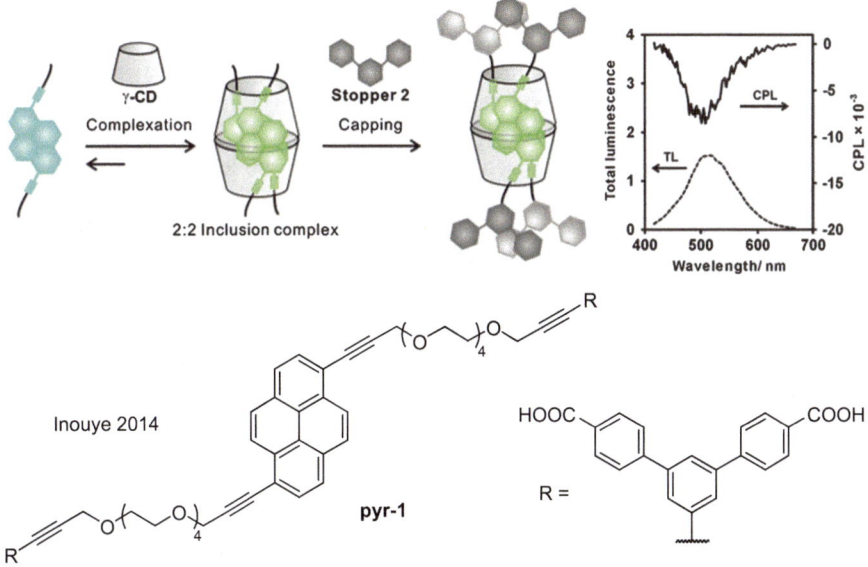

**Fig. 12.2** Top: adopted strategy for inclusion of pyrene dimers in γ-CDx and CPL/total luminescence spectra measured in $H_2O$ (pH 9.5, $C = 4.5 \times 10^{-5}$ M). Bottom: structure of the compound. Adapted with permission from reference [12]

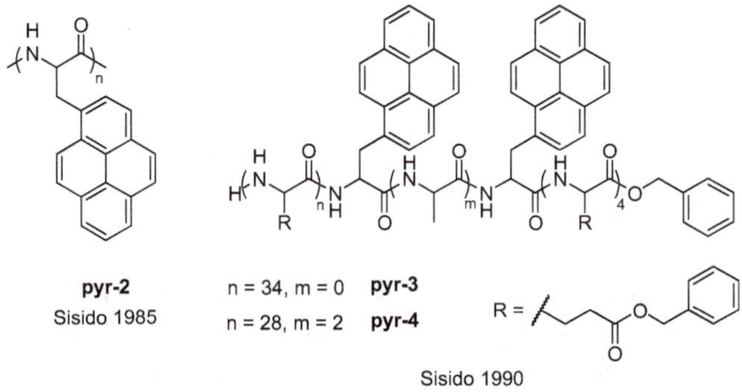

**Scheme 12.1** Structures of pyrene-decorated polypeptides developed by Sisido

This approach was followed for the first time by Sisido et al. in 1985 [13]. They synthesized a poly-pyrenylalanine (**pyr-2**, Scheme 12.1) and studied the CPL in polar solvents, such as dimethylformamide (DMF) at variable temperature $(6 \times 10^{-5}$ M). The $g_{lum}$ vs. wavelength plot showed a sign inversion upon changing the temperature (from 3 to 60 °C) and even a bisignate profile for intermediate temperatures. These features suggest the presence of two sources for the excimer transitions: one stems from an apolar contribution due to exciton resonance (Pyr∗-Pyr ↔ Pyr-Pyr∗), predominant at lower wavelengths, the second one comes from charge transfer resonance (Pyr⁺-Pyr⁻ ↔ Pyr⁻-Pyr⁺), predominant at longer wavelengths and stabilized in more polar solvents. These two components bring about different CPL signs, and the change of their relative contribution to the overall CPL spectrum at different temperatures thus explains the $g_{lum}$ profile variations. In this case $g_{lum}$ and $g_{abs}$ (measured on the first Cotton effect) are of the same order of magnitude ($|10^{-3}|$).

Later, Sisido's group synthesized two similar polypeptides, namely **pyr-3** and **pyr-4** (Scheme 12.1) [14]. In these examples, high $g_{lum}$ values of $|1.6 \times 10^{-2}|$ were detected associated with an excimer situation; no CPL being otherwise allied with monomer emission. The sign of the CPL was related to the handedness of the helix, which in turn depended on the length of the peptidic spacer: negative for **pyr-3** and positive for **pyr-4**. In these cases, the handedness is the same in both ground and excimer states, as indicated by the sign of the ECD band at 349 nm (negative for **pyr-3** and positive for **pyr-4**). For **pyr-3** and **pyr-4**, the ECD and CPL were also studied in DMF ($6 \times 10^{-5}$ M) and tetrahydrofuran (THF) at +20 and −40 °C [15]. As a first observation, the authors noticed that the $g_{lum}$ values were constant over the whole emission band, which is consistent with only a single excimer configuration being present in each polypeptide. As a second observation, the signs of $g_{lum}$ were opposite for **pyr-3** and **pyr-4** suggesting configurations with opposite screw sense in the two cases. For **pyr-3** specifically, a decrease of $g_{lum}$ value, from $|1.2 \times 10^{-2}|$ to $|0.4 \times 10^{-2}|$, was observed in DMF upon lowering the temperature from +20 to −40 °C; a similar behavior occurring in THF although less pronounced. This suggests that a major conformational change occurs in the −40 to +20 °C range.

More recently, other pyrene-decorated peptides bearing 1, 2, 3, or 4 pyrene moieties, **pyr-5**–**pyr-8**, were prepared and investigated by the group of Imai (Fig. 12.3) [17]. In CHCl$_3$ ($10^{-4}$ M), a low $g_{lum}$ value of $|1.9 \times 10^{-4}|$ was observed for **pyr-5** associated with the monomer fluorescence, while the dipeptide **pyr-6** showed the highest dissymmetry factor, $|1.1 \times 10^{-2}|$, allied with excimer emission. This latter $g_{lum}$ value is 240-fold higher than the corresponding $g_{abs}$ ($\sim|10^{-5}|$), while in the former case, in which no excimer occurs, the $g_{abs}$ and $g_{lum}$ are of the same order of magnitude.

Furthermore, Imai and collaborators prepared several other peptides bearing two pyrene moieties spaced by chains with different number ($n$) of methylene spacers, allowing the authors to study the relationship between chiroptical properties and spacer lengths (**pyr-9**–**pyr-16**, Fig. 12.3) [16]. The sign of CPL associated with excimer emission underwent a sign inversion going from $n = 1$ to $n = 2$ (**pyr-9** and **pyr-10**), showing a sort of odd–even effect, and from $n = 6$ to $n = 7$ (**pyr-14** and **pyr-15**, see Fig. 12.3). On the other hand, the first Cotton effect, as shown in the ECD spectrum, did not show any sign inversion, revealing a completely different geometry of the ground and excited excimer states, confirming again the complementary nature of ECD and CPL when excimer states are at play. The maximum $g_{lum}$ was recorded for $n = 3$ (**pyr-11**, $g_{lum} = 0.8 \times 10^{-2}$). In all instances, $g_{lum}$ values were 1 to 2 orders of magnitude higher than the corresponding $g_{abs}$ factors ($|10^{-3}|$–$|10^{-4}|$). With **pyr-12** ($n = 4$), CPL sign inversion was also observed going from relatively apolar chlorinated solvents, such as CHCl$_3$/CH$_2$Cl$_2$, to polar ones

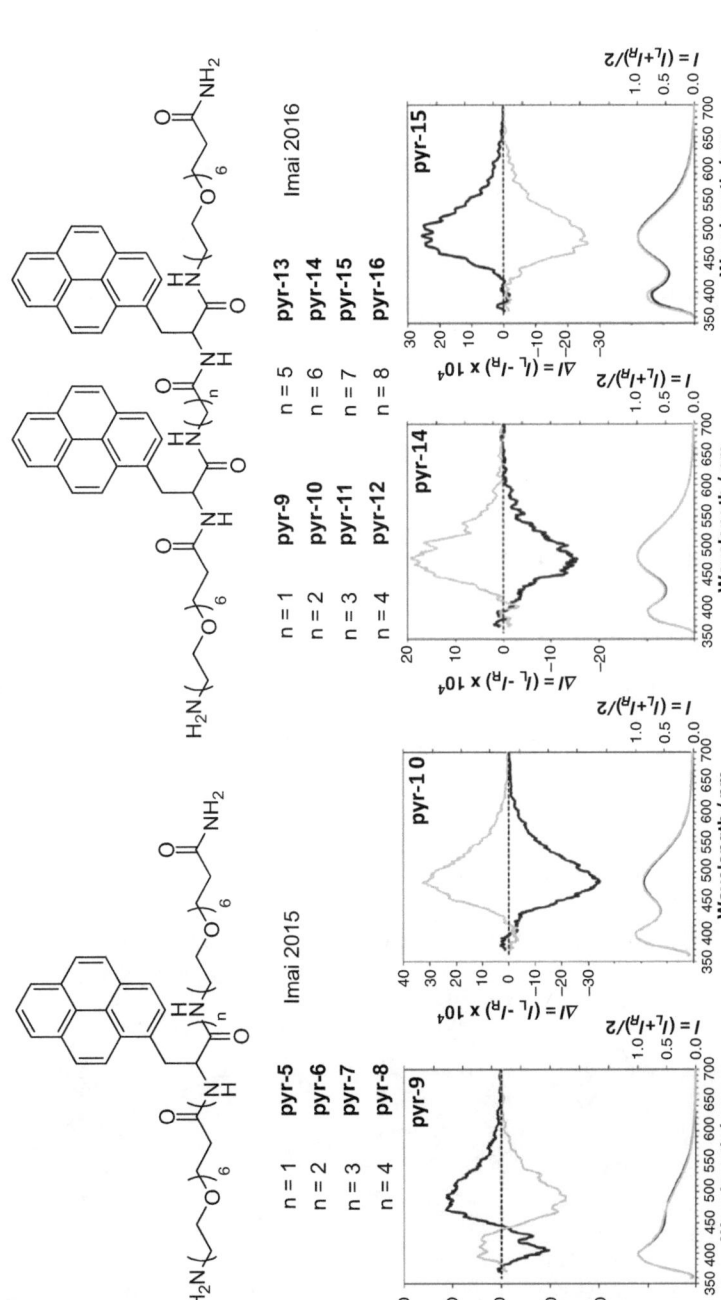

**Fig. 12.3** Top: structures of pyrene-decorated polypeptides developed by Imai's group. Bottom: selected examples of CPL and fluorescence (D-isomers in black, L-isomers in gray) measured in CHCl$_3$ ($10^{-4}$ M). Adapted with permission from reference [16]

[Methanol    (MeOH)/DMF/N-methyl-2-pyrrolidone    (NMP)/dimethylacetamide (DMAc)], while no major changes occurred in the ECD spectra [18].

## 12.2.2   Ground State Cryptochirality and Excimer CPL

As the polarization effects on excimer bands can be higher than the ones in absorption by orders of magnitude, it is possible in some cases to specifically detect a CPL band even when the ECD is extremely small or non-detectable. This behavior is clearly observed when the chirality is brought by stereogenic elements (centers or axes) positioned at a distance from the chromophore/fluorophore units. The ground-state of such systems can be then considered as "cryptochiral," for which the chirality cannot be (fully) detected through absorption-based chiroptical techniques, but the molecular asymmetry becomes evident upon emission polarization. In principle, this peculiarity could open a door to applications in cryptography and steganography [19].

Imai et al. prepared **pyr-17** bearing two pyrene units linked to a core binaphthyl moiety through ethoxyethyl chains and ester linkages (Scheme 12.2) [20]. When dissolved in $CHCl_3$ ($10^{-3}$ M), this molecule displayed a $g_{abs}$ of $|3.8 \times 10^{-5}|$ at 344 nm while the excimer associated $g_{lum}$ value was $|7.8 \times 10^{-4}|$ (20.5 times higher). The excimer CPL, in comparison with ECD, displayed thus a quite higher sensitivity in revealing a long-distance chiral environment.

A similar behavior was observed for **pyr-18–pyr-20** for which the two pyrene moieties were linked to the central binaphthyl unit through amide, ether, and ester bonds, respectively (Scheme 12.2) [19, 21]. In these examples, the measured $g_{abs}$ values were around $10^{-5}$, while the excimer associated $g_{lum}$ factors were around $10^{-3}$ ($C = 10^{-5}$ M in $CHCl_3$), with $g_{lum}/g_{abs}$ ratios ranging from 17.9 to 62.5. Density functional theory (DFT) structures showed that the two pyrene moieties assume a nearly achiral T-shaped arrangement in $S_0$ state, while in the excited state

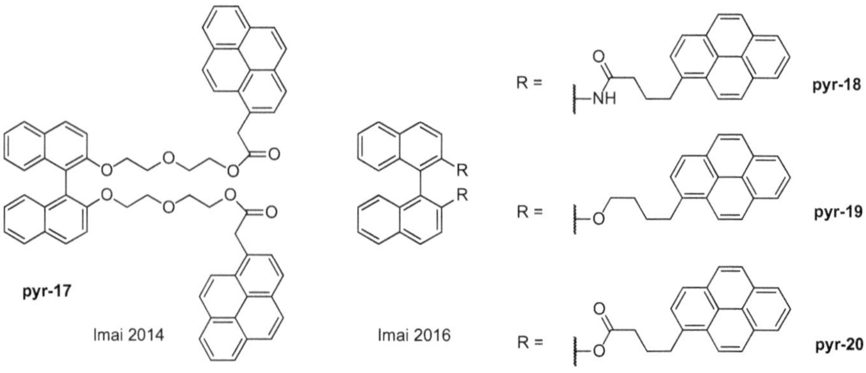

**Scheme 12.2** Structures of pyrene decorated binaphthyl systems

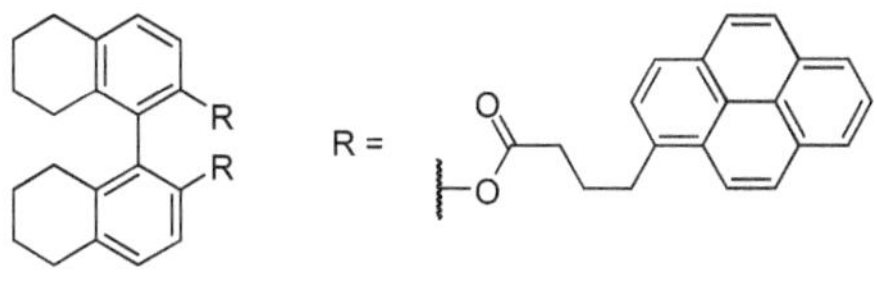

**pyr-21**

Imai 2018

**Scheme 12.3**  Structure of a pyrene-decorated octahydro-binaphthyl system

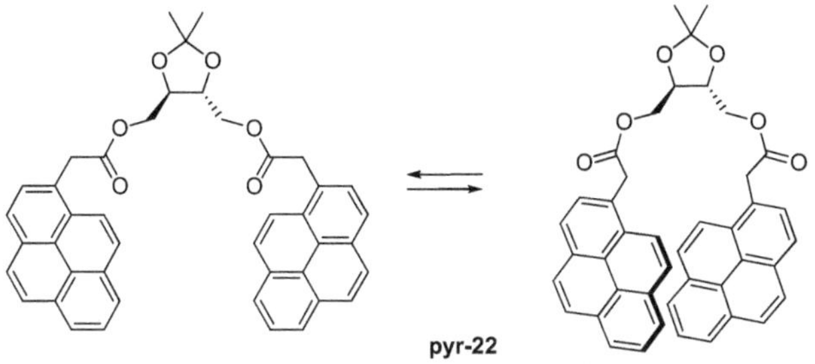

**pyr-22**

ground state            Imai 2015            excited state

**Scheme 12.4**  Structure of a pyrene-decorated dioxolane system

they acquire a π-stacked skewed geometry. The same properties are retained as well by partially hydrogenated compound **pyr-21** (Scheme 12.3), analogous to compound **pyr-20** where the binaphthyl moiety is replaced by an octahydro-binaphthyl moiety [22].

Imai's group developed a dioxolane, bearing two stereogenic centers, from which two pyrene moieties are connected through two $-CH_2OC(O)CH_2-$ ester chains (**pyr-22**, Scheme 12.4) [23]. The two pyrenes are thus located at the extremities of two floppy arms that dispose from a rather large conformational freedom. As such, π–π interactions between pyrene units could not be detected in the ground state; the aromatic moieties assuming overall a T-shaped arrangement, as shown by DFT calculations and hence non-detectable ECD. On the other hand, a weak but observable excimer CPL band was measured with a $g_{lum} = |8.9 \times 10^{-4}|$ ($C = 10^{-4}$ M in $CHCl_3$), demonstrating again that a chiroptically active geometry, bound by π–π stacking of the aromatic subunits, is obtainable in the excited state.

## 12.2.3  Excimer CPL in Chiroptical Switches

Thanks to the exceptional CPL activity associated with pyrene excimers, researchers have designed molecular systems that modulate their CPL response upon the occurrence of an external stimulus, such as light or the addition of ions. These systems are examples of so-called chiroptical switches.

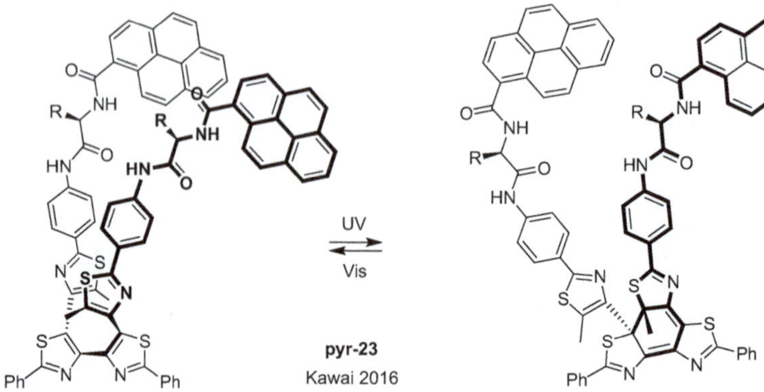

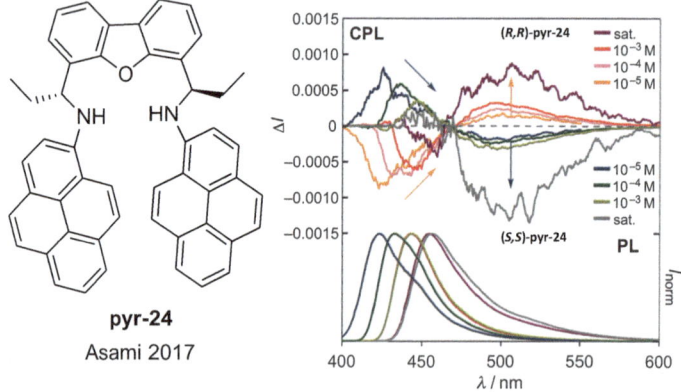

**Scheme 12.5** Tetrathiazole-based chiroptical switch reported by Kawai et al.

**Fig. 12.4** Concentration responsive pyrene-dibenzofuran molecule and CPL and total fluorescence as a function of **pyr-24** concentration in toluene. Adapted from reference [25] with permission from The Royal Society of Chemistry

Kawai's group reported a tetrathiazole-based compound which undergoes ring-opening/closure through a photoinduced reaction promoted by ultraviolet (UV) (for ring closure) or visible (for ring opening) light (**pyr-23**, Scheme 12.5) [24]. This transformation is accompanied by a large structural change. In the ring-open form, the two pyrene moieties are in close proximity giving rise to excimer-associated CPL with $g_{lum}$ around $10^{-2}$ ($C = 1.7 \times 10^{-4}$ M in CHCl$_3$). Upon photocyclization and the rigidification of the core three-dimensional (3D) structure, the two pyrene units are brought far apart, which results in a quenching of excimer CPL. Such light promoted CPL switching can be repeated over several cycles.

A molecule displaying concentration dependent CPL was developed by Asami and coworkers: [25] a dibenzofuran molecule linked to two pyrene moieties through a chiral chain (**pyr-24**, Fig. 12.4). Such compound shows a bisignate CPL profile, where monomer and excimer allied CPL have opposite signs (Fig. 12.4). Such sign

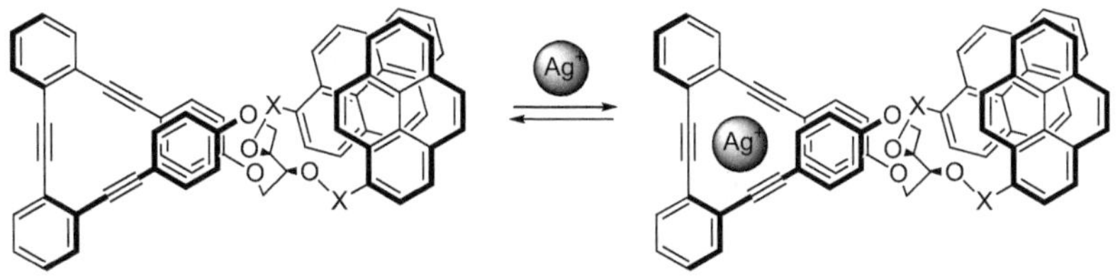

Cuerva 2018

X = CH$_2$    **pyr-25**
X = CO      **pyr-26**

**Scheme 12.6** Pyrene-decorated ortho-oligo(phenylene)-ethynylenes showing Ag-induced chiroptical switch behavior

difference allows one to disentangle monomer and excimer contributions from the overall CPL spectrum ($g_{lum}$ ~|10$^{-4}$| for the monomer and ~|10$^{-3}$| for the excimer). On the contrary, these two contributions are not clearly separable in the fluorescence spectra where the excimer is present as a broad tailing of the main monomer band (Fig. 12.4). Upon decreasing the concentration of **pyr-24** (from 8 × 10$^{-3}$ M to 1 × 10$^{-5}$ M in toluene, see Fig. 12.4), an increase in the monomer CPL $g_{lum}$ was observed (from |3.3 × 10$^{-4}$| to |6.9 × 10$^{-4}$|), while excimer allied $g_{lum}$ showed only minor variations (from |3.3 × 10$^{-3}$| to |3.9 × 10$^{-3}$|). This effect is probably due to different contributions to total luminescence and CPL from intra- and intermolecular excimer at different concentrations. On the other hand, ECD spectra at various concentrations did not show any variation of shape or intensity ($g_{abs}$ = | 6.2 × 10$^{-4}$|), indicating that the ground state remains unaffected.

Recently, Cuerva et al. reported *ortho*-oligo(phenylene)-ethynylenes [26] decorated with two pyrene moieties (**pyr-25** and **pyr-26**, Scheme 12.6) [27]. These systems displayed again a bisignated CPL spectra ($C = 10^{-5}$ M in CH$_2$Cl$_2$/acetone 95:5): the spectrum was dominated by the CPL emission associated to both phenylene-ethynylene moiety and to pyrene monomer at shorter wavelengths (~400 nm), while longer wavelength region (~500 nm) showed CPL allied to pyrene excimer emission with $g_{lum}$ ~|10$^{-2}$|. Upon Ag$^+$ addition, the CPL profile is modified and the relative intensities of shorter and longer wavelength components change. In this way it was possible to quantify Ag$^+$ concentration, thanks to the determination of the ratio between $\Delta I$ ($I_L - I_R$) at 400 nm and $\Delta I$ at 500 nm. A CPL-based ratiometric probe sensitive to Ag$^+$ was thus established. Moreover, considering the ECD of **pyr-26**, the authors noticed that, upon Ag$^+$ addition, only the band related to oligo (phenylene)-ethynylene scaffold was affected (345 nm), while the bands related to pyrene absorption (387 nm) remained unaffected.

A different strategy was followed by the Lacour's group [28]. Thanks to a versatile synthetic approach in two steps only from common precursors, a family of polyether and polyether-like macrocycles was prepared carrying a wealth of different fluorophores (Scheme 12.7) [29]. In particular, 18C6, 18C4 and 16C4-type scaffolds were decorated with pyrene units through amide bonds (**pyr-27–pyr-29**, Scheme 12.7). These molecules displayed very intense excimer emission

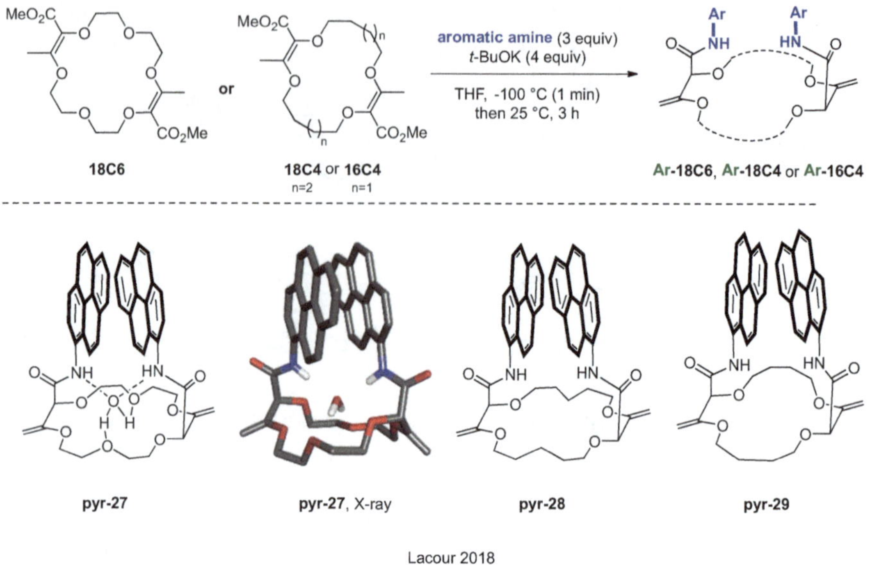

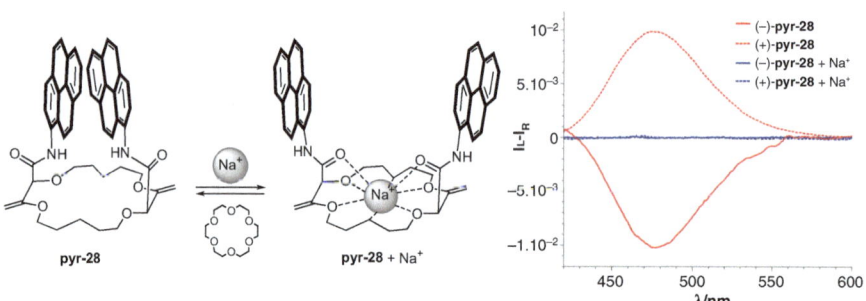

Lacour 2018

**Scheme 12.7** Straightforward synthesis of chiral functionalized macrocycles and structures of pyrene-functionalized macrocycles

**Fig. 12.5** Reversible complexation of sodium cations by **pyr-28** and associated CPL spectra recorded in $CH_2Cl_2$ ($10^{-5}$ M). Adapted from reference [28]. Published by The Royal Society of Chemistry

(compared to monomer band) with an associated $g_{lum}$ in the range of $|10^{-2}|$. The intensity of the excimer fluorescence is the result of the peculiar geometry of these systems, which brings the pyrene units in close spatial proximity (see the X-ray structure in Scheme 12.7). Moreover, upon cation addition (e.g., $Ba^{2+}$ or $Na^+$ in acetonitrile or $CH_2Cl_2$, respectively), profound conformational changes occur. In particular, the carbonyl groups of the amides turn inward and the two pyrene moieties part away from each other. It results in a loss of excimer fluorescence and in a total CPL quenching (Fig. 12.5). If the cation is removed, excimer CPL is fully recovered. In the ECD spectra instead, a signature inversion of most bands is visible upon cation addition. In contrast to the $g_{lum}$ factors, the $g_{abs}$ values measured on the first Cotton effect are in the range $|10^{-3}|–|10^{-4}|$.

### 12.2.4 Solid State Excimer CPL from Pyrene-Based Molecules

Some pyrene-based molecules display CPL activity even in solid state, both in bulk (as powders) or in polymeric matrix-based films. In these cases, a contribution of not only intra- but also intermolecular excimers may be present. For example, the already cited compound **pyr-17** displays exciter associated CPL in polymethyl methacrylate matrix ($g_{lum} = |3.6 \times 10^{-4}|$), even though with a $g_{lum}$ factor around half the one in solution and with sign inverted [20].

Intense excimer CPL was recently reported by Takaishi and Ema in quaternaphthyl structures bearing 6 or 8 pyrene moieties (**pyr-30** and **pyr-31**, Fig. 12.6) [30]. They showed that at least 6 pyrene units were necessary to achieve a rigid helical structure with skewed-arranged pyrenes (see DFT optimized structure in Fig. 12.6). In fact, for **pyr-30** and **pyr-31**, $g_{lum}$ values equal to $|3.7 \times 10^{-2}|$ and $|3.4 \times 10^{-2}|$ were recorded in $CH_2Cl_2$ ($10^{-6}–10^{-5}$ M). One order of magnitude lower $g_{abs}$ values was recorded for the first Cotton effect ($g_{abs} = |4.8 \times 10^{-3}|$ for **pyr-31**) in the ECD spectrum. Importantly, weaker but still relatively intense CPL emissions, $g_{lum}$ $|5.3 \times 10^{-3}|$ and $|5.6 \times 10^{-3}|$, respectively, were detected for the same molecules in solid state, as potassium bromide (KBr) pellet. It is worth mentioning that in the present example, solid state recorded $g_{lum}$ values are 6–7 times lower than in solution.

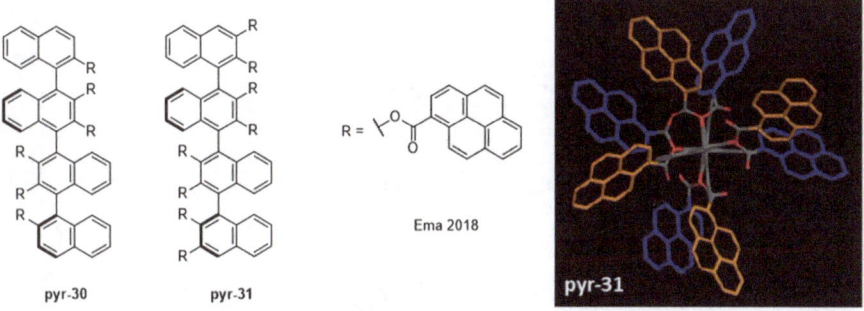

**Fig. 12.6** Pyrene-decorated quaternaphthyl structures reported by Takaishi and Ema, and DFT optimized structure of **pyr-31**. Reproduced from reference [30] with permission from The Royal Society of Chemistry

## 12.3   Perylene and Perylene Bisimide-Based Systems

Perylene and derivatives, such as perylene bisimides, are very well-established fluorophores, thanks to versatile access routes, very high quantum yields with emission shifted to long wavelengths. Traditionally, in these systems, the excimer formation is less efficient than with pyrenes. It is usually observed as a broad tailing along with the very intense and vibronically structured monomer-centered emission. Despite this limitation, some studies have been reported for which relatively intense excimer CPL signals have been measured, in particular in the case of designed intramolecularly-formed excimers.

### 12.3.1   Excimer CPL from Perylene-Based Systems

**Pery-1**, prepared in an analogous manner to **pyr-1** (vide supra) [11], displays circularly polarized excimer fluorescence from perylene upon inclusion into a γ-CDx and subsequent capping via Sonogashira reactions (Fig. 12.7) [31]. For such a system, CPL with a $g_{lum}$ of $|2.1 \times 10^{-2}|$ was recorded (Fig. 12.7). This value is comparable to that measured for **pyr-1** in similar conditions [11].

Following the same synthetic strategy exploited for compounds **pyr-27–pyr-29**, Lacour and collaborators prepared diperylene derivatives **pery-2–pery-4** (Scheme 12.8) [28]. These systems displayed excimer fluorescence either as a principal component of the fluorescence spectrum (**pery-2** and **pery-3**) or as a tailing of the very intense monomer emission (**pery-4**). In the CPL spectra, two components are clearly visible; one associated to excimer emission with $g_{lum}$ ca. $|10^{-3}|$ and one allied to monomer fluorescence with a $g_{lum}$ value, ca. $|10^{-4}|$, one order of magnitude lower. These two components have the same sign with **pery-2** and **pery-3**, but opposite signs for **pery-4**. In all these cases, CPL allows for a clear distinction between monomer and excimer contributions which is not obvious in fluorescence spectra. Upon $Ba^{2+}$ (in $CH_3CN$ for **pery-2**) or $Na^+$ (in $CH_2Cl_2$ for **pery-2–pery-4**) additions,

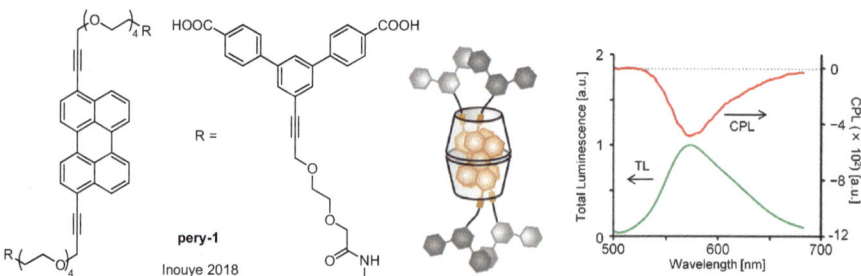

**Fig. 12.7** Structure of a substituted perylene included in γ-CDx and CPL/total luminescence spectra measured in $H_2O$ (pH 9.5, $C = 4.5 \times 10^{-5}$ M). Adapted with permission from reference [31]

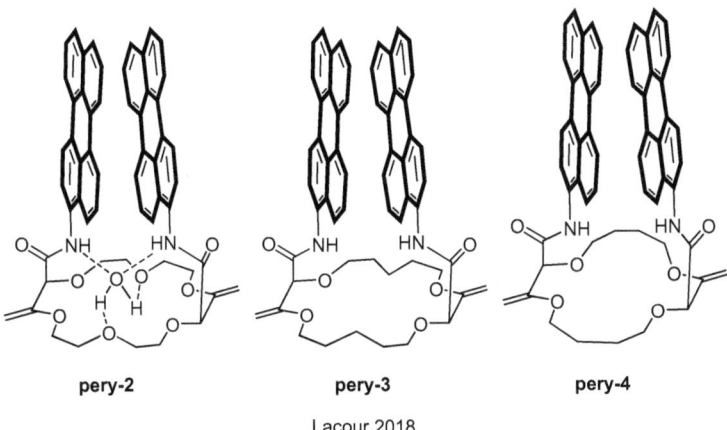

pery-2  pery-3  pery-4

Lacour 2018

**Scheme 12.8** Structures of perylene-functionalized macrocycles

only monomer emission and its allied CPL remain, while excimer fluorescence and CPL are quenched in all cases. ECD spectra instead showed an exciton couplet structures in the visible range, which is slightly blue-shifted upon cation addition.

## 12.3.2 Excimer CPL from Perylene Bisimide-Based Systems

Kawai's group reported 1,2-diaminocyclohexane-based molecules linked to two perylene bisimide moieties through a glycine linker (**pery-5**) or a β-alanine spacer (**pery-6**, Scheme 12.9) [32]. A clear odd–even effect depending on the length of linker is visible in the ECD profile which shows exciton couplets featuring different signs in the two cases. In the case of **pery-5**, in CHCl$_3$ as the solvent ($C = 10^{-5}$ M), CPL profile showed a single band around 540 nm with the same sign as the first Cotton effect in the ECD spectrum with similar $g_{lum}$ and $g_{abs}$ values ($g_{lum} \sim |6 \times 10^{-4}|$ and $g_{abs} \sim |4 \times 10^{-4}|$). This indicates that such contribution stems from the weakly coupled lowest excitonic state. Compound **pery-6** showed the same CPL band. However, it displayed in addition an intense and broad band with opposite sign around 630 nm originated from the excimer state. As expected, this band displayed a much higher $g_{lum}$ ($|8 \times 10^{-3}|$) than that associated to the excitonic state emission.

On the contrary, when perylene bisimide moieties were mounted directly on a binaphthyl scaffold (**pery-7**, Scheme 12.10) [33], only an emission from the lowest excitonic state was observed in toluene, associated with a $g_{lum}$ value of $|2 \times 10^{-3}|$–$|3 \times 10^{-3}|$ around 550 nm. No significant contribution from the excimer was visible, indicating that such structure does not allow for sufficient conformational freedom to obtain an effective overlap between the two fluorophores.

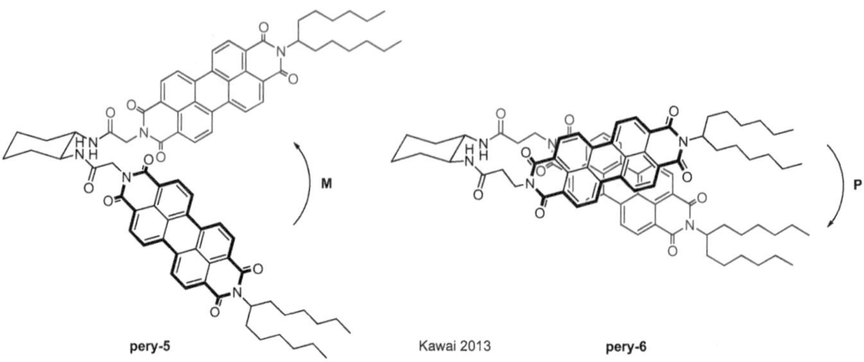

**Scheme 12.9** 1,2-Diaminocyclohexane/perylene bisimide adducts reported by Kawai et al.

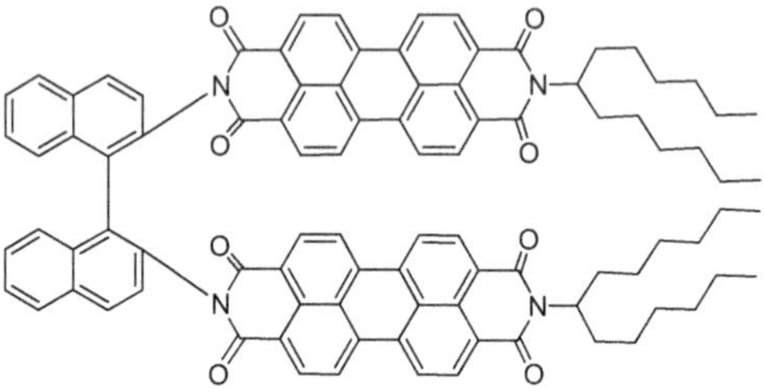

**Scheme 12.10** Binaphthyl/perylene bisimide system

## 12.4 1,8-Naphthalene Monoimide

Another candidate for CPL excimer observation is the 1,8-naphthalene monoimide core, abbreviated NMI, which usually displays emission maxima in the 450–500 nm range. There are fewer reported examples of inter- and intramolecular excimer CPL with NMI and related structures, such as naphthalene bisimide [34]. Below, examples of NMI-based molecules that are able to display CPL allied to intramolecular excimer situations are described.

As in many of the above examples, it was possible to benefit from a central binaphthyl core and connect NMI fluorophores. Cheng et al. decorated a binaphthyl with two naphthalene monoimides (**NMI-1** and **NMI-2**, Fig. 12.8), linked through chains bearing either 2 or 4 methylenes [35]. Intramolecular excimer allied CPL was observed in THF solutions ($C = 10^{-5}$ M) with $g_{lum}$ of $|5.5 \times 10^{-3}|$–$|6.1 \times 10^{-3}|$ for **NMI-1** and $|3.0 \times 10^{-3}|$–$|4.1 \times 10^{-3}|$ for **NMI-2**. Upon addition of water, the molecules undergo aggregation. Inverted CPL with possible contributions from both inter- and intramolecular excimers and lower $g_{lum}$ factors were observed. Similarly, when two NMI moieties were directly linked to a 1,2-diaminocyclohexane scaffold to obtain **NMI-3** and **NMI-4** (Fig. 12.8), the resulting derivatives showed excimer CPL in THF ($C = 10^{-5}$ M) with $g_{lum}$ values in the order of $|10^{-2}|$ [36]. Again, in

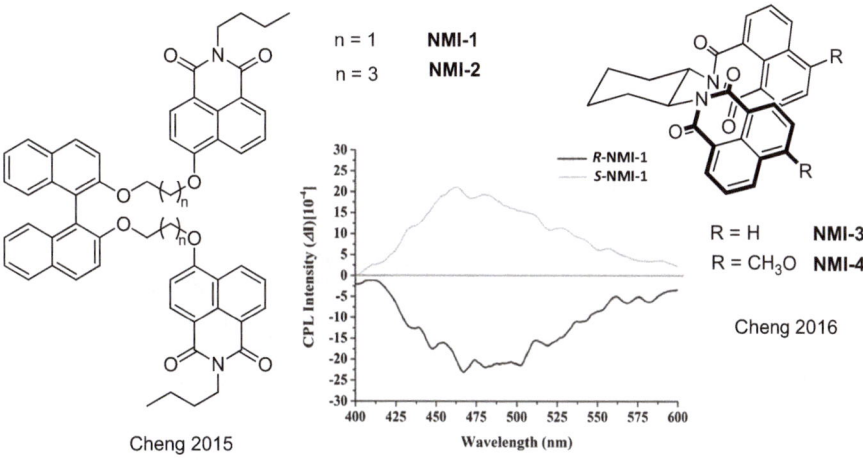

**Fig. 12.8** 1,8-Naphthalene monoimide derivatives developed by Cheng et al. CPL spectra of NMI-1 (THF, $C = 10^{-5}$ M). Adapted with permission from reference [35]

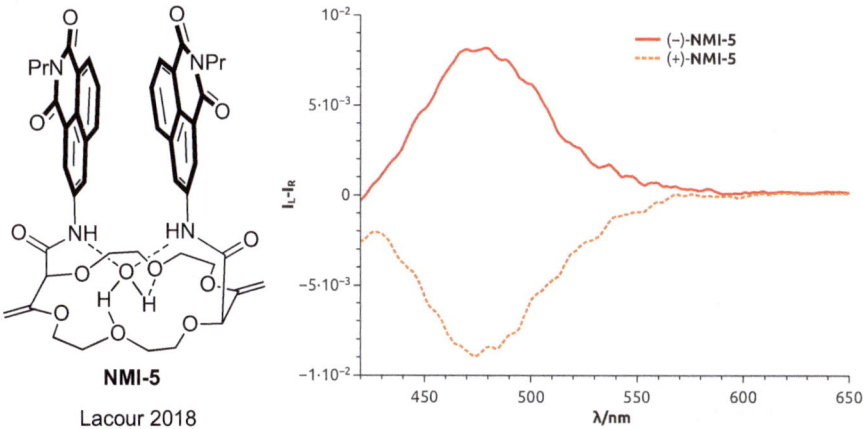

**Fig. 12.9** NMI-functionalized 18C6-macrocycle and CPL spectra recorded in CH$_3$CN, $C = 10^{-5}$ M. Adapted from reference [28]. Published by The Royal Society of Chemistry

THF/water mixtures where aggregation takes place, $g_{lum}$ factors remain of the same order of magnitude.

Finally, the Lacour's group was able to prepare **NMI-5** and observe excimer CPL with $g_{lum} \sim |8 \times 10^{-3}|$ (Fig. 12.9), to be compared with a $g_{abs} \sim |2.6 \times 10^{-4}|$. In this compound, the donor group on the NMI moieties is attached at position 3 of the naphthalene subunit (and not 4 as it is usually the case), and this heteroatom serves as a linker to the 18C6-scaffold [28].

## 12.5 Conclusions

Chiral molecules bearing two or more fluorophores able to interact in the excited state and generate an emitting excimer induce fluorescence often with high quantum yield allied with highly polarized emission. Through this strategy, it is thus possible to maximize polarization output not only in terms of dissymmetry factor, but also in terms of circular polarization brightness [6], to an extent which is generally unsurpassed by most single organic molecule (i.e., non-aggregated) emitters. This field will benefit from further experimental efforts to produce and investigate new compounds. Moreover, computational analysis of excimer allied CPL could provide in the future better insight into the mechanism behind such phenomenon and help to design tailored systems. Unfortunately, to the best of our knowledge, no computational work has been reported so far on the topic. A possible difficulty lies in the optimization of excimer geometry, considering that such geometry exists only in the excited state, therefore no direct experimental structure is available for a straightforward check of the calculated one. In such cases, in fact, a correct geometry is necessary to reproduce emission [37, 38], and even more CPL [39], in a trustable way.

## References

1. IUPAC, Compendium of chemical terminology, 2nd ed. the "Gold Book"
2. Birks J (1975) Excimers. Rep Prog Phys 38:903
3. Pucci A, Ruggeri G (2011) Mechanochromic polymer blends. J Mater Chem 21:8282–8291
4. Sánchez-Carnerero EM, Agarrabeitia AR, Moreno F, Maroto BL, Muller G, Ortiz MJ, de la Moya S (2015) Circularly polarized luminescence from simple organic molecules. Chemistry 21:13488–13500
5. Tanaka H, Inoue Y, Mori T (2018) Circularly polarized luminescence and circular dichroisms in small organic molecules: correlation between excitation and emission dissymmetry factors. ChemPhotoChem 2:386–402
6. Zinna F, Di Bari L (2018) Emerging field of chiral Ln(III) complexes for OLEDs. In: Martín-Ramos P, Ramos Silva M (eds) Lanthanide-based multifunctional materials. Elsevier, Amsterdam, pp 171–194
7. Lunkley JL, Shirotani D, Yamanari K, Kaizaki S, Muller G (2011) Chiroptical spectra of a series of tetrakis ((+)-3-heptafluorobutylyrylcamphorato) lanthanide (III) with an encapsulated alkali metal ion: circularly polarized luminescence and absolute chiral structures for the Eu (III) and Sm (III) complexes. Inorg Chem 50:12724–12732
8. Zinna F, Di Bari L (2015) Lanthanide circularly polarized luminescence: bases and applications. Chirality 27:1–13
9. Kumar J, Nakashima T, Kawai T (2015) Circularly polarized luminescence in chiral molecules and supramolecular assemblies. J Phys Chem Lett 6:3445–3452
10. Duhamel J (2012) New insights in the study of pyrene excimer fluorescence to characterize macromolecules and their supramolecular assemblies in solution. Langmuir 28:6527–6538
11. Kano K, Matsumoto H, Hashimoto S, Sisido M, Imanishi Y (1985) A chiral pyrene excimer in γ-cyclodextrin cavity. J Am Chem Soc 107:6117–6118

12. Inouye M et al (2014) A doubly alkynylpyrene-threaded [4] rotaxane that exhibits strong circularly polarized luminescence from the spatially restricted excimer. Angew Chem 126:14620–14624
13. Egusa S, Sisido M, Imanishi Y (1985) One-dimensional aromatic crystals in solution. 4. Ground-and excited-state interactions of poly (L-1-pyrenylalanine) studied by chiroptical spectroscopy including circularly polarized fluorescence and fluorescence-detected circular dichroism. Macromolecules 18:882–889
14. Inai Y, Sisido M, Imanishi Y (1990) Strong circular polarization in the excimer emission from a pair of pyrenyl groups linked to a polypeptide chain. J Phys Chem 94:2734–2735
15. Inai Y, Sisido M, Imanishi Y (1990) Excimer formation on a polypeptide carrying two pyrenyl groups in the middle of an α-helical main chain. J Phys Chem 94:8365–8370
16. Nishikawa T, Kitamura S, Kitamatsu M, Fujiki M, Imai Y (2016) Peptide magic: interdistance-sensitive sign inversion of excimer circularly polarized luminescence in bipyrenyl oligopeptides. ChemistrySelect 1:831–835
17. Nishikawa T, Tajima N, Kitamatsu M, Fujiki M, Imai Y (2015) Circularly polarised luminescence and circular dichroism of L-and D-oligopeptides with multiple pyrenes. Org Biomol Chem 13:11426–11431
18. Mimura Y, Kitamura S, Shizuma M, Kitamatsu M, Fujiki M, Imai Y (2017) Solvent-sensitive sign inversion of excimer origin circularly polarized luminescence in bipyrenyl peptides. ChemistrySelect 2:7759–7764
19. Nakanishi S, Nakabayashi K, Mizusawa T, Suzuki N, Guo S, Fujiki M, Imai Y (2016) Cryptochiral binaphthyl–bipyrene luminophores linked with alkylene esters: intense circularly polarised luminescence, but ultraweak circular dichroism. RSC Adv 6:99172–99176
20. Nakabayashi K, Amako T, Tajima N, Fujiki M, Imai Y (2014) Nonclassical dual control of circularly polarized luminescence modes of binaphthyl–pyrene organic fluorophores in fluidic and glassy media. Chem Commun 50:13228–13230
21. Nakabayashi K, Kitamura S, Suzuki N, Guo S, Fujiki M, Imai Y (2016) Non-classically controlled signs in a circularly polarised luminescent molecular puppet: the importance of the wire structure connecting binaphthyl and two Pyrenes. Eur J Org Chem 2016:64–69
22. Hara N, Yanai M, Kaji D, Shizuma M, Tajima N, Fujiki M, Imai Y (2018) A pivotal biaryl rotamer bearing two floppy pyrenes that exhibits cryptochiral characteristics in the ground state. ChemistrySelect 3:9970–9973
23. Amako T et al (2015) Pyrene magic: chiroptical enciphering and deciphering 1, 3-dioxolane bearing two wirepullings to drive two remote pyrenes. Chem Commun 51:8237–8240
24. Hashimoto Y, Nakashima T, Shimizu D, Kawai T (2016) Photoswitching of an intramolecular chiral stack in a helical tetrathiazole. Chem Commun 52:5171–5174
25. Ito S, Ikeda K, Nakanishi S, Imai Y, Asami M (2017) Concentration-dependent circularly polarized luminescence (CPL) of chiral N, N′-dipyrenyldiamines: sign-inverted CPL switching between monomer and excimer regions under retention of the monomer emission for photoluminescence. Chem Commun 53:6323–6326
26. Morcillo SP et al (2016) Stapled helical o-OPE foldamers as new circularly polarized lumines-cence emitters based on carbophilic interactions with Ag(i)-sensitivity. Chem Sci 7:5663–5670
27. Reiné P et al (2018) Pyrene-containing ortho-oligo (phenylene) ethynylene foldamer as a ratiometric probe based on circularly polarized luminescence. J Org Chem 83:4455–4463
28. Homberg A et al (2018) Combined reversible switching of ECD and quenching of CPL with chiral fluorescent macrocycles. Chem Sci 9:7043–7052
29. Vishe M, Hrdina R, Poblador-Bahamonde AI, Besnard C, Guénée L, Bürgi T, Lacour J (2015) Remote stereoselective deconjugation of α, β-unsaturated esters by simple amidation reactions. Chem Sci 6:4923–4928
30. Takaishi K, Takehana R, Ema T (2018) Intense excimer CPL of pyrenes linked to a quaternaphthyl. Chem Commun 54:1449–1452

31. Hayashi K, Miyaoka Y, Ohishi Y, Ta U, Iwamura M, Nozaki K, Inouye M (2018) Observation of circularly polarized luminescence of the excimer from two perylene cores in the form of [4] rotaxane. Chemistry 24:14613–14616
32. Kumar J, Nakashima T, Tsumatori H, Mori M, Naito M, Kawai T (2013) Circularly polarized luminescence in supramolecular assemblies of chiral bichromophoric perylene bisimides. Chem Eur J 19:14090–14097
33. Kawai T, Kawamura K, Tsumatori H, Ishikawa M, Naito M, Fujiki M, Nakashima T (2007) Circularly polarized luminescence of a fluorescent chiral binaphtylene–perylenebiscarboxydiimide dimer. ChemPhysChem 8:1465–1468
34. Salerno F, Berrocal JA, Haedler AT, Zinna F, Meijer E, Di Bari L (2017) Highly circularly polarized broad-band emission from chiral naphthalene diimide-based supramolecular aggregates. J Mater Chem C 5:3609–3615
35. Sheng Y, Shen D, Zhang W, Zhang H, Zhu C, Cheng Y (2015) Reversal circularly polarized luminescence of AIE-active chiral binaphthyl molecules from solution to aggregation. Chem Eur J 21:13196–13200
36. Sheng Y, Ma J, Liu S, Wang Y, Zhu C, Cheng Y (2016) Strong and reversible circularly polarized luminescence emission of a chiral 1, 8-naphthalimide fluorophore induced by excimer emission and orderly aggregation. Chem Eur J 22:9519–9522
37. Hestand NJ, Spano FC (2017) Molecular aggregate photophysics beyond the kasha model: novel design principles for organic materials. Acc Chem Res 50:341–350
38. Plasser F, Lischka H (2012) Analysis of excitonic and charge transfer interactions from quantum chemical calculations. J Chem Theory Comput 8:2777–2789
39. Zinna F, Bruhn T, Guido CA, Ahrens J, Bröring M, Di Bari L, Pescitelli G (2016) Circularly polarized luminescence from axially chiral BODIPY DYEmers: an experimental and computational study. Chem Eur J 22:16089–16098

# Chapter 13
# Design of Circularly Polarized Thermally Activated Delayed Fluorescence Emitters

Gregory Pieters and Lucas Frederic

**Abstract** This chapter focuses on the molecular designs of Small Organic Molecules (SOM) merging Circularly Polarized Luminescence (CPL) and Thermally Activated Delayed Fluorescence (TADF) properties. In Introduction, the benefits associated with the combination of these properties into SOM for their application as emitting materials to construct Circularly Polarized Organic Light Emitting Diodes (CPOLED) are presented. Next, the different molecular designs leading to CPTADF SOM are described depending on the nature of the chirality of these molecules (point, axial, or planar chirality). The synthesis, photophysical, and chiroptical properties of these molecules and the performance of related CPOLED devices are discussed.

## 13.1 Introduction

Among all the potential applications of Circularly Polarized Luminescent active Simple Organic Molecules (CPL-SOMs), their use as emitters in Organic Light Emitting Diodes (OLED) is one of the most exciting applications. Indeed, the development of CP-OLED has emerged as an interesting direction to substantially increasing the energy efficiency of conventional OLED displays in which 50% of the light emitted is suppressed using a polarizer and a quarter-wave plate to reduce the external light reflection [1, 2]. This high energy loss can be overcome by using CP-OLED which generates circularly polarized electroluminescence that can pass through these filters without any attenuation, resulting in higher display performances in terms of autonomy and contrast. Although this type of device presents undeniable advantages over conventional OLED, described examples are still rare in the literature [3]. Among those research works, chiral lanthanides [4, 5] and cholesteric-based [6–8] CP-OLED devices have shown high degree of circular polarized electroluminescence ($g_{El}$ up to 1 and 1.6 for lanthanide and cholesteric

G. Pieters (✉) · L. Frederic
Université Paris Saclay, SCBM, CEA Paris Saclay, Gif-sur-Yvette, France
e-mail: gregory.pieters@cea.fr; lucas.frederic@cea.fr

© Springer Nature Singapore Pte Ltd. 2020
T. Mori (ed.), *Circularly Polarized Luminescence of Isolated Small Organic Molecules*, https://doi.org/10.1007/978-981-15-2309-0_13

CPL emitters, respectively), but these devices were strongly limited in efficiency (EQE~0.05%, notably because of the low luminescence quantum yield of these types of chiral emitter). Impressive results have also been reported by Fuchter et al. [9], using phosphorescent organometallic helicenic complex as emitters which has allowed to generate a degree of CP-electroluminescence ($g_{El} = 0.4$) sufficient to give a 19% increased brightness compared to unpolarized OLED of similar performances. Nevertheless, the corresponding luminescence and power efficiencies were still low (EQE < 1%), highlighting the need of molecular engineering to design more efficient chiral emitters. In the field related to the development of OLED technology, the design of Thermally Activated Delayed Fluorescence (TADF) materials is an exploding area of research (For recent reviews see: [10–12]). Indeed, in such materials both singlet and triplet excitons can be harvested for light emission by a reverse intersystem crossing process thanks to a small energy gap between their singlet and triplet states ($\Delta E_{ST}$). This property has recently motivated numerous research works because of the possibility to develop, in theory, OLEDs with maximum efficiency. Therefore, the design of new molecular architectures presenting both TADF and CPL emission properties appears as a cornerstone for enhancement of the performances of OLED devices. In order to merge these two properties, several molecular designs have been imagined during last 4 years. This chapter covers the design principles allowing to combine TADF and CPL emissions in simple organic molecules, the synthesis and photophysical performances of such compounds and their ability to generate CP electroluminescence once incorporated in OLED devices.

## 13.2  Principles of TADF and Design Rules

Fluorescence and electroluminescence can be produced in organic molecules upon and photo- and electrical excitation, respectively. Generally, fluorescent small organic molecules exhibit prompt fluorescence (PF), and in this case, emission takes place following photon absorption ($S_0 \rightarrow S_n$) and excited-state relaxation to the lowest excited singlet state ($S_1$). On the other hand, delayed fluorescence can be generated via another mechanism involving the participation of the triplet-excited state. Once $S_1$ is attained, if the energy difference between the two lowest excited states $S_1$ and $T_1$ ($\Delta E_{ST}$) is sufficiently small ($\Delta E_{ST} < 100$ meV), the molecule undergoes intersystem crossing (ISC) to the lowest triplet-excited state. Then due to thermal activation, the molecule may go back to $S_1$ via a process named RISC for Reverse InterSystem Crossing and generates delayed fluorescence owning the same emission spectrum than PF but with a longer lifetime, generally in the micro- to millisecond timeframe (see Fig. 13.1). In the emissive layer of an OLED, electron-hole recombination in the organic emitters creates 25% singlet excitons and 75% triplet excitons. This means that in the case of an OLED using a classical fluorophore as emitter the maximal internal efficiency is limited to 25%. Because in TADF molecules both singlet and triplet excitons can be harvested for light emission, they

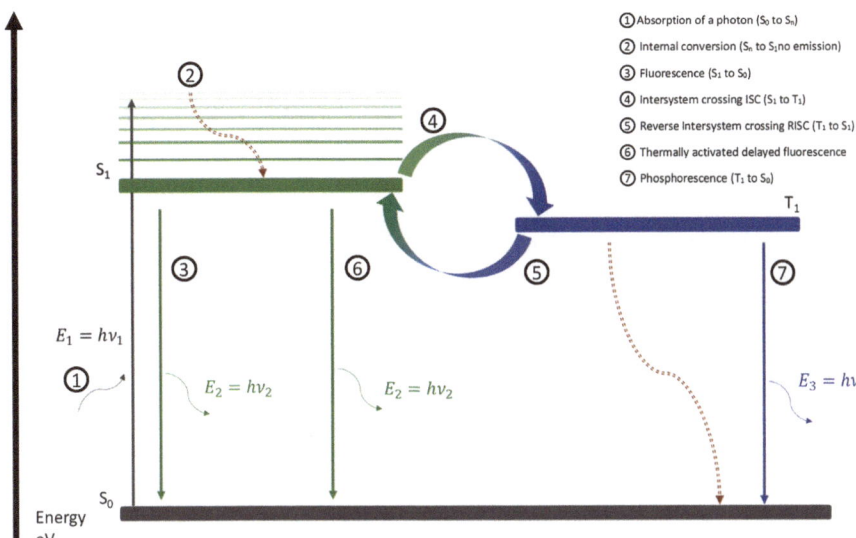

**Fig. 13.1** Simplified Perrin-Jablonski diagram

appear as a solution of choice to build more efficient OLED devices. But because of the atypical conditions that are required to generate such a delayed fluorescence: small $\Delta E_{ST}$, high ISC, and RISC probabilities along with rather long $T_1$ lifetime, efficient TADF emitters are difficult to design.

In order to minimize the $\Delta E_{ST}$ in small organic molecules, one of the most reliable approaches described so far involves the design of Donor–Acceptor (D-A) molecules in which a steric hindrance allows to twist D and A moieties around the D-A axis. The origin of this design rule is arising from the fact that such geometry in D-A molecules reduces the HOMO and LUMO overlap, which is proportional to the $\Delta E_{ST}$. We will see in the following lines that most of the chiral TADF molecules developed so far adhere to this design paradigm. However, this has the potential to reduce the oscillator strength of the transition and therefore to be detrimental to the fluorescence quantum yield ($\Phi_{PL}$). Please see references [10–12] for further information about the design of TADF molecules.

## 13.3 CPTADF Emitters: Design, Synthesis, Photophysical, and Chiroptical Properties

### 13.3.1 TADF Molecules Possessing Point Chirality

Shuzo Hirata and co-workers were the first to report the design and synthesis of a molecule merging both TADF and CPL properties (named CPTADF molecules for

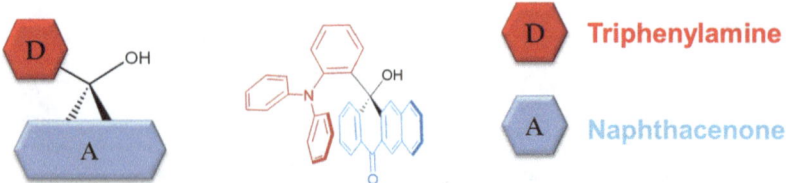

Chiral carbon sandwiched between a Donor and an Acceptor

**Fig. 13.2** Design of CPTADF molecules reported by Hirata et al

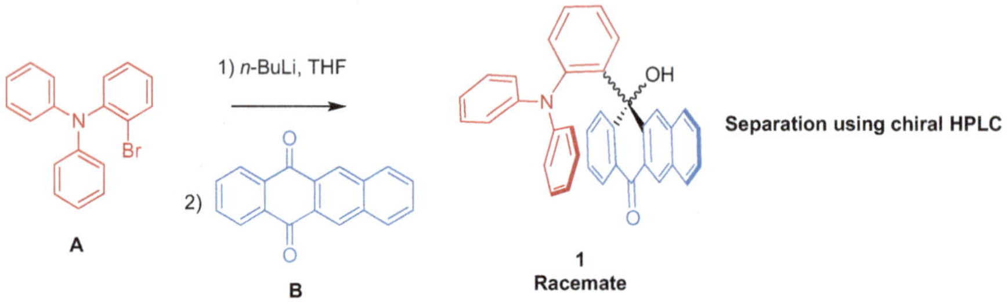

**Scheme 13.1** Synthesis of the 12-(2-(diphenylamino)phenyl)-12-hydroxynaphthacen-5(12H)-one

Circularly Polarized Thermally Activated Delayed Fluorescence molecules in the rest of the document) [13]. This first approach toward such molecules was based on the construction of a chiral carbon center bearing an electron donor, a triarylamine unit, and an acceptor unit, naphthacen-5(12H)-one (see Fig. 13.2).

The target molecule, 12-(2-(diphenylamino)phenyl)-12-hydroxynaphthacen-5 (12H)-one (DPHN), was obtained by the addition of the 2-litiumtriphenylamine, prepared by a classical halogen-lithium exchange on the 2-bromotriphenylamine **A**, to the 5,12-naphthacenequinone **B** using THF as solvent giving the target compound in 47% yield (see Scheme 13.1). Thereafter, the two enantiomers of the racemic mixture were separated using preparative chiral high-pressure liquid chromatography (HPLC).

Both enantiomers of compound **1** display green emission in toluene solution with a maximum wavelength centered on 513 nm. DFT and TD-DFT calculations at the B3LYP/6-31G(d,p) level have revealed some interesting insight concerning the photophysical properties of this molecule which can be resumed as follows: at the fundamental state ($S_0$), the transition that generates the first absorption band showed a weak charge transfer (CT) character with rather high spatial separation of the HOMO and LUMO. At the lowest singlet state ($S_1$), the character of the HOMO-LUMO transition was pure CT with a better spatial separation of the HOMO (located predominantly on triarylamine substructure) and LUMO (mainly located on the acceptor unit) compared to the $S_0$ state. The calculated energy difference between the lowest singlet state ($S_1$) and the lowest triplet state ($T_1$), $\Delta E_{ST}$, was found to be very small with a value of 0.07 eV indicating the potential of such molecular design to generate TADF. Unfortunately, this molecule possesses a relatively small quantum yield ($\Phi_{PF}$ = 4%) in toluene solution and the TADF emission in solution state was not demonstrated. The photophysical properties of compound **1** were then studied in film state by spin casting a chloroform solution containing 9% wt of compound **1** in N,N′-4,4′-dicarbazole-3,5-benzene (mCP) on quartz substrate. Interestingly at the solid state, higher quantum yields were obtained for prompt fluorescence ($\Phi_{PF}$ = 11%) and more importantly, TADF emission was observed ($\Phi_{DF}$ = 15%) with a lifetime in the order of milliseconds. This moderate photophysical performances can be explained by quite large $\Delta E_{ST}$ values (experimentally determined to be 0.19 eV) and the small fluorescent rate constant ($k_F$) resulting in a deactivation of excitons from $T_1$. Concerning chiroptical properties, such molecules possess interesting performances with a dissymmetry factor, $|g_{lum}|$ of $1.1 \times 10^{-3}$ measured in toluene solution, which is in the range of the value found for purely organic chromophores ($10^{-5}$ to $10^{-2}$). However, the possibility to use such molecules as dopants in CP-OLED was not demonstrated.

### 13.3.2 TADF Molecules Possessing Axial Chirality

The second example of CPTADF-SOM has been reported by Pieters et al. few months after the pioneer work from Hirata [14]. Conceptually, this other molecular design involves the tethering of a chiral unit (derived from BINOL) to an active TADF chromophore (see Fig. 13.3). Here, the proximity of the chiral unit is anticipated to induce chiroptical properties (induced circular dichroism and circularly polarized luminescence) to the TADF emitter, like in the chiral O-BODIPY dyes described by De la Moya et al. in 2015 [15]. The TADF unit used in this design is a donor–acceptor system composed of carbazoles as electron donor and a terephthalonitrile unit as acceptor. Such type of D-A systems has been previously used to construct very efficient TADF molecules by Adachi and coworkers [16].

The target enantiomers, (R)-**2** and (S)-**2**, were synthesized through a one-pot sequential procedure at room temperature involving commercially available

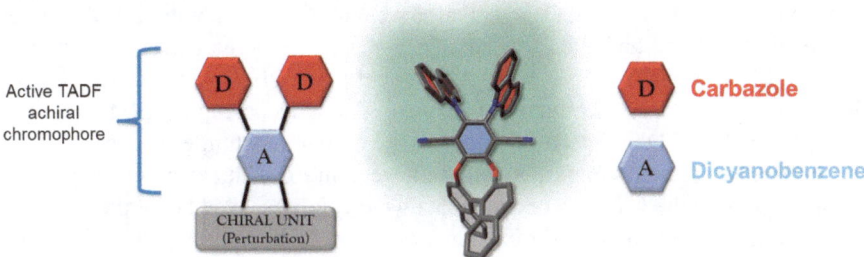

**Fig. 13.3** Design of CPTADF molecules reported by Pieters et al

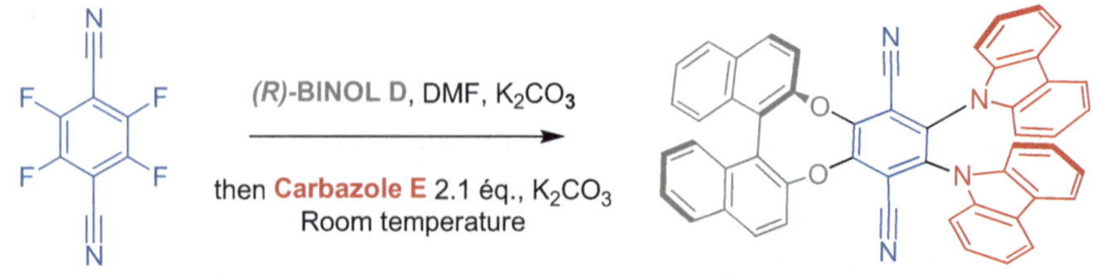

**Scheme 13.2** Synthesis of the CPTADF molecule reported by Pieters et al

compounds (Scheme 13.2). The first sequence consists in a highly selective desymmetrization of the starting tetrafluoroterephthalonitrile **C**, involving enantiopure BINOL **D** and $K_2CO_3$ as a base, which led to the formation of a chiral difluorinated intermediate. Then, carbazole **E** (2.1 equiv) and additional amount of base were added to the reaction mixture, leading to the targeted molecule **2** with high yield (83%) and enantiomeric excess (>99%, based on chiral HPLC) after purification over silica gel (see Scheme 13.2).

DFT calculations performed at the B3LYP/6-31G level have revealed some interesting information about such compounds. First, the spatial separation of HOMO and LUMO is significant with a HOMO mainly located on carbazolyl units and a LUMO centered onto the terephthalonitrile substructure. Second, at the ground state at least, no orbital coefficient has been found of the BINOL moiety. This means that the BINOL unit plays, at least at the ground state, a role of chiral perturbator for the potentially TADF active chromophore. Such compounds own quite interesting photophysical properties with a strong positive solvatochromism with $\Phi_{PF}$ values between 0.74 and 0.06 depending on the solvent polarizability. By contrast with the first example of CPTADF molecules described earlier, TADF properties were clearly observed in toluene solution with an important increase in the quantum yield in degassed toluene compared to the one in aerated solution (from 0.28 to 0.53). A bi-exponential fluorescence decay was observed with lifetime of 20 ns for prompt fluorescence and of 2.9 μs for the delayed component. The chiral perturbation from the BINOL unit to the TADF active emitter was found to be effective, both at ground and exited state, with CD signal located at the wavelengths corresponding to the Internal Charge Transfer band (ICT band) of the TADF emitter (centered at 420 nm) and a $|g_{lum}|$ value of $1.3 \times 10^{-3}$ recorded in degassed toluene solution ($C = 1$ mM). These compounds were successfully applied as emissive dopant in OLED devices built via Chemical Vapor Deposition (CVD). Using a nonoptimized OLED architecture, a maximal external quantum efficiency of 9.1% has been measured. Although the authors have confirmed the enantiopurity of the compounds after the CVD process, no data were communicated about the potential polarization light emanating from the device.

The potential of this molecular design to generate efficient emitters for construction of CP-OLED was later confirmed by Tang and coworkers [17]. Indeed,

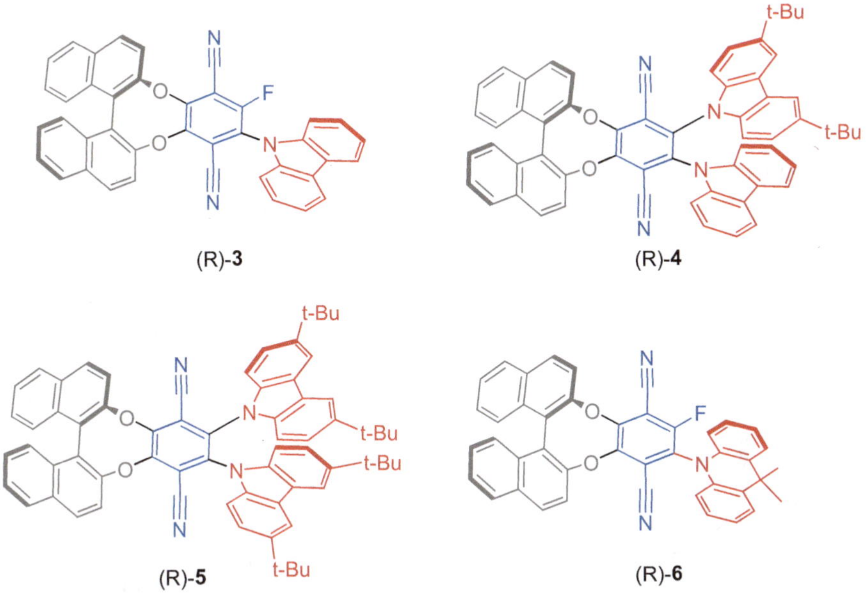

**Fig. 13.4** Structure of CPTADF emitters synthetized by Tang et al

**Table 13.1** Photophysical properties of compounds **3–6**

| | 3 | 4 | 5 | 6 |
|---|---|---|---|---|
| $\lambda abs_{CT}$ (nm) | 400 | 420 | 440 | 450 |
| $\lambda em_{max}$ (nm) | 493 | 530 | 531 | 585 |
| $\lambda em_{max}$ (nm) in neat film | 520 | 538 | 553 | 580 |
| Quantum yield in neat film | 0.4 | 0.2 | 0.2 | 0.05 |

following the same design rules, they have synthesized four novel chiral TADF emitters, involving carbazole, 3,6-ditertbutylcarbazole and 9,10-Dihydro-9,9-dimethylacridine as electron donors units (see Fig. 13.4). Those compounds were synthesized using the same synthetic procedure described for compound **2**, except for compound **6**, another couple base-solvent was used ($^t$BuOK as base and ACN as solvent). In terms of photophysical properties, molecules **3–6** exhibit broad and relatively weak absorption bands centered at 400, 420, 440, and 450 nm, respectively, which can be associated with intramolecular charge transfer (ICT) process derived from the Donor-Acceptor (D-A) active chromophore units. Emission spectra were measured in toluene solution and maximum emissions were observed between 495 nm and 595 nm depending on the nature of electron donor moiety (see Table 13.1). Low to good quantum yields from 0.05 for compound **6** to 0.4 for compound **3** were determined in neat thin film fabricated by CVD. Fluorescence decays were also investigated for each molecule in thin film. All compounds were found to be TADF emitters and show bi-exponential fluorescent decays with nanosecond-order prompt fluorescence (lifetimes from 11.3 to 162 ns) and microsecond-order delayed fluorescence (lifetimes from 0.04 to 24 μs). Chiroptical properties were also investigated in toluene solution. Compounds 3–6 possess $g_{lum}$ values in the same order of magnitude than compound 2 (from $0.5 \times 10^{-3}$ to $1.2 \times 10^{-3}$).

Importantly, Ben Zhong Tang and coworkers have demonstrated the Aggregation Induced Enhancement Emission (AIEE) properties of such chiral compounds by studying their fluorescence emissions in THF-$H_2O$ mixtures with different water fractions ($f_w$). Here, upon the formation of aggregates in water, the intramolecular rotational and vibrational motions occurring in solution state are limited and nonradiative decay processes from the excited states are decreased, leading to an important increase in the fluorescence upon aggregation.

The authors have also evidenced a strong amplification of the chiroptical properties of this kind of emitters in thin films. Indeed, analyzing CD responses of neat and doped films fabricated by CVD, a slight enhancement of Kuhn's dimensionless anisotropy factors ($g_{abs}$ multiplied by 2 up to 5 depending on the compound) was observed for all emitters 3–6. Surprisingly, the enhancement of the $g_{lum}$ values was found to be much higher, with an impressive 40-fold increase in the case of compound 3, exhibiting a $g_{lum}$ value of $1.3 \times 10^{-3}$ in toluene solution compared to a value of $4.1 \times 10^{-2}$ in neat film state (see Fig. 13.5). The authors have explained this important amplification of the dissymmetry factors by the formation of chiral aggregates during the formation of the thin film. Interestingly, this amplification also occurs when the molecules 3–6 are used as emissive dopants in a mCP matrix. In order to understand the origin of this amplification, more detailed characterization of the postulated chiral aggregates would be necessary.

Nevertheless, based on those impressive chiroptical properties in thin film state, CP-OLED were fabricated through CVD using neat and doped film involving molecules 3–6 as emissive layers. This has led to the fabrication of the most performant CP-OLEDs reported so far using CPL-SOMs in terms of circular polarization with $g_{el}$ up to 0.08 and External Quantum Efficiency (EQE) up to 3.5% for OLED using neat films as emissive layers.

The third example of molecular design allowing the synthesis of molecules merging CPL and TADF properties has been described by the group of Man-Keung Fung and Chuan-Feng Chen [18]. Their original approach relies on the linkage of two active TADF units, composed of a D-A-D system involving carbazoles as donors and an aromatic-imide as acceptors, by chiral unit derived from the commercially available trans-1,2-diaminocyclohexane (see Fig. 13.6).

Both enantiomers of this molecule were prepared in two steps starting from commercially available $(-)$-$(R,R)$- or $(+)$-$(S,S)$-diaminocyclohexane F (see Scheme 13.3). This straightforward synthesis involved the lactamization of 4,5-difluorophthalic anhydride G with the chiral diamine F in AcOH as a first step, followed by a nucleophilic aromatic substitution reaction on the fluoroaromatic imide intermediate H with carbazole anions formed in THF using NaH as base. Overall, the target molecule was obtained with a high enantiomeric excess (ee > 99% measured using chiral HPLC) in almost 40% yield over the two synthetic steps.

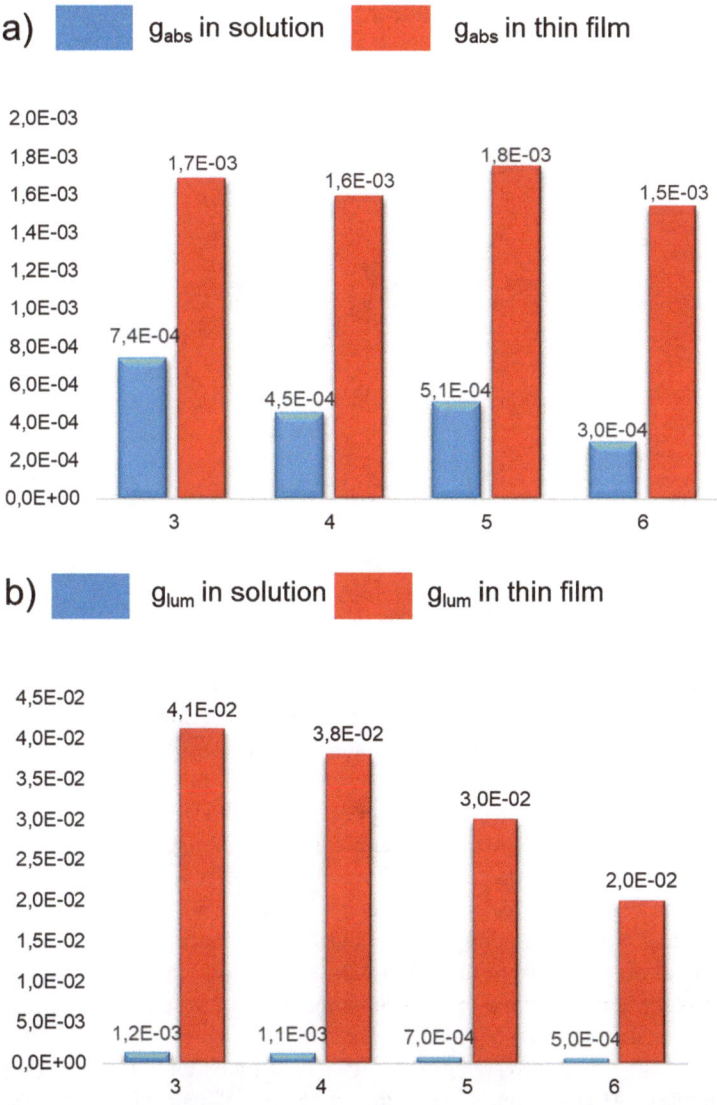

**Fig. 13.5** (**a**) $g_{abs}$ values in toluene solutions and in thin films measured at 405 nm for **3**, 436 nm for **4**, 440 nm for **5**, and 450 nm for **6**; (**b**) $g_{lum}$ values in toluene solutions and in thin films. Measured at 493 nm for **3**, 530 nm for **4**, 531 nm for **5**, and 585 nm for **6** in solution (wavelengths used to measure gabs and glum in thin film were not indicated by the authors)

In terms of photophysical properties, the absorption spectrum in THF of (+)-(*S, S*)-**7** exhibited an intense intramolecular charge-transfer absorption band at about 380 nm in THF. As expected for such D-A-D chromophores, this molecule exhibited a strong positive solvatochromism with maximum emission wavelengths ranging from 508 nm in hexane to 562 nm in acetonitrile. Interestingly, although the

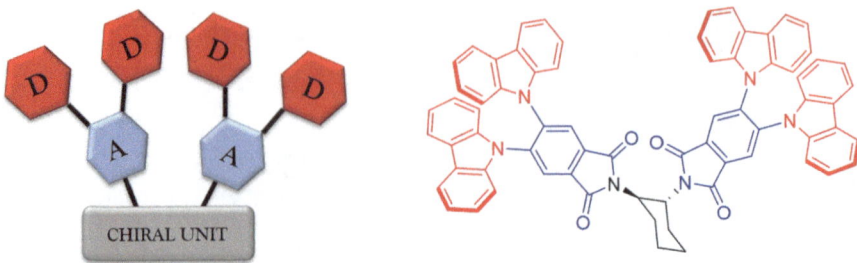

**Fig. 13.6** Design of CPTADF molecules described by Fung and Chen

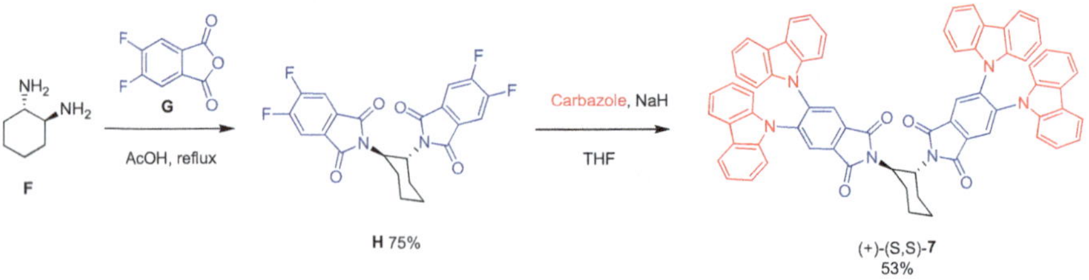

**Scheme 13.3** Synthesis of the CP-TADF molecules based on TADF aromatic imide

quantum yields were rather low in solution (0.18 in toluene), they were found to be much higher in thin film state, with values up to 0.78 in doped film using 3,3′-di (9H-carbazol-9-yl)biphenyl (mCBP) as the host matrix. By studying temperature-dependent transient photoluminescence characteristics of the co-doped films, the authors confirmed that these chiral compounds were TADF active and measured a very small $\Delta E_{ST}$ value of 0.06 eV. The transient photoluminescence decays of molecule **7** in the co-doped film under vacuum show an expected biexponential distribution. The fitting gave a short lifetime of 44 ns and a long one of 130 μs which confirms that **7** emits both prompt and delayed fluorescence. The efficient induction of chiroptical properties from the chiral 1.2-diaminocyclohexane unit to the TADF active chromophores was next confirmed by the study of the chiroptical properties in thin film state. Indeed, circular dichroism spectra show a clear CD signal for the band centered on 400 nm corresponding to the ICT transition of the TADF active chromophores. At the excited state, the induction of chiroptical properties was also demonstrated with the measurement of luminescence dissymmetry factors of $+/-1.1 \times 10^{-3}$. Using **7** as dopant in the emitting layers (15 wt % (+)-(S,S)-**7** or (−)-(R,R)-**7** in mCBP matrix), CP-OLED possessing a high maximum EQE of 19.7% and electroluminescent dissymmetry factors ($g_{EI}$) of $2.2 \times 10^{-3}$ (close to the photoluminescence dissymmetry factor measured in the thin film state, $1.1 \times 10^{-3}$) was fabricated. Based on the same molecular design, Chen's research group has also synthesized the red emitting CPTADF molecule **8** (see Fig. 13.7) [19].

In molecule **8**, the red emitting TADF active units are composed of a 1,8-naphthalimide moiety as electron acceptor (in blue Fig. 13.7) and a 9,9-dimethyl-9,10-dihydroacridine (DMAC, in red) as electron donor. The tethering

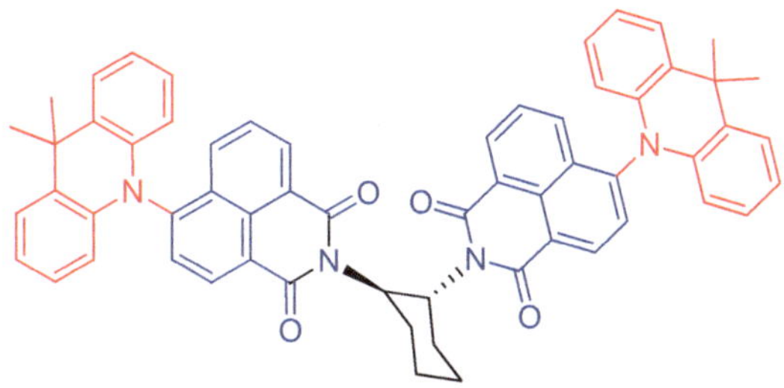

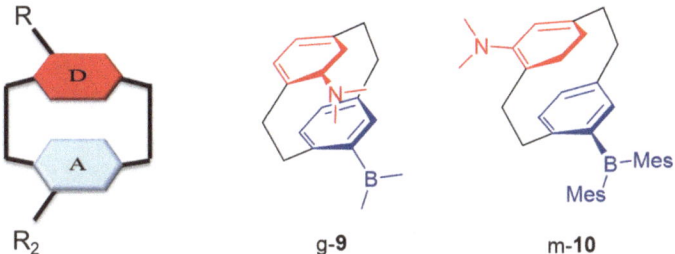

$(+)-(S,S)-\mathbf{8}$

**Fig. 13.7** Chemical structure of the red CPTADF emitter described by Chen et al

**Fig. 13.8** Paracyclophane based CPTADF molecules designed by Zhao et al

of these active chromophores by a chiral 1,2 diaminocyclohexane unit also allows to induce CD and CPL properties to the TADF emitters. As a matter of fact, enantiomers of those molecules exhibited a mirror-image CD and a CPL signals with $|g_{lum}|$ value of $9.2 \times 10^{-4}$ in thin film state. These molecules were also successfully used as dopant in the emissive layer of an OLED owning an orange-red emission (centered at 592 nm), and exhibiting a maximum EQE value of 12.4%. Although the authors have demonstrated the circular polarization of the emitters' photoluminescence in thin film state, no information was given concerning the potential circular polarization of the light emanating from the constructed OLED devices.

### 13.3.3 TADF Molecules Possessing Planar Chirality

At the end of 2018, a fourth molecular design allowing to merge TADF and CPL properties in a small molecule has been reported by Cui-Hua Zhao et al. [20] In the first three examples of CPTADF molecules described in this chapter, the HOMO-LUMO spatial separation requisite for TADF activity was attained using D-A chromophores possessing a highly twisted relative conformation. Here, in order to obtain a molecule owning a small exchange integral between the two frontier orbitals, they have embedded a Donor and an Acceptor units in a chiral[2.2]-paracyclophane skeleton (see Fig. 13.8).

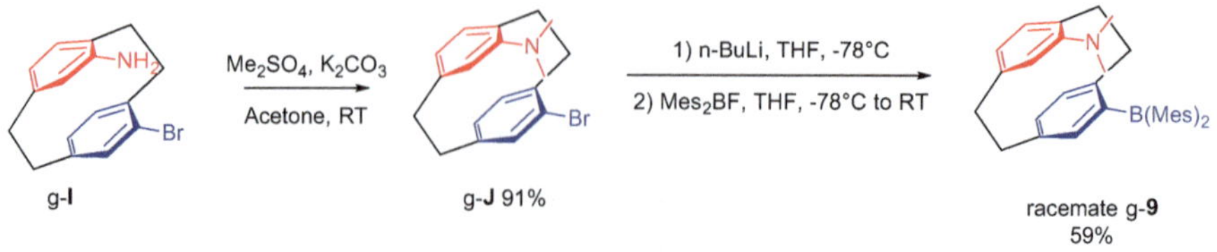

**Scheme 13.4** Synthesis of g-**9**

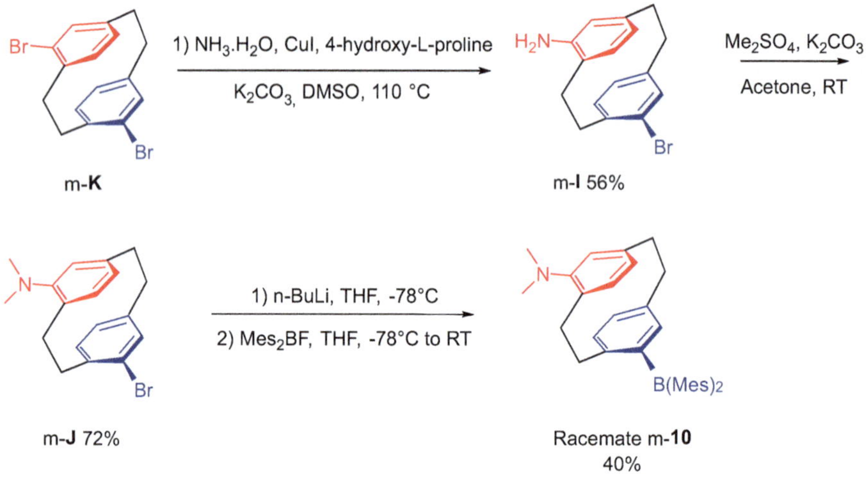

**Scheme 13.5** Synthesis of m-**10**

In these [2.2]paracyclophane derivatives, g-**9** and m-**10**, electron-donating dimethylamine group and the electron-accepting dimesytylboryl moiety are introduced at the pseudo-gem and pseudo-meta positions. In order to synthetize the emitter g-**9** (see Scheme 13.4), 4-bromo-13-amino[2.2]paracyclophane g-**I** was methylated using dimethyl sulfate as methylation agent in acetone using potassium carbonate as base. Then lithiation of the bromide g-**J** followed by addition of dimesitylboron fluoride was achieved to give g-**9** a 59% yield.

The synthesis of the m-**10** isomer was initiated by the amination of the 4,15-dibromo-[2.2]paracyclophane m-**K** via copper-assisted coupling reaction using hydroxyproline as ligand and aqueous ammoniac. Then, the same reaction sequence used for the synthesis of g-**9** gave racemate m-**10** with an overall yield of 16% over the three synthetic steps (see Scheme 13.5).

In terms of photophysical properties, compound g-**9** displays a broad and moderately intense absorption band centered at 374 nm, which was assigned using theoretical calculations to the intramolecular charge transfer transition from the HOMO, located on dimethylaminophenyl moiety, to the LUMO, localized on dimesitylborylphenyl unit. For m-**10**, the assignation of this ICT band was found to be more difficult because of the very small oscillator strength (0.0209) of the first excited state (corresponding to the ICT band). In terms of emissive properties, both isomers emit green fluorescence with a maximum emission wavelength of 521 nm for g-**9** and of 531 nm for m-**10**. Interestingly, the values of the quantum yields were found to be highly dependent on the presence of oxygen in the toluene solution. Indeed, g-**9** and m-**10** possess quantum yields of 0.46 and 0.34 in degassed toluene solution, respectively, whereas these values drop to 0.068 and 0.049 in presence of oxygen. This oxygen dependence on fluorescence efficiencies can be explained by the oxygen triplet state quenching and preludes therefore the possibility of exciton exchanges via efficient intersystem crossing. TADF properties were confirmed by the measurement of fluorescent decays in toluene solution. Biexponential decays were observed for both molecules with prompt fluorescence lifetimes of 17 ns for g-**9** and 22 ns for m-**10** and delayed fluorescence lifetimes of 0.38 µs for g-**9** and 0.22 µs for m-**10**. Those TADF active molecules show also good quantum yields at the solid state with a value of 0.53 for m-**10** and 0.33 for g-**9**. Because of its better photophysical performances, g-**9** was selected to study the chiroptical properties of such chiral TADF paracyclophanes. The enantiomers of g-**9** were synthesized starting from enantiopure pseudo-gem-substituted bromo amines g-**H** (see Scheme 13.4), which were obtained through optical resolution of its racemic form with (R)-(−)-10-camphorsulfonyl chloride. Both enantiomers display a small CD signal for the ICT band centered at 374 nm and a larger signal around 313 nm. The dissymmetry factor ($g_{abs}$) of absorbance at the longest wavelength (374 nm) was determined to be $1.44 \times 10^{-3}$. These compounds were also proved to be CPL active and possess a $g_{lum}$ of $4.24 \times 10^{-3}$, the highest value measured in solution so far for CPTADF molecules. Despite their promising photophysical and chiroptical properties, the authors did not report so far about the use of such chiral paracyclophanes as emitters to construct efficient CP-OLED.

## 13.4  Conclusion and Perspectives

After the first report about CPTADF molecules in mid-2015, only three other molecular designs were described in the literature so far (see Fig. 13.9).

This small number of effective compounds can be explained by the inherent difficulty related to the merging of CPL and TADF properties in a small organic molecule. Indeed, such molecules should possess a low $\Delta E_{ST}$ and a large magnetic dipole transition moment while maintaining a strong oscillator strength of the transition involved in the fluorescence emission. More systematic studies are needed in order to establish general guidelines for the design of efficient CPTADF emitters, as well as other approaches to combine CPL and TADF properties should be sought. So far, only two of these four efficient molecular designs have furnished molecules successfully used as emitters to construct CP-OLEDs. This highlights another important requirement to integrate in the future design of such molecules that is related their configurational stability. Indeed, in order to be incorporated in OLED stacks fabricated through CVD (the fabrication process affording the highest level of

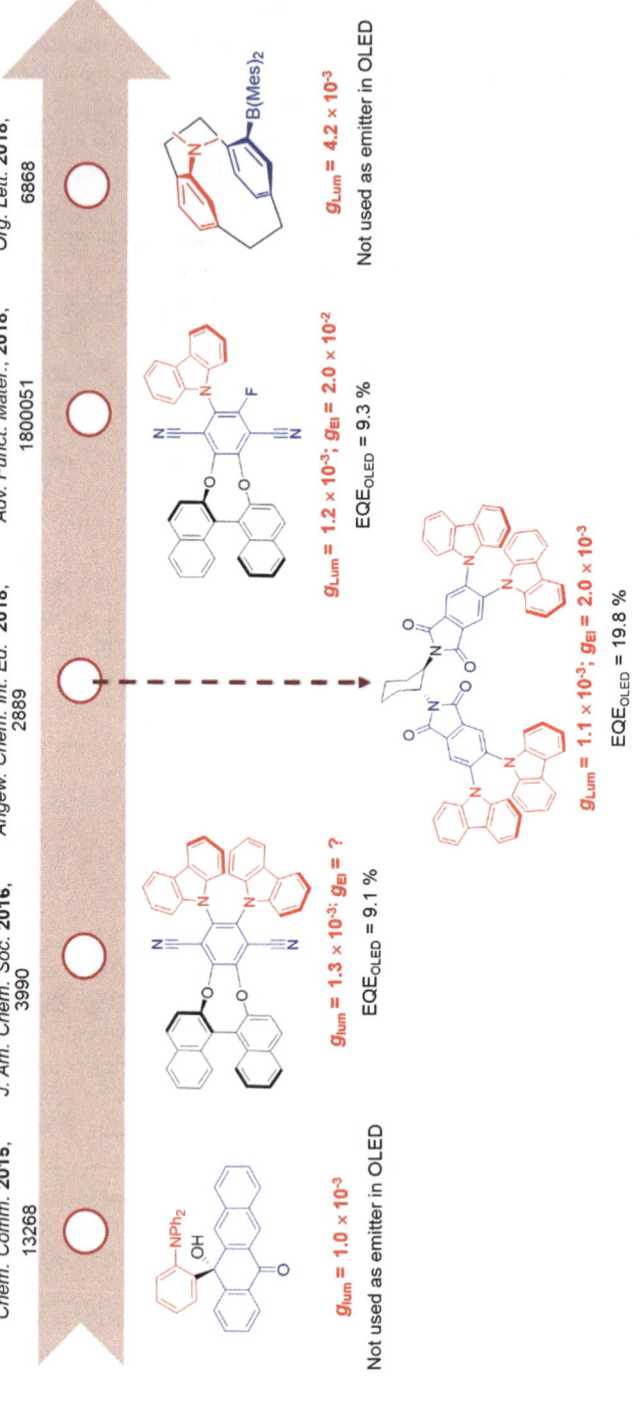

**Fig. 13.9** Evolutionary time scale related to the development of CPTADF molecules

performance so far), these chiral molecules should also possess very high racemization barriers. Because of the fundamental challenge that represents their molecular design and their high potential in terms of application in the context of display devices, there is no doubt that numerous future research works will be dedicated to the discovery of novel CPTADF molecules.

# References

1. Schadt M (1997) Liquid crystal materials and liquid crystal displays. Annu Rev Mater Sci 27:305–379
2. Singh R, Unni KNN, Solanki A (2012) Improving the contrast ratio of OLED displays: an analysis of various techniques. Opt Mater 34(4):716–723
3. Han J, Guo S, Lu H, Liu S, Zhao Q, Huang W (2018) Recent progress on circularly polarized luminescent materials for organic optoelectronic devices. Adv Opt Mater 6(17):1800538
4. Zinna F, Giovanella U, Di Bari L (2015) Highly circularly polarized electroluminescence from a chiral europium complex. Adv Mater 27(10):1791
5. Zinna F, Pasini M, Galeotti F, Botta C, Di Bari L, Giovanella U (2017) Design of lanthanide-based oleds with remarkable circularly polarized electroluminescence. Adv Funct Mater 27 (1):1603719
6. Peeters E, Christiaans MPT, Janssen RAJ, Schoo HFM, Dekkers HPJM, Meijer EW (1997) Circularly polarized electroluminescence from a polymer light-emitting diode. J Am Chem Soc 119(41):9909–9910
7. Geng Y, Trajkovska A, Culligan SW, Ou JJ, Chen HMP, Katsis D, Chen SH (2003) origin of strong chiroptical activities in films of nonafluorenes with a varying extent of pendant chirality. J Am Chem Soc 125(46):14032–14038
8. Lee DM, Song JW, Lee YJ, Yu CJ, Kim JH (2017) Control of circularly polarized electroluminescence in induced twist structure of conjugate polymer. Adv Mater 29(29):1700907
9. Brandt JR, Wang X, Yang Y, Campbell AJ, Fuchter MJ (2016) Circularly polarized phosphorescent electroluminescence with a high dissymmetry factor from PHOLEDs based on a Platinahelicene. J Am Chem Soc 138(31):9743–9746
10. Yang Z, Mao Z, Xie Z, Zhang Y, Liu S, Zhao J, Xu J, Chi Z, Aldred MP (2017) Recent advances in organic thermally activated delayed fluorescence materials. Chem Soc Rev 46:915–1016
11. Wong MY, Zysman-Colman EC (2017) Purely organic thermally activated delayed fluorescence materials for organic light-emitting diodes. Adv Mater 29(22):1605444
12. Czerwieniec R, Leitl MJ, Homeier HHH, Yersin H (2016) Cu(I) complexes – Thermally activated delayed fluorescence. Photophysical approach and material design. Coord Chem Rev 325:2–28
13. Imagawa T, Hirata S, Totani K, Watanabe T, Vacha M (2015) Thermally activated delayed fluorescence with circularly polarized luminescence characteristics. Chem Commun 51:13268–13271
14. Feuillastre S, Pauton M, Gao L, Desmarchelier A, Riives AJ, Prim D, Tondelier D, Geffroy B, Muller G, Clavier G, Pieters G (2016) Design and synthesis of new circularly polarized thermally activated delayed fluorescence emitters. J Am Chem Soc 138(12):3990–3993
15. Sanchez-Carnerero EM, Moreno F, Maroto BL, Agarrabeitia AR, Ortiz MJ, Vo B, Muller G, de La Moya S (2014) Circularly polarized luminescence by visible-light absorption in a chiral O-BODIPY dye: unprecedented design of CPL organic molecules from achiral chromophores. J Am Chem Soc 136(9):3346–3349
16. Uoyama H, Goushi K, Shizu K, Nomura H, Adachi C (2012) Highly efficient organic light-emitting diodes from delayed fluorescence. Nature 492:234–238

17. Song F, Xu Z, Zhang Q, Zhao Z, Zhang H, Zhao W, Qiu Z, Qi C, Zhang H, Sung HHY, Williams ID, Lam JWY, Zhao Z, Qin A, Ma D, Tang BZ (2018) Highly efficient circularly polarized electroluminescence from aggregation-induced emission luminogens with amplified chirality and delayed fluorescence. Adv Funct Mater 28(17):1800051

18. Li M, Li SH, Zhang D, Cai M, Duan L, Fung MK, Chen CF (2018) Stable enantiomers displaying thermally activated delayed fluorescence: efficient OLEDs with circularly polarized electroluminescence. Angew Chem Int Ed 57(11):2889

19. Wang Y-F, Lua H-Y, Chen C, Li M, Chen C-F (2019) 1,8-Naphthalimide-based circularly polarized TADF enantiomers as the emitters for efficient orange-red OLEDs. Org Electron 70:71–77

20. Zhang M-Y, Li Z-Y, Lu B, Wang Y, Ma Y-D, Zhao C-H (2018) Solid-state emissive triarylborane-based [2.2]paracyclophanes displaying circularly polarized luminescence and thermally activated delayed fluorescence. Org Lett 20(21):6868–6871

# Chapter 14
# Principles and Applications of Circularly Polarized Luminescence Spectrophotometer

Satoko Suzuki

**Abstract**  Materials that exhibit circularly polarized light characteristics are finding an increasing range of applications, such as liquid crystal display backlights, three-dimensional displays, holographic displays, plant growth control illumination, security systems for optical communications, and printing. In this chapter, we will describe the basic principles of circularly polarized luminescence (CPL) spectroscopy, the instruments used to measure CPL signals, including optical systems and signal acquisition methods in commercially available CPL instruments, and also methods for calibrating such equipment. Example measurements are presented for camphor and camphorquinone, which are long-established CPL samples, in addition to lanthanoid complexes that are promising CPL luminescent materials, biopolymers that are highly sensitive to circularly polarized light, and solid samples that have been attracting attention in recent years.

## 14.1  Introduction

Circular dichroism (CD) spectroscopy is widely used in the study of optically active substances. However, in recent years, circularly polarized luminescence (CPL) spectroscopy has also attracted a great deal of attention. The two methods are complimentary to each other. Materials that exhibit circularly polarized luminescence are currently being actively investigated for applications in fields such as liquid crystal display backlights, three-dimensional (3D) displays, holographic displays, light sources for controlling plant growth, and security systems for optical communications and printing. It is therefore important to identify molecules that emit one-handed circularly polarized light (large $g_{lum}$) with a high quantum yield. Consequently, there is a need for the development of highly sensitive CPL measuring instruments.

S. Suzuki (✉)
Jasco Corporation, Tokyo, Japan
e-mail: satoko.suzuki@jasco.co.jp

© Springer Nature Singapore Pte Ltd. 2020
T. Mori (ed.), *Circularly Polarized Luminescence of Isolated Small Organic Molecules*, https://doi.org/10.1007/978-981-15-2309-0_14

In this report, we introduce the basic principles of CPL measurements, which include the meaning of the signals obtained in CPL spectroscopy, the type of equipment used, the calibration methods, and measurement examples.

## 14.2    Introduction to CPL

CPL can be best described by comparing it to CD, which involves a difference in the amount of absorption of left- and right-handed circularly polarized light by an optically active substance, as shown in Fig. 14.1a. In contrast, as shown in Fig. 14.1b, CPL spectroscopy measures the difference in the intensity of left- and right-handed circularly polarized fluorescence during optical excitation of a sample.

The magnitude of the CD and CPL signals is expressed by Eqs. (14.1) and (14.2), respectively.

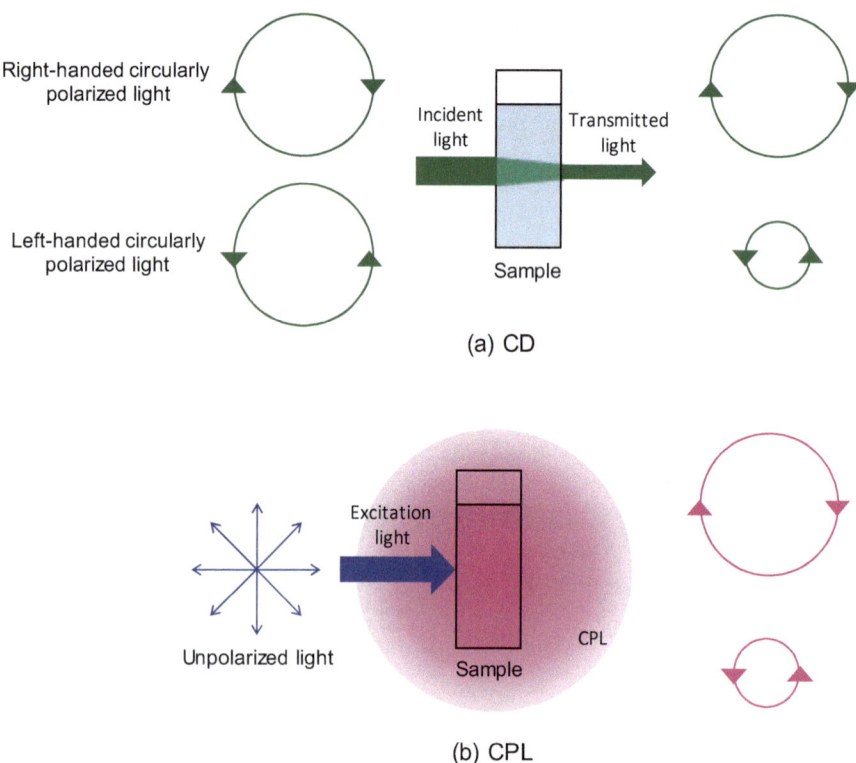

**Fig. 14.1** Comparison of CD (**a**) and CPL (**b**)

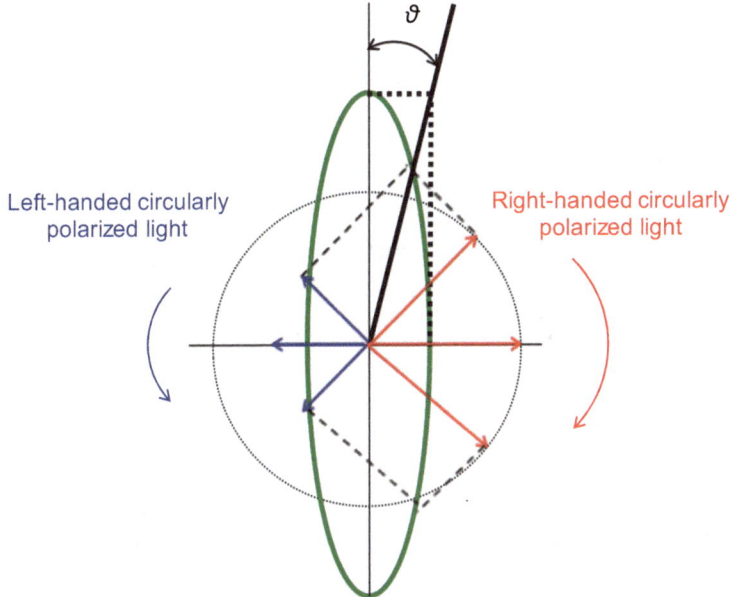

**Fig. 14.2** Ellipse angle ($\theta$) for CD and CPL signals

$$\Delta\varepsilon = \varepsilon_L - \varepsilon_R \qquad (14.1)$$

$$\Delta I = I_L - I_R \qquad (14.2)$$

$\varepsilon$ extinction coefficient for left- ($L$) and right-handed ($R$) circularly polarized light, $I$ fluorescence intensity of left- ($L$) and right-handed ($R$) circularly polarized light

CD and CPL signals are generally normalized by the averaged absorbance and the averaged fluorescence intensity for a sample, and are expressed in terms of $g_{abs}$ and $g_{lum}$ values:

$$CD : g_{abs} = \frac{2(\varepsilon_L - \varepsilon_R)}{\varepsilon_L + \varepsilon_R} \qquad (14.3)$$

$$CPL : g_{lum} = \frac{2(I_L - I_R)}{I_L + I_R} \qquad (14.4)$$

Both $g_{abs}$ and $g_{lum}$ take values between $-2$ and $2$. When only left-handed circularly polarized light is absorbed or emitted, the $g$ value is $+2$, while when only right-handed circularly polarized light is absorbed or emitted, it is $-2$.

The vertical axis in a CD or CPL spectrum is traditionally the ellipse angle $\theta$ [mdeg]. Figure 14.2 shows the overall polarization trajectory for a mixture of left- and right-handed circularly polarized light. If there is a difference in the intensity of the left- and right-handed polarized light, the overall trajectory is ellipse.

## 14.3  Instrumentation

### 14.3.1  Optical Layout in JASCO CD and CPL Spectrometer

Figure 14.3 shows schematics of the optics for CD and CPL spectroscopy. In a CD spectrometer, as illustrated in Fig. 14.3a, light from a light source is dispersed by a monochromator, and is linearly polarized by a polarizer. In contrast, a commercially available CD spectrometer (e.g., JASCO J-1000 series) is capable of not only monochromating light but also producing linearly polarized light using a quartz prism. This linearly polarized light is modulated at a frequency of 50 kHz using a photo-elastic modulator (PEM) to produce left- and right-handed circularly polarized light. This circularly polarized light is introduced to a sample, and the difference in the transmitted left- and right-handed circularly polarized light is measured using a detector. Figure 14.3b shows a schematic diagram of optics layout in CPL measurements. Light from a light source is first passed through a monochromator, and the sample is then irradiated by nonpolarized light. The left- and right-handed circularly polarized fluorescence emitted from the sample is alternately linearly polarized light at a frequency of 50 kHz using a PEM. A polarizer is placed at the back of the PEM, and the polarized fluorescence is transmitted synchronously with the modulation of the PEM, and is passed through a monochromator and then detected by a detector. The CPL signal corresponds to the difference in fluorescence intensity between left- and right-handed circularly polarized light.

### 14.3.2  Phase-Sensitive Detection of CD and CPL With Photo-Elastic Modulator (PEM)

As shown in Fig. 14.4a, a PEM is an optical device consisting of an isotropic optical material such as quartz, and a piezo actuator. The polarization of the incoming light

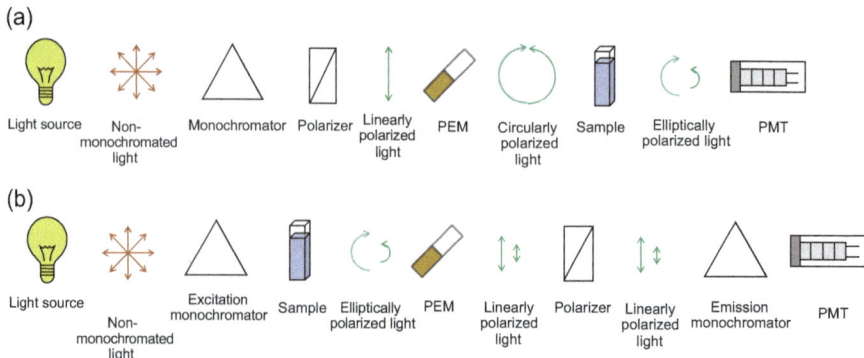

**Fig. 14.3** Schematic diagram of optics layouts in CD (**a**) and CPL (**b**) measurement principles

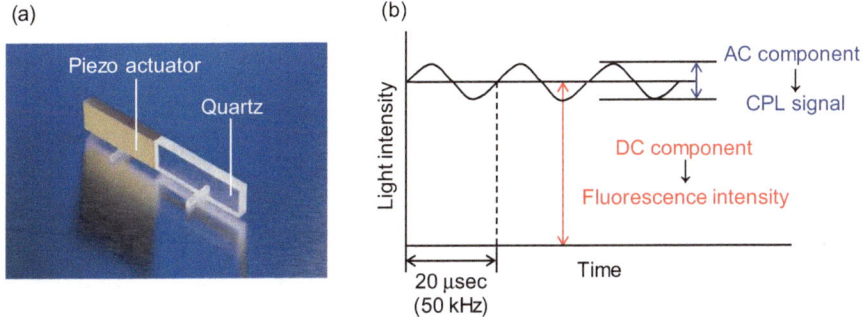

**Fig. 14.4** Phase-sensitive detection of CPL using PEM. (**a**) Appearance of PEM. (**b**) Time-domain signal of CPL and fluorescence

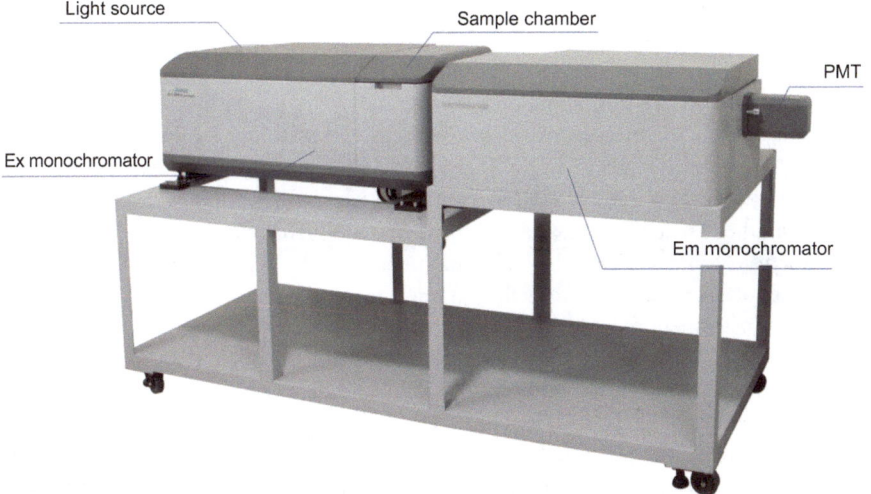

**Fig. 14.5** External view of CPL-300

can be changed by expanding and contracting the piezo actuator at a particular frequency, thereby applying a periodic stress to the quartz to produce birefringence. During CPL measurements, left- and right-handed circularly polarized fluorescence emitted by the sample is converted to linearly polarized light synchronously with the PEM frequency. This is transmitted through the polarizer and then detected. When a sample exhibits CPL, an alternating current (AC) signal is produced as shown in Fig. 14.4b, and its amplitude corresponds to the CPL signal intensity. The direct current (DC) component corresponds to the normal fluorescence spectrum.

We next describe the JASCO CPL-300 circularly polarized luminescence spectrophotometer (Fig. 14.5), as an example.

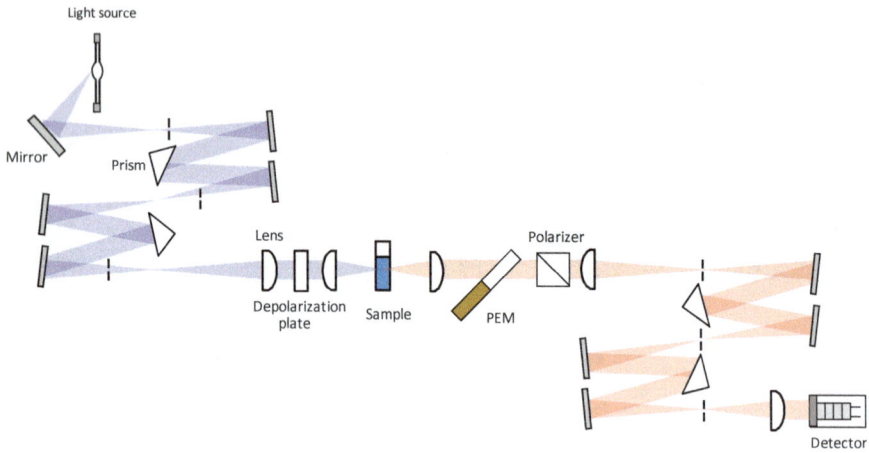

**Fig. 14.6** Optical system for CPL-300

Figure 14.6 shows a schematic of the optical system used in the CPL-300. The light source for CPL measurements can be a laser, a xenon or mercury xenon lamp, or a light-emitting diode (LED). In the CPL-300, a high-brightness xenon light source is installed by default and the excitation wavelength can be varied depending on the sample. The light is first passed through a monochromator that uses a prism or a diffraction grating as the dispersive element. This light is then depolarized by a quartz depolarizer and irradiates the sample. Since CPL signals are generally very weak, the instrument must be capable of high-sensitivity measurements that are free from artifacts. The CPL-300 uses a double-prism monochromator in order to achieve low stray light levels, avoid higher-order light due to a diffraction grating and light polarization due to Woods anomalies, and minimize the distortion of the CPL spectrum. Circularly polarized fluorescence emitted from the sample should be collected by the photo-detector placed in the angle of 90° or 180° toward the extinction direction. The light from the sample first passes through a PEM and polarizer, and is then passed through a monochromator and detected by a detector. In the 90° arrangement, although the detected light is less likely to be affected by excited scattered light, in the case of highly viscous or solid samples, there is a possibility that the CPL spectrum will be distorted by fluorescence anisotropy. To avoid this, the CPL-300 employs the 180° arrangement. A highly sensitive photomultiplier tube is used for the photo-detector.

## 14.4    Calibration of CPL Instrument

The most important aspect of CPL measurements is the accuracy of the spectrum. It is therefore important to have a reliable calibration method. In ultraviolet and visible spectrophotometers, fluorescence spectrophotometers, and circular dichroism spectrometers, the most common method of calibrating the wavelength is to use a bright line spectrum. In the CPL-300, a low-pressure mercury lamp is built into both the excitation and emission monochromators, and the wavelength is calibrated at the wavelength of 546.1 nm of a bright emission line. This is done first for the emission monochromator and then for the excitation monochromator.

A (1S)-(+)-10-ammonium camphorsulfonate aqueous solution is used to calibrate the CD intensity. The CD value at 290.5 nm for a 0.06% (w/v) aqueous solution of this substance is known to be +190.4 mdeg for an optical path length of 10 mm, and this is used for the scale standard for CD measuring instruments [1]. In the CPL-300, this solution is placed in the sample compartment, and the wavelengths of both the excitation and emission monochromators are set at 290.5 nm. Since the difference in the transmitted light intensity for left- and right-handed circularly polarized light is accurately known, this is used to calibrate the instrument, as described in detail below.

The CD value for (1S)-(+)-10-ammonium camphorsulfonate aqueous solution is mentioned above, indicating that absorption of left-handed circularly polarized light is larger than that of right-handed circularly polarized light, as shown in Eq. (14.1). This difference of transmitted left- and right-handed polarized light will be the clear standard of the intensity for CPL spectrometer. When both excitation and emission wavelengths of CPL spectrometer are then set to 290.5 nm, the sign of the signal is negative and the intensity is −190.4 mdeg according to Eq. (14.2).

## 14.5  Measurement Examples

Substrates that show CPL are widely authorized including organic compounds, metal complexes, fluorescent proteins, and colloidal molecules. The following sections describe examples involving camphor, camphorquinone, lanthanoid complex, and green fluorescent protein (GFP). Finally, CPL measurements of solid samples are discussed, since such applications have recently been attracting attention.

### 14.5.1  Camphor and Camphorquinone

These compounds have been the subject of CPL measurements for a long time. Figure 14.7a, b show CD, CPL, absorption, and fluorescence spectra of camphor and camphorquinone, respectively.

Because there are two carbonyl groups in camphorquinone, compared to just one in camphor, the structural difference between the ground and excited states for

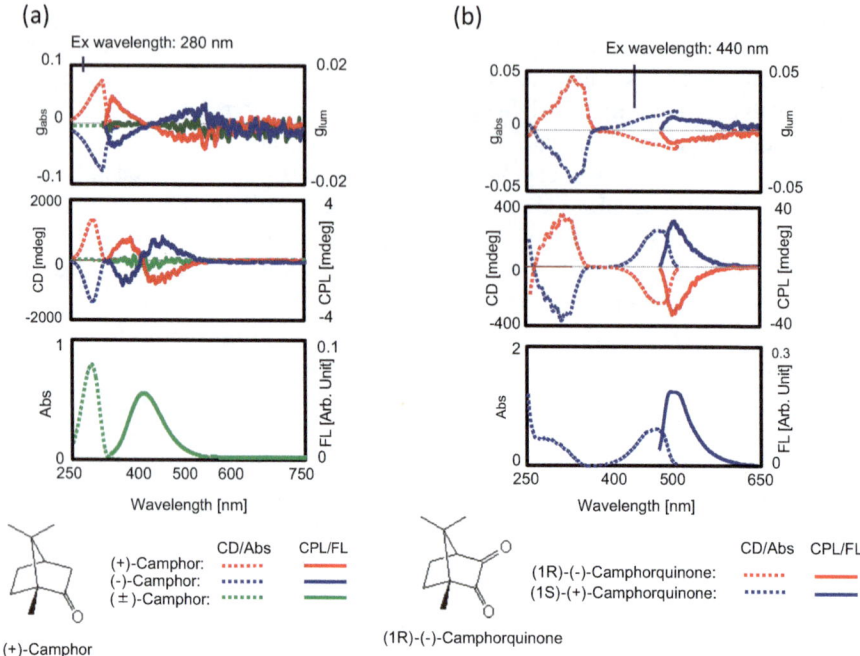

**Fig. 14.7** CD, CPL, absorption, and fluorescence spectra of camphor (**a**) and camphorquinone (**b**)

camphorquinone is predicted to be smaller because the molecule has a rigid structure. This is supported by the fact that although CD and CPL spectra associated with n–π* transitions in carbonyl groups have been observed for both camphor and camphorquinone, $g_{abs}$ and $g_{lum}$ for camphorquinone are closer than those for camphor [2].

## 14.5.2 Lanthanoid Complexes

Figure 14.8 shows CPL and fluorescence spectra of the lanthanoid complex (3-(trifluoromethylhydroxymethylene)-(+)-camphorate: Eu(facam)$_3$). Such complexes are widely used as LED materials because of their strong sharp emissions. They also potentially possess an ability in applications such as 3D displays and security markers. The synthesis and evaluation of lanthanoid complexes exhibiting CPL are underway. It is generally known that forbidden transitions such as n–π*, d–d, and f–f produce a larger $g_{lum}$ than allowed transitions such as π–π* [3]. Muller et al. reported europium complexes with a large $g_{lum}$ of +1.38 [4]. In Fig. 14.8, it can be seen that Eu(facam)$_3$ also exhibits a large $g_{lum}$. It should be noted that

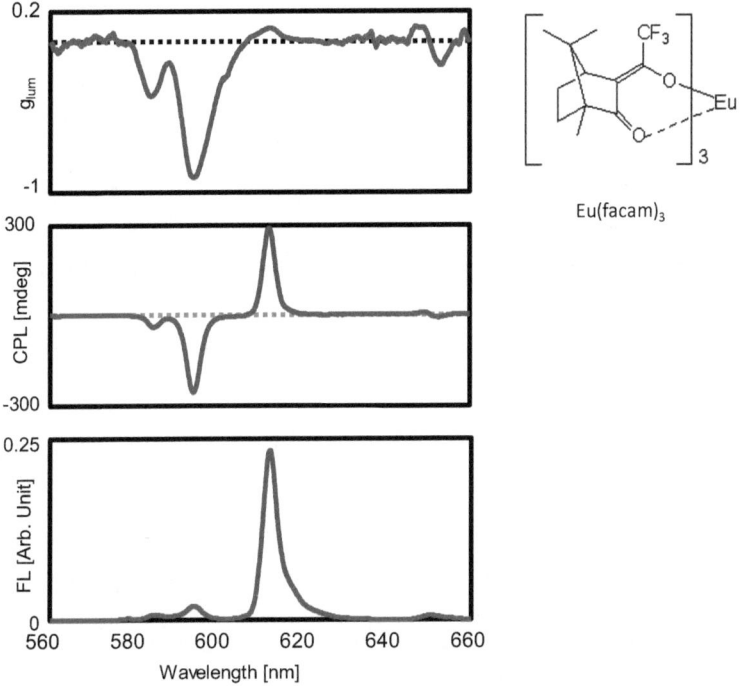

**Fig. 14.8** CPL and fluorescence spectra of 5.5-mM Eu(facam)$_3$ in dimethyl sulfoxide (DMSO) at room temperature

measurements of such sharp CPL spectra require careful setting of bandwidth for the quantitative evaluation of CPL intensity.

Figure 14.9a shows the change in the CPL spectrum of Eu(facam)$_3$ with fluorescence bandwidth. As the fluorescence bandwidth is increased, the CPL spectrum becomes broader, and $g_{lum}$ in the vicinity of 595 nm decreases slightly. Therefore, for compounds showing a very sharp CPL spectrum like Eu(facam)$_3$, i t s necessary to use a small fluorescence bandwidth to perform high-resolution measurements. Figure 14.9b shows a comparison of the fluorescence spectrum obtained using the CPL-300 with a fluorescence bandwidth of 3 nm and that obtained using the general-purpose FP-8300 fluorescence spectrophotometer with a fluorescence bandwidth of 2.5 nm. The good agreement between these spectra reveals that the CPL-300 is capable of high-resolution measurements. Since a trade-off relation exists between the bandwidth of the instrument and the signal-to-noise ratio (S/N), when measuring faint CPL signals at high resolution, cumulative measurements should be performed.

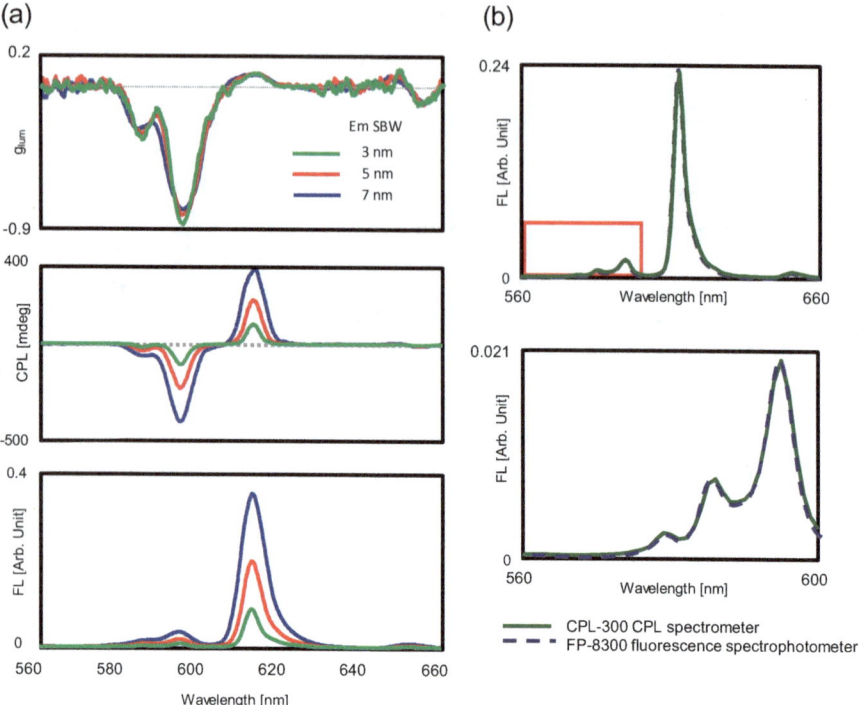

**Fig. 14.9** High-resolution CPL spectra. (**a**) CPL spectra in various bandwidth settings. (**b**) Comparison of CPL-300 (Em SBW = 3 nm) and general purpose fluorescence spectrophotometer (Em SBW = 2.5 nm)

## 14.5.3 Green Fluorescent Proteins

Many biological organisms respond to circularly polarized light. For example, the crustacean *Odontodactylus scyllarus* can detect circularly polarized light [5], and the growth rate of seaweed is promoted by right-handed circularly polarized light but is disturbed by left-handed circularly polarized light [6]. The reason why circularly polarized light affects biological communication of creatures and plant growth is considered to be the optical activity of the organism itself. We therefore performed measurements on green fluorescent protein (GFP) derived from a biological organism.

Figure 14.10a shows the CPL spectrum of wild-type GFP. Although the sample concentration was only 30 μg/mL and the optical path length was only 10 mm, the CPL signal could be clearly observed by increasing the number of accumulations. Although the S/N could be further improved by using a larger number of accumulations, proteins can become photodegraded under prolonged illumination. The occurrence of photodegradation of this GFP was therefore investigated using a small excitation bandwidth in order to decrease the excitation light intensity, and

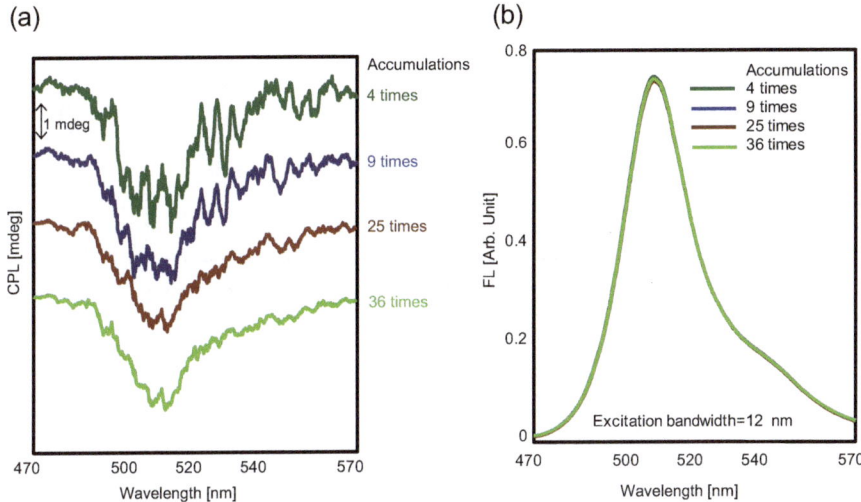

**Fig. 14.10** CPL and fluorescence spectra of wild-type GFP. (**a**) Accumulation dependence on CPL spectra of GFP. (**b**) Sample stability check during CPL spectra accumulation

the temporal change in the fluorescence spectrum was monitored. Figure 14.10b shows the 1st, 9th, 25th, and 36th fluorescence spectrum obtained during the CPL measurements. It can be seen that even after 36 accumulations, no change in the spectral shape occurred.

Figure 14.11 shows CD and CPL spectra of this GFP. The CD and absorption spectra were obtained with 30 μg/mL of the sample concentration and 20-mm path length. The CD originated from aromatic amino acid side chains and was observed in the near-ultraviolet region, whereas the optical absorption and CPL originating from functional groups emitting green fluorescence appear in the visible region. Since $g_{abs}$ and $g_{lum}$ are similar to each other, structural differences between the ground and excited states of the fluorescent functional groups are considered to be very small.

Figure 14.12 shows CD and CPL spectra of enhanced GFP, which offers higher-intensity emission with respect to wild-type GFP. In this case also, $g_{abs}$ and $g_{lum}$ are small and similar to each other, indicating that structural differences between the ground and excitation states are very small.

### 14.5.4  CPL Measurements of Solid Sample

Finally, we introduce a method for measuring CPL for a solid sample. When circularly polarized light-emitting materials are used for displays and illumination, these materials are expected to be not liquids but solids. Therefore, it is necessary to

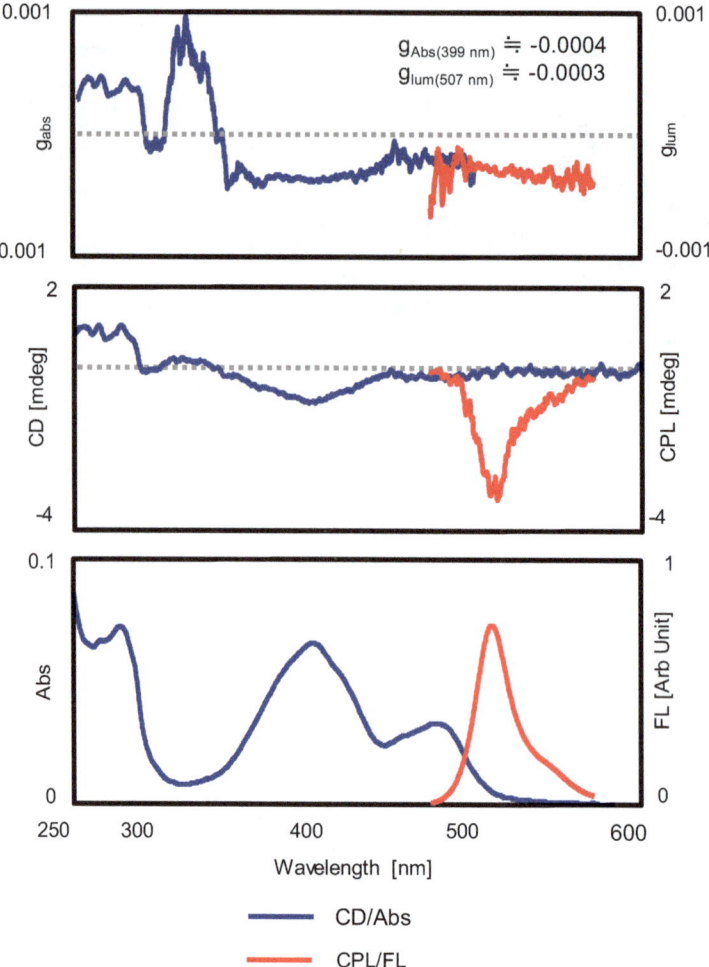

**Fig. 14.11** CD, CPL, absorption, and fluorescence spectra of wild-type GFP

have a means of evaluating CPL in materials in solid form like powder and film. We herein introduce a method for measuring powder samples.

In the Nujol mull method, a powder sample is coated on a quartz plate, and the KBr pellet method, commonly utilized in infrared spectroscopy, can be used. The CPL-300 is designed to detect transmitted fluorescence through samples. Thus, it can detect fluorescence emitted from the back of a pellet while suppressing the effect of fluorescence anisotropy (Fig. 14.13). In the case of a solid sample containing a powder, the spectrum may be affected by artifacts due to birefringence. To determine whether such artifacts are present, it is necessary to measure enantiomers to obtain symmetrical spectra. Furthermore, the confirmation of the isotropic CPL spectra by the sample rotation to the optical axis of a light polarizer is necessary. Such measurements can be

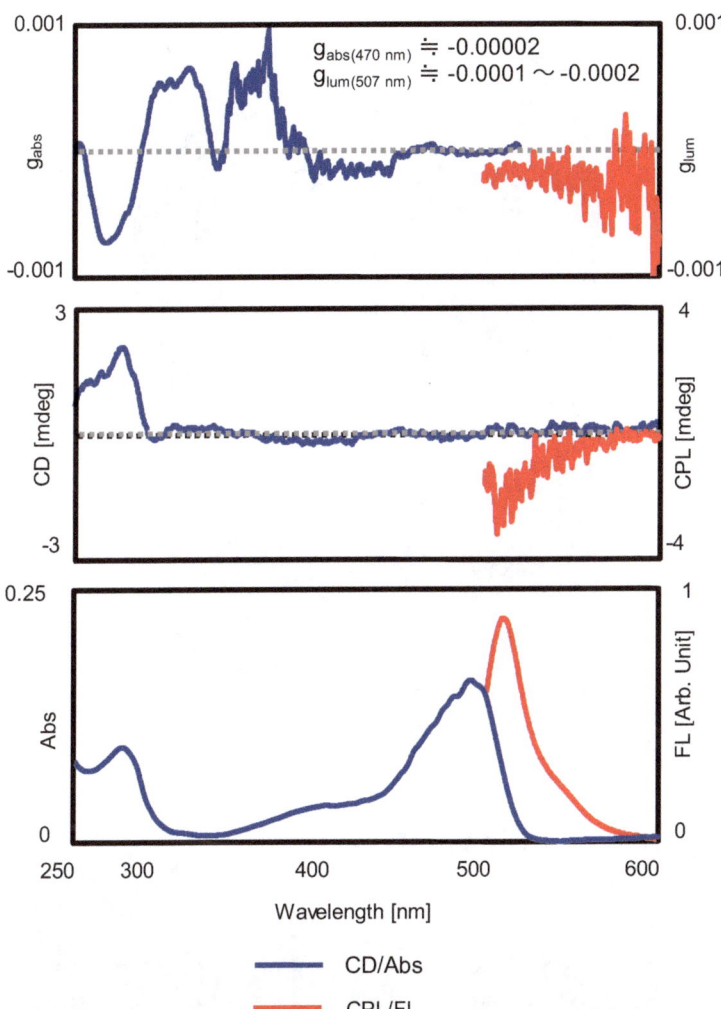

**Fig. 14.12** CD, CPL, absorption, and fluorescence spectra of enhanced GFP

performed using the pellet holder shown in Fig. 14.14a, which has a scale indicating the rotation angle. Figure 14.14b shows a KBr pellet installed in the holder.

Figure 14.15 shows spectra of powdered $Eu(facam)_3$ in a KBr pellet measured using this pellet holder. The spectra were obtained at rotation angles of $0°$, $45°$, and $90°$. There is no significant change in spectral shape with rotation angle, indicating that there is little effect of anisotropy.

Since $g_{lum}$ for europium complexes is known to be strongly affected by the type of solvent used, it was expected that $g_{lum}$ determined using the KBr pellet method would be different to that for a solution. Figure 14.16 shows CPL spectra of Eu $(facam)_3$ in different solvents and in a KBr pellet. $g_{lum}$ values obtained using the KBr

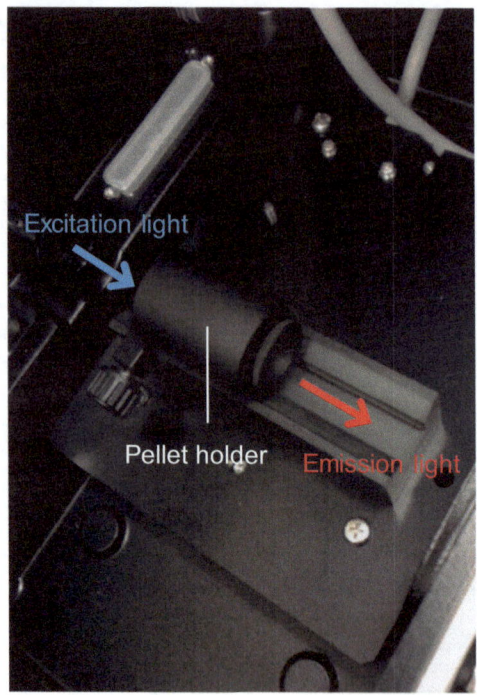

**Fig. 14.13** Sample compartment of CPL-300 with installed pellet holder

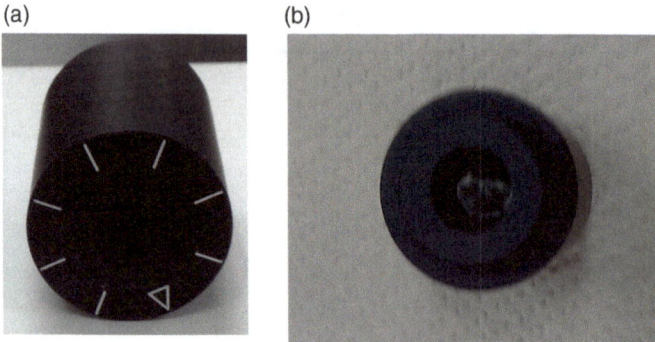

**Fig. 14.14** External view of pellet holder. (**a**) Front view of pellet folder. (**b**) KBr pellet installed in pellet holder

pellet method are much smaller than those obtained using dimethyl sulfoxide (DMSO) and *N,N*-dimethylformamide (DMF) as a solvent, and the value around 590 nm is positive.

In addition to inorganic and organic molecules, including lanthanoid complexes, there is a great deal of research into materials that exhibit the CPL characteristics of supramolecular assemblies, and polymers and colloids that are aggregates of

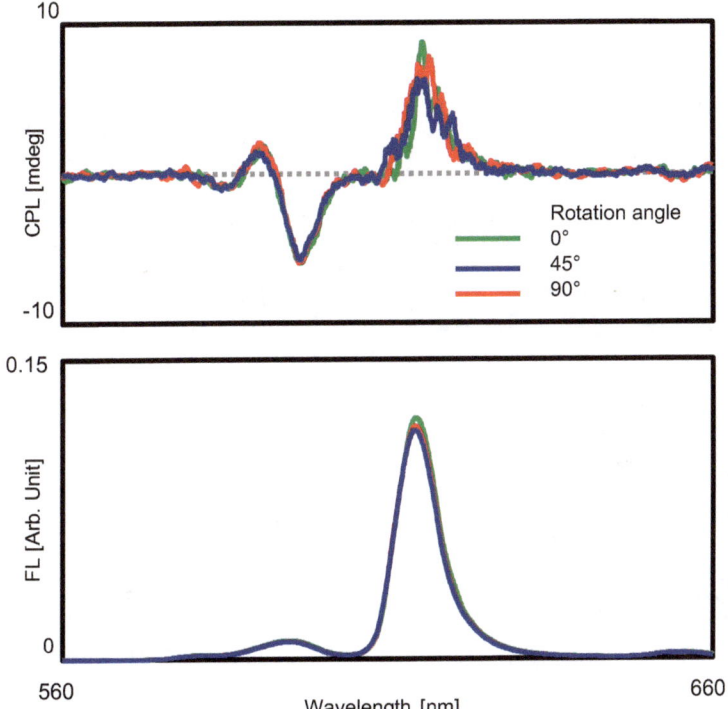

**Fig. 14.15** CPL and fluorescence spectra of Eu(facam)$_3$ in KBr pellet

molecules. This latter group of materials offers advantages in that characteristics such as the emission wavelength and sign of the $g$ value can be easily tailored. In addition, aggregation-induced emission (AIE) [7], first discovered by Tang et al. in 2001, has applications to materials that are capable of light emission in aggregate form, and can be used in organic LEDs (OLEDs) and cell dyeing agents. These materials are often measured in thin-film form. However, it is necessary to ensure that CPL measurements of solid sample are not affected by the artifacts described above.

Furthermore, recent results concerning the design of CPL materials based on quantum chemistry calculations, and the subsequent synthesis of such a material [8], highlight the possibility of a method for developing novel CPL materials that does not require an empirical approach. Given the wide range of potential applications of circularly polarized light-emitting materials, the ability to perform CPL measurements in the solid state is highly significant.

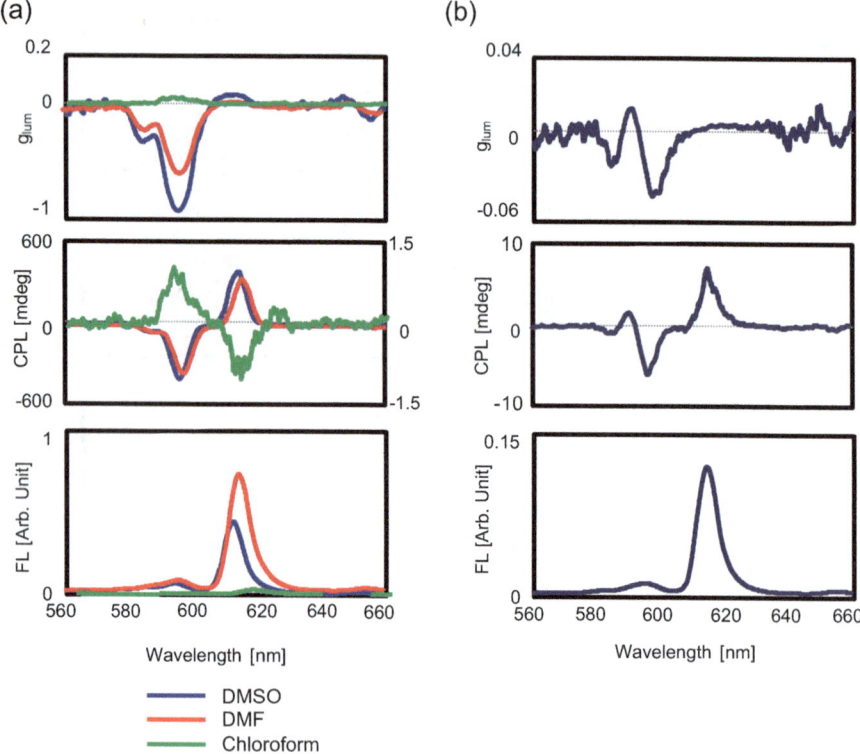

**Fig. 14.16** CD, CPL, absorption, and fluorescent spectra of Eu(facam)$_3$. (**a**) Eu(facam)$_3$ in various solvents. (**b**) Eu(facam)$_3$ in KBr pellet

## 14.6   Summary

In this chapter, we introduced the principles of CPL measurement instruments, calibration methods, and measurement examples. We expect that research into CPL materials will accelerate, and that these materials will find a wide range of practical applications such as 3D displays and security ink.

## References

1. Takakuwa T, Konno T, Meguro H (1985) New standard substance for calibration of circular dichroism: ammonium d-10-camphorsulfonate. Anal Sci 1:215
2. Chun Ka Luk FS, Richardson J (1974) Circularly polarized luminescence spectrum of camphorquinone. Am Chem Soc 96:2006–2009
3. Fujiki M (2014) Front-line polymer science: gigantic enhanced chiral photonics. Polymers 63:468–472

4. Lunkley JL, Shirotani D, Yamanari K, Kaizaki S, Muller G (2008) Extraordinary circularly polarized luminescence activity exhibited by cesium tetrakis(3-heptafluoro-butylryl-(+)-camphorato) Eu(III) complexes in EtOH and CHCl3 solutions. J Am Chem Soc 130:13814–13815
5. Vignolini S, Rudall PJ, Rowland AV, Reed A, Moyroud E, Faden RB, Baumberg JJ, Glover BJ, Steiner U (2012) Pointillist structural color in Pollia fruit. Proc Natl Acad Sci U S A 109:15712–15715
6. MacLeod GC (1957) The effect of circularly polarized light on the photosynthesis and chlorophyll a synthesis of certain marine algae. Limnol Oceanogr 2:360–362
7. Luo J, Xie Z, Lam JWY, Cheng L, Chen H, Qiu C, Kwok HS, Zhan X, Liu Y, Zhu D, Tang BZ (2001) Aggregation-induced emission of 1-methyl-1,2,3,4,5-pentaphenylsilole. Chem Commun 18:1740–1741
8. Tanaka H, Ikenosako M, Kato Y, Fujiki M, Inoue Y, Mori T (2018) Symmetry-based rational design for boosting chiroptical responses. Commun Chem 1:38

# Chapter 15
# Transient Circular Dichroism Approach to Chirality Detection in Dark Photo-Excited States

**Yasuyuki Araki**

**Abstract** A transient circular dichroism (CD)-based approach, called time-resolved circular dichroism (TRCD) spectroscopy, serves as an alternative to circularly polarized luminescence (CPL) spectroscopy for the investigation of excited-state chirality. The experimental apparatus for TRCD measurements was designed and is presented here. The Jones calculus approach to assessing experimental observables is also reported. Finally, to demonstrate the feasibility of TRCD, the CD spectrum of the photo-excited state of [6]carbohelicene is obtained.

## 15.1 Introduction

We may recognize circularly polarized luminescence (CPL) and circular dichroism (CD) as well-known fingerprints of the chirality possessed by chiral molecules [1–3]; thus, the spectroscopy of CD and CPL are considered useful, unique, and irreplaceable tools to obtain the chiral molecular picture in many problems in chemical and biological fields [4–6].

At the same time, studies of electronically excited states are usually performed by fluorescence and absorption spectroscopy. Moreover, transient fluorescence and absorption measurements often reveal the dynamic behavior of the electronically excited states, and such experiments give us direct insight into the excited states, e.g., the lifetime of the electronically excited state. Such characteristics of the electronically excited states obtained by fluorescence lifetime and transient absorption studies are useful for the material development of light-emitting materials at high quantum yield; therefore, we presume that transient CPL and CD experiments could clarify the dynamics of the excited "chiral" states, and such knowledge contributes to the further development of efficient CPL-emitting materials. In a strict sense, transient CPL and CD do not provide the same information. CPL only

Y. Araki (✉)
Institute of Multidisciplinary Research for Advanced Materials, Tohoku University, Sendai, Japan
e-mail: y.araki@tohoku.ac.jp

© Springer Nature Singapore Pte Ltd. 2020

327

T. Mori (ed.), *Circularly Polarized Luminescence of Isolated Small Organic Molecules*, https://doi.org/10.1007/978-981-15-2309-0_15

indicates the emitting electronically excited state; however, transient CD may show evidence of any type of chiral transient species, in principle. Hence, the transient CD approach would be considered as a versatile way to access the chirality of the excited state. Moreover, transient CPL studies are scarce [7, 8], whereas there have been several reports of transient CD, initiated by the pioneering work by Kliger and coworkers [9]. Our group has also recently developed a transient CD apparatus and reported the CD spectrum of small organic molecules in the photo-excited triplet state [10, 11].

In this section, we describe the measurement technique for transient circular dichroism measurement of electronically excited states.

## 15.2 Road to Transient Circular Dichroism (TRCD) Measurement

CD may be defined as the difference in the molar extinction coefficient for left and right circularly polarized light ($\Delta\varepsilon = \varepsilon_L - \varepsilon_R$), and CD measurement is essentially a method to detect this difference. To achieve sensitivity in CD detection, we use a phase-sensitive technique, which is actualized by the phase-sensitive detection of the transmitted intensity of left and right circularly polarized light, transformed from linearly polarized light with a photo-elastic modulator (PEM), operated at thousands of cycles per second. Because of the use of phase-sensitive detection, the time resolution of a usual CD detection depends on the operating frequency of the PEM. The practical time resolution of the manufactured products is considered to be on the order of milliseconds. Therefore, several efforts have been made for implementing CD measurement technology without using PEM.

Time-resolved circular dichroism (TRCD) spectroscopy with broadband and reasonable time response, applicable to a great number of photochemical events in the ns to μs time regimes, was initiated by the Kliger group [9]. Their approach relied on CD detection with elliptically polarized light rather than circular-polarized light. Their approach has been continuously improved and applied to photo-biological events [12–15]. Along with these ideas, TRCD measurement is a promising method for detecting the structural dynamics of chiral molecules and molecular ensembles, ensuring external impulses such as light excitation.

Notwithstanding recent TRCD developments with sophisticated techniques in ultrafast events (1–1000 ps), the applications were limited to the biological field, and the applications of TRCD measurement to the photo-excited state detection of small inorganic and organic molecules are relatively few [16, 17].

Recently, our group also has developed the steady-state CD apparatus, shown in Fig. 15.1, which employs elliptically polarized light as a probe light to detect CD. Our method is conceptually the same as the Kliger method [9]; however, we found the conditions for easier control of light ellipticity by the precise azimuth control of the retarder, yielding the high reproducibility of the CD signals in the

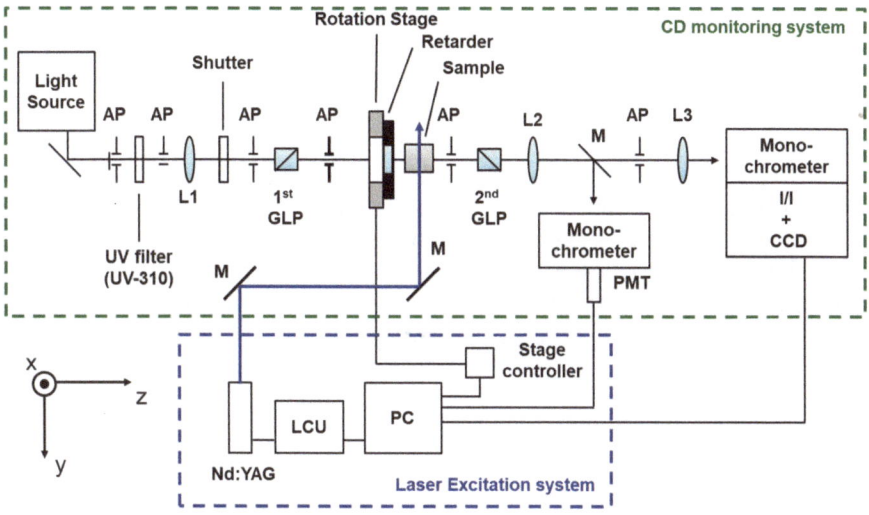

**Fig. 15.1** Top view of the setup for TRCD measurement system. Experimental coordination axis is also shown in the figure. Light source: Xe lamp (Energetiq Technology Inc. LDLS EQ-99) for CD spectrum measurement. Light source: He–Ne laser (OptoSigma 05-LHP-121) for time-course CD measurement. L1, L2, L3: Lens. AP: Aperture. GP: Glan-laser prism (THORLABS Inc., GL15-A). M: Mirror. Retarder: quartz plate: 2 mm width, diameter 30 mm. Rotation stage: THORLABS Inc., Ø1″ Motorized precision rotation stage, PRM1/MZ8. I/I+CCD: Image-intensifier and Multichannel analyzer (Unisoku, MD200). Monochrometer: Acton Research Corporation, SpectraPro-150, slit width 0.5 mm. PC: Personal computer. LCU: Laser control unit. Nd:YAG: Nd:YAG laser (continuum, Surelite, 6 nm FWHM, 355 nm, 10 Hz, 0.01 J). Stage controller: K-cube brushed DC servo motor controller (THORLABS Inc., KDC101)

steady-state CD measurement of Δ- and Λ-Ru complex [11]. Here, we describe our recent progress for TRCD experiments with elliptically polarized measurements.

## 15.3   Jones Vector Description of CD Measurement with Elliptically Polarized Light

The CD measurement apparatus discussed here is schematically shown in Fig. 15.1.

In Fig. 15.1, light from Xe lamp is directed to the first glan-laser prism to yield linearly polarized light. To produce elliptically polarized light, linearly polarized light is directed into the home-built retarder, in which a $\theta$-stage is mounted to control the azimuth ($\theta$). Note that the azimuth is defined as the angle between the fast axis of the home-built retarder and polarized plane of the introduced linearly polarized light proceeding in the counterclockwise direction. Usually, the polarized plane is set vertically; thus, the long axis of the elliptical polarization is also in the vertical direction. We can produce left- and right-handed elliptically polarized light by controlling the azimuth $\theta$, in particular $+\theta$ for left-handed polarized light and $-\theta$

for right-handed polarized light. Passing through chiral media, which exhibit molar circular dichroism ($\Delta\varepsilon$), the ellipse undergoes some variation, namely, light intensity in the vertical and/or horizontal axis of the ellipse. To estimate such a change, the second glan-laser prism, which is in crossed Nicol directions with respect to the first glan-laser prism, is placed to detect the horizontal intensity of the elliptically polarized light. Finally, the light is collected by a diode array, equipped with an image-intensifier to obtain horizontal intensity as a function of $\theta$.

Such an experimental situation is best formulated by the Jones vector approach. First, we show the relationship between elliptically polarized light and molar circular dichroism ($\Delta\varepsilon$), described by the Jones calculus shown below [9].

Polarized light $J_{\text{init}}$ is generally expressed by the two-component vector

$$J_{\text{init}} = \begin{bmatrix} E_x\, e^{i(\varphi_x + 2\pi\, \nu\, t)} \\ E_y e^{i(\varphi_y + 2\pi\, \nu\, t)} \end{bmatrix}, \tag{15.1}$$

where $E_x$ and $E_y$ denote the complex magnitudes of electric vectors of the observation light along the x- and y-axis in right-handed coordinate system, respectively, $z$ is the direction of light propagation, $\varphi_x$ and $\varphi_y$ are the phases of those vectors, and $\nu$ is the light frequency. This apolarized light from the light source is converted to linearly polarized light through the first linear polarizer (LP$_x$) in Fig. 15.1. The linearly polarized light is generated by the first linear polarizer, followed by the conversion to an elliptically polarized light by the retarder ($\alpha(\theta, \delta)$), which has two variables—retardance ($\delta$) and azimuth ($\theta$). The Jones matrices of the first linear polarizer and $\alpha(\theta, \delta)$ are represented as

$$LP_x = \begin{bmatrix} 1 & 0 \\ 0 & 0 \end{bmatrix} \tag{15.2}$$

and

$$\alpha(\theta, \delta) = \begin{bmatrix} \cos^2\theta \cdot e^{i\frac{\delta}{2}} + \sin^2\theta \cdot e^{-i\frac{\delta}{2}} & 2i\cos\theta\sin\theta\sin\dfrac{\delta}{2} \\ 2i\cos\theta\sin\theta\sin\dfrac{\delta}{2} & \cos^2\theta \cdot e^{-i\frac{\delta}{2}} + \sin^2\theta \cdot e^{i\frac{\delta}{2}} \end{bmatrix}. \tag{15.3}$$

Retardance is defined as the difference in phase shift of its slow axis against its fast axis. Now, the fast axis is defined from top to bottom in the home-built retarder (the pressing direction of the quartz plate by the screw, which is the $-y$ direction in Fig. 15.1), and the azimuth is also defined as depicted in Fig. 15.2.

The performance of the retarder is shown in Fig. 15.3. Figure 15.3a is the front view of our home-built retarder. In the home-built retarder, the 2-mm thick quartz plate was mounted on the $\theta$-stage and pressed slightly from the top of the plate.

Owing to the pressure, the quartz plate possesses birefringence. Figure 15.3b shows the two-dimensional map of the quartz plate birefringence. The fast axis of the

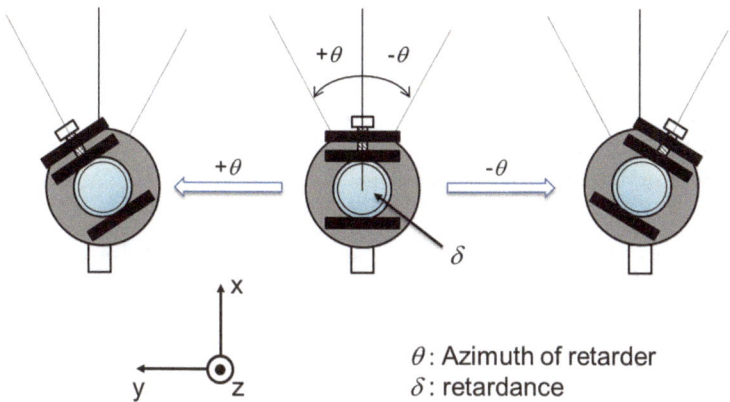

**Fig. 15.2** Illustration of home-built retarder view from detector (I/I) direction. Experimental coordination axis and $\theta$ definition are also shown in the figure

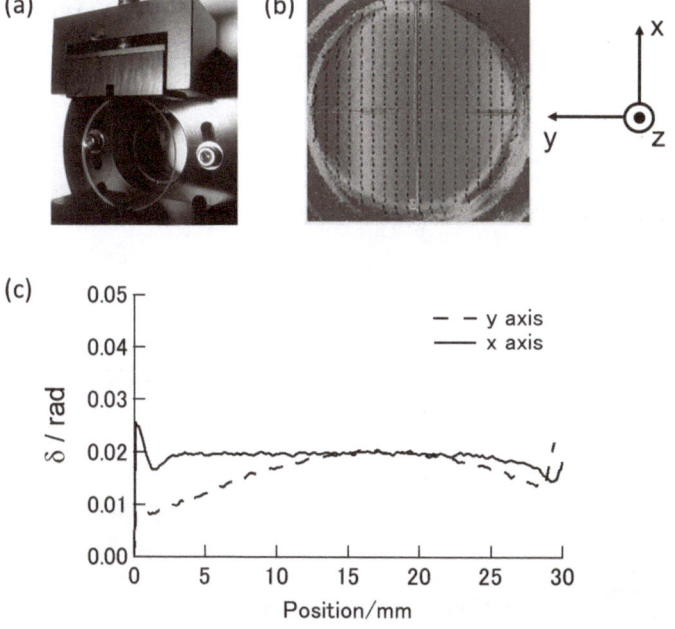

**Fig. 15.3** (**a**) A picture of home-built retarder. (**b**) Birefringence mapping of quartz plate evaluated by 2D birefringence measurement system (PA-110, Photonic lattice, Inc.). Birefringence was monitored at 520 nm. The fast axis is observed as *x*-axis. (**c**) Retardance profile of the cross section of birefringence mapping in *x*- and *y*-axis

plate is defined by the pressing direction; namely, the y-axis from top to bottom. Figure 15.3c is the retardance plot against the position of the quartz plate. A uniform retardance of $2 \times 10^{-2}$ rad was realized in the central position of the plate; thus, we can use that value for the elliptically polarized light in this experiment.

The matrices of the absorption of left and right circular-polarized light ($\beta$ and $\gamma$) by the chiral sample are described as

$$\beta = \frac{1}{2} \begin{bmatrix} k' + 1 & i(k' - 1) \\ -i(k' - 1) & k' + 1 \end{bmatrix}, \tag{15.4}$$

and

$$\gamma = \frac{1}{2} \begin{bmatrix} k + 1 & -i(k - 1) \\ i(k - 1) & k + 1 \end{bmatrix}, \tag{15.5}$$

where the transmittance of left- and right-handed circular-polarized light is $k^2$ and $k'^2$, respectively. Note that $\beta$ and $\gamma$ commute; thus, the Jones matrix for circular dichroism ($M_{CD}$) can be written as their sum [7], as follows:

$$M_{CD} = \frac{1}{2} \begin{bmatrix} k + k' & -i(k - k') \\ i(k - k') & k + k' \end{bmatrix}. \tag{15.6}$$

The second linear polarizer is described by

$$LP_y = \begin{bmatrix} 0 & 0 \\ 0 & 1 \end{bmatrix} \tag{15.7}$$

Finally, the transmitted light ($J(\theta, \delta)$) in Fig. 15.1 after the second linear polarizer is calculated by the sequential multiplication of Eqs. 15.6 and 15.7:

$$J(\theta, \delta) = LP_y \cdot M_{CD} \cdot \alpha(\theta, \delta) \cdot \zeta_x \cdot J_{init}$$
$$= \frac{1}{2} \begin{bmatrix} 0 \\ E_x \left\{ \left( \rho \cos \frac{\delta}{2} + (2k - \rho) \sin 2\theta \sin \delta \right) + i\rho \cos 2\theta \sin \frac{\delta}{2} \right\} e^{i\left(\varphi_x + \frac{\pi}{2} + 2\pi\nu t\right)} \end{bmatrix}. \tag{15.8}$$

The observed light intensity as a function of $\theta$ and $\delta$ ($I(\theta, \delta)$) is the square of $J(\theta, \delta)$:

$$I(\theta, \delta) = |J(\theta, \delta)|^2$$
$$= \frac{1}{4} I_0^2 \left\{ \rho^2 \cos^2 \frac{\delta}{2} + \rho(2k - \rho) \sin 2\theta \sin \delta + (2k - \rho)^2 \sin^2 2\theta \sin^2 \frac{\delta}{2} + \rho^2 \cos 2\theta \sin^2 \frac{\delta}{2} \right\} \tag{15.9}$$

where we defined the parameter $\rho$ as

$$\rho = k - k' = k\left(1 - \frac{k'}{k}\right) \cong k\left(1 - e^{1.15\,\Delta\varepsilon C\,l}\right) \tag{15.10}$$

and $I_0$ is equal to the square of the complex amplitude of the initial light source.

Now, we can calculate the following $S_{CD}$ value:

$$S_{CD} = \frac{I_R - I_L}{I_R + I_L} = \frac{I(\theta,\delta) - I(-\theta,\delta)}{I(\theta,\delta) + I(-\theta,\delta)}. \tag{15.11}$$

When $\Delta\varepsilon\,C\,l = (\varepsilon_L - \varepsilon_R)\,C\,l << 1$,

$$\rho \cong -1.15\,k\,\Delta\varepsilon c l, \tag{15.12}$$

and

$$S_{CD} = \frac{I_R - I_L}{I_R + I_L} = \frac{1.15\,\Delta\varepsilon\,C\,l}{\sin 2\theta\,\tan\frac{\delta}{2}}, \tag{15.13}$$

where $C$ and $l$ are the molar concentration of the sample and optical path length of the sample cell, respectively.

Equation 15.13 gives the relationship between the experimentally observed values ($S_{CD}$) and molar circular dichroism ($\Delta\varepsilon$). In Eq. 15.13, we realize that two parameters, $\theta$ and $\delta$, need to be known to evaluate $\Delta\varepsilon$.

The angle between the polarization plane of the initial linearly polarized light after the first prism and the first axis of the retarder is defined as $\theta$. In Fig. 15.1, the retarder is mounted on the $\theta$-stage; thus, $\theta$ is well known when the experiment is performed.

To determine $\delta$, we should understand the effect of $\delta$ on the polarization of transmitted light through all optics without any light absorber (sample solution) in Fig. 15.1. By calculation similar to that mentioned above, we have an equation indicating the polarized light intensity as follows:

$$I(\theta,\delta) = I_0 \sin^2 2\theta \sin^2 \frac{\delta}{2}. \tag{15.14}$$

Thus, we can estimate $\delta$ by the following equation:

$$\delta = 2\sin^{-1}\sqrt{\frac{I(\theta,\delta)}{I_0 \sin^2 2\theta}}. \tag{15.15}$$

Through Eq. 15.13, we account that $S_{CD}$ value may have a $\theta$-value dependency; moreover, $S_{CD}$ increases with decreasing $\theta$-value. Such a characteristic feature is represented in Fig. 15.4.

**Fig. 15.4** Circular
dichroism spectra of [6]
helicene in toluene obtained
by our apparatus. (Top) $\theta$-
dependence of $S_{CD}$. Inset:
CD intensity at 327 nm
plotted against $(\sin(2\theta))^{-1}$.
(Bottom) CD spectrum by
our apparatus. CD spectrum
obtained by *J*-820 is also
shown for comparison

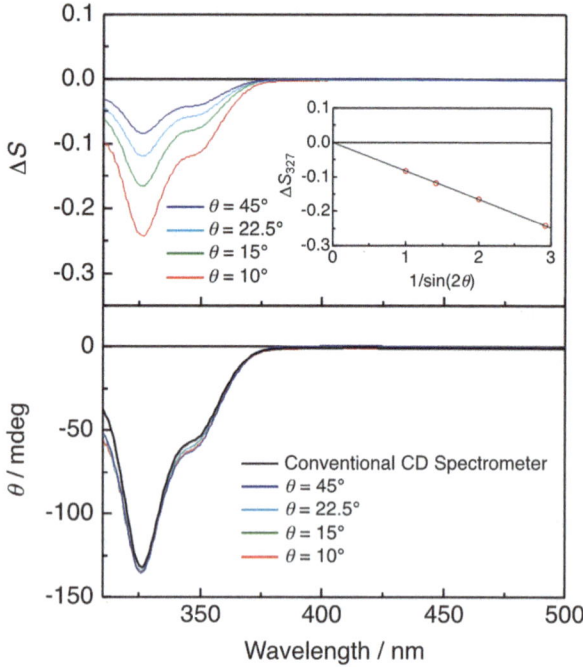

In Fig. 15.4, the CD spectrum of (*M*)-[6]carbohelicene ((*M*)-[6]CH) by our setup
is presented. (*M*)-[6]CH shows negative CD signal at 327 nm [18]. Note that
negative CD signal increased when the smaller $\theta$-value was set. The increase in
$S_{CD}$ was proportional to the $(\sin(2\theta))^{-1}$, as suggested in Eq. 15.13. Furthermore,
signals shown in the top part of Fig. 15.4 converged to the CD spectrum measured by
conventional CD polarimeter, within experimental error. This finding allows us to
obtain CD signals without lock-in detection by PEM usage.

At the end of this section, let us briefly comment on the circular birefringence
(CB) contribution on this CD measurement. Knowledge of the CB contribution is
needed for observation in CD measurements with elliptically polarized light
[14, 19]. Here, we described the CB signal by Jones calculus to give the zeroth-
order picture of this issue.

CB is described by the following Jones matrix:

$$M_{CB} = \begin{pmatrix} \cos\chi & \sin\chi \\ -\sin\chi & \cos\chi \end{pmatrix}, \tag{15.16}$$

where $\chi$ is the product of sample thickness ($l$) and the phase shift due to CB, defined
by the asymmetric refractive index of media as

$$\chi = \frac{\pi(n_L - n_R)l}{\lambda}. \tag{15.17}$$

The results of $S_{CB}$ are as follows:

$$S_{CB} = \frac{\sin 2\theta \, \cos\chi \, \sin^2\frac{\delta}{2}}{\sin\chi(\cos^2\frac{\delta}{2}\,\sin\chi + \sin^2\frac{\delta}{2}\,\cos 2\theta)}. \tag{15.18}$$

At first glance, the most important difference between $S_{CD}$ and $S_{CB}$ is the factor $\Delta\varepsilon$ in the quantity of the former. Otherwise, all the other quantities contribute to both $S_{CD}$ and $S_{CB}$. We observe that $S_{CD}$ and $S_{CB}$ in the case $\delta$ approaching to zero approach their corresponding values in the completely opposite direction, namely, $S_{CD}$ increases nonlinearly when $\delta$ goes to zero due to the $(\tan(\delta/2))^{-1}$ factor, while $S_{CB}$ approaches zero due to the $(\tan(\delta/2))^2$ factor. CD and CB should have the same amplitude because they are connected by the Kramers–Kronig relationship; thus, the ratio, $S_{CD}/S_{CB}$, in the small $\delta$ case is increased just by a factor of $(\tan(\delta/2))^{-3} \sim (\delta/2 + 1/6(\delta/2)^3)^{-3}$. In Fig. 15.3, we used $\delta$ as small as 0.02 rad; thus, $S_{CD}/S_{CB}$ is on the order of $10.^6$ This zeroth-order estimate suggests us that the CB contribution to CD measurement with elliptically polarized light provided by the retarder with small retardance is vanishingly small. As a matter of course, we should note that this issue should be explicitly treated by $N$-matrix formalism of Jones calculus [20].

## 15.4  Examples of Transient CD Spectrum: CD Spectrum of [6]Carbohelicenes in Solution [17]

We employ the above described system to carry out TRCD experiments, by combining with ns-laser flash photolysis system, in a time interval of up to 1 μs (Fig. 15.1). We selected a target molecule, optical active [6]carbohelicene ([6]CH), which has photophysical properties and has been well studied and documented [21–25]. Photoexcitation of [6]CH yields the excited triplet state ($^3$[6]CH$^*$) efficiently; thus, early work suggested the CD spectrum of $^3$[6]CH$^*$ derivatives in frozen 2-methyltetrahydrofuran at 98 K with careful data analysis to eliminate the linear dichroism (LD) component originating from the photo-selection in the frozen media [26]. In this work, we aimed at observing the circular dichroism spectrum of $^3$[6]CH$^*$ in solution, and at the μs timescale, we essentially observed an LD-free CD spectrum because of the free rotation of $^3$[6]CH$^*$ and compare our results with the previous ones.

Therefore, here we consider it worthwhile to show the transient CD spectra of ($P$)- and ($M$)-$^3$[6]CH$^*$ in toluene at room temperature (RT) and the time-trace analysis of transient absorption and circular dichroism measurement.

At this stage, we can evaluate the $\Delta S$ spectra of ($P$)- and ($M$)-$^3$[6]CH$^*$ at the initial time (1–11 μs) observed as a function of the wavelength with the image-intensifier/diode array. They showed a 620-nm peak signal with positive $\Delta S$ for ($P$)-$^3$[6]CH$^*$

**Fig. 15.5** (a) Anisotropy ($g* = \Delta\varepsilon/\varepsilon$) factor and (**b**) TRCD spectra accumulated during 1–11 µs interval after laser excitation of (*P*)- (red) and (*M*)-$^3$[6]CH$^*$ (blue) in toluene at RT monitored with an image-intensifier. Azimuth of retarder $\theta = 15°$. (**c**) Transient absorption spectrum of $^3$[6]CH$^*$ in toluene at RT (reprinted from Kuronuma et al. [11])

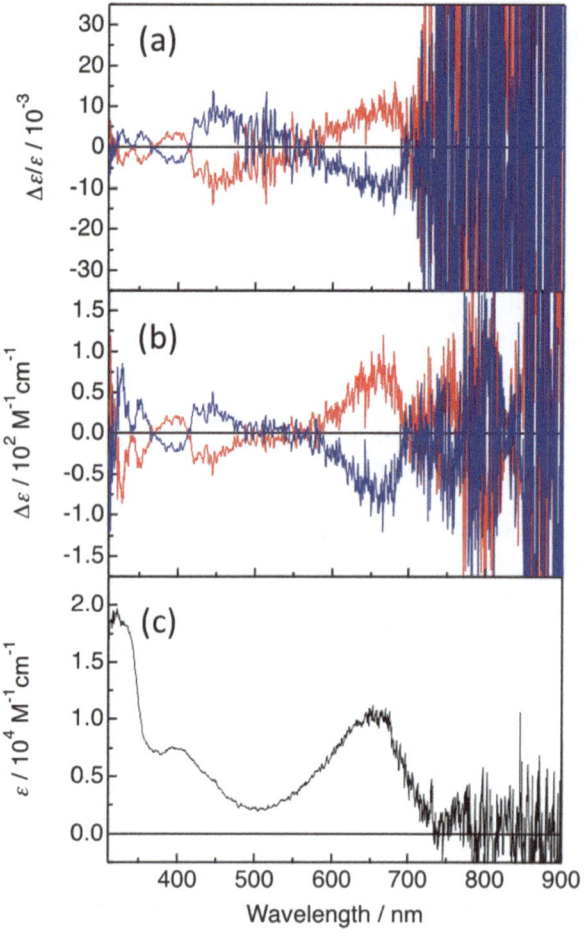

and negative $\Delta S$ for (*M*)-$^3$[6]CH$^*$ in addition to the weak signal near 420 nm with opposite $\Delta S$ signs with respect to the 620-nm peak. The $\Delta S$ values can be transformed to $\Delta\varepsilon$ values by Eq. 15.13. To evaluate the initial concentration of $^3$[6]CH$^*$, the $\varepsilon$ values of $T_n \leftarrow T_1$ absorption with the comparison method, which gives the $\varepsilon$ value at 660 nm to be $(1.1 \pm 1.5) \times 10^4$ M$^{-1}$ cm$^{-1}$ [11].

The transient absorption spectrum of $^3$[6]CH$^*$ in the 100–150 ns region measured with an image-intensifier is shown in Fig. 15.5 (bottom), in which the $T_n \leftarrow T_1$ absorption peak appears at 650 nm [19], with weak peaks at 400 nm, in addition to the 330-nm peak in the ultraviolet region. The average concentration of $^3$[6]CH$^*$ in the 1–11 µs interval was evaluated as $1.3 \times 10^{-6}$ M in this time regime from the estimated $\varepsilon_{660}$ (= $(1.1 \pm 1.5) \times 10^4$ M$^{-1}$ cm$^{-1}$) value. Employing the $^3$[6]CH-$^*$-concentration in the 1–11 µs interval, the CD spectrum in $\Delta\varepsilon$ is depicted, as shown in Fig. 15.5 (middle). The main CD peak at 620 nm exhibits a positive sign for (*P*)-$^3$[6]CH$^*$ and a negative sign for (*M*)-$^3$[6]CH$^*$ in a mirror image. For (*P*)-$^3$[6]CH$^*$, several CD peaks appear; i.e., negative 420-nm CD, positive 380-nm CD,

**Fig. 15.6** (**a**) Time-profile of $\Delta S$ of $(P)$-$^3$[6]CH$^*$ (light red dot) and $(M)$-$^3$[6]CH$^*$ (light blue dot) in toluene at RT monitored at 632 nm from He–Ne laser as a stable monitor light, which was detected with photomultiplier tubes after passing Gran-laser prism, retarder, and quartz cell. Azimuth of retarder $\theta = 15°$. Solid red and blue lines; fitting curves. (**b**) $\Delta$Abs-decay of $^3$[6]CH$^*$ at 632 nm. Orange solid line; fitting curve (reprinted from Kuronuma et al. [11])

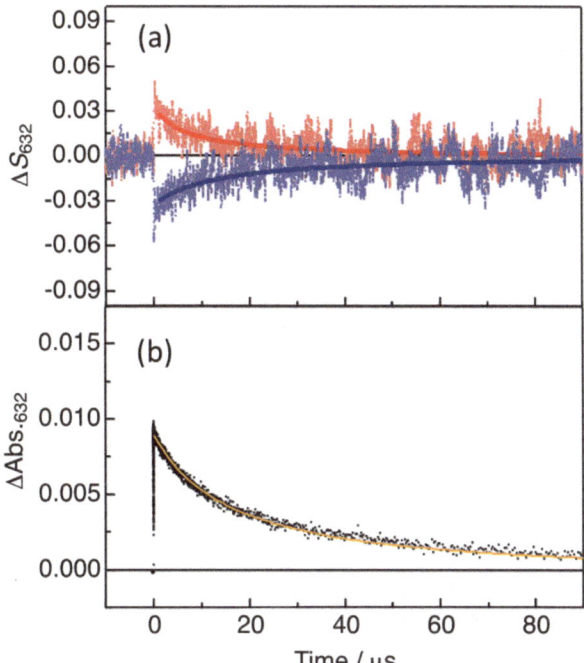

positive 340-nm CD, and 310-nm positive peaks, which are all mirror images for $(P)$-$^3$[6]CH$^*$. In the transient absorption spectrum, only two absorption bands may be identified, whereas each CD spectrum clearly shows four distinct transitions. It should be noted that these TRCD spectra shown here are essentially the same as those in the previous report for the 600–700 nm range, which demonstrates the validity of our measurement [26].

The observed temporal changes of $\Delta S(t)$ are shown in Fig. 15.6 (upper panel), in which the decay of the positive region is seen for $(P)$-$^3$[6]CH$^*$ with a half-decay time of ca. 10 μs, while the decay in the negative region is seen for $(M)$-$^3$[6]CH$^*$ as the mirror image with almost the same half-decay time.

We have reached the conclusion that we have successfully observed the CD spectrum of $^3$[6]CH$^*$ in solution by means of TRCD spectroscopy. The CD spectrum of $(P)$-$^3$[6]CH$^*$ in 600–700 nm has positive Cotton effects, while $(M)$-$^3$[6]CH$^*$ exhibits negative effects with the same spectral shape, one of the most distinguishing characteristics of the CD spectrum. Our experimentally observed CD spectrum in solution was almost identical to the previously reported CD spectrum carefully evaluated under the photo-stationary state of [6]CH derivatives in the frozen media, which means our observation has reached the criteria of validation.

We hope such information may be helpful in developing circularly polarized light-emitting devices. Furthermore, electronic interaction between chromophores in chiral orientation is given by CD spectroscopy from the analysis of the exciton interaction method. Thus, from this point of view, transient CD technique will receive much attention in the field of analysis of interacting photo-excited systems, such as singlet fission and triplet-triplet annihilation photon up-conversion processes.

# References

1. Nakanishi K, Kuroyanagi M, Nambu H, Oltz EM, Takeda R, Verdine GL, Zask A (1984) Pure Appl Chem 56:1031–1048
2. Murphy WS (1975) J Chem Educ 52:774–776
3. Smith HE, Fontana LP (1991) J Org Chem 56:432–435
4. Kelly SM, Jess TJ, Price NC (2005) Biochim Biophys Acta 1751:119–139
5. Sugimoto N, Nakano S, Katoh M, Matsumura A, Nakamuta H, Ohmichi T, Yoneyama M, Sasaki M (1995) Biochemistry 34:11211–11216
6. Ranjbar B, Gill P (2009) Chem Biol Drug Des 74:101–120
7. Schauerte JA, Schlyer BD, Steel DG, Gafni A (1995) Proc Natl Acad Sci U S A 92(2):569–573
8. Schauerte JA, Steel DG, Gafni A (1992) Proc Natl Acad Sci U S A 89:10154–10158
9. Kliger DS, Lewis JW (1987) Res Chem Intermed 8:367–398
10. Murakami M, Araki Y, Sakamoto S, Hamada Y, Wada T (2013) Chem Lett 42:261–262
11. Kuronuma M, Sato T, Araki Y, Mori T, Sakamoto S, Inoue Y, Ito O, Wada T (2019) Chem Lett 48:357–360
12. Stadnytskyi V, Orf GS, Blankenship RE, Savikhin S (2018) Rev Sci Instrum 89:033104
13. Lewis JW, Tilton RF, Einterz CM, Milder SJ, Kuntz ID, Kliger DS (1985) J Phys Chem 89:289–294
14. Lewis JW, Goldbeck RA, Kliger DS, Xie X, Dunn RC, Simon JD (1992) J Phys Chem 96:5243–5254
15. Björling SC, Goldbeck RA, Paquette SJ, Milder SJ, Kliger DS (1996) Biochemistry 35:8619–8627
16. Chen E, Wood MJ, Fink AL, Kliger DS (1998) Biochemistry 37:5589–5598
17. Thomas YG, Szundi I, Lewis JW, Kliger DS (2009) Biochemistry 48:12283–12289
18. Nakai Y, Mori T, Inoue Y (2012) J Phys Chem A 116:7372
19. Hiramatsu K, Nagata T (2015) J Chem Phys 143:121102
20. Kliger DS, Lewis JW, Randall CE (1990) Polarized light in optics and spectroscopy. Academic, San Diego, pp 133–150
21. Niezborala C, Hache F (2008) J Am Chem Soc 7:12783–12786
22. Weigang OE Jr, Turner JA, Trouard PA (1966) J Chem Phys 45:1126
23. Vander Donckt E, Nasielski J, Greenleaf JR, Birks JB (1968) Chem Phys Lett 2:409
24. Rhodes W, Amr El-Sayed MF (1962) J Mol Spectrosc 9:42
25. Sapir M, Vander Donckt E (1975) Chem Phys Lett 36:108
26. Tetreau C, Lavalette D, Balan A (1985) J Phys Chem 89:1699